Logic, Mathematics, Philosophy: Vintage Enthusiasms

THE WESTERN ONTARIO SERIES IN PHILOSOPHY OF SCIENCE

A SERIES OF BOOKS IN PHILOSOPHY OF MATHEMATICS AND NATURAL SCIENCE, HISTORY OF SCIENCE, HISTORY OF PHILOSOPHY OF SCIENCE, EPISTEMOLOGY, PHILOSOPHY OF COGNITIVE SCIENCE, GAME AND DECISION THEORY

VOLUME 75

David DeVidi · Michael Hallett · Peter Clark
Editors

Logic, Mathematics, Philosophy: Vintage Enthusiasms

Essays in Honour of John L. Bell

Springer

Editors
David DeVidi
University of Waterloo
Department of Philosophy
200 University Ave West
Waterloo, Ontario N2L 3G1
Canada
ddevidi@uwaterloo.ca

Michael Hallett
McGill University
Department of Philosophy
855 Sherbrooke St. West
Montreal, Québec
Leacock Bldg.
Canada H3A 2T7
michael.hallett@mcgill.ca

Peter Clark
University of St. Andrews
Department of Philosophy
Edgecliff, The Scores
St. Andrews, Fife, UK, KY16 9AR
pjc@st-andrews.ac.uk

ISSN 1566-659X
ISBN 978-94-007-0213-4 e-ISBN 978-94-007-0214-1
DOI 10.1007/978-94-007-0214-1
Springer Dordrecht Heidelberg London New York

Library of Congress Control Number: 2011920962

Printed on acid-free paper

Springer is part of Springer Science+Business Media (www.springer.com)

John and Mimi Bell in Oxford in 1970, following the conferral of John's D. Phil.

To the memory of $-M^2$

Preface

If things of Sight such Heavens be
What Heavens are those we cannot see?
Andrew Marvell

This collection of essays was put together to celebrate John Bell's sixtieth birthday on the 25 March 2005. The list of contributors signals some of the important stations of John's career as a mathematician, teacher, colleague and friend: the student days in Oxford; the years of the young Lecturer, pacing the rooming houses of Londinium's Bedsit Land; the years of the Reader, ensconsed in more sedate accommodations; the years of the Canadian Professor in London, Ontario, no longer a philosopher in a mathematics department, but now a mathematician in a philosophy department. And all of them years of books, records and beloved recordings, from low-fi to high-fi, the sounds of Beethoven and Schoenberg, Heifetz and Gould, Parker and Powell, conversations (monologues?) late (very late!) into the night—vintage years of undying, revivifying enthusiasms, not least among them, the enthusiasm for vintages.

The contributions are in no way intended to be commentaries on John's work; they are to be seen rather as presents from some of the people John has influenced, been inspired by or inspired, encouraged, amazed and amused through the years. They include contributions from former fellow students, former and current colleagues, former pupils, collaborators and joint authors, and from friends and admirers. The papers are grouped into a small number of broad categories, though the breadth of topics gives some indication of the range of John's interests and influence, running from mathematics to aesthetics, and from philosophy of science to political theory. What unites the material in this volume is what is characteristic of John's work: when it is mathematical, the topics are chosen because of their philosophical interest; when it is philosophical, it takes, where appropriate, full advantage of illumination from relevant work in mathematics and formal logic.

The attentive reader will no doubt have noticed that a considerable time has elapsed between March of 2005 and the date of publication. Some of the lapse has been in the interests of the quality of the volume itself, as we waited for some of our eminent, and correspondingly busy, contributors to complete their papers. The Editors, less eminent but perhaps not less busy, accept full responsibility for the rest of the delay. We wish to thank the contributors, not just for their papers, but for their patience and efficiency in the face of delays, pesterings and quibbles. We also wish to thank John, Bill Demopoulos (the General Editor of the Western Ontario Series in the Philosophy of Science, in which this volume appears) and the publisher for their tolerance and understanding. We also owe an enormous debt to the meticulous work of Oran Magal on the penultimate draft of the manuscript, and for compiling the Name Index; his diligence and care has saved us from numerous infelicities.

Because of the various delays in its completion, John began referring to the book waggishly as "the memorial volume." But the joke, meant entirely good-heartedly, has now, sadly, a cruel sting. We had intended to dedicate the volume to the fixed point in John's "perpetual motion," Mimi, or $-M^2$ as her dedicatory sobriquet became. However, Mimi died from cancer on 20 November 2009; she had been John's constant companion for over 40 years, from their early days together at LSE.

Two of the editors first got to know the Bells at LSE in the early 1970s. Thanks to John's (literally) prodigious talent, which led to an Oxford Scholarship at the age of 15, he was officially our senior as teacher and supervisor, and indeed, it seemed then, as intellectual and cultural being, although he was scarcely older. We picture John and Mimi in the large LSE Refectory (the standard meeting place, and John's constant resort), at one of the long formica tables, perhaps with a bowl of the stodgy spaghetti or one of the glutinous curries, or just mugs of coffee or tea and a biscuit or a cheese roll. The sense of fun was intense, as was the Bells' delight in the ridiculous, of which the Refectory, with its food and habitués (including ourselves), provided a constant supply. John, of course, was the outwardly dominant one, but Mimi was firmly in charge, a fact which became clearer as one got to know them better. John was never allowed lasting dominance, and Mimi wouldn't permit the conversation to be swamped by John's obsession *du jour*. After a first rebuke, John would try reviving the topic on which he had fixed, especially if there was some *aperçu* which had occurred to him and he was itching to get out; this would be followed by a somewhat exasperated second rebuke, and the cycle would repeat itself. Eventually, Mimi would exclaim "Oh John, you're so irri*tat*ing," and the cycles would be at an end. It was hard work, but Mimi always won, as she did at Scrabble. Many eccentric and odd characters would turn up, and (of course!) gravitate to the Bell circle, if sometimes only briefly. John encouraged this, and delighted in it; in fact, for him the odder and more eccentric, the better. Mimi, however, would usually remain aloof, composed, dignified and mildly sceptical, a rock of reassurance in the unpredictable political and emotional turbulence.[1]

Eventually, the centre of this semi-communal social life shifted from the LSE to the Bells' rented flat in Alexandra Grove, where the lads of humble English origin, having grown up in the monochrome aftermath of the Second World War, were first exposed to new worlds of intellectual and culinary creation. The only occasion when

[1] A fine sense of many of the LSE characters encountered by the Bells at this time is conveyed by the chapter "London, 1968–73" in John's memoir *Perpetual Motion: My First Thirty Years*, available from his website.

at least one of the Editors saw John genuinely silenced and overawed was when their son, Alex, was born. After witnessing Mimi give birth, John came to a dinner party to which he had been previously invited. He hardly spoke a word all evening and sat amazed by what he had experienced; when he did speak, it was to express his admiration for Mimi and his complete devotion to her and their new son.

Evolving social and academic commitments and careers dispersed us somewhat, and, as the years went by, there was less communal life and social gatherings became rarer; new circles were formed of which we were no longer part. At the end of the 1980s, the persuasive powers of some of John's future colleagues at the University of Western Ontario, following on a decade of Thatcherism, convinced the Bells to swap Londinium, as it became known for purposes of disambiguation, for London, Ontario. This was a massive disruption, involving significant adaptation, no matter how willingly undertaken. Academic life remained relatively familiar, and, for John, the major difficulty now was being surrounded by students and colleagues with philosophical, rather than mathematical, background. But Canada generally, and London in particular, involved considerable cultural bemusement for both the Bells. John greatly enjoyed regaling visitors with lists of pros, cons and constants. John was eventually able to reconcile himself to the strange ways of Canadians; one suspects that he would thrive anywhere, but it has been the good fortune of Canadian philosophy that he has been thriving in Canada. The Bells came to admire greatly Canada's well-run, multi-cultural mix, largely untroubled as it is by the hatreds that had scarred old London in the thirties and which were, by the time the Bell family left, beginning to reassert themselves in the darker corners of British political and social life.

Once they'd arrived in Canada, and escaped a particularly miserly landlord (a source of new obsessions!) by moving to their own home, the Bells kept a busy social schedule, one that included frequent visits from graduate students. It was during this transition period that the third editor got to know the Bells, again first as John's Ph.D. student, then as friend. For the shyer or more reserved of the Canuck students, a visit to the Bell household promised intimidation, what with John's boisterousness on home soil added to his brilliance. But, once arrived at the house, Mimi (again firmly in charge) ensured that everyone recognized how welcome they were. This third editor, who doesn't think of himself as a shrinking violet, fondly remembers spending a good portion of his first party at the Bells in the kitchen helping Mimi prepare the food, a respite from the storm of conversation in the living room. That job involved, among other things, the peeling of vegetables the editor had never seen before, and was a first hint of the fact that every visit to the Bells involved marvellous food, prepared with flair and imagination.

While John pursued his usual academic life, one of enormous productivity and Oxonian "effortless superiority," Mimi's endeavours in new London were more varied. For years, she worked at a local shelter for victims of domestic violence, putting her deeply felt (if not often voiced) political commitments into action; while she didn't talk about this work often, at least at social gatherings, the work was essential and potentially dangerous. In more recent years, she was involved in helping new immigrants and refugees get settled in Canada, steering them through the formidable bureaucracy involved in finding adequate housing, language classes, training and jobs. She was also eventually persuaded to share her culinary expertise with circles outside the guests at her house. For years she taught cooking classes, and she set up and ran her own catering business. She wrote a cookery book (which is partly responsible for the difficulty one of the editors has in keeping his weight under control). She wrote poetry and painted, including work that found its way into exhibitions. All this in addition to raising the Bells' son, Alex, and doing the bulk of the work keeping a frenetic household running.

We cannot, sadly, dedicate this book to Mimi, as John so often did with his own books. Instead, we dedicate it to her memory, with wonderfully fond recollection.

Waterloo (Canada) — David DeVidi
Montreal (Canada) — Michael Hallett
St. Andrews (UK) — Peter Clark

Contents

Contributors

Peter Aczel Former Professor of Mathematical Logic and Computer Science, University of Manchester

Frank Arntzenius Sir Peter Strawson Fellow in Philosophy, University College, Oxford

Joel Bennhall Musicologist and Composer

Ken Binmore, CBE, FBA Emeritus Professor of Economics, University College, London

Peter Clark Proctor and Provost of St Leonard's College, Vice-Principal and Dean of Graduate Studies; Professor of Philosophy, University of St. Andrews

William Demopoulos Professor of Philosophy, Killam Research Fellow, University of Western Ontario

David DeVidi Professor of Philosophy, University of Waterloo

Max Dickmann Member of the Équipe de Logique Mathématique, Université Paris VII, Paris, France; Associate at the Projet: Topologie et Géométrie Algébriques, Institut de Mathématiques de Jussieu, Paris

Robert DiSalle Professor of Philosophy, University of Western Ontario

Sir Michael Dummett, FBA, D.Litt Wykeham Professor of Logic Emeritus, University of Oxford

Richard Feist Dean and Associate Professor of Philosophy, St Paul University, Ottawa

Michael Hallett John Frothingham Chair of Logic and Metaphysics, McGill University

Wilfrid Hodges, FBA Former Professor of Mathematics, Queen Mary University of London

Colin Howson Professor of Philosophy, University of Toronto

Daniel Isaacson University Lecturer in Philosophy of Mathematics, University of Oxford; Fellow of Wolfson College, Oxford

Elaine Landry Associate Professor of Philosophy, University of California, Davis

F. William Lawvere Professor of Mathematics, State University of New York, Buffalo

Moshé Machover Emeritus Professor of Philosophy, King's College, London

J.P. Mayberry Research Fellow in the Department of Philosophy, University of Bristol

Alberto Peruzzi Professor of Theoretical Philosophy, University of Florence

Graham Priest Boyce Gibson Professor of Philosophy, University of Melbourne and Distinguished Professor of Philosophy, Graduate Center, City University of New York

Michael Redhead Centennial Professor, London School of Economics and Emeritus Professor of History and Philosophy of Science, Cambridge University

Adam Rieger Senior Lecturer in the Department of Philosophy, University of Glasgow

Giovanni Sambin Professor of Mathematical Logic, University of Padua

George Wilmers Lecturer in Mathematics and Member of the Mathematical Logic Research Group, University of Manchester

Elie Zahar Reader Emeritus in Logic and Scientific Method, London School of Economics

Part I
History and Philosophy of Mathematics and Logic

Chapter 1
On Logicist Conceptions of Functions and Classes

William Demopoulos

1 Introduction

John Bell arrived in "new" London in 1989, a refugee from the academy under Margaret Thatcher. We soon became good friends, and during the early years of our friendship we collaborated on two papers (Bell and Demopoulos, 1993, 1996). The first of these collaborations was a paper on the foundational significance of results based on second-order logic and Frege's understanding of his Begriffsschrift; the second was on various notions of independence that arise in connection with elementary propositions in the philosophy of logical atomism. I retain fond memories of both collaborations; they proceeded quickly and almost effortlessly. In this contribution to John's Festschrift, I propose to revisit our paper on Frege. That paper was occasioned by (Hintikka and Sandu, 1992), which questioned whether Frege's understanding of second-order logic corresponded, in his framework of functions and concepts, to what we would now regard as the standard interpretation, the interpretation that takes the domain of the function variables to be the full power-set of the domain of individuals. Hintikka and Sandu maintained that it did not on the basis of a number of arguments, all of which they took to show that Frege favored some variety of non-standard interpretation for which the domain of the function variables is something less than the characteristic functions of *all* subsets of the domain over which the individual variables range.

Although Hintikka and Sandu based their contention on many different considerations, the one to which they assigned the greatest weight was that Frege lacked the concept of an arbitrary correspondence from natural numbers to natural numbers (or from real numbers to real numbers).[1] Hintikka and Sandu contrasted their view of

W. Demopoulos (✉)
Professor of Philosophy, Killam Research Fellow, University of Western Ontario,
London, ON, Canada
e-mail: wgdemo@uwo.ca

[1] The textual considerations Hintikka and Sandu advance in support of this interpretive claim were shown in (Burgess, 1993) to rest on simple misreadings of Frege's *Venia docendi*, and in the case of Frege's relatively late paper, "What is a function?," on passages that are far from unequivocal.

D. DeVidi et al. (eds.), *Logic, Mathematics, Philosophy: Vintage Enthusiasms*,
The Western Ontario Series in Philosophy of Science 75,
DOI 10.1007/978-94-007-0214-1_1,

Frege on the concept of a function with that of Dummett who, in *FPL*,[2] had maintained that Frege's notion of a function coincides with the concept of an arbitrary correspondence:

> And it is true enough, in a sense, that, once we know what objects there are, then we also know what functions there are, at least, so long as we are prepared, as Frege was, to admit all "arbitrary functions" defined over all objects (*FPL* p. 177, quoted by (Hintikka and Sandu, 1992, p. 303)).

However in *FPM* Dummett reversed himself on this question, and suggested that Frege implicitly assumed the adequacy of a substitutional interpretation of his function variables:

> ... Frege fails to pay due attention to the fact that the introduction of the [class] abstraction operator brings with it, not only new singular terms, but an extension of the domain. ... [I]t may be seen as making an inconsistent demand on the size of the domain D, namely that, where D comprises n objects, we should have $n^n \leqslant n$, which holds only when $n = 1$, whereas we must have $n \geqslant 2$, since the two truth values are distinct: for there must be n^n extensionally non-equivalent functions of one argument and hence n^n distinct value ranges. But this assumes that the function-variables range over the entire classical totality of functions from D into D, and there is meagre evidence for attributing such a conception to Frege. His formulations make it more likely that he thought of his function-variables as ranging over only those functions that could be referred to by functional expressions of his symbolism (and thus over a denumerable totality of functions), and of the domain D of objects as comprising value-ranges only of such functions (*FPM*, pp. 219–220).

In *FPM* [Chapter 17, *How the serpent entered Eden*] Dummett put forward a further consideration that can be seen as lending support to Hintikka and Sandu's view that Frege favored some sort of non-standard interpretation. In the course of his discussion of the attempted consistency proof of *Gg* § 31, Dummett suggested that Frege erroneously supposed that the consistency of certain restricted interpretations of the function variables extends to the system of *Gg*'s theory of functions and classes in the context of full second-order logic. This unjustified, and unjustifiable, assumption is what Dummett means by "Frege's amazing insouciance regarding the second-order quantifier" [*FPM* p. 218]. Subsequent developments have shown that there is a systematic consideration in favor of Dummett's claim and, a fortiori, in favor of Hintikka and Sandu's suggestion that Frege assumed a non-standard interpretation: some restricted interpretations yield consistent fragments of *Gg*.[3] Hence it might be that Frege was blind to the inconsistency of his system because he considered the question of consistency only from the perspective of such a restricted interpretation and failed to ask whether what holds of it, holds in general.

But although Dummett shares Hintikka and Sandu's conclusion that Frege tended toward a non-standard interpretation, his analysis does not support Hintikka and

[2] I use the following mnemonic abbreviations: *FPL* for (Dummett, 1981), *FPM* for (Dummett, 1991), *Grundlagen* for (Frege, 1884), *Gg* for (Frege, 1903), *FoM* for (Ramsey, 1990), *Principia* for (Whitehead and Russell, 1910), *Tractatus* for (Wittgenstein, 1922).

[3] See for example the system **PV** discussed in (Burgess, 2005, § 2.1).

Sandu's evaluation of Frege's foundational contributions. If we follow Dummett, Frege missed the fact that the consistency of *Gg*, relative to a nonstandard interpretation, does not necessarily extend to its consistency when the logic is given a full interpretation. This is certainly an oversight, but it is not the oversight that is appealed to in those of Hintikka and Sandu's criticisms of Frege that so offended some of their critics, as for example, whether, without having isolated the notion of a standard interpretation, Frege could have even conceptualized results like Dedekind's categoricity theorem.

Apropos of Frege and categoricity, in their response to Hintikka and Sandu, (Heck and Stanley, 1993) observed (what (Heck, 1995) elaborates in detail) that Frege *proved* an analog of Dedekind's theorem using his own axiomatization of arithmetic (one that is only a slight variant of the Peano-Dedekind axiomatization). And in our response, John and I argued that the relevance of the dependence of this and other similar foundational results on the standard interpretation is not entirely straightforward since the actual arguments which support them have the same character, whether one is working in second-order logic or in a suitably rich first-order theory such as Zermelo-Fraenkel set theory. Hintikka and Sandu's claim that Frege could not even have formulated (let alone appreciated) these results because of their dependence on the standard interpretation is therefore incorrect both historically and methodologically. It is incorrect historically because Frege successfully proved a categoricity theorem like Dedekind's. And it is incorrect systematically because essentially the same argument establishes the categoricity of second-order arithmetic in any of the usual systems of set theory. And surely it is implausible that only someone familiar with the categoricity of the Peano-Dedekind Axioms as a theorem of second-order logic has really grasped the theorem or its proof. At most, Frege might be charged with having missed a subtlety concerning the distinction between formal and semi-formal systems; but this is hardly surprising for the period in which he wrote.

In our paper, John and I accepted Dummett's view in *FPL* and based our claim that Frege's interpretation of the function variables was the standard one on the premise that Frege's concept of a function coincides with the set-theoretic notion of an arbitrary correspondence, in which case the domain of the function variables is in one-one correspondence with the power-set of the domain of the individual variables. Our thought was that whatever covert role the neglect of Cantor's theorem might have played in the inconsistency of *Gg*, it is unlikely that Frege sought to ignore the theorem by assuming that the totality of functions, like the totality of expressions, is countably infinite. But we sided with Dummett in *FPM* and supposed that Frege might very well have been misled into assuming that what holds for certain countable interpretations of the function variables holds in general; hence we agreed with Dummett's evaluation of the sense in which Frege missed the significance of the possibility of different interpretations for his program.

More recently, reflection occasioned by reading (Sandu, 2005) has convinced me that the equation of Frege's concept of a function with the notion of an arbitrary correspondence should be reconsidered, and that it might be fruitful to reconsider it from the perspective of Ramsey's interpretation of *Principia*'s propositional functions.

Let us call *classes* the extensions of functions in the *logical* sense of "function," a notion I will explain in the sequel. Then we wish to explore whether the classes determined by such functions correspond to only a fragment of the sets associated with arbitrary correspondences. The assumption that something like this is correct evidently underlies Ramsey's proposed reinterpretation of the first edition *Principia*'s notion of a propositional function in his essay, *FoM*. That essay is usually cited for its separation of the paradoxes and its isolation of the simple hierarchy of *Principia*'s functions of lowest order. It is generally assumed that by this hierarchy Ramsey understood a hierarchy of functions interchangeable with the standard hierarchy of sets, modulo the condition that *Principia*'s hierarchy is stratified rather than cumulative. In fact, Ramsey regards *Principia*'s simple hierarchy of functions of lowest order as inadequate, and he devotes a central chapter of his essay [*FoM*, Chapter IV] to criticizing *Principia*'s notion of propositional function and arguing for its extension. He achieves this extension, and more, by the introduction of the notion of a propositional function in extension, or what I will call an *extensional propositional function*. The consideration of Ramsey's criticisms of *Principia* has led me to the view that the interest of Hintikka and Sandu's paper has less to do with standard vs. non-standard interpretations of second-order logic than with Frege's concept of a function. As I now see it, the chief interest of their paper, although not perhaps their principal aim, is the suggestion that the difference between logical and set-theoretic notions of *function* parallels the well-known contrast between the logical notion of *class* and the mathematical notion of *set*.

My strategy for the balance of the paper is to proceed anti-historically by reviewing Ramsey's notion of an extensional propositional function, its difference from what Ramsey calls the predicative propositional functions of *Principia*, and the uses to which Ramsey put the notion in his defense of a modified logicist position. To motivate the notion of an extensional propositional function, I will begin by briefly recounting the relevant Tractarian background to Ramsey's thought. I will then describe two reconstructions of the Axiom of Infinity, one with, and one without, the notion of an extensional propositional function. Both reconstructions trace back to Ramsey. The reconstruction involving the notion of an extensional propositional function was elaborated in his *FoM*; my discussion of it constitutes the principal novelty of the present paper. Having clarified the predicative and extensional notions, I will return to Frege's functions. I will argue that they have a feature in common with *Principia*'s propositional functions, and that it is plausible to argue, on the basis of this, that they should be distinguished from the notion of an arbitrary correspondence. However the situation is not entirely straightforward, and we will see that there are also reasons to suppose that Frege's notion can be regarded as falling in line with the mathematical (or "extensionalist") tradition.

2 The Tractarian Background to Ramsey's Extensional Propositional Functions

There are three key ideas of the *Tractatus* that form the background to Ramsey's notion of an extensional propositional function: (i) A proposition is significant only insofar as it partitions all possible states of affairs into two classes. This is what (Potter, 2005) has called "Wittgenstein's big idea." It implies that tautologies cannot be regarded as significant propositions, and insofar as the propositions of logic are all tautologies, the propositions of logic are not significant propositions; in particular, they cannot be merely more general than the truths of zoology or any other special science as Russell claimed (in (Russell, 1919, p. 169)). (ii) Objects, which constitute the substance of the world, are constant across alternative possibilities; so also therefore is their number. Potter puts the point well when he says that "what changes in the transition between [possibilities] is how objects are combined with one another to form atomic facts; what the objects are does not change because they are the hinges about which the possibilities turn and hence are constant" (Potter, 2005, p. 72). It follows that there cannot be a significant proposition regarding the number of objects. At best language can "reflect" the cardinality of the world. (iii) Since atomic propositions are logically independent of one another, statements of identity cannot express significant propositions, for if they did, they would establish relations of logical dependence among atomic propositions.

It is a consequence of Wittgenstein's rejection of identity, a rejection with which Ramsey concurred, that the notion of class implicit in Russell's theory of classes must be an "accidental" one, that is, one according to which it is possible that every property might be shared by at least two individuals, with the consequence that there would be nothing to answer to the number 1 [*FoM*, p. 213]. If we could express significant propositions with the use of identity, then among the properties of *a* there would be the property of *being identical with a*, and this would settle the question of the existence of the number 1 on Russell's theory. But acceptance of the third Tractarian thesis argues against a solution which, like this one, treats identity as a possible constituent of a significant proposition. Hence if we follow the *Tractatus* on identity, Russell's theory cannot recover the notion of class that its account of number—and indeed, of the whole of mathematics—requires. This is an objection that is in some ways more fundamental than the standard objection to the apparent contingency of the Axiom of Infinity, since even if that axiom could in some way be made acceptable, according to the present objection, it is still *possible* on Russell's theory that there might not be enough numbers to go around. Hence, for this and other reasons we will soon come to, mathematics cannot be based on *Principia*'s theory of classes.

Ramsey's notion of an extensional propositional function emerged from his attempt to address this and other defects in *Principia*'s theory of classes, such as those surrounding its account of Choice. But perhaps the most important consideration in favor of the notion was that it proved essential to a formulation of Infinity that accords with the first and second Tractarian theses. For, if we follow Russell and simply take Infinity to assert that for every inductive cardinal number n, there is a class whose cardinality is given by n, then we treat the axiom as an accidental truth concerning the number of things of a particular sort. As a consequence, we miss the peculiar character of claims regarding the cardinality of objects—in the present case, of individuals—enunciated by (ii): The existence of objects is a precondition for significant propositions and cannot be a subject with which such propositions deal; the same is true of their number. In order to see how Ramsey's notion of an extensional propositional function leads to a formulation of Infinity that addresses these issues, it will be useful to begin with a formulation of the axiom that has all the ingredients of the formulation of *FoM* *except* for the notion of an extensional function.

3 Infinity Without Extensional Propositional Functions

Potter (2005) discusses an early unpublished fragment of Ramsey's ("The number of things in the world") which contains what Potter calls "Ramsey's transcendental argument for Infinity." Potter discusses this argument at length, and canvasses a number of questions of Ramsey-interpretation which the fragment raises. The transcendental argument evidently precedes Ramsey's discovery of extensional propositional functions. For my purposes, the argument's most interesting feature is the formulation of Infinity it suggests, rather than the considerations in favor of the axiom that it advances. I will refer to this as "Ramsey's early formulation of Infinity"; however, my discussion is not based on Ramsey's unpublished fragment, but on Potter's reconstruction of it. The historical accuracy of Potter's account is of course completely irrelevant to its usefulness for motivating the formulation of Infinity that Ramsey gives in *FoM*. That formulation is my main concern, and I will take it up in the next section.

Let φx be any propositional function, and let Tx be $\varphi x \vee \neg\varphi x$. Then to say that there are at least n individuals write

$$p_n =_{Df} \exists x_1 \ldots \exists x_n T x_1 \ \& \ldots \& \ T x_n,$$

where, here and elsewhere, Ramsey assumes Wittgenstein's "nested variable convention," which says that whenever a variable occurs within the scope of another variable, it is to be assigned a different individual as its value.[4] Thus, for example, "$\exists x \exists y \ldots$" is read, "there is an individual x and *another* individual y ...." Then if there are n individuals $a_1, \ldots, a_n$, this is reflected by the tautology $Ta_1 \ \& \ldots \& \ Ta_n$.

---

[4] What I am calling Wittgenstein's nested variable convention is explained in detail in (Wehmeier, 2008) in his discussion of "W-logic." Wehmeier (2009) discusses a variant of W-logic ("R-logic") which originated from Ramsey's initial misunderstanding of Wittgenstein's convention.

Now let p_ω be the "logical product" of the p_n, for all finite n. Then p_ω is Ramsey's early formulation of Infinity, where what Ramsey means by the logical product of the p_n can be gleaned from what he says about logical sums:

> A logical sum is not like an algebraic sum; only a finite number of terms can have an algebraic sum, for an "infinite sum" is really a limit. But the logical sum of a set of propositions is the proposition that these are not all false, and exists whether the set be finite or infinite (*FoM* p. 219, n. 1).

Ramsey clearly takes this to apply mutatis mutandis to the notion of a logical product, in which case we are to understand by p_ω the proposition that every member of the p-series is true. This is expressed by the infinite conjunction of the p_n, which is what Ramsey means by the proposition that every member of the p-series is true. Hence, if there are infinitely many individuals, this is reflected by the tautology Ta_1 & ... & Ta_n & ..., which asserts, or appears to assert, the infinity of the number of objects of the type of individuals, and is to be contrasted with a claim about the number of things there are of some particular kind.

To construct an "Axiom of Infinity" regarding some sort of thing, we also proceed by first describing a series of propositions such as

$$q_1 =_{Df} \exists x(x \text{ is a hydrogen atom})$$
$$q_2 =_{Df} \exists x \exists y(x \text{ and } y \text{ are hydrogen atoms})$$
$$\vdots$$

where once again Wittgenstein's convention regarding nested variables is assumed. In this case, each q_n is an ordinary empirical proposition concerning not objects in general but the number of things of a particular kind, and q_ω (the product of all the q_n) says that there are infinitely many of them. The *Tractatus* imposes no prohibition against the significance of a proposition like q_n since it is a genuine possibility that there are at least n hydrogen atoms, even if there are in fact only m of them for m less than n. It is likewise a genuine possibility that there are *infinitely many* hydrogen atoms. All such claims mark genuine possibilities in the sense demanded by the Tractarian notion of a significant proposition.

There is a well-known distinction of Carnap's that bears on the evaluation of Ramsey's early proposal. Carnap (1931, p. 41) distinguished between two types of logicist reduction. Let us call a reduction *type (a)* if it defines all concepts of a mathematical theory in terms of those of logic, and *type (b)* if, in addition to providing definitions of a mathematical theory's concepts in terms of logical concepts, it derives the axioms of the theory from purely logical axioms. It is worth remarking that although Russell may have sometimes expressed the view that a type (a) reduction suffices for logicism, Frege was always clear on the need to secure both type (a) and type (b) reductions.

In *FoM* [pp. 166–167], which was written after the fragment containing Ramsey's early formulation of Infinity, there is an anticipation of Carnap's distinction between type (a) and type (b) reductions. Ramsey gave the distinction an interesting interpretation when, under the clear influence of Wittgenstein, he remarked that while a type (a) reduction might illuminate the *generality* of mathematics, a type (b) reduction would address what is truly distinctive about it: its *necessity*. But if Ramsey and Wittgenstein are the source of the view that necessity is the chief characteristic of mathematical propositions that requires explanation, it is important to remember that there is not a trace of an interest in necessity in Frege's *Grundlagen* or in the first edition *Principia*. Both works stress the generality of mathematics and the possibility that this might be illuminated by assimilating it to the generality of logic. Ramsey's aim was to illuminate the necessity of mathematics by assimilating it to the necessity of logic. But he also saw that there are at least two ways in which this might be accomplished. We can try reducing the propositions of mathematics to those of logic after the fashion of a type (b) reduction of the sort Frege tried, unsuccessfully, to achieve. Alternatively, we might attempt to show that the propositions of mathematics have the same kind of necessity that we find in the propositions of logic, *not* by deriving mathematics from logic, but by an analysis of the necessity mathematical propositions exhibit. Ramsey sought to explicate this characteristic by proposing that the propositions of mathematics, like those of logic, are "tautologies," an idea that evidently has its basis in the *Tractatus*. However, by a tautology Ramsey did not mean something that is *truth-table decidable*, as is clear from his discussion of the Axiom of Choice where he mentions with approval the possibility of "a tautology, which could be stated in finite terms, whose proof was, nevertheless, infinitely complicated and therefore impossible for us" [*FoM*, p. 222]. We will return to Ramsey's notion of *tautology*.

How do these considerations bear on Ramsey's early formulation of Infinity? There are two objections to this formulation, one more telling than the other. The less telling objection is that the existential claims expressed by the propositions of the p-series fall short of possessing the necessity of logical propositions. The fact that Tx_1 & ... & Tx_n becomes a tautology when there are individuals $a_1, \ldots, a_n$ which accord with Wittgenstein's convention does not show that p_n is a tautology, and therefore does not show that the axiom is necessary. In a domain of $m < n$ objects, p_n is simply false, a fact which on Ramsey's formulation is represented by the absence of values for the variables of p_n that accord with Wittgenstein's nested variable convention. But to this objection Ramsey can respond that it was not his intention to show that p_n, let alone p_ω, is a tautology. Rather, the point of the formulation was to capture the idea that if there are n individuals, then p_n is *witnessed* by the tautology, Ta_1 & ... & Ta_n. The objection mistakenly assumes that the only way to establish the logical necessity of Infinity is to show that it holds in every "universe of discourse." This is one sense in which a proposition may be seen to be a logical proposition, but it is not the sense Ramsey's formulation was intended to capture. A proposition can be one whose truth is witnessed by a tautology, and thus be a logical proposition in Ramsey's sense, without holding in every universe of discourse. And this is precisely the case with Infinity.

The second objection is harder to articulate and will only become clear after we have examined the formulation of Infinity in *FoM* that is based on Ramsey's notion of an extensional propositional function. As a first approximation, the difficulty is that the tautological character of the propositional function T_x which figures in Ramsey's p-series is not integrated into the theory of functions and classes as it would be on a formulation of Infinity that was derived from an analysis of *class* and *propositional function*. For this reason, Ramsey's formulation carries little conviction as an account of the necessity that a fundamental axiom like Infinity is supposed to exhibit. Since, according to Ramsey's reading of the *Tractatus*, the explanation of this characteristic is the central task of a philosophy of mathematics, it seems likely that the source of his dissatisfaction with his early proposal, and the reason he never published it, was the recognition that it fails to present a convincing account of Infinity's necessity.

4 Infinity with Extensional Propositional Functions

In *FoM* Ramsey rejected the notion of propositional function that we associate with the 1910 *Principia*. Such functions are, in Ramsey's phrase, "predicative" in the sense that the proposition φa which the propositional function φ assigns to a says or *predicates* the same thing of a as the proposition φb, which φ assigns to b, does of b. This connection with predication is essential to any *logical* notion of the functions which determine classes, and it stands in contrast with the idea that a function is an *arbitrary* correspondence. According to Ramsey's new conception, the way to accommodate the broader notion of set which is implicit in the idea of an arbitrary correspondence is by conceiving of a propositional function as an arbitrary mapping, in the present case from individuals to propositions, with φa and φb merely the values of φ for a and b as arguments. Here "arbitrary" means both that the mapping is not constrained to preserve a property in the sense that the propositions φa and φb are not required to say the same thing of a and b, *and* that it allows all combinatorial possibilities of functional pairings of individuals with propositions. On Ramsey's view, it is only *propositions* that involve properties which are predicable of individuals; the *classes* which propositional functions determine should not be constrained by the demand that their elements share a common property.[5]

When this purely extensional notion of a propositional function is adopted, there is, of course, no difficulty with the existence of unit classes and the number 1, since

[5] By an abuse of notation which uses the same symbol φ in two very different ways—both for a mapping from individuals to propositions and for a predicate of the language—a predicative propositional function φ takes an individual a to a proposition of the form φa, i.e., to a proposition which is expressed by a sentence consisting of the concatenation of the predicate φ with the constant a. The class determined by φ is the class of all a such that φa is true. If φ is extensional but not predicative, the class determined by φ is the class of all individuals a which φ maps to truths. Under an extensional understanding of propositional functions, there is not in general a correspondence between propositional functions and predicates of the language, so that the association with propositions is in this sense "arbitrary." It is also arbitrary in the stronger sense of allowing all possible pairings of individuals with truth values.

every object has a mapping to propositions that is unique to that object.[6] Notice however that, of the two features of extensional propositional functions, it is only the second, namely the fact that such functions exhaust all combinatorial possibilities, that the solution to this difficulty depends upon, and this can readily be accommodated within a framework that considers only pairings of individuals with truth values. The "propositional" aspect of propositional functions, the fact that they map individuals to propositions, doesn't enter into the solution of the problem raised by the possibility that every property is shared by at least two individuals.

Ramsey says of the Axiom of Choice that given his notion of an extensional propositional function, "it is the most evident tautology ... [and not something that] can be the subject of reasonable doubt" [*FoM*, p. 221]. He then proceeds to show how, on *Principia*'s understanding of *propositional function* and *class*, the axiom, though not a contradiction, is also not a tautology. Ramsey does not pause to explain why on his account the axiom is an evident tautology, but as John once observed in conversation, after the existence of singletons has been shown to follow from the concept of an extensional propositional function, Choice, in the form "If K is a family of disjoint non-empty classes, then K has a choice class," is clearly true. For if classes are determined by extensional functions, it is evident that every class in K can be shrunk to a singleton; the sum of all such singletons is the required choice class. Like the difficulty with the number 1, the foregoing justification of Choice depends only on the fact that extensional propositional functions exhaust all combinatorial possibilities.

Clearly, what Ramsey means by the tautologousness of Choice is not captured by the idea that it holds in all universes of discourse, still less that it is truth-table decidable. Ramsey perceived that there are "interpretations" of *Principia*, by which he meant understandings of the notion of *propositional function*, of which the one favored by Whitehead and Russell is just one example, under which Choice can be shown to be false. And in a remark made in the course of a discussion of Infinity and the possibility of saying something about the cardinality of the world given his adherence to the second Tractarian idea to which we called attention in Section 2, Ramsey shows his appreciation of the possibility of falsifying a fundamental axiom by "imagining a universe of discourse, to which we may be confined, so that by 'all' we mean all in the universe of discourse" [*FoM*, p. 224]. From this we may conclude that Ramsey's understanding of the "tautologousness" of Choice is that relative to his extensional understanding of "propositional function" and the intended meaning of "all," and relative perhaps as well to the intended meanings of the propositional connectives, Choice is evidently true.[7] Ramsey expresses this by saying that, under these conditions, the axiom is "an evident tautology," to draw attention to the fact that its obvious demonstration in the finite case proceeds by inspecting all the com-

---

6 For example the singleton of a is determined by a propositional function which maps a to an arbitrarily selected truth and maps every other individual to a falsehood.

7 Sandu misses this point about the tautologousness of Choice when he criticizes Ramsey's contention that Choice is a tautology and argues that since "... there are models of set theory in which the Axiom of Choice is false, ... it cannot, therefore, be a tautology" (Sandu, 2005, p. 252).

binatorially possible relations between classes and their elements. In the general case, we are incapable of carrying out such an inspection, but this does not affect the truth of the principle on an understanding of *propositional function* that admits all combinatorial possibilities.

The situation is different with Ramsey's analysis of the Axiom of Infinity. Here, in addition to the fact that Ramsey's functions exhaust all combinatorial possibilities, the propositional character of extensional propositional functions is an indispensable component of the solution to the problem the axiom presents. To understand Ramsey's account of Infinity, suppose we try expressing the idea that there are at least two things by the proposition

$$\exists x \exists y \neg (\varphi)(\varphi x \equiv \varphi y), \qquad (*)$$

where it is assumed that propositional functions are to be understood predicatively, after the manner of Whitehead and Russell. Now consider a universe of discourse U containing precisely two individuals a and b which have all their properties in common. For Ramsey there is nothing absurd in the idea that two things might share all their properties and thus be indistinguishable but distinct; and since he rejects identity he will not allow an appeal to the property, *being identical with a* to insure the truth of $(*)$ in U under such a predicative understanding of propositional functions. Under these circumstances, $(*)$ will fail to reflect the fact that a and b comprise a two-element universe. But for Ramsey that it should even be *possible* that $(*)$ can fail in this way shows that, under its predicative interpretation, $(*)$ purports to express a general truth, and hence, a significant proposition; it cannot therefore be the correct expression of the idea that there are at least two things.

Let us now consider what happens when propositional functions are understood extensionally. Assuming Wittgenstein's nested variable convention, $(*)$ *must* be true in any two-element universe such as U. For if we consider all possible mappings φ, there must be one among them that assigns a to the negation of whatever proposition it assigns b. Hence the function $\neg(\varphi)(\varphi x = \varphi y)$ will map (a, b) to the negation of a contradiction, and therefore on Ramsey's understanding of *propositional function*, the truth of $(*)$ in U will be witnessed by a tautology. This contrasts with the predicative interpretation under which $(*)$ can fail in a two-element universe, so that even in cases where it holds, it does so under a predicative understanding of *propositional function* only in virtue of a contingent fact about individuals and their properties. It also contrasts with Potter's reconstruction of Ramsey's early formulation, since the fact that $(*)$ is witnessed by a tautology is not an ad hoc stipulation, but a consequence of Ramsey's extensional understanding of the nature of a propositional function.

In light of the foregoing considerations, let us define the propositional function

$$T(x, y) =_{Df} (\varphi)(\varphi x = \varphi y).$$

As we have seen, when propositional functions are understood extensionally, the function $T(x, y)$ maps to a tautology when the values of x and y are the same and to a contradiction otherwise. Thus if propositional functions are understood extensionally, then if there are two individuals, and we understand $\exists x \exists y T(x, y)$ to say, "There is an x and *another* y such that T(x,y)," then $\exists x \exists y T(x, y)$ reduces to a contradiction, while $\exists x \exists y \neg T(x, y)$ reduces to a tautology. This suggests a *new* p-series defined as follows:

$$p_1 =_{Df} \exists x T(x, x)$$
$$p_2 =_{Df} \exists x \exists y \neg T(x, y)$$
$$p_3 =_{Df} \exists x \exists y \exists z \neg T(x, y) \ \& \ \neg T(x, z) \ \& \ \neg T(y, z)$$
$$\vdots$$

For each n, p_n is true in a universe of n individuals. The witnesses to the propositions of the series alternate between a tautology, when a member of the series is true, and a contradiction, when it is false. Ramsey's new formulation of the Axiom of Infinity is given by the logical product $p_{\aleph_0}$ of the propositions of this series. Observe that if there are $\aleph_0$ individuals, $p_{\aleph_0}$ is witnessed by a product of tautologies, and if fewer than $\aleph_0$, its falsity is witnessed by a contradiction.

To appreciate Ramsey's achievement, consider how his proposal differs from the following reduction of truths to tautologies. (For simplicity, we restrict ourselves to quantifier-free molecular formulas, since this suffices to illustrate the point of difference with the alternative reduction to which I wish to call attention.) Replace a quantifier-free molecular formula φ by its equivalent disjunctive normal form. Next replace any literal of φ's disjunctive normal form which is true by $x = x$, and replace a literal by $x \neq x$ if it is false. Then the resulting formula is a truth function of "tautologies" and "contradictions," and it reduces to one or the other according to whether it is true or false. This is clearly artificial since it enables us to "reduce" to tautologies and contradictions many propositions which are merely true or false. But *FoM*'s proposal regarding Infinity does not simply replace truths with tautologies, and falsehoods with contradictions; it *derives* the tautologous or contradictory character of a witness to a proposition of the p-series from an analysis of the notion of a propositional function: a proposition of the p-series is witnessed by a tautology or contradiction as a consequence of the cardinality of the domain *and the extensional character of propositional functions*. In this respect Ramsey's reduction procedure stands in marked contrast with one which merely stipulates the tautologousness of the witness to the truth of a proposition.

Does the formulation of Infinity based on Ramsey's reinterpretation of the notion of a propositional function provide a logical justification in the usual sense? It does not. Ramsey's achievement consists in showing how the witness to the truth of Infinity reduces to a product of tautologies when the cardinality of the domain of individuals is infinite. In conformity with the first of the Tractarian theses noted earlier, the truth of the axiom is not expressed by a genuine proposition; rather, the cardinality of the world is "shown"—or as I prefer to say, "witnessed"—by a tautology. This is not the provision of a logical justification for Infinity that purports to show that it is true in every universe of discourse, but an explanation of how, *as a consequence of the notion of an extensional propositional function*, a witness to the truth of the axiom reduces to a product of tautologies in any domain in which the axiom holds, and how a witness to its falsity reduces to a contradiction in any in which it does not hold. By integrating the idea that Infinity, if true, is witnessed by a tautology into the notion of a propositional function, it addresses the second of the two objections we noted in our discussion of the formulation of the unpublished fragment.

5 Extensional Propositional Functions and Logicism

Hintikka and Sandu's discussion of Fregean concepts shares a number of similarities with Ramsey's discussion of the 1910 *Principia*'s propositional functions: Like Ramsey, Hintikka and Sandu regard the notion of *class* on which the early logicists relied as inadequate by comparison with the extensionalist tradition's notion of *set*; and both Hintikka and Sandu and Ramsey base their analyses of the inadequacy of classes on a failure of the early logicists to conceive of a function as an arbitrary correspondence. But there are also significant differences. By contrast with Hintikka and Sandu, Ramsey took himself to be contributing to a more defensible version of logicism. He hoped to show that his formulation of an appropriate extensionalist replacement of *Principia*'s predicative propositional functions would address an internal difficulty with the work, one surrounding the adequacy of its account of mathematical propositions.

Ramsey's extensional propositional functions attempt to marry two seemingly incompatible ideas: they preserve the letter of the logicist thesis that classes are determined by functions, but they also invoke the combinatorial notion of set by representing a propositional function as an arbitrary functional pairing of individuals with propositions. As a consequence, one gives up the Russellian idea that a propositional function determines a class in terms of an antecedently available property. But Ramsey's extensionalism is in some respects congenial to Frege. This is because Frege's functions satisfy a condition of extensionality in the sense that functions whose courses-of-values coincide are not distinguished, and because Fregean functions form a simple hierarchy which is not constrained by ramification. Despite these points of agreement with an extensionalist viewpoint, Frege's assimilation of concepts to functions which map into truth values is usually understood to share with the 1910 *Principia* the idea that the correspondence is not arbitrary, but is constrained by the principle that if a function maps two objects to The True, they must fall under a common concept. Hence Frege's generalization of the function concept to include those that map to truth values is, in Ramsey's sense of the term, a generalization to a notion of function that is just as *predicative* as Russell's.

If I have understood him, Sandu's suggestion in (Sandu, 2005) is that his paper with Hintikka should be understood as arguing that the predicativeness of Frege's functions is sufficient to show that they do not exhaust all mappings between objects and truth values, and that this lends support to his and Hintikka's claim that in general Frege's functions are not arbitrary correspondences. But it is at this point that I think a difference between *Principia*'s propositional functions and Frege's concepts may be important. For *Principia*, a function maps to the truth values only by "passing through" a proposition; this, after all, is why they are called *propositional* functions. But Frege's concepts map directly to the truth values. To be sure, the informal explanation of the Fregean idea that concepts are a kind of function invariably proceeds by saying that a function maps an object to The True just in case the object falls under the associated concept. But as pedagogically natural as this explanation may be, it is also misleading, since Frege's assimilation of concepts to functions does not exclude the possibility that a concept might be no more accessible to us than the correspondence which a mapping to truth values establishes. In such a case, there would be little to choose between a Fregean function and the extensional notion of a function as an arbitrary correspondence between objects and truth values.

What of Hintikka and Sandu's question, "What was Frege's concept of *function*?" Frege explains our acquisition of the functions associated with concepts—functions which express a judgeable content—by our acquisition of their linguistic expression, something we achieve by an analysis of sentences. For Frege this may well have been an essential component of his conception of his theory of functions and concepts. But like our informal explanation of how to understand functions which map to truth values, perhaps it, and the predicative interpretation it suggests, can be regarded as a consideration which merely motivates the notion. The question is whether, by allowing for functions as arbitrary correspondences between objects and truth values, we do as much violence to the Fregean idea of a concept or function as Ramsey's extensional functions do to the propositional functions of *Principia*. As often happens, settling on an answer to an interpretive question such as this one is less important than achieving greater clarity on the systematic issues the question raises. It seems clear that both *Principia*'s propositional functions and Frege's concepts admit of extensionalist interpretations—in the strong sense considered here, of arbitrary mappings between *Principia*'s individuals and propositions, or between Frege's objects and truth values. Such interpretations tend to undermine the motivation for both ideas, a fact that is particularly evident in the case of Russell's propositional functions, where the extensionalist interpretation seems more properly regarded as a wholesale *replacement* of the notion.

The situation is less clear in the case of Fregean functions and concepts because they lack the explicit association with propositions that is characteristic of propositional functions; an extensionalist interpretation of a Fregean concept as an arbitrary mapping of objects to truth values is arguably still a Fregean concept. However its utility for Frege's theory of classes is unclear. According to a theory like Frege's, concepts provide the principle which gives classes their "unity," and they also serve the epistemological function of providing the principle under which a collection of objects can be regarded as a separate object of thought. A class that is generated by an arbitrary pairing of individuals with truth values might be one that is "determined by a concept," but the concept which determines it seems no more epistemically accessible than the collection itself. Even if it can be convincingly argued that such concepts sustain the unity of the classes they determine, it can hardly be maintained that they are capable of playing the epistemological role which the predicative interpretation can claim for its functions and concepts.[8]

Bibliography

Bell, J. and Demopoulos, W. (1993). Frege's theory of concepts and objects and the interpretation of second-order logic. *Philosophia Mathematica*, 1:139–156.

Bell, J. and Demopoulos, W. (1996). Elementary propositions and independence. *Notre Dame Journal of Formal Logic*, 37:112–124.

Burgess, J. (1993). Hintikka et Sandu versus Frege in re arbitrary functions. *Philosophia Mathematica*, 1:50–65.

Burgess, J. (2005). *Fixing Frege*. Princeton University Press, Princeton, NJ.

Carnap, R. (1931). The logicist foundation of mathematics. In Putnam, H. and Benacerraf, P., editors, *Philosophy of Mathematics: Selected Readings*, pages 41–52. Cambridge University Press, Cambridge, second edition. Transl. E. Putnam and G. Massey.

Dummett, M. (1981). *Frege, Philosophy of Language*. Harvard University Press, Cambridge, second edition.

Dummett, M. (1991). *Frege, Philosophy of Mathematics*. Harvard University Press, Cambridge.

Frege, G. (1884). *The Foundations of Arithmetic: A Logico-Mathematical Enquiry into the Concept of Number*. Northwestern University Press, Evanston, second revised edition. Transl. J.L. Austin. English edition 1980.

Frege, G. (1893–1903). *The Basic Laws of Arithmetic: Exposition of the System*. University of California Press, Berkeley, CA. Two volumes. Partial translation by M. Furth, English version 1964.

Heck, R. and Stanley, J. (1993). Reply to Hintikka and Sandu: Frege and second-order logic. *Journal of Philosophy*, 90:416–424.

Heck, R. (1995). Definition by induction in Frege's *Grundgesetze der Arithmetik*. In Demopoulos, W., editor, *Frege's Philosophy of Mathematics*, pages 295–333. Harvard University Press, Cambridge.

Hintikka, J. and Sandu, G. (1992). The skeleton in Frege's cupboard: The standard versus nonstandard distinction. *Journal of Philosophy*, 89:290–315.

Potter, M. (2005). Ramsey's transcendental argument. In Lillehammer, H. and Mellor, D., editors, *Ramsey's Legacy*, pages 71–82. Oxford University Press, Oxford.

Ramsey, F. (1990). The foundations of mathematics. In Mellor, D., editor, *F.P. Ramsey: Philosophical Papers*, pages 164–224. Cambridge University Press, Cambridge. Ramsey's essay was originally published in 1925.

Russell, B. (1919). *Introduction to Mathematical Philosophy*. Allen and Unwin, London.

Sandu, G. (2005). Ramsey on the notion of arbitrary function. In Frapolli, M., editor, *Ramsey: Critical Reassessments*, pages 237–256. Continuum Studies in British Philosophy, London.

Wehmeier, K. F. (2008). Wittgensteinian tableaux, identity, and co-denotation. *Erkenntnis*, 69: 363–376.

[8] The present paper borrows from a longer study, "The 1910 Principia's theory of functions and classes," which will be published at a later date. I wish to thank Gerry Callaghan, Greg Lavers and Peter Koellner for their insightful comments on earlier drafts. I am indebted to Michael Hallett for suggestions which led to many improvements. Thanks to John for many hours pleasantly spent discussing the issues dealt with here. Support from the Killam Foundation and the Social Sciences and Humanities Research Council of Canada is gratefully acknowledged.

Wehmeier, K. F. (2009). On Ramsey's "silly delusion" regarding Tractatus 5.53. In Primiero, G. and Rahman, S., editors, *Acts of Knowledge: History, Philosophy and Logic: Essays Dedicated to Goran Sundholm*, tributes series. College Publications, London.
Whitehead, A. and Russell, B. (1910). *Principia Mathematica*, Volume 1. Cambridge University Press, Cambridge.
Wittgenstein, L. (1922). *Tractatus Logic-Philosophicus*. Routledge Classics, New York, NY, 2001 edition. Transl. D. Pears and B. McGuinness.

Chapter 2
Metaphysical Necessity

Michael Dummett

Saul Kripke deserves great credit for fastening the minds of the philosophical community on the existence of different types of possibility and, correlatively, of necessity. The type of necessity upon which analytical philosophers had been accustomed to concentrate was analyticity; for clarity, I will replace this term by "a priori knowability." We need not stop to argue over a precise characterisation of this familiar notion: I will say that a statement, considered as made by the utterance, actual or hypothetical, of a meaningful assertoric sentence by a particular person at a specific time, is a priori knowable if the speaker is or would be able to recognise its truth simply from the meanings of the words composing the sentence, together, when needed, with some deductive argument which he devised or with which he was presented. When it is not knowable a priori that a given statement is not true, let us say that that statement is possible a priori. And I shall say that a statement is knowable only a posteriori if the speaker is able to recognise its truth only on the basis of some observations he has made or of reports made to him by others of observations they have made. No doubt there are many holes in these formulations, considered as definitions, but they will serve present purposes.

Kant of course distinguished between the analytic and the synthetic a priori, but his distinction is little to the question here at issue. More relevant is the distinction Aquinas made in discussing the ontological argument for the existence of God. The proposition that God exists is, he said, *per se nota* but not *nota quoad nos*. By "*nota quoad nos*" he meant essentially "knowable a priori" in the sense explained, although he most probably would not have classified "I am here" under this head; he was distinguishing *propositions*, not utterances of sentences. Since the proposition that God exists was not *nota quoad nos*, there could not be a valid argument which, like the ontological argument, derived it without appeal to any facts that could be established only by observation: that is why Aquinas's own proofs of the existence of God all proceed from very general premisses about how things are in the world. What, then, did Aquinas mean by "*per se nota*"? Plainly, this notion has to do with the sort of thing that makes the proposition in question true, in contrast with the

M. Dummett, FBA (✉)
Wykeham Professor of Logic Emeritus, University of Oxford, Oxford, UK

D. DeVidi et al. (eds.), *Logic, Mathematics, Philosophy: Vintage Enthusiasms*,
The Western Ontario Series in Philosophy of Science 75,
DOI 10.1007/978-94-007-0214-1_2,

means by which we can come to know it. Why would one who believes that there is a God be reluctant to add, "But there might not have been"? Not because he thinks that the existence of God is capable of being proved by purely logical means, but because there is nothing of which it would make sense to say, "Were it not for that, there would not have been a God."

If it were not so gross an anachronism, we might compare Aquinas's distinction with that between the derivability in some formal system of a formula of second-order logic and its being true in all models, or with that between Frege's definition of analyticity and Bolzano's. Frege defined a statement to be analytic if it was logically derivable from fundamental logical laws; Bolzano if it held true under every replacement of its non-logical expressions by syntactically similar ones. Mathematical realists, usually called platonists, consider that a mathematical statement may be true even if it is beyond our means to prove. They would consider that Goldbach's celebrated conjecture that every even number greater than 2 is the sum of two primes might be one of these. It would then be *per se nota*, since it would hold good in all models of second-order arithmetic (or all standard models of the first-order theory), but not *nota quoad nos*.

It is, then, plain that there is a notion of necessity distinct from that of being knowable a priori; perhaps there is more than one such notion. Kripke labels the notion he contrasts with knowability a priori "metaphysical necessity"; the function of the word "metaphysical" here is to indicate that necessity of this type is not explained in terms of our capacity to know the proposition in question. Knowability and possibility a priori may therefore also be termed "epistemic" necessity and possibility; the reference is not to what we do know, but to what, given only an understanding of the language and a capacity to reason, we can know. The clue Kripke follows in characterising the metaphysical modalities is our use of "might" and "might have": something is metaphysically necessary if it is not the case that it might have been otherwise. But care is needed. Suppose that someone recounts some incident involving Anderson which took place in London; I say, "How fortuitous! Anderson might not even have been in London that day," but the other replies, "Oh, no, he had to be in London for the meeting of the committee." This reply does not express the metaphysical necessity of Anderson's presence in London on the day in question. Wittgenstein said that the modal verb "can" is always tacitly qualified by a phrase of the form "as far as ... is concerned": possibility is always relative. Possibility a priori is intended to be an absolute notion, however, needing no such qualification; and the same holds good for metaphysical possibility.

Quine distinguished three grades of modal involvement. At the first grade, we classify true statements as necessary or contingent, and false statements as impossible or possible: "necessarily true" and "possibly true" are metalinguistic predicates. At the second grade, we apply necessity and possibility operators to sentences of our object-language to form new sentences of our object-language; these new sentences are subject to other sentential operators such as negation and conditionalisation. The two grades are connected by a linking principle:

(L1) A is necessarily true iff $\ulcorner$Necessarily $A\urcorner$ is true.

When "A is possibly true" is explained as meaning "$\ulcorner$Not $A\urcorner$ is not necessarily true," and $\ulcorner$Possibly $A\urcorner$ is defined to mean $\ulcorner$Not necessarily not $A\urcorner$, the dual linking principle:

(L2) A is possibly true iff $\ulcorner$Possibly $A\urcorner$ is true

follows. These linking principles may be regarded as holding irrespectively of the direction of explanation. If we start at the first grade of modal involvement and advance to the second, the direction of explanation is from left to right. The linking principles then lay down the condition for the truth of any sentence built up by means of modal operators and ordinary sentential operators in which no modal operator stands in the scope of any other: they do not lay down that of a sentence in which any modal operator stands in the scope of another. The third grade of modal involvement is reached when the modal operators "Necessarily" and "Possibly" are allowed to stand within the scope of quantifiers: modalised statements can now be quantified into. When we advance from the second grade to the third, the foregoing linking principles do not suffice to explain the truth-condition of a sentence in which a modal operator stands in the scope of a quantifier. For that we need the first-grade notion of a predicate's being necessarily true of an object, with that of its being possibly true of an object explained in terms of the former notion in an analogous way. It may then appear open to us to lay down the linking principles:

(L3) $A(x)$ is necessarily true of the object denoted by a iff $\ulcorner$Necessarily $A(a)\urcorner$ is true

(L4) $A(x)$ is possibly true of the object denoted by a iff $\ulcorner$Possibly $A(a)\urcorner$ is true.

We shall need to look at (L3) and (L4) again later. We may rank as a fourth grade of modal involvement that modal operators should be freely allowed to stand within the scope of other modal operators; but this fourth grade will not here concern us.

Kripke does not explain the modal operators expressing metaphysical necessity and possibility by the transition from left to right of the linking principles (L1) and (L2). He does not start by specifying the condition for a statement to possess the property of being metaphysically necessary or being metaphysically possible, and then give a partial explanation of the corresponding modal operators by appeal to the linking principles. Rather, he adopts the opposite direction of explanation. A statement A is metaphysically possible if $\ulcorner$Possibly $A\urcorner$ is true, where $\ulcorner$Possibly $A\urcorner$ is interpreted as meaning $\ulcorner$It might have been the case that $A\urcorner$, as this form of words is understood in ordinary discourse; correlatively, A is metaphysically necessary if $\ulcorner$It could not have been the case that not $A\urcorner$ is true, where "could not have been" is the negation of "might have been."

The distinction between the metaphysical modalities and the epistemic ones is then drawn, at least in the first place, by invoking the phenomenon of modal rigidity. By drawing attention to the phenomenon of rigidity, Kripke undoubtedly made a signal contribution to philosophical logic. There is not only modal but also temporal rigidity. If I say, "It is always cold where I am," I mean that, for every place s and every time t, if I am at s at t, then it is cold at s at t. But if I say to someone on the telephone, "It is very cold where I am," and my interlocutor replies, "It is always cold there," he does not mean that I am always in a cold place, but that, if I am, for example, in Dundee at the time, then it is always cold in Dundee. The adverb "there" is temporally rigid, whereas the phrase "where you are" can be used as temporally flexible. A news correspondent might say, "It is very noisy where Mr. Blair is"; if someone comments, "It is always very noisy where Mr. Blair is," he means that Mr. Blair is constantly surrounded by a hubbub. Suppose now that we introduce the expression "Blairabout" to bear the same relation to "where Mr. Blair is" as does "there" to "where you are" or "here" to "where I am." Then the statement "It is always noisy Blairabout" means that if Mr. Blair is now at s, then, for every time t, it is noisy at s at t. At any time, the condition for the truth of "It is noisy Blairabout" coincides with that for the truth of "It is noisy where Mr. Blair is"; but when quantification over times is introduced into both sentences, the truth-conditions diverge.

Let us say that a sentence *A* is *presently true* at any given time if by uttering *A* at that time the speaker would make a true statement. And let us say that it is *perpetually true* if, at any time, it is presently true. Then the sentence "Mr. Blair (if now alive) is where Mr. Blair is" is perpetually true; and, since it has the same truth-condition, so is "Mr. Blair (if now alive) is Blairabout." We may, however, introduce the notion of a sentence's being *eternally true* by appeal to a linking principle, applied from right to left:

(L5) A is eternally true iff ⌜Always *A*⌝ is true.

We may say that *A* is *temporarily true* at any given time if it is true, but it is not eternally true. Then, while "Mr. Blair (if now alive) is where Mr. Blair is" is eternally true (provided that the phrase "where Mr. Blair is" is understood as temporally flexible), the sentence "Mr. Blair (if now alive) is Blairabout" is not eternally true: it is only temporarily true.

Kripke's distinction, when it rests on the distinction between epistemic and metaphysical possibility, is similar. The sentence "Mr. Blair is where he is" is epistemically necessary, since, whenever it is uttered, a true statement is made (and anyone who understands the words knows this). But it is not metaphysically necessary, because Mr. Blair might have been somewhere else: the phrase "where he is," even if temporally flexible, appears to be modally rigid. The point seems delicate. There is an ambiguity, akin to Russell's joke answer to "I expected you to be taller than you are"—"Of course I am not taller than I am." It is at any rate clear that "Mr. Blair is Blairabout" is epistemically but not metaphysically necessary, just as it is perpetually but not eternally true. The point was glimpsed by G. E. Moore when he replied to Russell's observation that "This exists" is necessarily true by remarking that nevertheless "This might not have existed" may be true; but he never followed up the clue.

A disagreement between Saul Kripke and me concerned whether pairs of sentences such as "It is cold here" and "It is cold where I am now" could be said to express the same proposition. This question depends upon the answer to the prior question whether the notion of rigidity is relevant only to sentences containing modal (or temporal) operators. Consider the two sentences

(i) "Mr. Blair is wherever he is" and
(ii) "Mr. Blair is Blairabout";

the phrase "wherever Mr. Blair is" is certainly modally flexible. Kripke argued that neither sentence contains a modal operator, and yet they differ in modal status: both sentences are epistemically necessary, but only (i) is metaphysically necessary. The difference arises from the modal rigidity of "Blairabout", as opposed to the modal flexibility of "wherever Mr. Blair is": hence modal rigidity is relevant to sentences not containing modal operators. The modal status of a sentence, he argued, depends upon the modal status of the proposition it expresses: hence the two sentences must express different propositions.

This argument appears to me to be back to front. We have been given no criterion for the modal status of a proposition save the modal status of a sentence expressing it; and we have been given no criterion for the metaphysical modal status of a sentence save the truth or falsity of the sentence that results from applying a modal operator to it. The explanation was from right to left of the linking principles (L1) and (L2). It is in virtue of the different ways in which "Blairabout" and "wherever Mr. Blair is" behave when in the scope of a metaphysical modal operator that sentences (i) and (ii) are accorded different metaphysical modal status: without reference to that, such a difference would be imperceptible. People who had a language lacking modal operators (or modal auxiliary verbs or modal adverbs) could have no idea of such a difference in modal status, nor, indeed, of modal status. Modal rigidity directly affects only the truth-conditions of modalised sentences. This difference may then be projected back on to the unmodalised sentences to which the modal operators were attached in the form of a difference in modal status; but this difference merely reflects the difference in the truth-conditions of the modalised sentences, and could not be perceived save by appeal to it.

It might be retorted that speakers of a language devoid of modality could still recognise the difference in what we might call temporal status between the two sentences (one is eternally true, the other only temporarily true). That is correct; but it is more tendentious to claim that sentences differing in temporal status must express distinct propositions.

We are therefore free, if we wish, to regard the proposition a sentence expresses as Frege did, namely as determined by the condition for it to be true. Understanding an assertoric sentence involves knowing its significance when uttered on its own, and, in addition, knowing how it contributes to the sense of a complex sentence built up from simpler sentences of which it is one. The significance of a sentence when uttered on its own on a particular occasion to make an assertion may be called its *assertoric content*: it is what determines the difference that is made to the hearer's picture of the world if he accepts it as correct. Frege's truth-conditional characterisation of sense gives an account of the assertoric content of a sentence when uttered on a particular occasion. Just as he took the sense of the sentence when so uttered to be a thought, so we should take the proposition expressed by uttering the sentence as depending solely on its assertoric content and the occasion of utterance. More may need to be known about a sentence than its assertoric content in order to understand how it contributes to determining the assertoric content of a more complex sentence of which it is a constituent: we may call this its *ingredient sense*. Any semantic property of the sentence not strictly relevant to its assertoric content belongs to its ingredient sense, which bears solely upon its role as a subsentence of more complex ones. Among such properties is the presence in the sentence of temporally or modally rigid terms.

It may be objected that it is quite unfair to say that metaphysical modality is explained solely by reference to behaviour within the scope of modal operators, on the ground that there is a direct explanation in terms of truth at possible worlds. But the notion of a possible world is explicable only by the use of modal operators: a possible world is a way the world might have been. It is urged in reply that to understand a sentence demands an ability to judge of its truth-value in a variety of circumstances, which means, it is said, a grasp of which possible worlds it is true in and which it is false in: the notion of a possible world is therefore already implicit within the understanding of any unmodalised sentence. Understanding a sentence indeed demands an ability to judge of its truth-value whatever circumstances may obtain, but it does not require the capacity to envisage hypothetical circumstances and consider what its truth-value would be in them. Furthermore, the notion of truth or falsity invoked in the explanations of the modal operators in possible-worlds semantics is that of the truth-value *at* a possible world; for example, ⌜Possibly A⌝ is true at a world w if A is true at some world possible relatively to w. The truth-value of a sentence *at* a world is the truth-value with respect to that world that we ascribe to a sentence understood as uttered *in the actual world*. It is not to be equated with that of its truth or falsity *in* the possible world, that is, the truth-value ascribed to the sentence, when uttered *in that world*, by the inhabitants of that world; this notion plays no role in the semantics. For instance, if uttered when Mr. Blair is (in the actual world) in Milan, the sentence "Mr. Brown will be Blairabout tomorrow" will be true *at* a world w if in w Mr. Brown is the next day in Milan; but if, in w, Mr. Blair is today not in Milan but in Edinburgh, the same statement will be true *in* w if Mr. Brown is, in w, in Edinburgh the next day. It is the presence of modally rigid terms that brings about the divergence between truth *at* a world and truth *in* it. But the understanding of a sentence demands the ability to judge of its truth-value *in* whatever circumstances in fact obtain. A grasp of the condition for the truth of an unmodalised sentence *at* any given possible world, as this notion is used in possible-worlds semantics, is therefore quite irrelevant to a grasp of its assertoric content. To know this, we need to know how the reference of any term it contains is determined in the world in which it is uttered or considered as being uttered. The inhabitants of a possible world would need, in order to understand the sentence, as uttered on its own in that world, to know how the reference of a term occurring in it was determined *in* that world. It may be argued that they should know how it would be determined in other worlds; but they need know nothing about how its reference was determined *at* any other world.

The fact that an epistemically necessary sentence may not be metaphysically necessary is solely due to its containing a modally rigid term. The standard examples of epistemically necessary sentences that are not metaphysically necessary all depend upon the rigidity of some term. The existence of sentences which, conversely, are metaphysically but not epistemically necessary does not, however, follow from the phenomenon of rigidity alone. "Solomon was a son of David" is, according to Kripke, metaphysically necessary, and its being so depends upon the rigidity of the proper name "Solomon": "The king who built the first Temple in Jerusalem was a son of David" is not metaphysically necessary, because, after all, the Temple might have been built by a successor of Solomon. But, to apprehend the necessity of "Solomon was a son of David," we must know more than that "Solomon" is modally rigid: we must know that being David's son was an essential property of Solomon's, whereas building the Temple or being the last king of undivided Israel was not. Essential properties are properties of *objects*, not of ways of referring to them: the direction of explanation in (L3) and (L4) is from left to right. This feature of metaphysical modality therefore does not turn primarily upon the behaviour of particular linguistic items, but upon the principles that determine what is metaphysically possible.

The doctrine of essential properties, as Kripke understands it, creates some difficulty for modal logic. The statement "David begot Solomon" is, on Kripke's view, metaphysically necessary by the criterion that it is true in all possible worlds in which both David and Solomon exist, since having been begotten by David was an essential property of Solomon: hence "Necessarily, David begot Solomon" is true, in accordance with the linking principle (L1). But we cannot now conclude, by (L3), that begetting Solomon was an essential property of David's; in fact, it was one he was morally wrong to have had. For it to have been one of David's essential properties, the statement "David begot Solomon" would have had to be true in every possible world in which David exists, whereas it is not true in those worlds in which Solomon does not exist. It follows that (L3) and (L4) are unsound. In fact, in order to express the fact that satisfying "$F(x)$" is an essential property of an object **a**, we cannot use the *sentential* operator "necessarily": we need an operator that is specifically attached to one-place predicates. If we write the result of applying such an operator to a predicate "$F(x)$" as yielding another predicate $\ulcorner \mathrm{N}[\lambda y.F(y)](x)\urcorner$, we can assert

$$\mathrm{N}[\lambda y.\text{David begot } y](\text{Solomon})$$

but also

$$\text{Not: } \mathrm{N}[\lambda y.y\text{begot Solomon}](\text{David}).$$

Modal logic is obviously greatly complicated by the need to have such an operator upon predicate-abstracts as well as one operating on sentences.

This leads us on to the metaphysical necessity of all true identity-statements. Since it is not necessarily true of David that he was the father of Solomon, "David was the father of Solomon" had better not be an identity-statement. Hence "the father of Solomon" must be a definite description rather than the result of applying a functor to a proper name, and definite descriptions must be interpreted in accordance with Russell's Theory of Descriptions. But "Phosphorus is the same planet as Hesperus" is metaphysically necessary, even though it is not knowable a priori. Why? Because "Phosphorus," being modally rigid, must designate in every possible world what it designates in the actual world, and likewise for "Hesperus": so the identity-statement is true in every possible world.

The doctrine that all true identity-statements are metaphysically necessary is founded on the doctrine that all proper names are modally rigid. If we say that Einstein might never have published his famous papers, we are speaking of that very man, even though *we* identify him as the author of the Special and General Theories of Relativity. It does not matter how little we know about the bearers of the names we use in hypothetical suppositions. Even though we did not know that Eric Blair and George Orwell were the same man, and even though we therefore identified them in different ways, the two names would designate the same man in all possible worlds in which they designated anyone; even if we imagined a world in which Eric Blair never took up writing, and therefore adopted no pen-name, "George Orwell" would still designate Eric Blair in that world.

I think we have to accept this. I do not see how it can be denied that proper names are modally rigid; and the rest follows. But when we contemplate the metaphysical necessity of "George Orwell was Eric Blair," we must not think that we are recognising some deep philosophical truth: we are merely acknowledging a mechanical consequence of the modal rigidity of proper names.

But have we not learned something of philosophical importance about psychophysical identity? We used to think that the proposition that sensations are *identical to* stimulations of certain nerves, rather than merely *correlated with* them, could be dismissed out of hand, since the two things are identified in quite different ways. Now, however, the theory of metaphysical necessity has taught us that it does not impugn the truth of an identity-statement that we identify the bearers of the two names in different ways. But we knew that already: that is the whole point of Frege's famous Phosphorus/Hesperus example. What interests us is whether these psychophysical identities are true (or even can be true—the "can" here expressing possibility a priori). The notion of modal rigidity adds only that, provided the two terms of an identity-statement are modally rigid, if the statement is true, it is necessarily true; but once we are satisfied of the truth of a psychophysical identity, its (metaphysical) modal status will not concern us greatly.

Yet the notion of modal rigidity cannot dispel the requirement that an identity-statement cannot hold good unless the two terms designate things of the same sort. Phosphorus and Hesperus are the same *celestial body*; George Orwell and Eric Blair were the same *man*. But are there not identities that violate this principle? For example, are not sounds waves of compression and rarefaction in the air or other medium? If sounds and sound-waves are things of different sorts, they *cannot* be identical. Cross-sortal identifications form an interesting topic, deserving of close study. I think they are usually proposals that we should do without things of one sort in favour of those of the other. Whatever is to be said about them, I do not see that either the notion of modal rigidity or the doctrine of the metaphysical necessity of true statements of identity is of any great help in giving an account of them.

Given the metaphysical necessity of true identity-statements, it seems that we can derive by L3 that being identical with Hesperus is an essential property of the planet Venus. It then appears that being identical with Hesperus is the *same* property as being identical with Venus, or, equally, with Phosphorus. It may be objected that it is a priori knowable that Hesperus has the property of being identical with Hesperus, but not that it has that of being identical with Phosphorus, and hence that these two properties cannot be the same as each other or as that of being identical with Venus. The Kripkean must answer that properties are what *objects* have, however they may be referred to, whereas a priori knowability is a feature of *sentences*; we ought therefore to speak of its being a priori knowable, not that being identical with Venus is a property of Venus, but that the *sentence* "Venus is identical with Venus" is true. It is, he must say, for one and the same reason that it is metaphysically necessary that Venus has the property of being identical with Venus, and that it has that of being identical with Hesperus: hence these properties are the same. What, then, is a property? From every standpoint, it is something expressible by a one-place predicate. But, the Kripkean must say, it is not the *sense* of that predicate, which reflects *our* understanding of the predicate: it is what, in any possible world, makes it true of an object in that world that the predicate applies to it. Whether the notion of what *makes* something true is sustainable is a question into which we need not here go further.

The notions of rigidity and of essential properties are linked in the following way. A term t is modally rigid if, when the truth-value of a sentence in which it occurs is evaluated at a possible world w, the denotation of t is taken as the same as its denotation in the actual world, not its denotation in w; its denotation in w is what any speakers of our language (English) existing in w would take it to denote. (Perhaps we should say: what the speakers in w of a language syntactically similar to English would take to be denoted by that term of their language which corresponded to t.) The essential properties of whatever t denotes in the actual world supply the criteria which determine whether anything existing in any world w is the same as what t denotes in the actual world: if t denotes **a** in the actual world, then, if **b** exists in w, and has, in w, all the essential properties of **a**, then **b** is the same as **a**. The notion of an essential property thus governs the *application* of the notion of rigidity.

It may be objected that Kripke denies that we need any criterion of transworld identity. He says that when we envisage hypothetical circumstances, we use no criterion to identify hypothesised objects with actual ones: we simply *stipulate* of some actual object that it plays some role in the envisaged circumstances. That is true. But of course the question arises whether the circumstances we envisage are genuinely possible ones (metaphysically possible, that is). People are apt to say things like, "If Lewis Carroll had been born 35 years later, he would have become a great logician." They are then stipulating that the person they are talking about is C.L. Dodgson (Lewis Carroll) just as if they had said, "If Lewis Carroll had visited Germany, ..." But anyone born 35 years after the birth of Dodgson would have had a different mother from Dodgson, and thus, on Kripke's view, would have lacked one of Dodgson's essential properties; he would therefore not have been Dodgson, according to Kripke. The situation envisaged in the antecedent of the counterfactual is not a metaphysically possible one on Kripke's view, and the counterfactual is therefore void.

There is a parallel point. Kripke is surely right in saying that, when we are interested in how the life of a particular human being would have gone in circumstances other than the actual ones, we usually make clear that we are talking about that individual simply by using his or her name. But such specification by name cannot exhaust the way in which a particular individual can get into a possible world or set of hypothetical circumstances. Suppose that, in describing the hypothetical circumstances he has in mind, the speaker refers to another individual than the one in whom he is principally interested, and describes that other individual only in general terms. And suppose that the description characterises him as having some, and lacking none, of the essential properties of some actual individual. Then the speaker must be talking about that actual individual, even if he has never heard of him. Kripke has as much need of a criterion of transworld identity as any other possible-worlds theorist; and it is supplied by the doctrine of essential properties.

What, then, is an essential property of an object? If a king abdicates, he ceases to be a king, but he still exists and thereby remains the same man as before. When we say of him, "He was a king once," the pronoun does not exemplify deferred ostension. Remarkable transformations occur in nature without loss of identity or cessation of existence: a tadpole becomes a frog, and a caterpillar eventually becomes a butterfly, but each is still the same creature after the metamorphosis. But a prince cannot become a frog, a man a beetle, a woman a pillar of salt, a hunter a stag, a girl a laurel bush or a man or woman a constellation. If any such transformations were apparently to take place, they would mark the end of the life, and so of the existence, of the human beings they occurred to: if it was said of a bush that it was once a beautiful woman, this would be analogous to saying the same of a mummified corpse. There is nothing that was a woman and is now a corpse, only something that was the body of a woman and is now a corpse. So it is an essential property of every human being that he or she is a human being: a man, woman or child cannot lose that property and continue to exist.

Essential properties of this kind may be called *existentially necessary properties*. Such a property attaches to an object in virtue of the sort of object it is: it could not continue to exist while ceasing to be of that sort. Possession by an object of an existentially necessary property has to do with the identification of the object over time. But are there *individually essential properties*—properties which an individual of a given sort must have in order to be *that* individual? In all the cases just considered, except transformation into a constellation, the body of a living person became something else, in most cases the body of something else (a bush is not normally said to have a body, although it can die). If we were to illustrate an individually essential property in the same way, we should need a case in which the body of one person became the body of another. People who supposed that a rite of passage—a puberty ritual, or reception into a religious elite—literally marked the end of one person's existence and the beginning of another's would find no difficulty in this. In Western culture, it is hard to give plausible examples. We may classify all essential properties derivable from criteria for identification over time as *persistence properties*. Kripke recognises individually essential properties, but not as persistence properties: his examples have to do with the identification of an individual across possible worlds rather than over time.

It is clear that, stated in terms of possible worlds, the essential properties of an object are those which it possesses in every possible world in which it exists: unless the range of possible worlds is constrained in such a way as to ensure that there are such properties, there are no essential properties of objects save the persistence properties. Examples such as that of Solomon's being a son of David illustrate the well known necessities of origin: among the essential properties of any object are the circumstances of its having come to be, according to Kripke's theory. An object's having come into existence in given circumstances is not a persistence property: it is a consequence of identity over time, not a criterion or necessary condition for it. If at a particular time it is true of a given man that he had certain parents, it follows logically from the identification of someone at a later time as that same man that the individual so identified had those parents: we could not discover that he had those parents *before* making the identification. Someone's losing the property of having had the parents that he had is a conceptual impossibility. Why, then, do the circumstances in which an object came into existence constitute an essential property of it? The answer lies in Kripke's conception of a possible world. Any possible world exactly resembles the actual world up to a certain time, and thereafter develops differently, though in accordance with the persistence properties and the natural laws obtaining at the moment of divergence.

Now it is certainly the case that many counterfactual suppositions leading to the conclusion that something or other might have held good take the form that Kripke regards as canonical, namely to imagine things as having gone as they went in fact up to a certain moment, and as having thereafter diverged from what actually happened. Let us call a counterfactual based on such a supposition a "Kripke counterfactual" if the moment of divergence occurred at a time when all individual persons, animals or things considered in the counterfactual were already in existence. Does a restriction to Kripke counterfactuals underpin the necessities of origin? It is obvious that formally it does. Practical applications may depend upon resolving problems or disputes about when a person or object does first come into existence. At what stage does an identifiable painting or other work of art begin to exist? At what point did a named human being first exist? I find it impossible to accept the answer "at the first moment of conception," understood as the moment of impregnation, for the decisive reason that the possibility that twins would form was then open, and it would be senseless to ask which twin was the individual whose existence had begun with impregnation. Kripke's doctrine of metaphysical necessity is powerless to resolve questions like these. It can only say that an actual individual—person or thing—may be considered as involved in hypothetical circumstances from any moment from that individual's coming into being, whenever that was.

In any case, the restriction to Kripke counterfactuals does not seem to me compelling. I do not find it difficult to attach a sense to speculations about how things would have been if some individual had been born much earlier or much later than he in fact was, or had had different parents. It could be said that in such cases all that is being imagined is the birth at a different time or to different parents of someone very like, but not identical to, the individual in question; but I cannot say that I see any compelling reason to maintain this, nor, indeed, to combat it. I doubt that there is any clear truth of the matter: I doubt that it makes any difference which we decide to say. When it comes to things other than human beings or animals, such as cities or institutions, the idea that their origin is essential to their identity seems far less compelling. The supposition that the city of Venice, for example, was founded 30 years later than in fact it was does not strike me as the entertainment of an impossibility. The objection that a city founded at such a date would not *be* Venice appears quite unconvincing: what would make it Venice would be its being located where Venice is now, and its having for the past many centuries looked as the city we know by that name had looked in those centuries, and the things that have happened in that time in and to that city having happened in and to it.

If, like David Lewis, one is a realist about possible worlds, thinking that they are as much part of reality as the actual world, then I suppose that the questions whether objects to be found in the actual world also inhabit any other worlds, and, if so, what are the criteria that render an object in one world identical with one in another world, are genuine questions with determinate answers. But Saul Kripke is not a realist about possible worlds. He nevertheless believes that individual objects, and natural kinds, have essential properties. There can be no possible world in which some individual or natural kind exists but lacks any of its essential properties; but, for any consistent set of inessential properties of any given individual or natural kind, there will be a possible world in which it has just the properties in that set. One of the tasks of metaphysics is to determine which types of properties that individual objects and natural kinds possess are essential to them; it is a task for empirical investigation to discover which specific properties those are.

This sounds indeed a grand conception. It remains that it rests on a slender linguistic basis: the rationale for the whole apparatus of essential properties and possible worlds is given by our practice of using modal auxiliaries like "might have" and the counterfactual conditionals on which we base statements involving such auxiliaries. It must be agreed that a statement to the effect that something might have been so is most usually defended by citing circumstances in which it *would* have been so, that is, by putting forward a counterfactual conditional. But I do not think that the use of counterfactual conditionals is regimented to anywhere near the extent that Saul Kripke supposes. When such a statement is made, it will usually be apparent which features of the actual world, that is, of reality, the speaker intends to be understood as holding constant; but sometimes it is obscure, and then the speaker can be challenged to make the antecedent of the conditional more precise. There is no *general* criterion by which the truth of an arbitrary counterfactual conditional is to be judged.

We have lighted on three types of essential property: self-identity; persistence properties; and circumstances of origin. But, in seeking others, what are we looking for? Is there any general criterion by which we can judge, of any given property that an object may have, that it is essential to that object (or to any object that has it)? I do not find any general principle underlying Kripke's classification of certain properties as essential. Rather, his category of essential properties seems to me a ragbag, with properties being awarded this status on very heterogeneous grounds. Self-identity properties were reckoned essential because of the denotations in possible worlds assigned to modally rigid terms; circumstances of origin were deemed essential on the basis of a doctrine about what possible worlds there are. In both these cases the direction of explanation in the principle

(L6) $\lambda x[A(x)]$ is an essential property of **a** iff $\ulcorner A(\mathbf{a}) \urcorner$ is true on every possible world in which *a* exists

is from right to left. But essential properties of other types are not counted as such because of any argument that they must be possessed in every possible world by any object that exists in that world. Rather, they are first classified as essential on intuitive grounds, and then from this it is deduced that any object possessing any of them must possess it in all possible worlds in which it exists. The range of possible worlds is delimited by the prior classification of these properties as essential; the direction of explanation is from left to right. The next subclass of essential properties I shall consider is that of internal structure, of substances and of animals of the same species: the essential properties of natural kinds.

Chemical compounds and elements have, as we know, highly determinate internal structures; the notion of essential properties therefore fits them very well. It appears to justify a notion of the *real*, rather than the merely nominal, essences of such things: these are to be discovered by scientific investigation of them, not of our uses of the words denoting them. Nevertheless, the doctrine of the real essences of these types of substance has prompted a correlative analysis of such words. The word "water," for instance, is held to have an indexical component together with a tacit reference to the real essence: on this analysis, it means "whatever has the same internal structure as the liquid found in terrestrial streams and rivers," and has always meant that, even before we knew anything about its chemical composition. Hence the Twin Earth paradox.

Yet a great many of the substances to which we habitually refer in everyday life are not chemically pure at all: wool, leather, silk, wood, air and mud, for example. Many of those substances are identifiable by their origin rather than their internal composition: there is no wood that was not once part of a tree. When we ask what must be true of some material for it to be wood or wool, we are not asking a metaphysical or a scientific question about its essential properties, but about the meanings of the *word* "wood" or "wool"; we are asking after its nominal essence. It is by no means obvious why it should be different with those substances which we have discovered to have determinate internal structures. Now that we know about the chemical composition of water, we may henceforward so use the word "water" as to make it essential to anything's being called "water" that it is a compound whose molecules contain two atoms of hydrogen and one of oxygen. But before anyone had our notions of elements and of molecules, the word "water" surely did not involve a tacit reference to an as yet unknown inner structure. The Twin Earth paradox will arise only while the speakers on both planets do not yet know the chemical compositions of the substances which those on one and those on the other call "water," but, in using that term, make a tacit reference to the chemical composition. Once they know the composition, and make it part of the sense of that term, there is no paradox; nor is there any paradox if, before they know the composition, they do not understand the reference of the term "water" as determined by the as yet unknown composition. Until the molecular composition of material substances was recognised, there could be no conception of the inner structure of those substances comparable to that which we have now; for anyone who believed that a substance continuously filled the whole volume that it occupied, there could be no distinction between a compound and a mixture. It is a matter for discovery that a substance for which we have a single name has a definite internal structure, and what that structure is: there can be no presumption in advance that it has one. It is therefore unlikely that, before such a discovery, the common understanding of any substance-term would have included a tacit reference to a possible but as yet unknown internal structure.

We are here concerned, of course, with everyday substance-terms, with the word "water," for example, as employed in ordinary colloquial speech. In scientific discourse, the chemical composition of a named substance is part of its nominal essence: the name applies solely in virtue of that composition, and so there is no divergence here between epistemic and metaphysical necessity. With animals, the everyday criterion is that of descent: the offspring of a bull and a cow, however much of a freak, is a calf, and a bull or cow can only have been sired by a bull upon a cow. If creatures resembling terrestrial tigers, however closely, had been discovered on Mars, they would not have been tigers (although they would certainly have been called "Martian tigers"), unless they were proved to have been, by some extraordinary means, descended from terrestrial tigers. The everyday criterion for being an animal of a particular kind would, if universally maintained, yield a proof that every kind of animal had a history going back to the origin of life on earth, which few people believe; but even if new species can arise out of pre-existing ones, the principle that the classification of animal kinds is to be based on descent and genetic relatedness holds good in scientific biology as much as in popular thinking. Thus it is in accordance both with everyday and with scientific criteria to say that white ants (termites) are not true ants because they are not genetically related to genuine ants (being in fact of a different order). What makes internal structure relevant to the identification of the kind to which an animal belongs is that it has been found to be a far more faithful guide to genetic relatedness than external appearance; but it is still genetic relationship that is the controlling criterion. The selection of the internal structure of an animal as one of its essential properties is thus an amalgam of what belongs to the nominal essence of the kind to which it belongs, and is thus a criterion for its being of that kind, namely its descent, with a scientific symptom of its having had that descent. No doubt a layman would be unskilled at identifying the specific features of the internal structure which a biologist would take as being decisive; but the general idea would accord well with his understanding of the question. Thus what Kripke takes as an essential property of an animal is very near the nominal essence of the kind to which the animal belongs, being a scientifically identified symptom of the individual animal's conforming to that nominal essence. There is a much smaller divergence between the scientific and the everyday conception of kinds of animal than between the scientific and the everyday conception of kinds of substance.

Are individual objects and natural kinds the only things to which essential properties can be ascribed? The tight connection between essential properties and rigid designation suggests that they are not, since dates and terms for lengths are supposed to be rigid designators: the standard metre rod—that very rod—might have been more than a metre long. What are the essential properties of a metre or of a difference of temperature of 1 °C? It is not apparent that such things have any essential intrinsic properties: they have only essential relations given by scientific laws. What are the essential properties of any particular year? It was the period during which the Earth completed an orbit about the Sun; but that was part of its nominal essence. Apart from that it appears to have only necessary relations: it necessarily succeeded what was in fact the previous year, and necessarily preceded what was in fact the subsequent one.

In the notion of what is metaphysically necessary and that of essential properties, we do not have hold of any firm concepts. We are, indeed, exploring a loose one. Many of our uses of modal auxiliaries such as "might have" are indeed not explicable in terms of epistemic possibility, and a good proportion of them are amenable to a Kripkean account. But the truth-conditions of statements involving them are not definite; and in my opinion there is no determinate notion of metaphysical necessity to be sought.

Has the notion of metaphysical necessity, whether a sharp one or not, been of use as a philosophical tool? The notion of analyticity, or of a priori knowability, has long been such a tool. It is a legitimate and a useful one, since it bears on the *senses* of expressions, and thus upon the concepts they express. It is far from clear that the notion of metaphysical necessity is of similar philosophical utility. It has become a standard notion among analytical philosophers; but what clarification has it enabled us to make? Kripke's notion of metaphysical necessity is not to be equated with Aquinas's notion of the *per se nota*. For him whatever is *nota quoad nos* must also be *per se nota*: he had no room for the contingent a priori. The concept of metaphysical necessity can tempt philosophers to use it where they should be using that of aprioricity. For example, in a lecture which I heard on the subject of colour an argument was put forward against the so-called dispositional analysis of colour-concepts, in terms of how things appear to normal observers in normal circumstances. It was said that, if the analysis were correct, then the following modalised biconditional must hold of any object *a*:

Necessarily, *a* is red iff *a* would appear red to any normal observer in normal circumstances.

This modalised biconditional was confuted on the ground that there is surely a possible world in which indeed *a* is red, but nevertheless appears yellow to all observers in all circumstances. Whether or not there is such a possible world, the argument is beside the point. If we are concerned with analysing the concept of something's being red, and thus the sense of the adjective "red," the question is not whether the biconditional is metaphysically necessary, but whether it is knowable a priori. The addition of metaphysical necessity to our repertoire can have a confusing effect on philosophical analysis of concepts; that it is in any way helpful is seriously to be doubted.

The formulation of the notion of rigidity (modal or temporal) was a fruitful addition to the philosopher's toolkit. The introduction of the concept of metaphysical necessity has led only to irrelevance and confusion.

Chapter 3
Ibn Sīnā and Conflict in Logic

Wilfrid Hodges

1 To John

In autumn 1965 I became one of John Bell's first students. He and I were both registered at Oxford University to work for MPhil—later upgraded to DPhil—under the supervision of John Crossley (though for reasons I don't remember, I began as a supervisee of Michael Dummett). John Bell and Alan Slomson (another supervisee of John Crossley) had put together a clear and elegant first course on model theory, concentrating on ultraproducts and the construction of elementary embeddings one element at a time, in the style being pioneered at the time by Chang and Keisler. That course was my introduction to model theory. A fuller version, with some extra material from George Wilmers, became the famous *Bell and Slomson: Models and Ultraproducts*. It was published two years earlier than the definitive tome of Chang and Keisler, and made a lot of people happy.

I particularly want to thank John for another thing that he taught me during our time in Oxford. In 1966 C. C. Chang himself came to Oxford with a copy of Jack Silver's dissertation on indiscernibles and large cardinals. This was novel stuff and we all found it challenging. But John lectured on it, and what I learned from his lectures led directly to my own DPhil thesis, and in turn to much of what I did later in the field.

In 1967 I was invited to the Philosophy Department at UCLA at short notice. The scheduled lecturer in Philosophy of Religion, the logician John Lemmon, had died of a heart attack while climbing in the San Bernadino Mountains just a few weeks before the beginning of term, and I was brought in as an emergency replacement. I think it was during my year in California that John Bell was appointed to LSE in London. For a while, he and Moshé Machover were the London mathematical logicians. There were a number of people in London at the time who knew something about mathematical logic and took a positive view of it—Daniel Cohen, Paul Cohn, Chris Fernau, Ivor Grattan-Guinness, Clive Kilmister, Geoffrey Kneebone,

W. Hodges FBA (✉)
Herons Brook, Sticklepath, Okehampton, Devon EX20 2PY, England
e-mail: wilfrid.hodges@btinternet.com

D. DeVidi et al. (eds.), *Logic, Mathematics, Philosophy: Vintage Enthusiasms*, The Western Ontario Series in Philosophy of Science 75,
DOI 10.1007/978-94-007-0214-1_3,

Imre Lakatos, Peter Landin, David Wiggins, and probably others I missed. But none of these were trained specialists in the field. So John and Moshé, who made a splendid team together, filled an important slot. In 1968 I moved to London too, to lecture at the now deceased Bedford College, so I graduated from being John's pupil to become his colleague.

An aside: as my memory serves me, John told me once that according to his mother, his surname translates into Russian as *zvonok*. I was appalled. A *zvonok* is a doorbell—John Doorbell! I saw him much more as a *kolokol*, one of those tuneful and authoritative church bells that ring across the countryside and are imitated by bluebells in the spring.

That brings me to translations. Though the Oxford logicians were barely aware of it, John Bell in the late 1960s was building for himself the foundations of a possible second career as a scientific translator. His translation (with Michael Woods) of Jean Nicod's "Geometry and Induction" was published in 1970 and is still cited. So I hope John will allow me to honour him with a translation of some hitherto untranslated work of another logician, the eleventh century Arabic-Iranian scholar Ibn Sīnā.

I thank Amirouche Moktefi and Dimitri Gutas for helpful comments; in particular Amirouche read through the Arabic with me and made many improvements to the translation. Of course any mistakes are my responsibility.

2 De Interpretatione Chapter 14

2.1 The Text Translated Below

During the 1020s Ibn Sīnā wrote a commentary in Arabic on the logical works of Aristotle, as part of his encyclopedia *Al-Šifā'* (*The Cure*). The commentary runs to some 2180 pages in the recent Cairo edition; this figure includes his commentary *Madkal* on Porphyry's *Eisagōgē*, which he counted as an introduction to Aristotle's work. Apart from the *Madkal* which was translated into Latin in the 12th century, and the section of *Qiyās* dealing with propositional logic, barely any of Ibn Sīnā's commentary has been translated into any western language.

Ibn Sīnā refers to Aristotle's texts as "The First Teaching" (as in [2.5.7], [2.5.9] below). Their first five books are the subjects of the second to sixth volumes of the *Šifā'* respectively. After *Madkal* which comments on the *Eisagōgē*, there come *Maqūlāt* (on Aristotle's *Categories*), *ᶜIbāra* (on *Perì Hermēneías*), *Qiyās* (on *Prior Analytics*), *Burhān* (on *Posterior Analytics*) and *Jadal* (on *Topics*). Ibn Sīnā also commented on the *Sophistical Refutations*, *Rhetoric* and *Poetics*, but we won't use these commentaries. We will cite two later works of Ibn Sīnā: the *Easterners* and the *'Išārāt*. The numerous references to the works of Ibn Sīnā will be by title in the text. References to the work of other authors follow the standard format for this volume.

The passage translated in §6 below is Section 2.5 from Ibn Sīnā's commentary *ᶜIbāra* on Aristotle's *Perì Hermēneías*, the book that the Latins knew as

De Interpretatione. Ibn Sīnā is commenting on Section 14 of the *Perì Hermeneías*; henceforth I abbreviate this to PH14. The commentator Stephanus—who probably taught in Alexandria around AD 600—said of this passage

> The enquiry now undertaken is certainly not Aristotle's, but is written as an exercise. That is why Porphyry, writing a lengthy commentary on [*Perì Hermēneías*], did not judge this section worthy of the thought needed to clarify it. ((Charlton, 2000, p. 185). (Ammonius, 1897) says similar things in more detail.) (3.1)

In fact most commentators have found it difficult to fit the passage convincingly into *Perì Hermēneías*, or even to make sense of it on its own. This is helpful for us in two ways. First, the passage is a misfit, so we can reasonably take it separately from the rest of *Perì Hermēneías*. Second, it will serve to illustrate how Ibn Sīnā deals with challenging material.

My title—"conflict in logic"—has two meanings. First we will see that Ibn Sīnā takes up the view of some earlier commentators, that Aristotle's passage is about conflict as a notion related to logic. Second, Ibn Sīnā was notorious for treating other logicians with disdain, and one of his livelier paragraphs here is a vivid demonstration of how to wipe the floor with people you despise. I think Ibn Sīnā is not just being obnoxious; he has a significant point to make about how to do logic.

2.2 Ways of Teaching

The Aristotelian commentators set themselves the task of making Aristotle consistent with himself. They developed a battery of excuses for the contradictions and obscurities that they found in him. It became a cardinal point that Aristotle intentionally made himself difficult to understand. Thus Ibn Sīnā's predecessor Al-Fārābī:

> ... the person who researches the Aristotelian sciences, studies his books, and applies himself with perseverance to them, knows full well the different methods he used to render things inaccessible, cryptic, and intricate, despite his express intent to expound and explain. Among these [methods] are the following: (3.2)
>
> i. In many of the syllogisms ... he omits the necessary premiss. (Etc. through five methods.)

Dimitri Gutas argues that Ibn Sīnā accepted what he took to be Aristotle's reasons for this policy, and sometimes copied Aristotle by adopting a similar obscurity. Thus Ibn Sīnā writes:

> Whatever I am able to bring to light I will do so either openly, or from behind a veil which acts as a useful kind of stimulus and drill ... ((Gutas, 1988), p. 228 for the Al-Fārābī quotation and p. 307 for the Ibn Sīnā.) (3.3)

In fact Ibn Sīnā does assume that the obscurities of PH14 are deliberate, but he has another explanation for them. In *Qiyās* he makes the following remarks:

> Teaching is of two kinds. [First there is] teaching which supplies knowledge of something that wasn't already known in the nature of things; as when one teaches that the three angles of a triangle are equivalent to two right angles. [Second there is] teaching that consists of reminding (*taḏkīr*) and facilitating (*'i^{c}dād*). Reminding is what causes a thing to come into the mental processor (*bāl*) when the thing was already known. ... Facilitation is that a thing comes into the mental processor together with other things that have similar properties to it. Each of these other things, when it is known, gives no further knowledge beyond itself; but when [the first thing] is brought into the mental processor in the vicinity of this thing, the two of them [together] supply new knowledge. ... *Perì Hermēneías* mostly consists of reminding and facilitating, though some of it is argument and reasoning. (*Al-Qiyās* 15.13–17.1. On "mental processor" see §3.2 below and the notes on [2.5.13].) (3.4)

In paragraphs [2.5.9] and [2.5.10] of his commentary on PH14, Ibn Sīnā states his view that this section of *Perì Hermēneías*, or at least a major part of it, was written as a facilitation. (He says *tawṭi'a*; this is a near synonym of *'i^{c}dād*, and there is no reason to think that he means anything different by it.)

The notion of facilitation as a style of teaching seems to be Ibn Sīnā's own. I think it entitles him to a mention in the history of cognitive science. There is more to be said on this, but not here.

The thing to take home here is that on Ibn Sīnā's account, Aristotle is making points not for their own sake, but so that they can serve as catalysts for the student to develop other related pieces of insight. So there is no need for PH14 to hang together. Its overall structure is not what the student should be learning from it. Rather it gives the student a collection of bullet points that can facilitate other knowledge in analogous situations.

Ibn Sīnā writes his commentary on PH14 in a similar style. He makes points that should be followed up but aren't. He jumps forwards and backwards through Aristotle's text for no apparent reason. With the exception of the remarkable paragraph [2.5.16], he gives many suggestions but few arguments. He is surprisingly casual about whether his explanations are accurate to Aristotle's intentions. In the context described above, one can see why he could have thought that this was an appropriate way to treat a historical text; today we would regard it as less than professional.

2.3 Summary of Aristotle's Text

Ibn Sīnā would have had available to him the excellent Arabic translation of Aristotle's text made around the year 900 by 'Isḥāq ibn Ḥunain. This translation survives, but Ibn Sīnā tends to quote so loosely that it's hard to be sure exactly what he is quoting. (For an example see the note on 131.12 "Contraries" below.)

The following summary of Aristotle's argument imposes a shape that may go some way beyond what Aristotle himself intended. Incoherent arguments are very hard to summarise without imposing some kind of coherence. Ackrill (1963) translates the passage with a commentary.

At 17b4 earlier in the *Perì Hermēneías*, Aristotle has explained that the two sentences "Every A is an X" and "No A is an X" are contraries (*enantíai*) of each other. This was by stipulation; in the present passage, Aristotle aims to give a heuristic argument to justify the stipulation on the basis of more fundamental principles.

The principles that he assumes are as follows:

1. For every proposition p there is a proposition q such that p and q are contraries of each other. (Implicit in 23a27–30.)
2. If p, q, r are propositions and both q and r are contraries of p, then q and r express the same belief. (Implicit in 23a32–35.)
3. If p is a contrary of q then p and q can't both be true. (24b9.)

Step 1 (23a32–39). Aristotle begins by reducing the question to one about beliefs. I guess that this is in order to replace the equivalence relation in Principle 2 by identity. But in fact he continues his argument in terms of the propositions expressing the beliefs.

Step 2 (23a40–23b7). Next he considers the case of singular propositions. An affirmative singular proposition has the form

$$A \text{ is an } X. \tag{3.5}$$

and a negative singular proposition has the form

$$B \text{ is not an } X. \tag{3.6}$$

Aristotle claims that if the subjects A and B are distinct, then in general nothing prevents (3.5) and (3.6) from both being true, even with the same X; and the same holds between two affirmatives or two negatives. So we conclude that two singular propositions that are contrary to each other must have the same subject.

Step 3 (implicit in 23b7–9). Next he points out that if (3.5) is contrary to a singular proposition q, then q must entail

$$A \text{ is not an } X. \tag{3.7}$$

Otherwise Principle 3 is violated.

Steps 4a, 4b, 4c. Having established this, he gives three arguments why the contrary of (3.5) must be (3.7) (up to identity of belief). The first argument (Step 4a, 23b7–27) is very obscure; maybe Aristotle intended it to be more than one argument. It includes two further notions: that of a thing being true essentially as opposed to accidentally, and that of a belief being further from the truth than another belief.

The second argument (Step 4b, 23b27–32) is that for some particular values of A and X the only candidate for a contrary of (3.5) is (3.7). But the generality in Principle 1 implies that there is some uniform formula for reaching the contrary. The third argument (Step 4c, 23b33–24a3) is broadly similar, but in the opposite direction: the contrary of (3.7) can only be (3.5), and so the contrary of (3.5) is (3.7) by the symmetry in Principle 1.

Step 5 (24b1–6). Finally Aristotle reaches his conclusion as follows. By analogy with singular propositions, the contrary of

Every A is an X. (3.8)

must have the same subject as (3.8). Now the subject of (3.8) is A, taken universally; so the same must hold for its contrary. Also by analogy with singular propositions, the contrary of (3.8) should conclude "... is not X." Putting these together, the contrary of (3.8) is

Every A is not an X. (3.9)

As required, this expresses the same belief as "No A is an X."

3 Ibn Sīnā's Semantics

3.1 Ideas

For Ibn Sīnā, logic is about ideas (he calls them "things," "*ašyā*"). More precisely it's about how ideas are derived from other ideas by definition or deduction. Ideas are objective entities, as distinct from the representations of ideas in your mind or my mind. In fact Ibn Sīnā believes that there is a divine intellect that holds the stock of ideas. But he has a strong antipathy to mixing logic and metaphysics, so he never mentions this point in his logical writings.

A typical idea is the meaning of a word or meaningful phrase of a natural language. Here we need some notation. I use quotes to name words and phrases:

"not a horse" (3.10)

and semantic quotes (in the style of (Jackendoff, 1990) and others) to name the meanings of words and phrases:

[NOT A HORSE] (3.11)

Ibn Sīnā himself has neither of these notations. For the first he would write the equivalent of

the phrase not a horse. (3.12)

He has no expression for semantic quotes, but there are a number of phrases that serve as a cue that he is talking about meanings. For example in [2.5.5] below he says the Arabic equivalent of "as for it isn't good itself"; he is talking about the meaning [NOT GOOD]. (See the note on 125.17 below.) A more extravagant example is

not-a-horse-ness insofar as it is not-a-horse-ness (3.13)

(*Al-Maqūlāt* 242.3), which refers to [NOT A HORSE].

A typical idea like [HORSE] has at its core a principle for classifying actual or possible entities into two sorts, those which satisfy it and those which don't. This principle is (near enough for present purposes) what Ibn Sīnā calls the "nature" (*tabī*^{c}*a*) of the idea.

Some ideas are atomic and come to us direct. Others are compound; their nature is built up from the natures of simpler ideas. A typical compound idea like [HORSE] has a feature which records how the nature is built up. This feature is called the "essence" (*ḏāt*) or "whatness" (*māhiyya*) of the idea. The idea also has a "definition," which is a linguistic expression that reports the essence in a canonical form. The simpler ideas from which an idea is formed are said to be "internal" (*dākil*) to the idea or the definition. For example [ANIMAL] is internal to [HORSE], and [NECESSARY] is internal to [POSSIBLE].

The essence of [HORSE] actually contains [ANIMAL] as a conjunct: to check that a thing is a horse, you need to check among other things that it's an animal. We can express this by saying that [ANIMAL] is "constitutive" (*muqawwim*) for [HORSE], or that it's a constitutive of [HORSE]. Unfortunately we often meet loose vocabulary in this area. For example when Ibn Sīnā says that I is "essential" for J, he sometimes seems to mean that it's internal, and sometimes that it's constitutive. Constitutive implies internal but not vice versa; for example [NECESSARY] is certainly not constitutive for [POSSIBLE].

This confusion arises from an endemic false assumption of Aristotelian logic, which I discuss in (Hodges, 2009) under the name of Top-Level Processing. Briefly, the assumption is that logical processing never reaches below the top syntactic level. Ibn Sīnā deserves credit for making the assumption explicit, but (at least in the West) we have to wait till Frege to see a serious assault on it.

The way it shows up in connection with "internal" is that definitions are required to express intersections; for example

> So the definition has to be composed from the genus and the differentia ... as when we define "human" by saying "Human is animal that is rational." (*Al-Madkal* 48.17–19) (3.14)

> ... composition in the form of restriction, which is what happens when we obtain concepts through definitions ... (*Al-*^{c}*Ibāra* 31.16f) (3.15)

(In all ages of Aristotelian logic, $a \cap b$ is thought of as b restricted to a. As Boole puts it, "The mental operation represented by the adjective ... is that of selecting from a certain class as subject all the individuals which together answer to a given description. ... the subject class is expressed by that word or combination of words to which the adjective is prefixed." (Boole, 1997, p. 5)) Thanks to Top-Level Processing, the theory of definitions tends to fasten on the top-level constitutives and overlook features that are lower down in the syntactic structure, such as negations on subphrases.

3.2 Attachments

We can form the idea [HUMAN]. We can also form the ideas [RATIONAL HUMAN] and [YOUNG HUMAN]; let us express this by saying that the ideas [RATIONAL] and [YOUNG] are "attachments" of the idea [HUMAN]. There is a crucial difference between these two attachments: [RATIONAL] is (for Aristotelians) constitutive for [HUMAN], but [YOUNG] certainly isn't. Those attachments that are not constitutive are said to be "accidents" (*ᶜaraḍ*). The distinction between constitutives and accidents is one of the fundamental principles of Aristotelian philosophy, though not always in this terminology.

Aristotle himself said a number of things relevant to the distinction between constitutives and accidents, and the early commentators added more. By the late fourth century AD, the Phoenician philosopher Porphyry decided that the range of views on this and related topics had become a barrier to beginners in the field, and so he wrote his *Eisagōgē* to draw some lines in the sand.

Porphyry divides attachments into three groups, namely non-accidents (i.e. constitutives), inseparable accidents (*ᶜaraḍ ġair mufāriq* in the Arabic translation) and separable accidents (*ᶜaraḍ mufāriq*) (*Eisagōgē* 12.25ff, (Barnes, 2003, p. 12)). He classifies them by a device called "removal," *rafᶜ* in Arabic. Given an idea I and an attachment J, we first form the idea NOT-J. (At this point an aristotelian would want to distinguish between removing J and affirming NOT-J. For simplicity I suppress this distinction.) Are there any things that satisfy I and NOT-J? If there aren't, then J is a constitutive of I. If there are in the real world, then J is a separable accident of I. If there aren't any in the real world, but we can imagine one, then J is an inseparable accident of I. Thus [RATIONAL] is constitutive for [HUMAN] and [YOUNG] is a separable accident of [HUMAN]. Porphyry offers [BLACK] as an inseparable accident of [CROW].

Ibn Sīnā reviews this classification in his *Madḵal*. In places he merely reports what Porphyry said. Elsewhere (e.g., 86.4ff) he gives a textbookish critique of the classification, scolding Porphyry for his careless formulations. But the full extent of Ibn Sīnā's disagreement with Porphyry comes to light when he forgets Porphyry and sets out his own views, in *Madḵal* and elsewhere.

Like Porphyry, Ibn Sīnā divides attachments into three groups, but the groups are not Porphyry's. Ibn Sīnā distinguishes (1) the constitutives, (2) the inherents (*lāzim*) and (3) the rest. The *rafᶜ* test plays no part in defining these groups. The constitutives of an idea I are those ideas which are conjuncts of the essence of I, as we saw above. The inherents are those attachments that are not constitutive but follow necessarily from the essence of I.

> Sometimes [an idea] has inherents that follow from it because of its whatness, though the whatness is established first and then these things follow from it. Thus [EVEN] follows from [TWO] .... (*Al-Madḵal* 34.10f) (3.16)

> A demonstration gives an inherent ... whereas a definition gives something in the constitutive essence. An inherent [of an idea] is not internal to the definition of the idea. ... For example there is a demonstration telling us that a triangle has angles equal to two right angles; this meaning is external to the definition of [TRIANGLE]. (*Al-Burhān* 199.15–18) (3.17)

The third class consists of those attachments that don't follow from the essence of I.

How does this correlate with *rafᶜ*? Strictly there are two forms of *rafᶜ*, namely removal in the real world and removal in the imagination. Ibn Sīnā mentions the first, but he has very little interest in it. Logicians don't concern themselves with the actual world. As for removal in the imagination, he is clear that this can't happen with constitutives:

> It's impossible to introduce into the mental processor both an idea and one of its constitutives, so that the idea is actually present and the constitutive is negated from it; this would destroy the conceptualisation of the whatness. (*Al-Madkhal* 34.21–35.3) (3.18)

But which accidents can be removed in the imagination? Ibn Sīnā's answer has two ingredients, one logical and one cognitive.

The logical part is that a proof of J from I can be long or short, depending on what I and J are. For Ibn Sīnā the crucial dividing line is whether I entails J immediately, or whether you need to introduce an intermediate (*mutawassiṭ*) idea in order to deduce J from I. For example, part of the statement of Proposition I.32 of Euclid's *Elements* is that the internal angles of a (plane) triangle ABC add up to the sum of two right angles. Euclid proves this by first extending the side BC to BD, then adding a line segment CE parallel to BA:

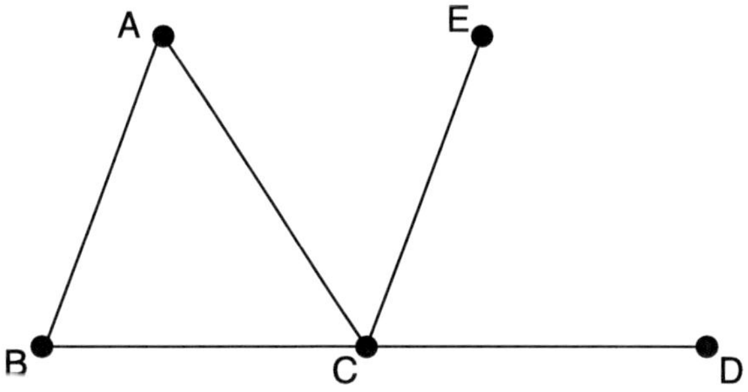

(3.19)

Then by earlier propositions in the *Elements*, the angles ACE and ECD are respectively equal to BAC and ABC, and the theorem follows at once. The theorem shows that [ANGLES 180°] is an inherent of [TRIANGLE]. But at least by this proof, it's not an immediate inherent. Ibn Sīnā thinks that the possibility of extending the side BC to BD is an immediate inherent of [TRIANGLE], but the segment CE adds a whole new idea (*Al-Madkal* 35.20–36.3).

This theorem has a history. Proclus in the fifth century AD (Proclus, 1970, 296ff) mentions two other proofs of it, but both of them involve drawing extra lines.

Mancosu (1996, p. 13ff) quotes a tract of the sixteenth century author Pereyra, in which Pereyra uses the *raf*ᶜ test to show that the extendibility of BC to BD is an accidental property of the triangle, and hence that the proof is not "scientific." (For Ibn Sīnā *all* scientific proofs are proofs of accidents—why should one bother to prove a definition? Also we will see in a moment why he believes that the *raf*ᶜ test doesn't give the answer that Pereyra claims.)

We turn to the cognitive part. According to Ibn Sīnā, our *bāl* (which I translated as "mental processor" in (3.4) above) is the part of our mind where reasoning gets processed. From what he says about it, apparently it has a very limited range of actions: we send two ideas into the *bāl*, and it either finds some overlap or similarity between the ideas, or it tells us they are incompatible. (So the *bāl* is basically a unification engine. In syllogistic reasoning it serves essentially the same role as unification in the resolution calculus. Obviously there has to be some other mental element that chooses the appropriate ideas to send into the *bāl*; this raises some issues which are important for understanding Ibn Sīnā, but they are not our concern here.)

Ibn Sīnā believes that the *raf*ᶜ test involves a simple act of comparison between two ideas, with no further reasoning. So we do the test by a single pass through the *bāl*. This is enough for the *bāl* to recognise the incompatibility of I and NOT-J when J follows from I immediately:

> There are accidents that are inherents of the whatness by a primary and clear entailment that is not mediated by any other accident. So when the entailment is not via some intermediate, it is impossible to negate the accident from the whatness at the same time as affirming the whatness, having them both enter the mental processor together. An example is [TRIANGLE] and [CAN IMAGINE A LINE OF THE TRIANGLE EXTENDED]. (*Al-Madkal* 35.18–20) (3.20)

But in a single pass, with no memory and no internal controls, the *bāl* has no way of recognising an incompatibility that depends on some intermediate that it hasn't seen:

> It can sometimes happen that the holding of the accident is through something intermediate, so when this intermediate thing doesn't come into the mental processor, one can negate [the accident]—for example [one can negate] that any two angles of a triangle are [together] less than two right angles. (*Al-Madkal* 36.1–3) (3.21)

In short, the test of *raf*ᶜ (in the imagination) separates an idea I from its accident J if and only if J is not an immediate consequence of I.

Somehow this theory is a little glib. There are signs that Ibn Sīnā himself has reservations about it. In *Al-Burhān* 38.3–8 he says that if a person didn't realise that all humans are rational, it would be possible for him to imagine there are humans with no sense of humour. Now Ibn Sīnā believes that [HAS A SENSE OF HUMOUR] is a consequence of [IS CAPABLE OF BEING SURPRISED], which in turn is a consequence of [RATIONAL] (e.g. *Al-Madkal* 30.1f). This is a two-step inference, so by the theory we have been reviewing, Ibn Sīnā should believe that it's possible, as things are, to imagine that there is a human with no sense of humour. So why does he introduce the desperate premise that the imaginer doesn't realise that humans are rational?

Nevertheless Ibn Sīnā is robust in his belief that there are non-constitutives that can't be removed in the imagination:

> Pay no attention to the theory that says that non-constitutives can legitimately be removed in the imagination. (*Al-'Išārāt* i.12 p. 49) (3.22)

3.3 Meanings of Sentences

Semantics is a theme that runs throughout the logical part of the *Šifā'*. But Ibn Sīnā is probably not intending to make a contribution to general linguistics. His context is that we use logic to analyse arguments—other people's or our own. These arguments reach us in the form of sets of sentences. The sentences can even be written in a book, separated and in the wrong order, and with bits missing or added (*Al-Qiyās* 460.4–8). So we have to try to reconstruct the author's intentions. In other words, we have to reconstruct the "reading" (*'akḏ*, literally "taking") that the author put on his words. In this setting, Ibn Sīnā often refers to the kinds of information that we need to supply, and the ways open to us for finding them.

Of course this includes disambiguating both words and syntactic constructions. But as Ibn Sīnā often emphasises, it can also involve adding things that weren't explicit in the sentence:

> ... a time, or a place, or a comparison of how things are, or an implied event, or an action or an experience, or some consideration of possible versus actual, or some consideration associated with an agent or an experiencer .... (*Easterners* 48.6–8) (3.23)

(The question how much interpretation we are entitled to add to a text was a hot topic in Qur'ānic exegesis at the time.)

Thus every sentence has an indefinite number of possible readings. In general some of these readings will be true and some will be false. This is probably how we should understand Ibn Sīnā's notion of two sentences being "true together," or "agreeing in truth" (and likewise "false together"). Thus a sentence p and a sentence q are "true together" if there are a reading of p, and a corresponding reading of q, such that p and q are both true under these readings. This notion obviously depends on what we count as a "corresponding" reading. But in fact Ibn Sīnā normally uses this notion of "true together" for two sentences that are very close syntactically; for example they may differ only in their quantifier. If the reading provides references for the names in p, together with a place and time for the whole sentence, then the same references and place and time carry over directly to the other sentence.

Another notion in the same general area is *'īhām*, what a sentence suggests but doesn't in fact say.

> But with propositions one should focus not on what they suggest but on what they mean in themselves. (*Al-*^{c}*Ibāra* 55.13f) (3.24)

Later in the ^{c}*Ibāra* (p. 104) Ibn Sīnā uses this principle to dismiss the Grices of his time. He doesn't seem to have noticed that what the sentence suggests may be exactly the reading intended by the speaker or writer. This becomes highly relevant in paragraph [2.5.16] below; see the note on line 130.16.

4 Conflict and Contrariety

4.1 Types of Opposition

The notions of two ideas "agreeing" or being "opposed" are two of the primitive notions of Aristotelian logic. Since these notions are primitive, they tend to get short shrift in terms of definitions and explanations. But we do meet classifications of different types of opposition.

For Ibn Sīnā the general term for "opposed" is *muqābil*, literally "facing." He does offer a definition of this in *Al-Maqūlāt* 241.7f:

> We say: opposing pairs are those which don't combine in a single subject from a single aspect at a single time together. (3.25)

Unfortunately we have to guess the meaning of "combine"; but almost certainly he means "are true together" as in §3.3 above. We also need to know what "aspects" are; I ignore this here. I think we have to read "single subject" in a rather forced way as "the same singular subject," for example "Zayd." Thus [MOVING] and [AT REST] are opposed, because for example the following sentences can't both be true at the same time with the same Zayd:

> Zayd is moving. Zayd is at rest. (3.26)

Ibn Sīnā's definition seems to apply only to noun-type ideas; but we can extend it to sentence-type ideas by regarding these as classifiers of times and/or circumstances. This reduction of sentence-type ideas to noun-type ideas runs throughout Ibn Sīnā's logic, though as a heuristic principle rather than a formal reduction.

Ibn Sīnā mentions three main types of opposed pairs: negations (*salb*), contradictions (*naqīd*) and contraries (*ḍidd*). Negations are got by adding or removing a particle of negation, provided of course that it applies to the whole idea.

> When the predicate is negated on its own without negating the quantity with which it is predicated [i.e. the quantifier on the subject], then the negation is not a negation of what was affirmed. (*Al-ᶜIbāra* 94.4–6) (3.27)

In practice contradictories are the same thing as negations, though Ibn Sīnā may intend the contradictory of p to be by definition a sentence which is true exactly when p is false (i.e. under the same readings).

There is some discussion of contraries in book 7 of *Maqūlāt*. Ibn Sīnā mentions a number of pairs which are generally considered contrary. They include [HEAT] and [COLD], [HEALTH] and [SICKNESS], [MOVEMENT UP] and [MOVEMENT DOWN]. He discusses which of these pairs can be described by saying that some particular idea is present on one side and absent on the other. If he has a general definition for "contrary" here, I haven't found it. But in ^{c}*Ibāra* he gives the classical definition for the case of sentence-type ideas:

> Let us call this opposition 'contrariety,' where the two opposing things don't ever agree in truth, but they do agree in falsehood. ... Contraries can't be true together, but they can be false together, as you know. (46.16–47.2.) (3.28)

Just before giving the definition, he has quoted the example pair

> Each person is a writer. No person is a writer. (3.29)

He says the second sentence is got from the first by negating the predicate; so he is not distinguishing the second sentence from "Each person is not a writer."

4.2 Aristotle's Question in PH14

Ibn Sīnā opens his commentary on PH14 with a remark that the passage is about the relation "in greater conflict with" (*'ašadd* ^{c}*inād*). This reading can be traced back at least to Ammonius (*mâllon mákhesthai*, (Ammonius, 1897, p. 252, 202 r120), who led the Alexandrian school in around AD 500, and it survives down to Whitaker ("more violently opposed," (Whitaker, 1996, p. 172)). But the evidence to support it is vanishingly small. In both the original Greek and 'Isḥāq's Arabic translation, the comparatives are limited to just a few lines (23b20–24, a small part of what I called Step 4a in §2.3 above). But for reasons given in §2.2 above, I don't think Ibn Sīnā is much concerned about whether he has interpreted this particular passage as Aristotle intended it.

We don't know specifically that Ibn Sīnā knew Ammonius' commentary. But it's likely that he did, since some half a century earlier Yaḥyā ibn ^{c}Adī [1988, pp. 321–323] regarded the views of the "Alexandrian commentators" on PH14 as something to discuss between friends. I thank Peter Adamson for this reference.

The relation "in greater conflict with" is a rather strange one to find in an Aristotelian logical treatise. It has three variables: X is in greater conflict with Z than Y is. Most Aristotelian logicians found it beyond their capabilities to handle even a relation with two variables. Ibn Sīnā himself has an ambivalent attitude to variables in relations. Throughout his logical writings he calls attention to them, and particularly to the need to supply the assumed parameters when we interpret sentences. But his formal logic avoids them completely. There has to be a reason for this discrepancy. I think it's Top-Level Processing again; for further details see (Hodges, 2009).

The word ^{c}*inād* "conflict" never appears in 'Isḥāq's translation of PH14, though *muqābil*, *salb*, *naqīḍ* and *ḍidd* "contrary" are all frequent ((Jabre, 1999)—^{c}*inād* does appear at 21a38, earlier in the *Perì Hermēneías*). I suspect Ibn Sīnā chooses a word distinct from all of these four, and uses it throughout his commentary, in order to make it clear that he wants to visit a new question. In PH14 one easily gets the impression that Aristotle is reworking issues that he has forgotten he settled earlier.

("The body of the chapter ... upsets the distinction between contraries and contradictories which was drawn in Chapter 7," (Ackrill, 1963, p. 153).) The notion of ^{c}*inād* turns up again quite often in *Al-Jadal*, Ibn Sīnā's commentary on Aristotle's book about debate. Ibn Sīnā says in [2.5.1] that the topic of PH14 has more to do with debate than with logic; but I don't think the question of "greater conflict" reappears in *Jadal*.

4.3 The Quickest Contradiction

Ibn Sīnā's interpretation of Aristotle's text seems to rest on two intuitions, which we can call Canonicality and Immediacy.

Canonicality is the intuition that if every proposition has a unique most conflicting opposite proposition, then there has to be a recipe for finding this most conflicting opposite which is uniform across all propositions.

A simple application of this intuition is to think of an idea I in terms of its extension $\bar{I}$, which is the class of things that satisfy I. Suppose Ω is the relevant universe of things. An idea J is contrary to I if and only if $\bar{J}$ is a subclass of Ω that is disjoint from $\bar{I}$. Just from the given data, how can we define a particular contrary J for I? There are only two options: $\bar{J}$ must be either the complement of $\bar{I}$ in Ω, or the empty class. Aristotelian logicians are unsure whether there are ideas with empty extension; and in any case there are undesirable consequences if two different ideas have the same most conflicting opposite. This leaves only the idea NOT-I, whose extension is the unique contrary of $\bar{I}$ that includes the extensions of all other contraries of I. This is the argument of [2.5.5], and similar thoughts lie behind [2.5.4], [2.5.9], [2.5.14] and [2.5.17].

Immediacy is the intuition that if the incompatibility of I and K can be proved more briefly than the incompatibility of J and K, then I conflicts with K more strongly than J does.

The simplest example of this is where the only way to prove the incompatibility of J and K is by using the incompatibility of I and K as a lemma. Ibn Sīnā isolates this case at [2.5.6]. Also if the incompatibility of I and K can be proved using only ideas in the essences of I and K, whereas to prove the incompatibility of J and K one needs to go outside the essences, then prima facie one expects that the proof will be shorter for I than for J. This seems to be the intuition behind [2.5.11]–[2.5.13].

In §3.2 above we saw how Ibn Sīnā has reasons for being interested in lengths of proofs. The idea of grading contradictions as more or less lethal has some echoes in modern work; it can matter in databases.

Another area where people have considered false theories and classified them in terms of distance from truth is the philosophy of science, as for example in (Oddie, 2007). I thank Sjoerd Zwart for making me aware of this possible parallel, though I have to add that I think it would take some ingenuity to make Ibn Sīnā's suggestions relevant to the questions discussed by Oddie.

5 The "Well-Known Commentator"

5.1 Ibn Sīnā's Outrageous Analysis

In paragraph [2.5.16] Ibn Sīnā attributes the following argument to an unnamed commentator:

> A contradictory belief can be found for a true belief in each case. This fact is something essential, since an essential thing is a thing that is found in all cases. (3.30)

Ibn Sīnā seems to take the argument as follows. Write

Cx: x is the property of a thing, that true beliefs about it have contradictories.
Ex : x is essential. (3.31)
Ax : x is found in all cases.

Then Ibn Sīnā reads the syllogism as follows:

Every C is an A. Every E is an A. Therefore every C is an E. (3.32)

I am guessing at some details of (3.31). But the formal argument (3.32) can be read off from Ibn Sīnā's discussion.

This argument (3.32) is obviously invalid. Ibn Sīnā could demonstrate this in a couple of lines by giving interpretations of C, E and A that make the premises true and the conclusion false, following Aristotle's procedure in the *Prior Analytics*. But this is not what he does.

What follows calls for some knowledge of syllogisms and their metatheory. A good source for this material is the first three chapters of (Thom, 1981). Note that for Ibn Sīnā a syllogism strictly consists of just the two premises; hence he has only three figures, which are distinguished by where the middle term lies.

Ibn Sīnā assumes that (3.32) is intended as a syllogism. He reasons, assuming for contradiction that the syllogism is valid:

(a) (130.3) The conclusion is universally quantified.
(b) By (a), the syllogism is not valid as a third figure syllogism.
(c) (130.6) The conclusion is affirmative.
(d) By (c), the syllogism is not valid as a second figure syllogism.
(e) (130.4) By (b) and (d) the syllogism must be in first figure.
(f) By (e), the subjects of the premises are respectively the middle term and the minor term.
(g) (130.5) By (f), E must be either the middle term or the minor term.
(h) (130.1f) If E is the middle term then it isn't in the conclusion.
(i) (130.7) E is in the conclusion.
(j) By (h) and (i), E is not the middle term.
(k) (130.7f) By (e), if E is the minor term then it is the subject of the conclusion.
(ℓ) (130.8) E is the predicate in the conclusion.
(m) By (k) and (ℓ), E is not the minor term.
(n) By (g), (j) and (m), contradiction.

Ibn Sīnā leaves to the reader the steps for which I give no line reference. In fact we could have stretched out the argument still further by including the metatheorems that Ibn Sīnā uses but doesn't state. Recall §2.2!

But Ibn Sīnā has only just started his demolition. He moves on to consider a possible repair of the syllogism:

(o) (130.10) Suppose we replace the premise "Every C is an A" by "Every A is a C."
(p) (130.12) Then we get a valid syllogism with conclusion "Every E is a C."
(q) (130.11) But our new premise is false, and our new conclusion is not what was claimed.
(r) (130.13) In any case the commentator has forgotten to include the universal quantifier on this premise.

Next he reverts to the original syllogism.

(s) The middle term is predicate in both premises.
(t) (130.16) By (s), the syllogism is in second figure.
(u) (130.15) Both premises are affirmative.
(v) No valid syllogism in second figure has both premises affirmative.
(w) By (t), (u) and (v), the syllogism is invalid.

Ibn Sīnā has already shown this result, but this time he is using a different metatheorem (v).

Finally he tries another repair:

(x) (130.16) Suppose that we replace the premise "Every E is an A" by "Every A is an E."
(y) (130.16) Then we are using an obviously false premise.

On the basis of the facts that Ibn Sīnā gives us, this final repair is pretty clearly what the unnamed commentator intended in the first place. So a final twist of the knife is that Ibn Sīnā doesn't even bother to acknowledge that the resulting syllogism is a straightforward instance of the valid first figure mood *Barbara*.

The premise "Every A is an E" says "A thing that is found in all cases is essential." This is a tolerable statement of the *Eisagōgē* view of essence and accidents; see §3.2 and the note on line 130.16 below. Ibn Sīnā rejects this reading because he rejects the *Eisagōgē* view. But this could be a difference of terminology between Ibn Sīnā and the unnamed commentator, not a mistake by the commentator. Ibn Sīnā also refuses to consider the possibility that the commentator's final clause could be read as an equivalence between C and E; see the note on 130.16. So Ibn Sīnā's whole attack on the commentator is almost certainly based on a wilful misinterpretation of what the commentator said. This was hardly a new technique of debate, but the crudity of Ibn Sīnā's use of it can only be seen as undisguised personal malice.

5.2 Why?

Ibn Sīnā's procedure in demolishing the unnamed commentator does have another striking feature. He is conducting a philosophical debate by using a number of *metatheorems* of syllogistic logic. More precisely he is using metatheorems of the form "Any valid syllogism must satisfy the following condition ...." Let us call these NCV (Necessary Conditions for Validity) metatheorems.

The chief accepted use of logic in Ibn Sīnā's time was to check arguments by reducing them to syllogistic form. For this one would need to know the rules of syllogisms. But here Ibn Sīnā is using not those rules but higher-level rules about them. Perhaps he was the only logician of his age who was capable of deploying these metatheorems to make a philosophical point; if he believed he was, he certainly wouldn't have wanted to keep the fact to himself. But we have no reason to deny him credit for his observation that *NCV metatheorems are useful.*

The metatheorems that he invokes here are not very novel—I think they can all be found in Aristotle. Among the other known results in Ibn Sīnā's time, the only other NCV metatheorem that I can offer for comparison is Theophrastus' *peiorem* ("worse") rule, which said that on any of several measures of strength, the conclusion of a valid syllogism can never be stronger than the weaker of the two premises (Thom, 2003, pp. 23, 76). But this rule was imprecise and it didn't generalise beyond categorical syllogisms—as noted in Thom's book, there is a counterexample in Ibn Sīnā's modal syllogisms. (Ibn Sīnā gives his own version of the rule at *Al-Qiyās* 108.9, where *'aḥsan* "better" is clearly a corruption of *'aḵass* "worse"!) By contrast the NCV metatheorems that Ibn Sīnā uses in his commentary are ones that he had checked for himself in a range of extensions of categorical syllogisms.

Within a couple of centuries of Ibn Sīnā, versions of the laws of distribution started to appear in the West. These laws are NCV metatheorems. The best versions were more precise than the *peiorem* rule, but still there were problems about generalising them beyond categorical syllogisms. These problems were resolved only in the late twentieth century. Meanwhile in 1906, Frege, in one of his most perceptive papers (Frege, 1906), had argued that the logic of his time had no sound basis for proving general NCV metatheorems, and that such a basis would need to be found if Hilbert's arguments for the independence of the parallel postulate were to be put in an acceptable form.

5.3 Who Was the "Well-Known Commentator"?

Ibn Sīnā often discusses the views of other commentators, but he rarely names them except by epithets like "well-known." In the present case Fritz Zimmermann (personal communication) has suggested that it might be Abū al-Faraj Ibn al-Ṭayyib, a contemporary of Ibn Sīnā who did write a commentary on the *Perì Hermēneías*.

If Zimmermann's suggestion is right, the passage we are discussing falls neatly into line with an episode that Gutas discusses (Gutas, 1988, pp. 64–72). The episode took place in 1030, some half dozen years after Ibn Sīnā wrote his own commentary on *Perì Hermēneías*. According to an anonymous disciple of Ibn Sīnā, someone suggested to Ibn Sīnā that he might not be up to date with contemporary philosophy. Ibn Sīnā took umbrage, and sent one of his friends off to Baghdad with instructions to buy the latest books in the field. There Ibn al-Ṭayyib had some commentaries for sale; but when Ibn al-Ṭayyib heard that it was Ibn Sīnā who wanted the books, he "asked an exorbitant price," which was duly paid. Later Ibn Sīnā let it be known that in spite of paying the price, he had already formed a low opinion of Ibn al-Ṭayyib. In his account of the episode, the anonymous disciple goes on to include Ibn al-Ṭayyib among people who

> never acquired the habit of dealing with [the forms of syllogisms] and they never suffered the pains of analyzing the details of problems so that they may gain a syllogistic habit; their sole reliance, instead, is upon ideas not subject to rules. (Gutas, 1988, p. 69) (3.33)

If Ibn al-Ṭayyib was indeed Ibn Sīnā's "well-known commentator," then we can easily understand why he didn't want Ibn Sīnā to buy copies of his later writings.

The anonymous disciple lists those of Ibn al-Ṭayyib's commentaries that "became available to us," and he includes the *Perì Hermēneías* commentary. If this wording means that Ibn Sīnā didn't have Ibn al-Ṭayyib's *Perì Hermēneías* commentary before 1030, then Ibn al-Ṭayyib can't be the well-known commentator. But the wording might only mean that the list contains those commentaries that were available after the 1030 purchase, including some that Ibn Sīnā already had.

If we had the commentary then we could see whether it contains the argument that Ibn Sīnā complains of. Unfortunately it survives only in three copies of an epitome, all in India; (Lameer, 1996, p. 96) reports their coordinates. This epitome is our best hope for settling whether Ibn Sīnā's target was Ibn al-Ṭayyib. My attempts to see one of the copies (Calcutta, Būhār Library, Arab. Logic 283, fols. 44–79[32]) haven't so far borne fruit.

But did Ibn al-Ṭayyib in fact accept Porphyry's account of essence and accidents? Apparently yes. We have his commentary on Porphyry's *Eisagōgē*, and in it we find for example a description of accidents as things which

> when they are removed (*irtafaᶜat*, from *rafᶜ*) don't affect the essence of the thing (Ibn al-Ṭayyib, 1975, p. 139 l. 11). (3.34)

But probably there were other philosophers who followed Porphyry in this. The anonymous disciple quotes Ibn Sīnā as naming two other contemporary philosophers who "adhere more closely than others to the [traditional] transmission of certain books" (Gutas, 1988, p. 68).

6 Translation

The text is taken from Section 2.5 of (Ibn Sīnā, 1970), a volume of the Cairo edition of Ibn Sīnā's *Šifā'*. Reference 124.5 means line 5 on page 124. The division into paragraphs [2.5.1] etc. is my own. For transliteration of Arabic words I follow Wehr's Dictionary.

An explanation of whether the opposition between affirmative and negative is greater, or the opposition between two affirmatives whose predicates are contraries. 124.3

[2.5.1] It has been customary to conclude this part of Logic with something that logicians as such don't need. In fact it is more closely related to investigations in the form of debate. Namely: when a predicate is predicated of a subject— 124.5
and this predicate has a contrary—which is in greater conflict with [the predication], the affirmation of the contrary, or its negation which is its contradictory opposite?

For example when someone says

Zayd is just. (3.35)

and we say

Zayd is unjust. (3.36)

is it (3.36) that is in greater conflict with (3.35), or is it the sentence

[Zayd] is-not just? (3.37)

And if we say

Everybody is just. (3.38)

is it the contrary of this when we say

Everybody is unjust. (3.39)

or is [the contrary] what was mentioned earlier, namely 124.10

Not a single person is just? (3.40)

This is stuff for sects to quarrel about, but the truth of it is that his being unjust is in the nature of things more strongly in conflict with his being just than is his not being just.

124.12 [2.5.2] As concerns [the conditions for] assenting [to a proposition], and the force [of the proposition], regardless of whether [the proposition is taken to be] a belief or a verbal expression: the negative form [of the proposition] conflicts most strongly with the affirmative form and is furthest from matching it in truth and falsehood. The present investigation is about the force, and the force can be [taken] either as a
124.15 phrase or as a belief (because the phrase follows the belief). So let us carry out the investigation of these conflicting [propositions] in terms of beliefs.

125.1 [2.5.3] Consider a belief about something good, namely that it is good, and a belief that it is not good, and a belief that it is bad. You should know that if the belief is associated with two contrary [predicates], as when we believe that Moses is good and that Pharaoh is bad, or with two mutually contradictory [predicates], as when we believe that Moses is good and Pharaoh is not good, this doesn't imply that the two beliefs conflict with each other. For the two beliefs to negate each other, they would have to be about a single subject.

125.5 [2.5.4] So suppose we consider the truth about one subject, namely that he is good. If it is believed that he is bad, and it is also believed that he is not good, which of the two beliefs is in itself more strongly in conflict [with his being good]? The only thing that makes it impossible to believe that [the subject] is both good and bad is that a bad thing is not good. If in place of "bad" we have "thing that is not good and not bad," then it would still be impossible to believe that [the subject] is good and not good. In fact there are many things that are neither good nor bad. It's clear that the conflict in the case of the first belief [(that the subject is bad)] is not because the two believed things are contrary to each other, but because the contents of the
125.10 two beliefs deny each other. Denial is in the first instance between an affirmation and [the corresponding] negation.

125.11 [2.5.5] They say: Another piece of evidence for this is that when a thing is good and just, some affirmatives are true of it, for example that it is praiseworthy and preferable, and so are some negatives, for example that it is not blameworthy and not loathsome; and some affirmatives are false of it, for example that it is loathsome and blameworthy, and some negatives are false of it, for example that it is not praiseworthy and not preferable. Now being a genuine contrary [of X] is not something that one can impose on everything that disagrees [with X] this way or
125.15 that. In fact a single thing has just one genuine contrary. It follows that the contrary must be one of these [disagreeing things] that includes them all. So it includes all the affirmatives and the negatives which say falsely of the good thing that it is not good. Any [idea]—whether it affirms or negates—which is legitimately taken as [NOT

GOOD] is inconsistent [with [GOOD]], and [NOT GOOD] itself is inconsistent
with [GOOD] even if it is not considered to be one of those [disagreeing things]. 126.1

[2.5.6] Suppose a thing X can be distinguished from a thing Z without needing 126.1
another thing Y, whereas Y can't be distinguished [from Z] without [using the distinctness of] X [from Z]. Then X has a distinctness [from Z] which is prior [to the distinctness of Y from Z]. A thing that has a distinctness [from Z] that is prior [to all other distinctnesses from Z] is most strongly in conflict [with Z]. So the negation [of Z] is most strongly opposed [to Z], and what is most strongly opposed [to a thing] is the contrary [of the thing], so the negative [of Z] is the contrary [of Z].

[2.5.7] It's plausible that these two paragraphs in the First Teaching were not 126.4
intended to be an argument at all, and that the aim in the first of the two was just to 126.5
indicate that a contrariety in things doesn't itself entail a contrariety in beliefs, but rather a contradiction in things is a necessary condition for having a contrariety in beliefs.

[2.5.8] And the aim of the second paragraph is to indicate also that when beliefs 126.7
are mutually incompatible, even though the beliefs can't be true together, it doesn't show that they are contraries. So, to spell this out, there are infinitely many things which are legitimately denied of a person who is good and just. For example he is not a bird or a stone or the sky; to assert any of these [of him] is false. Also there are infinitely many things which it would be legitimate to assert of him, for example
that he is white and he is sitting and he is acting, so it is false to deny that these 126.10
things could be true of him. It's impossible for infinitely many things to be true of him, but infinitely many things are false of him. It's not appropriate to consider in each case whether or not the belief is contrary to the belief that he is good—this applies to infinitely many beliefs.

[2.5.9] But this enquiry is just about things which already had some befuddle- 126.14
ment in them, and this befuddlement lies just in what a thing can become. Thus,
granting that a good person is not a bird, and is also not bad, so that both [BIRD] 126.15
and [BAD] are false of him, still he could become one of these things, but he couldn't become the other. Of the two things that are opposed [to [GOOD]], the one that he
can become is [BAD], and the one that he can't become is [BIRD]. The befuddle- 127.1
ment is just about things opposed to [GOOD], like [BAD] and [UNJUST], namely whether the belief that he is just is contrary to the belief that he is bad and unjust. This fits what is said in the First Teaching, which aims to make a facilitation and to indicate that beliefs which deny [other beliefs] are not always opposed [to them] in the sense of contrariety. Otherwise we would be dealing with the befuddlement that
the belief that Zayd is just will be contrary to the belief that he is a bird—and in 127.5
fact contrary to infinitely many other beliefs.

[2.5.10] It's plausible that the aim of the First Teacher was what we indicated. 127.6
What he put at the beginning of this topic was put there not as an argument but as a facilitation. After finishing this statement of his intentions, he starts by claiming

that he has established [firstly] that the contrariety of two things doesn't itself make the two [corresponding beliefs] contrary to each other, and [secondly] that the fact that two beliefs negate each other doesn't force the two [corresponding] things to negate each other. So he has to undertake an investigation of the former point which is more specific than the investigations of these two points.

127.10

[2.5.11] So we say: In fact when we say "it is good" of whatever is good, we
127.10 speak truly, and when we say "it is not bad," we speak truly. But the truth that we express about [whatever is good] by saying "It is good" is a self-contained truth in the essence (of [GOOD]), whereas the truth that we express about it by saying "It is not bad" is not in the essence (of [GOOD]). [GOOD] is in the essence of [GOOD]. But as for [NOT BAD], that is an accident of [GOOD] [which becomes known] when [GOOD] is opposed to something that is not its essence and is distinct from its essence, namely [BAD], so that [BAD] is denied of [whatever is good]. So the assertion that [whatever is good] is good is completed for [GOOD] through the essence of [GOOD], while its denial is completed for it only through something else.
127.15 You already know that the negative inherents in cases like this are not internal to the essence. And corresponding to these two truths are two falsehoods. It is false that
128.1 [whatever is good] is not good, and this is a falsehood that is opposed to [GOOD] in its essence. And it is false that [whatever is good] is bad, and this is a falsehood that is opposed to something that is an accident of it. And when the belief that [something good] is good is true in an essential feature, [and] while the belief that [something good] is not bad is true in an accidental feature, a belief that [something good] is not good is false in an essential feature. Falsehood about something in the essence is more opposed to truth about something in the essence than falsehood about some accident is. This is how one should say it.

128.4 [2.5.12] About the belief that one of two [falsehoods can be] more strongly false
128.5 than the other: this [belief] is incorrect. There is no truth that is more strongly true than some [other] truth, nor is there any falsehood that is more strongly false than some [other] falsehood. However, some truths are more permanent and some are less permanent. Also some truths are about an essential matter while others are about a nonessential one. A falsehood about an essential matter is more strongly in conflict [with the truth].

[2.5.13] This could give rise to another argument that should be understood as follows. Suppose there is a just person that I know, and after checking it explicitly I believe that he is good. Then there is no need to believe at the same time that he
128.10 is not bad, since this is not in his essence; rather it is an accident of him. But for a thing in the essence to come into the mental processor, there is no need at all for it to refer to a relation to something external [to the essence]. Rather, the essential truth simply coalesces as a result of the subject and the predicate coming into the mental processor, whether or not anything else is brought into the mental processor as well. And if I were to introduce two [other] beliefs that oppose this belief, namely that he is bad and that he is not good, I would find that the belief that he is bad is

not complete for me unless it contains [the belief] that he is not good. A falsehood
which is opposed to an accidental truth is completed only by an essential falsehood 128.15
coming into the mental processor. So if it doesn't come into my mental processor,
about the just person whom I know to be good, that he is not good, then it is not
possible for me to believe that he is bad. And this is because I know and believe that
this just person is good, and that this is true. If I think of him as bad, so as to test 129.1
this opposition, there comes into my mental processor the compulsion to negate this
truth about him—[though conversely] when the negation of this truth comes into my
mental processor, it doesn't have to have already come into my mental processor that
he is bad. [Aristotle's] indication has to be understood in this forced way. Although
it is not quite right, it is close to what we said in the first place; it amounts to the
same thing.

[2.5.14] Here is another argument. All propositions have opposites in the form of 129.4
their contradictory negations. But not every proposition has an affirmative opposite 129.5
that expresses its contrary. In fact when we say "Such-and-such is square," facing
it we [may] find that it's not square, though we don't find that it's some other kind
of shape which is contrary to square. In this case the thing that conflicts [has to be]
the negative, and not an affirmative predicate which is contrary [to "square"]. When
a proposition has an affirmative contrary, it still has a negation that conflicts with
it. Thus every affirmative proposition has a negation that conflicts with it, while not
every affirmative proposition has a conflicting [proposition] that is affirmative. Just
by being affirmative, an affirmative proposition has a conflicting proposition that
is negative; the other conflicting [proposition] arises incidentally and not from the 129.10
affirmativeness [of the first proposition].

[2.5.15] But someone might well say: We are not discussing whether for every 129.11
affirmative there is an affirmative that conflicts with it in the way that [REST]
conflicts with [MOVEMENT] taken absolutely. Rather it can be assumed that the
negation gives the most general and greatest conflict. But take the case where
an affirmative which is contrary to a given predicate is narrowed down so as
to stay affirmative. Is there a contrary which results from narrowing down [the
predicate] over against it, and which is more strongly contrary to it? Thus when
[MOVEMENT] is specified to be [DOWNWARDS MOVEMENT], the contrary to
it [(namely [UPWARDS MOVEMENT])] is a movement which conflicts with it
more strongly than [REST] does.

[2.5.16] But consider the case of a certain well-known commentator and all those 129.15
people who come close to his level of deficiency. He supports this argument by the
following unsound syllogism: 130.1

> A contradictory belief can be found for a true belief about any thing.
> This is an essential thing, since an essential thing is a thing that is (3.41)
> found in all cases.

See the mistake he made in the syllogism. He produced the phrase since an essential thing is found in all cases (3.42)

as a premise for a syllogism that concludes:

This is an essential thing. (3.43)

This goal of his is universally quantified with a singular subject, [and the conclusion is affirmative], so that it can only follow in the first figure. So he can only put "the
130.5 essential" in (3.42) as either the middle term or the minor term, because it is the subject in this premise [[...]]. If he put it as the middle term, then it shouldn't occur in the conclusion, but he did make it occur there. If he put it as the minor term then the conclusion would be

The essential is such-and-such. (3.44)

not that such-and-such is essential. So in fact "the essential" has to be the minor term in the syllogism, not the major.

—

When we take into account the other premise, we find that what this premise shares [with the first-mentioned premise] is the property "found in all cases." Suppose we
130.10 make that the subject there, so that the inference is as follows:

The essential is what is found in all cases.
What is found in all cases is that a true belief has a contradictory opposite belief. (3.45)

Then—disregarding the fact that the major premise is false when "found in all cases" is taken in it in the same sense as in the minor premise—it is entailed that the essential is such-and-such, not that such-and-such is essential. Besides being false, the [major] premise was misstated; for the syllogism to have a valid conclusion, the premise is taken as universally quantified, not as unquantified.

—

And if he made "found in all cases" the predicate rather than the subject, as indeed he should, then [the syllogism would be]

That a true belief has a contradictory opposite belief is a thing found in all cases. (3.46)
The essential is what is found in all cases.

130.15 so he would have made a deduction from two affirmatives in the second figure!

—

If he had converted [the first premise], he would have made it

Anything found in all cases is essential. (3.47)

But this is obviously false.

[2.5.17] After this argument there comes a strong argument. It says that when something is not good and we believe that 130.17

It is not good. (3.48)

the only other beliefs (of the kind relevant to this discussion) that we can introduce in contrast to (3.48) are the beliefs that

It is bad, (3.49)

that

It is not bad (3.50)

and that

It is good. (3.51)

But there are many cases where a belief that it is bad can be true together with the 131.1
belief (3.48), so (3.49) will not be an absolute opposite of the belief (3.48). Also our belief (3.50) that it is not bad can be true [together with (3.48)]. In fact we [can] find an individual thing, for example an infant, that is neither good nor bad. Likewise [we can find something that is] intermediate [between good and bad]. The remaining case is that the belief which conflicts with [its being not good] is (3.51) that it is good. Therefore the belief that it is good conflicts with the belief that it
is not good, and is the genuine contrary of it. A contrary is the contrary of its own 131.5
contrary. So what conflicts with the belief that it is good is that it is not good. In fact it's impossible for X to be the genuine absolute contrary of Y when Y is contrary to something other than X.

[2.5.18] If we put the [same] question about a universally quantified sentence, 131.7
we will be asking whether what conflicts with the sentence

Every human is not good. (3.52)

is the sentence

Every human is bad. (3.53)

or the sentence

Every human is not bad. (3.54)

or the sentence

Every human is good. (3.55)

The contrary of (3.52), in the sense [of "contrary"] that we have explained, is (3.55). So the contrary of the sentence (3.52) is the sentence (3.55), whereas the contrary
131.10 of the sentence (3.55) is the sentence

No person is good. (3.56)

which denies of each individual that he is good. This same account applies to both singular sentences and universally quantified ones. But as for unquantified sentences, how could they be contrary to each other, given that they are true together? Likewise with two existentially quantified sentences. Contraries, even though they can be denied together and false together, can't be true together.

7 Notes

[2.5.1] This comments on *Perì Hermēneías* 23a27–32.

124.8 "is-not just," Arabic uses a single word for "is not." This precludes the reading "is not-just," which Arabic would express differently.

124.10 "mentioned earlier": Probably this refers to those places where Ibn Sīnā has said that a definition is a conjunction of genus and differentiae, for example (3.14). It seems absurd to invoke this fact, when the differentiae can contain any number of negations. The explanation is almost certainly the principle of Top-Level Processing, §3.1 above.

[2.5.2] This comments on *Perì Hermēneías* 23a32–39.

124.12 "force" (*ḥukm*): This very common word has a rather diffuse meaning. Possible translations range between "judgment," "content," "force," "logical properties" and "the rules for using it."

124.12 "negative form" (*sālib*): Ibn Sīnā sometimes writes as if the negation of an idea *X* is the same idea *X* but taken "negatively." Add to this that he sometimes speaks of noun-type ideas as "affirming" or "negating," as at 125.16. The outcome is that his words for "negation" and "negative" don't always translate smoothly into modern logical idiom.

[2.5.3] This comments on *Perì Hermēneías* 23a39–23b7.

[2.5.4] Here Ibn Sīnā gives his own broad take on the question.

125.9 "the two believed things": Apparently not the two beliefs mentioned in 125.6, but the first of those beliefs together with the supposition (from 125.5) that the subject is good.

[2.5.5] This doesn't fit anything in *Perì Hermēneías* closely, though there might be a reminiscence of 23b7–13. But the "They say" at the beginning is probably meant to indicate that Ibn Sīnā is commenting on some other commentator. A likely source is (Al-Fārābī, 1986, 201.21–202.6). Thus:

If the contrary is to be an affirmation, it must be the one that embraces (*taštamil* ^{c}*alā*) all the false affirmations; and if a negation, the one that embraces all other false negations (Zimmermann, 1981, p. 195). (3.57)

125.15 "includes" (*ya*^{c}*umm*): One of the primitive notions of Ibn Sīnā's logic, so that he never defines it. What he says about it is consistent with the following definition: idea A includes idea B if and only if for every idea C, if B is true of C then so is A. Al-Fārābī has *'a*^{c}*amm* in a corresponding place (Al-Fārābī, 1986, 202.3,5).

125.17 "[NOT GOOD]" (*'ammā laisa bi-ḵair nafsuh*): The expression *'ammā* "as for" is normally followed by a noun phrase in the nominative. So it is here, when we note that *nafsuh* and similar expressions can serve as indicators that Ibn Sīnā is referring to an idea. Then *bi-nafsih* later in the line picks up the other relevant idea, namely [GOOD].

126.1 "it is not considered": Ibn Sīnā has said first that [NOT GOOD] is itself an idea that is inconsistent with [GOOD], and second that it includes every such idea. That seems to complete the argument, so perhaps we should delete the pointless clause "even if ...." A suspect feature of the clause is that it finishes with *tilka*; this is uncommon but not unique, see for example *Al-Burhān* 20.19 and 22.4.

[2.5.6] This again is pure Ibn Sīnā. It seems to complete the argument of paragraph [2.5.4].

126.1 "a thing X" : Here we see the three variables of the relation "X is more strongly in conflict with Z than Y is." But none of them appear as variables in Ibn Sīnā's text. Thus Z is mentioned only by implication; X and Y are introduced briefly as "the thing" and "the other," and there is just one anaphoric pronoun referring back to "the thing." After a career teaching logic, I lay a heavy bet that only a fraction of Ibn Sīnā's readers reconstructed all the cross-references correctly.

Why did he do this? Even if he didn't want to use variables here, he could have clarified matters hugely by introducing a "first thing," a "second thing" and a "third thing" and cross-referencing properly. One suspects he wanted to make a point. Maybe it was that logic takes no prisoners. More likely it was that in interpreting anybody's statements you need to supply what the speaker assumes about other entities besides those that are mentioned explicitly; cf. (3.23) in §3.3 above.

[2.5.7] This comments on *Perì Hermēneías* 23b13.

126.4 "the first": The "first paragraph" is 23a32–23b7. But Ibn Sīnā has already extracted a sound point from this paragraph in his own paragraph [2.5.2].

[2.5.8] This comments on *Perì Hermēneías* 23b7–13.

126.7 As proposed by Moktefi, read *tanāfin* for *yunāfī*, and *wa-in* for *wa-an* at the beginning of the next line.

126.11 "impossible for infinitely many things":

> Also it is said [in the First Teaching] that only finitely many predicates are internal to the whatness of a thing, because these [predicates] are internal to the defining of things, ... So if definitions were to reach the point that there were infinitely many things in them, then it wouldn't be possible for us to define anything. But definitions do exist, since things get conceptualised. So they must have finitely many principles. (*Al-Burhān* 168.6–9) (3.58)

This paraphrases Aristotle *Posterior Analytics* A22 82b37–39.

[2.5.9] This comments on *Perì Hermēneías* 23b13–15, and perhaps also 23b21–23.

126.14 "befuddlement": Aristotle himself mentioned befuddlement (or more strictly deception, *apátē*) and becoming (*genéseis*). It seems that nobody has a convincing explanation of what Aristotle had in mind here. (Ackrill, 1963) doesn't attempt to explain it. Apparently the idea of mentioning something that Zayd couldn't become (namely a bird) is Ibn Sīnā's own. Readers will certainly notice that [BIRD] conflicts with Zayd's essence, whereas [BAD] doesn't. From the arguments to come in [2.5.11] and [2.5.12], this should suggest that [BIRD] is a better candidate for the contrary of [GOOD] than [BAD] is. But this conclusion doesn't fit with anything else in PH14, so Ibn Sīnā leaves it to the reader to spell out.

127.3 "facilitation": The word is *tawṭi'a*, as in line 127.7 below. But here Ibn Sīnā adds "indication" (*tanbīh*). This word belongs to one of the standard excuses for Aristotle, namely that sometimes he gave only hints, so as to protect his teachings from the intellectual riffraff. Cf. (Gutas, 1988, pp. 307–311), noting that *tanbīh* is the word translated as "reminder" on his p. 310.

[2.5.10] This comments on *Perì Hermēneías* 23b2–7.

127.9 "[secondly] ...": The point seems to be a gratuitous contradiction of 126.6f above. Probably the text is faulty.

[2.5.11] This comments on *Perì Hermēneías* 23b15–21.

127.10 "So we say": This is Ibn Sīnā's standard formula to indicate that he has been laying out a problem and he is about to give his own solution. This paragraph doesn't seem to reflect Ibn Sīnā's own views any more than, say, [2.5.4]. Maybe he is signalling that he thinks the argument in this paragraph is the heart of the matter.

127.10 "it is good": Aristotle talks of beliefs about "the good," which he makes neutral (*estìn agathòn*). Arabic has no neuter case, so Ibn Sīnā has a choice between reading the corresponding word (*al-ḵair*) as "the good person" or "the good thing," or simply as "the good." (cf. (Ackrill, 1963, p. 154) for a related comment on 23a32.) He will try "the good person" in paragraph [2.5.13]; in paragraph [2.5.11] he leaves it ambiguous.

Now Aristotle's argument is about the essence of the good (thing/person/...), which he says contains the good. In Ibn Sīnā's framework objects don't have essences; ideas do. So we have to suppose that we are thinking about the good (thing/person/...) through some idea that identifies it, and [GOOD] is in the essence of this idea. But how is this to work?

Suppose for example that we are thinking about Nelson Mandela; then Aristotle's argument assumes that [GOOD] is constitutive for [NELSON MANDELA]. But how could it be? Isn't it clear that we could imagine even Nelson Mandela turning evil, polluting the environment and depriving old ladies of their pension funds? Ibn Sīnā accepts the *rafᶜ* test in this direction.

The same problem applies if we take "it" in Aristotle's "the good" to stand for any one particular good entity—unless perhaps we believe in Platonic ideas, which Ibn Sīnā didn't. I think this leaves the argument of [2.5.13] as unrescuable, and that seems to be Ibn Sīnā's conclusion too. But we can more or less rescue paragraph [2.5.11] by reading "the good" generically, like "the human" or "the horse." Then to say that the good is good is in effect to assert the meaning [EVERYTHING GOOD IS GOOD]. I have translated it on that assumption.

128.2 With several manuscripts, read *ḥīna kāna iᶜtiqād* in place of *muqābilun lil-iᶜtiqād*.

[2.5.12] This comments on *Perì Hermēneías* 23b21f.

128.5 "more strongly true": Ibn Sīnā is quietly reprimanding Aristotle for a careless statement. Aristotle had said:

> The more true belief about anything is the one about what it is in itself. (23b17, trans. (Ackrill, 1963, p. 66)) (3.59)

For example the belief that gold is a metal is more true than the belief that gold has high market value. This is incorrect, says Ibn Sīnā; both beliefs are simply true, but Aristotle could have made his point by saying that the belief about the essence is more permanent than that about an accidental property. "Permanent" should probably be understood as in §3.3: a belief is more permanent if it is more resistant to changing truth value when one changes it by changing the "reading" of the time or circumstance that it applies to.

[2.5.13] This is a second attempt at *Perì Hermēneías* 23b15–21, which Ibn Sīnā has already tackled in [2.5.11].

128.11f "mental processor" (*bāl*): Ibn Sīnā has a number of words for different aspects of mind. One of the most specific is *bāl*. It's the part or aspect of the mind where rational processing takes place. His normal idioms for it are *yakṯir bil-bāl* ("it comes into the mind") and *'ukṯir bil-bāl* ("it is brought to the notice of the mind"). This notion of *bāl* need not be an intrusion of psychology into logic. It's better seen as part of a high-level description of the algorithms needed for reasoning.

The extent to which these algorithms probe the structure of the relevant sentences gives an upper bound on the proving power of Aristotelian logic. For example, with a few minor reservations, the algorithms never penetrate an NP-VP sentence further than separating the NP from the VP. (This is one formulation of Top-level Processing, §3.1). Line 128.12 is a typical illustration of this, with the NP and VP (or strictly their meanings) called respectively "subject" and "predicate."

128.11 "essential truth": I.e. truth about the essence. As in the notes on [2.5.11], Ibn Sīnā must mean here the essence of the idea of this just individual, for example [NELSON MANDELA]. Here he touches on the hoary problem of the meanings of proper names. At *Al-Madkal* 31.9 he suggests that the essence of the idea of an individual contains "whatever [the individual] is individuated by." But I don't know if he succeeds in taking this semantic question any further.

129.3 "forced way" (*takalluf*): One of Ibn Sīnā's commonest criticisms of other commentators is that their explanations are forced. In this case there is an implied criticism of Aristotle himself—that his argument can only be explained in a forced way. Presumably the forced point is the assumption that the essence of (the idea of) any individual person can contain [GOOD]; see the notes on [2.5.11].

[2.5.14] This comments on *Perì Hermēneías* 23b27–32.

129.10 "not": following the reading *lā* rather than *lahā*.

[2.5.15] This looks like a comment on something in the literature, but we don't know what.

[2.5.16] See §5.3 for the source.

129.15 "his level of deficiency" (^{c}*ahdah*): The noun ^{c}*ahd* has several meanings, none of them clearly appropriate here; so if the text is sound, it's likely that some idiom hasn't come down to us. But since what follows is specific to an individual commentator, probably Ibn Sīnā's phrase is meant to insult this commentator rather than to describe some other people. An ^{c}*uhda* is a claim you have against a person who has sold you substandard goods, and hence it comes to be used metaphorically for an attribute in which one is below standard. Commercial metaphors appear also at *Easterners* 24.18 (^{c}*uhda* again) and *Al-'Išārāt* 67.8 (*ḥawāla*).

129.15 "this argument": Paragraph [2.5.14] lines 129.8-9, "every affirmative proposition has a negation that conflicts with it." The commentator "supports" it in the sense of using it as a premise for a further argument.

130.4 "and the conclusion is affirmative": The clause is missing from the text, but the argument needs it. Lo and behold, there it is in the text at 130.6 (marked [[...]] in the translation) where it makes no sense at all. Presumably Ibn Sīnā himself, editing the passage, saw that the clause was needed and added it in the margin. The copyist wasn't concentrating and managed to incorporate it at the wrong place. There are many places in Ibn Sīnā where one suspects that something like this has happened, but this is a particularly clear example (thanks to the constraints of formal argument).

130.4 "singular": Literally "specialised." Ibn Sīnā's point may be that the subject is "the property of the thing, viz. that in all cases ...," which is singular, but the argument is made no less valid if we replace this by "every property of the thing such that in all cases ...." So without loss we can consider the conclusion as universally quantified.

130.10 "the major premise is false": The major premise is the second premise in (3.45). Ibn Sīnā reads it as implicitly universally quantified; so it says that the only thing true in all cases is that a belief has a contradictory opposite belief. But the implied subjects—true beliefs—have lots of other properties that hold in all cases, for example being true beliefs.

130.16 "Anything found in all cases": According to Ibn Sīnā's account at 130.1, the commentator had said something of the form "An *A* is a *B*." His argument clearly calls for this to imply "Every *B* is an *A*." But "An *A* is a *B*" can quite naturally be read as meaning "The *A*s are the same thing as the *B*s," which does have the required implication. Maybe Ibn Sīnā counts this reading as one of those things that are merely "suggested" by the commentator's sentence, as in §3.3 above. But professionalism should have warned him to start with the reading most likely to have been intended.

That much is from the form of the argument. When we turn to the content, the commentator seems to be claiming that if an idea I has an attachment J that holds "in all cases," then J is constitutive for I. If "all cases" includes imagined cases as well as real-world ones, then the claim is simply Porphyry's claim that non-constitutives can be separated off by *rafᶜ* in the imagination. As we saw in §3.2, Ibn Sīnā himself rejected this claim—so it's no surprise that he labels it as "obviously false."

[2.5.17] This comments on *Perì Hermēneías* 23b33–24a3.

130.18 "relevant": The restriction to these forms is from Aristotle. Ibn Sīnā makes no attempt to justify it. Maybe he has in mind some canonicality argument as in §4.3.

[2.5.18] This comments on *Perì Hermēneías* 24b1–9.

131.7 "universally quantified": These are sentences of either of the forms "Every A is a B," "No A is a B."

131.10 "is good" (second occurrence): This is reading *huwa ḵair* with two manuscripts. The majority reading *laisa bi-ḵair* "is not good" makes logical nonsense.

131.11 "account": Singular sentences are of either of the two forms "A is a B," "A is not a B," where A names an individual. The account applying to both these and universally quantified sentences is that we get the contrary by swapping "B" and "not a B"—where "No A is a B" is counted as "Every A is not a B."

131.12 "unquantified": For Ibn Sīnā these are sentences of the forms "The A is a B." "The A is not a B."

131.12 "Contraries": This last sentence picks up the final sentence of Aristotle's text. The common core of the two sentences is (in 'Isḥāq's translation) *fa-'ammā al-ḍāddān fa-laisa yumkin 'an yūjadā maᶜan*, and (in Ibn Sīnā) *wal-'aḍdād fa-laisa yajūz 'an taṣduq maᶜan*. Both mean "Contraries can't be true together," but Ibn Sīnā replaces most of the significant words by other words that have the same meaning in context. Thus he replaces the dual *ḍaddān* by a plural; *yumkin* and *yajūz* both mean "it's possible," and *yūjadā* and *taṣduq* both mean "are true." Differences like these could be evidence that Ibn Sīnā is using a different translation from 'Isḥāq's. But they are par for the course in Ibn Sīnā, and a more likely explanation is that he prefers to assert his independence by saying everything in his own words.

Bibliography

Ackrill, J. (1963). *Aristotle's Categories and De Interpretatione, Translated with Notes and Glossary*. Clarendon Press, Oxford.

Al-Fārābī (1986). *Commentary on Aristotle's De Interpretatione*. Dar el-Machreq, Beirut. W. Kutsch and S. Marrow, eds.

Ammonius (1897). *In Aristotelis De Interpretatione Commentarius*. Reimer, Berlin. Adolfus Busse, ed.

Barnes, J. (2003). *Porphyry: Introduction*. Clarendon Press, Oxford.

Boole, G. (1997). The nature of logic (1848). In Grattan-Guinness, I. and Bornet, G., editors, *George Boole: Selected Manuscripts on Logic and its Philosophy*, pages 1–12. Birkhäuser, Basel.

Charlton, W. (2000). *"Philoponus" on Aristotle's On the Soul 3.9–13 with Stephanus on Aristotle's On Interpretation*. Cornell University Press, Ithaca, NY.

Frege, G. (1903/1906). Grundlagen der Geometrie. *Jahresbericht der Deutschen Mathematikervereinigung*, 12/15:319–324, 368–375/ 293–309, 377–403, 423–430.

Gutas, D. (1988). *Avicenna and the Aristotelian Tradition: Introduction to Reading Avicenna's Philosophical Works*. Brill, Leiden.

Hodges, W. (2009). Traditional logic, modern logic, and natural language. *Journal of Philosophical Logic*. Volume in honour of Johan van Benthem, edited by Hans von Ditmarsch and Larry Moss.

Ibn al-Ṭayyib (1975). *Commentary on Porphyry's Eisagoge*. Dar el-Machreq, Beirut. Kwame Gyekye, ed.

Ibn Sīnā (1970). *Al-ᶜIbāra*. Dār al-Kātib al-ᶜArabī lil-Ṭabāᶜ wal-Našr, Cairo. Ibrahim Madkour et al., eds.

Jabre, F., editor (1999). *Al-Naṣṣ al-Kāmil li-Manṭiq 'Arisṭū*, volume 1. Dār al-Fikr al-Libnānī, Beirut.

Jackendoff, R. (1990). *Semantic Structures*. MIT Press, Cambridge, MA.

Lameer, J. (1996). Review of "Glosses and commentaries etc., ed. Charles Burnett". *Journal of the American Oriental Society*, 116:90–98.

Mancosu, P. (1996). *Philosophy of Mathematics and Mathematical Practice in the Seventeenth Century*. Oxford University Press, New York, NY.

Oddie, G. (2007). Truthlikeness. Stanford Internet Encyclopedia of Philosophy. plato.stanford.edu/entries/truthlikeness/.

Proclus (1970). *A Commentary on the First Book of Euclid's Elements*. Princeton University Press, Princeton, NJ. Glenn R. Morrow, ed.

Thom, P. (1981). *The Syllogism*. Philosophia Verlag, Munich.

Thom, P. (2003). *Medieval Modal Systems*. Ashgate, Gateshead.

Whitaker, C. (1996). *Aristotle's De Interpretatione: Contradiction and Dialectic*. Clarendon Press, Oxford.

Yaḥyā ibn ᶜAdī (1988). *Maqālāt Yaḥyā ibn ᶜAdī al-Falsafiyya*. Amman. Sahban Khalifat, ed.

Zimmermann, F. W. (1981). *Al-Farabi's Commentary and Short Treatise on Aristotle's De Interpretatione*. British Academy and Oxford University Press, Oxford.

Chapter 4
A Minimalist Foundation at Work

Giovanni Sambin

What is the nature of mathematics? The different answers given to this perennial question can be gathered from the different theories which have been put forward as foundations of mathematics. In the common contemporary vision, the role of foundations is to provide safe grounds a posteriori for the activity of mathematicians, which on the whole is taken as a matter of fact. This is usually achieved by reducing all concepts of mathematics to that of set, and by postulating sets to form a fixed universe, which exists by itself and is unaffected by any human activity. Thus it is a static vision.

A dynamic view is also possible, in which mathematics, as well as all of science, is the result of human interaction with the environment. In this view, mathematics is, as it has been traditionally, the science of abstract concepts of number, figure, measure, deduction, etc., and all these concepts are created by human beings through a process of abstraction which is a part of human evolution (Sambin, 2002).

To understand the true nature of such concepts, one should first of all avoid reducing them to a single one, such as that of set, since this would bring in undesired assumptions. More generally, one should aim at looking at reality, imposing as little as possible of one's preconceived notions and expectations, so that one can more easily accept different views as aspects of the dynamic process.

This general attitude corresponds fairly well to a precise foundational theory, which we have called the *minimalist foundation of mathematics* (Maietti and Sambin, 2005). Broadly speaking, it is obtained from a constructive type theory, such as Martin-Löf's, by avoiding the identification of logical propositions with sets and by distinguishing an intensional level (a sort of high-level programming language) to deal with computations from an extensional level to deal with abstract mathematics. The result is a foundation from which apparently all others can be obtained by the addition of some principles (essentially, a combination of: the axiom of choice, the law of excluded middle, the powerset axiom).

Rather than pondering over the conceptual advantages of a minimalist foundation (as is partly done in (Sambin, 2003; Maietti and Sambin, 2005)), after a short

G. Sambin (✉)
Professor of Mathematical Logic, University of Padua, Padua, Italy
e-mail: sambin@math.unipd.it

D. DeVidi et al. (eds.), *Logic, Mathematics, Philosophy: Vintage Enthusiasms*,
The Western Ontario Series in Philosophy of Science 75,
DOI 10.1007/978-94-007-0214-1_4,

introduction to the idea, my aim here is to illustrate how it works when it is put into action on a specific piece of mathematics. Thus I provide arguments showing that this way of conceiving foundations is not only possible and reasonable, but actually also relevant and useful in doing mathematics in a technical sense.

1 Foundations as a Choice of Abstraction Level

It is commonly believed that the choice of some principles stricter than some others is due to the desire for safer foundations for one and the same mathematics. This is a defensive and static view. More positively, the role of stricter principles is to allow looking at reality with a finer grid of distinctions.

In fact, to be able to pass from the chaotic nature of reality to mathematical thought, one has to abstract from details and contingencies, that is, to disregard what under a certain perspective appears to be irrelevant. Only by "forgetting" some information can one obtain some abstract concepts, treat them "mathematically," that is, with no mention of the reality they come from, and finally apply them back to reality successfully.

This is probably taken for granted by most. However, one tends to overlook the fact that there is no single and necessary method of doing it. Actually, different foundations can be seen, under this perspective, as different choices as to what should or should not be considered as relevant, that is, as different principles to be used to abstract from reality and idealize it.

In a certain sense, a foundational theory is a specification of the sense organs, or "lenses," through which one perceives the world. Different choices of "lenses" lead to different kinds of mathematics, with different degrees of abstraction and of effectiveness.

The most common foundation, namely the classical foundation, is based on classical logic and on axiomatic set theory, such as ZFC. Both existence of entities—which means only sets—and truth of propositions are identified with their potentiality, that is absence of contradictions. All that is consistently conceivable automatically exists and is true. So the task of mathematicians is reduced to discovering what is assumed to be there already. In particular, propositions are formally reduced to set-theoretic formulae, and their truth is fixed and independent of our knowledge, and hence must coincide with absence of falsity.

To be able to justify this incredibly static view, it comes as a consequence that one is led to postulate the existence of a world beyond sensory perceptions, in which all of mathematics has always been, and will forever be, as it is now. No wonder then that any residual effectiveness in mathematics is regarded as a lucky miracle, or even as unreasonable.

1.1 Different Constructive Foundations and Their Incompatibility

To free mathematics from the immobility of the classical conception, which was perhaps suitable in Plato's times but is now historically childish, one necessary step seems to be that of overcoming the rigidity of classical logic. In intuitionistic logic the notion of truth is linked with proof, and necessity, rather than with consistency, or possibility. This means that the Law of Excluded Middle

LEM: for every proposition A, $A \vee \neg A$ is true

is not universally valid, and this immediately opens up some space for movement. All foundations based on intuitionistic logic are called constructive, in the broad sense. Here only two of these will be considered, which we call the geometric and the computational vision of constructivism.

The geometric vision is expressed formally by topos theory (by which here and from now on we mean the theory of the internal language of a generic topos, as explained in (Bell, 1988; Maietti, 2005)). An explicit aim of toposes, as originally introduced by A. Grothendieck in the 50s, is to enrich geometry with a generalized notion of space. Their axiomatizable version, the notion of elementary toposes described by F.W. Lawvere and M. Tierney, can be seen as a possible universe of sets.

While truth of propositions is left open (since LEM is not assumed), in a topos the notion of proposition, and hence that of propositional function and of subset, is assumed to be fixed. This seems necessary to be able to justify the Power Set Axiom:

PSA: for every set X, the totality $\mathcal{P}X$ of all subsets of X also forms a set

which is assumed to hold in every topos. In fact, since $\mathcal{P}X$ is a set, one can freely apply quantification over subsets to produce propositions. So in particular one can define a "new" subset of X by making reference to the totality of subsets of X. This is an impredicative definition or, speaking frankly, a patent vicious circle. The only way to make it conceptually harmless seems that of postulating, in each topos, the notion of subset to be fixed and hence fully static. Indeed, only in this way one can say that the "new definition" is just a new name for a subset which was there anyway.

This is the reason why topos theory has no tools to discern, as a theory, sets which are given effectively. Treatment of effectivity is restricted to numerical functions and must remain at the level of semantics—as for instance in the effective topos (Hyland, 1982).

In the computational vision, as put forward in (Bishop, 1967, 1970); (Martin-Löf, 1970), each piece of mathematics is considered meaningful as long as it has an effective, or even numerical, content. It is then natural to assume the validity of the Axiom of Choice, in the form:

AC: every total relation contains a function, $\forall x \exists y R(x, y) \rightarrow \exists f \forall x R(x, fx)$

(where x is an element of the set A, y of the set B, and then f is a function from A to B).

A rigorous formulation of the computational view of constructivism relies on Martin-Löf's intuitionistic type theory, here shortened to ML type theory (see (Martin-Löf, 1984), and its refinement (Nordström and Petersson, 1990)). It can be described as a systematic development of the Brouwer-Heyting-Kolmogorov interpretation of intuitionistic logic, through the propositions-as-sets paradigm, that is, the identification of (inference, or deduction rules for) logic with (construction, or induction rules for) set theory.

In this way not only is AC justified conceptually, it actually becomes internally provable in ML type theory. In fact, by the propositions-as-sets paradigm, the characterization of the quantifier $\exists$ coincides with that of the set theoretic constructor Σ of disjoint sums. Thus a proof of $\exists y R(x, y)$ automatically contains both a witness a and a proof b that $R(x, a)$ holds, and so when $\forall x \exists y R(x, y)$ holds, such witness and proof are automatically produced by a function. In other terms, what is brought to a proof of AC is the very meaning given to quantifiers, and hence ultimately the identification of propositions with sets.

Effectivity of ML type theory lies in the fact that every entity can be understood as computable. This is expressed formally by the fact that ML type theory[1] is compatible with the internal Church thesis:

CT: every function on natural numbers is recursive, $\forall f : \mathbb{N} \to \mathbb{N}\ \exists n\ f = \varphi_n$,

where φ_n is the recursive function with index n.

As with human affairs, contact between rigid principles produces reciprocal rejection. Indeed, one can see that the principles of all propositions forming a set (hence PSA) and of each proposition being the same as a set (propositions-as-sets paradigm, hence AC) are constructively incompatible, in the sense that when put together they return us to classical logic. Formally, one can prove that

$$T + \mathsf{PSA} + \mathsf{AC} \quad \vdash \quad \mathsf{LEM}$$

where T can be either topos theory or ML type theory.[2] The conclusion seems to be that the geometrical and the computational view of constructivism are incompatible.

1.2 The Minimalist Foundation of Constructive Mathematics

The most practical and simple expression of the computational view is to say that it should be possible to implement any piece of (meaningful) mathematics in a computer. This cannot be understood in its weak form, with the computer acting only as a spell checker (are expressions well formed according to the given grammar?) and as proof checker (do proofs apply only inferences rules of the given deductive system?); in fact, any piece of mathematics can be implemented in this sense, as long as it is formalizable in an axiomatizable theory. Even a highly non-effective

[1] More precisely—as explained to me by Milly Maietti—ML type theory deprived of the reduction rule inferring $b = c$ from $\lambda x.b = \lambda x.c$, and of universes.

[2] This is essentially Diaconescu theorem saying that that every topos satisfying AC is boolean (see (Bell, 1988) for a simple proof). Its version for type theory says more precisely that PSA, or more generally extensional quotients, are incompatible with AC, since the choice function does not respect extensionality (Maietti and Valentini, 1999; Maietti, 1999).

theory like ZFC certainly is axiomatizable, so this kind of "light" implementation is not enough to characterize computational mathematics.

Informally, one could say that proper, or "deep" implementation requires that, besides checking syntactical correctness of formalized mathematics, the computer can "understand" it and that all theorems become valid under this "interpretation." That is, every definition in the given language for mathematics should correspond to some construct in the computer's internal language, and every mathematical proof should produce a computer program. In particular, every function between natural numbers in the language for mathematics has to produce a function in the computer's language, which automatically means a computable (or recursive) function. This is called the *proofs-as-programs* paradigm.

In the end, the proofs-as-programs principle should be equivalent to the more rigorous requirement that the theory in which mathematics is developed admits a recursive realizability interpretation. This in turn should be even stronger than the requirement of consistency with $\mathsf{AC} + \mathsf{CT}$ mentioned before (and put forward in (Maietti and Sambin, 2005)). However, the digression on computers reveals a second difficulty, which appears at first even worse than the constructive incompatibility discussed in the previous section.

In fact, any computer language is intrinsically intensional; this means, in particular, that a function is determined only through the instructions, or program, to compute it. ML type theory (Nordström and Petersson, 1990) is intensional in this sense, and this makes it possible to consider it as an abstract programming language. On the other hand, in an extensional theory a function is uniquely determined by its behaviour: the principle of extensionality for functions says that two functions producing equal outputs on equal inputs are equal:

ExtFun: extensionality for functions, $\forall x(fx = gx) \rightarrow f = g$.

This might look so trivially valid that it may seem worthless to spell it out. So does in fact the standard set-theoretic approach to mathematics, in which a function is *defined* as a total and single valued relation. And since equality of relations is tacitly understood as extensional (that is, R and R' are equal whenever, for every x, y, $R(x, y)$ holds if and only if $R'(x, y)$ holds), extensionality for functions is implicitly present in their usual definition.

The two definitions of equality between functions, as mentioned above, underlie two different conceptions of functions themselves. In the former a function is identified with its instructions, in the latter with its behaviour. Knowing about the intrinsic difficulty of going back from behaviour to instructions which determine it, it seems difficult to reconcile them. Actually, their mutual inconsistency can be proved rigorously. In fact, as remarked above, an intensional theory to deal with the computational content of mathematics should be consistent with AC and CT. But one can prove ((Troelstra and van Dalen, 1988); see (Maietti and Sambin, 2005) for a simple proof) that the conjunction of ExtFun, AC and CT is contradictory, symbolically

$$\mathsf{ExtFun} + \mathsf{AC} + \mathsf{CT} \quad \vdash \quad \bot.$$

So one can see that the apparent contrast is not just an alternative between two views on constructive mathematics, which one could consider as a matter of subjective choice, but rather between two needs whose accomplishment is assumed to be a matter of fact. On one hand, mathematics traditionally wants to deal with "objects" independently of their presentation, and this is just a way of saying that it must be extensional. On the other hand, the use of computers to develop mathematics is by now also an accepted reality, which has brought important achievements and which is absorbing more and more resources. It is out of question to stop at this point, in particular if one is not ready to abandon the computational interpretation of mathematics and expects that it can be applied also when doing mathematics in practice, that is in an extensional way.

The pressure of necessity always acts as a strong spring pushing towards a solution. And a solution is found in what looks, a posteriori, the most simple-minded of all possible ways: contrary to a common expectation, both among mathematicians and computer scientists, (the level of abstraction of) mathematics just cannot coincide with (that of) its proper implementation. That is, the language in which mathematical proofs are written, even if formulated in a computer, just cannot coincide with the computer's internal language, in which programs are written. So the solution, as proposed in (Maietti and Sambin, 2005), is an approach to foundations in which two different levels of abstraction live together and interact. There will be an intensional type theory to deal with computations, together with an extensional theory in which (extensional) mathematics can be developed and formalized.

The principal idea underlying the design of the intensional theory, as different from ML type theory, is that logic should be introduced independently of set theory. A specific formal system, called *minimal type theory*, or mTT, has been introduced in (Maietti and Sambin, 2005) and expanded in (Maietti, 2009). In mTT, every proposition is a set, namely the set of its proofs in the formal system, but not conversely. This means that the propositions-as-sets paradigm is not followed. As a result, the inference rules for the existential quantifier $\exists$ are weaker than the rules for disjoint sums of sets Σ. A consequence of this is that AC is in general not valid.

In all other respects, when restricted to sets minimal type theory essentially coincides with ML type theory. As with the latter, it is designed to allow mechanical checking of correctness of all judgements; to this end, equality is always intensional. So it can be seen as a specification for a functional programming language. For this reason, in this setting the implementation of mathematics can be identified, for theoretical purposes, with its formalization in minimal type theory.

The core of the extensional theory is essentially many-sorted intuitionistic logic, in which sorts can be sets or collections and both allow passing to quotients. Sets, which are the domains of first-order variables, are the same as in the intensional theory, except for equality which is extensional and is obtained by abstraction. In particular, if every proof of a proposition A is considered to be equal to any other, that is, if one abstracts from proofs, one obtains the judgement that a proposition A is true. This is an important tool for obtaining extensional concepts (and explains the technical reasons for logic to be independent from set theory). The main example is that of subsets; a subset D of a set X is a propositional function $D(x)$ $prop$ $(x \in X)$ and $x \in X$ is said to be an element of D, written $x \epsilon D$, if the proposition $D(x)$ is true, abstracting from its proofs. Impredicative definitions are avoided by restricting variables in D to range only over sets. The notion of relation (in a set, or between sets) is defined in a similar way, except for the number of arguments.

Besides abstraction, a further specification of the interaction between the two theories is required. In fact, to be able to keep the computational interpretation, it should be possible, for every piece of mathematics developed in the extensional theory, to formalize it in the intensional ground theory. To satisfy this aim, one has to take care that in the process of abstraction only those pieces of information are left out which can be restored later for implementation, or which are not essential for it. Against appearances, this requirement is not self-contradictory since it is possible to restore information by a meta-mathematical analysis of proofs. One can show (see (Maietti, 2009), which is based on an idea in (Sambin and Valentini, 1998)) that it can be achieved once and for all, by only introducing extensional constructions under a certain discipline.

In conclusion, the extensional and the intensional aspects of mathematics can both be present in a single framework, provided they are seen as living at two different levels, which interact through abstraction and implementation.

The choice of two levels of abstraction also has a second, equally important motivation: it makes it possible to see the different options for foundations as parts of a unitary and consistent picture. This is obtained directly by showing that such options are obtained by adding some principles. Roughly speaking, one adds propositions-as-sets, or directly AC, to mTT to obtain ML type theory at the intensional level, while one can add PSA to obtain topos theory at the extensional level (Maietti, 2005). By adding both assumptions one obtains a classical foundation (though one must give up to compatibility with CT).

All these reasons explain why it has been natural to call this approach to foundations minimalist, as was first proposed in (Maietti and Sambin, 2005). Its extensional level is meant to be a common base for developing mathematics constructively. In fact, all of mathematics developed in it can be understood as it is (with no need of translations) by all mathematicians, whatever their foundational view. In particular, it makes communication possible between different traditions in constructive mathematics; this has also a practical relevance, which should not be underestimated.

We have seen that the computational (type theoretic) and the geometrical (topos theoretic) view of constructivism are incompatible when formulated in the usual way. One tends to extend this incompatibility from their formulations to the views themselves, and in fact they are felt to be incompatible in actual life: it often happens that a mathematician who works, usually for contingent or subjective reasons, in one of the two traditions is not very familiar with the other. Sometimes awareness of incompatibility is not sharp enough; sometimes, owing to the fact that different languages are used, it is felt as insurmountable and hence the two options are perceived as mutually exclusive.

When the two views are expressed inside the minimalist framework, the aims inspiring their principles acquire a different perspective. Then one can realize that both aims can be achieved in the same foundation, if they are formulated in a weaker, less rigid (and perhaps more realistic) way.

In the minimalist foundation, at both levels one has sets and collections. To introduce a set, one must specify a finite number of effective rules generating all its elements. All set constructors preserve effectivity of sets; in other words, every set is inductively generated. On the other hand, collections lack any effectivity. One can deal both with sets and with collections, while keeping them distinct, because logic is given independently. In particular, the collection of propositional functions with one argument in a set is present in the formal system for the intensional level (Maietti, 2009); by taking quotients over propositional equivalence, one obtains the collection of subsets in the usual sense, at the extensional level. In general, the decidability of equality in a set or collection at the intensional level is lost when passing to one of its extensional quotients.

So, while in the computational view "even the most abstract mathematical statement has a computational basis" (Bishop and Bridges, 1985, p. 6), in the minimalist view this applies only to statements regarding sets, both at the intensional *and* at the extensional level (because its implementation is always possible). However, some idealistic aspects are also present, and they are represented by collections. So in the minimalist view any computational content is carefully preserved, and yet computations are not the only source of meaning.

On the other hand, the validity of PSA, as in topos theory, would destroy the difference between a set, which is inductively generated, and the collection of its subsets, which is not (and can never be). In the minimalist foundation, ideal aspects of mathematics are possible and freely accessible, but they are kept distinct from the computational ones, since sets are distinct from collections. In particular, the power of a set is fully accessible, but the absence of PSA means that it is not a set.

More generally, the use of ideal language is intentionally designed so as to avoid producing any fictitious computational reality. A typical example is found in the notion of function. One has to distinguish a notion of *operation*, or function in the intensional sense, from A to B, that is a "finite routine" (Bishop, 1967) producing an element of B as output when an element of A is given as input, from that of *function* as a total and single valued relation between A and B. This means, in particular, that the so-called axiom of unique choice

AC! every function is given by an operation, $\forall x \exists ! y R(x, y) \rightarrow \exists f \forall x R(x, fx)$

is in general *not* valid, even if it is valid in all the other views (this shows, incidentally, that the minimalist foundations is not a naked compromise between existing foundations). In fact AC! is valid in ML type theory, also at the extensional level (see the so-called setoids), since this follows from validity of AC at the intensional level (Martin-Löf, 2006). It is derivable also in topos theory (Bell, 1988; Maietti, 2005), and in set theory it even looks like a tautology, by the very definition of function; the truth is that both these theories lack any notion of operation, that is function which allows one to *compute* the output.

Failure of AC!, which goes together with the distinction between operation and function, allows one to introduce a properly idealistic notion as such Brouwer's free choice sequences (see (Dummett, 1977)) without spoiling or damaging the computational aspects. A choice sequence of nodes in a tree can be described mathematically as a subset α of nodes of the tree which contains the root and which satisfies: for every node $a \in \alpha$, there exists exactly one node b which is a successor of a and such that $b \in \alpha$.[3] This is a statement of the form $\forall\exists!$, so by AC! we would always have an operation f doing the same job as α. That is, if AC! holds, every sequence is lawlike.

The notion of choice sequence was introduced by Brouwer mainly to be able to give meaning to the principle called Bar Induction. One can prove that if all sequences are lawlike and AC! + CT holds, then Bar Induction fails. In the minimalist foundation, owing to the lack of AC! one can prove that one can have a proper notion of choice sequence without the loss of computational aspects, since it can be shown that Bar Induction is consistent with CT (see (Maietti and Sambin, 201x)). Viewed in this way, Bar Induction becomes a very intriguing reducibility (or conservativity) principle, saying that a certain ideal property holds (existence of a bar) only if it corresponds to an effective property (the bar is inductively defined).

2 A Minimalist Foundation at Work

The short explanation of good reasons for a minimalist foundation, and its first results, have, I hope, provided evidence that it is interesting enough and worthy of further attention. However, rather than plunging into technical details of formal systems or lingering over arguments of purely philosophical nature, I believe that here it is better to illustrate the benefits of a minimalist foundation, and compare it with other foundations, by looking at how they work in practice on a specific, significant example in mathematics. This seems appropriate also because the minimalist approach was born from actual research in constructive mathematics, in the field started in (Sambin, 1987) and now known as formal topology.

It is a widespread belief that there is only one kind of mathematics, which is that developed in the classical foundation, and that the aim of constructivism is to develop as much as possible of it while avoiding "strong" but unreliable principles, such as the law of excluded middle or impredicative definitions. From this perspective, the excitement of discovery appears to be exclusively in the hands of classical mathematicians. Then it is no surprise that so few people are attracted by research in constructive mathematics: the effort of developing mathematics in a different foundation does not seem to be worthwhile if the outcome is only already known mathematics, with no novelty except perhaps a sharper treatment.

A more careful reflection brings the opposite conclusion. If a foundation is a choice of the criteria by which one abstracts from reality to obtain mathematics, then each choice should bring a different kind of mathematics. Indeed, in my opinion the

---

[3] More precisely, choice sequences can be defined as formal points over a suitable formal topology on the tree.

most attractive and important aspect of research in constructive mathematics is that it reveals *new* structures and *new* ways of mathematical thought.

So the mission of constructivism is not to accept mathematics as given and redeem it from its classical sins, but to produce mathematics of a different kind: a mathematics inaccessible from a classical standpoint, although compatible with it. Every constructive foundation, when put in action in the practice of research, should offer some specific contributions. This applies also to the minimalist foundation: if it does not help to deliver something really new in mathematics, then it would have little to say to working mathematicians.

The gain of constructive foundations, over the classical one, is usually to be found in improved computational meaning or clearer conceptual structures. The motivations of the minimalist foundation can be summarized in one sentence: provide a language for developing mathematics which can be understood and used by mathematicians of whatever creed, and yet such that each piece of mathematics developed in it is automatically formalizable in computer language.

We have seen above how this has been achieved. However interesting, it is still information on known mathematics reached through a meta-mathematical study. To create *new* mathematics, a foundation must be put to work and actually used in *doing* mathematics. This has happened with the minimalist foundation (whose substance was pretty clear well before its rigorous formulation was given in (Maietti and Sambin, 2005)) in the development of formal topology beginning over 10 years ago. It has brought the emergence of a new conceptual structure which underlies topology; its development gives rise to an extension and generalization of topology, which has been called the Basic Picture (Sambin, 2003, 201x). To be able to appreciate its novelty, a short summary of the basic picture is needed. After that, it will become easy to observe what had remained invisible from the perspective of other foundations, as well as the reasons for this. This will provide a specific illustration of the idea that, even starting from the same reality, different foundations produce different mathematics.

2.1 Dynamics Between Two Sets

The common definition says that a topological space is a pair $(X, \mathcal{O}X)$, where X is a set and $\mathcal{O}X$ is a topology on X, that is, a family of subsets of X which is closed under arbitrary unions and finite intersections. Elements of X are called points, and subsets in $\mathcal{O}X$ are said to be open. Given in this way, the definition has little computational value and thus cannot be satisfactory for a constructivist. The simplest way to obtain a topology on X in an effective way is to start from a second set S and a family

$$\mathsf{ext}\,(a) \subseteq X \ (a \in S)$$

of subsets of X indexed by S. Open subsets are obtained by union of these, that is, they are of the form

$$\mathsf{ext}\, V \equiv \cup_{b \epsilon V} \mathsf{ext}\,(b)$$

for some $V \subseteq S$. The union of any family of open subsets $\mathsf{ext}\, V_i \subseteq X$ $(i \in I)$, indexed by a set I, is automatically also an open subset, since one can easily show that $\cup_{i \in I} \mathsf{ext}\, V_i = \mathsf{ext}\,(\cup_{i \in I} V_i)$. In particular, $\mathsf{ext}\, \emptyset = \emptyset$ is open. One can show that closure under finite intersections is obtained by requiring two conditions on ext, that is

B1: $\mathsf{ext}\, U \cap \mathsf{ext}\, V = \mathsf{ext}\,(U \downarrow V)$
B2: $\mathsf{ext}\, S = X$

for every $U, V \subseteq S$, where by definition $a \downarrow b \equiv \{c \in S : \mathsf{ext}\, c \subseteq \mathsf{ext}\, a \cap \mathsf{ext}\, b\}$ for every $a, b \in S$, and $U \downarrow V \equiv \cup_{a \epsilon U} \cup_{b \epsilon V} a \downarrow b$ for every $U, V \subseteq S$.

A function from S into subsets of X, as ext, can equivalently be expressed as a relation between X and S, defined by

$$x \Vdash a \equiv x \epsilon \mathsf{ext}\, a$$

for every $x \in X$, $a \in S$. So one can see that the basic ingredients are just two sets X, S linked by any relation $\Vdash$. A structure like this has been called a *basic pair*. When it satisfies the conditions B1-B2, it is called a *concrete space*. However, knowing whether B1-B2 hold or not is not necessary for introducing topological notions on X.

The presence of the second set S alongside the set of points X immediately produces movement. The notions of open and of closed subset of X can indeed be explained, as we are going to see, purely as a result of the dynamic interaction between X and S, through the relation $\Vdash$. One may think of X and S as two persons, or domains, who are eager to communicate with each other about their own views, that is about subsets of their domain, which they interpret as regions, concepts, etc. S can simulate a region D of X by putting questions to X. But since $\Vdash$ is the only mean of communication, and in general it is not a function, when S asks about an element $a \in S$, all that X understands is $\mathsf{ext}\,(a)$, and then a (positive) answer can be of two different kinds: a is inside D, if $\mathsf{ext}\, a \subseteq D$ and a touches D, if $\mathsf{ext}\, a \between D$. Note that here, and from now on, $\between$ denotes the relation of overlap between two subsets D, E, which is defined by $D \between E \equiv \exists x(x \epsilon D \;\&\; x \epsilon E)$ (the importance of overlap is not visible in a classical approach, since $D \between E$ is there equivalent to $\neg\forall x \neg(x \epsilon D \;\&\; x \epsilon E)$, that is $D \cap E \neq \emptyset$). Accordingly, S will simulate D in two different ways:

$$\Diamond D \equiv \{a \in S : \mathsf{ext}\, a \between D\} \quad \exists\text{-simulation of } D,$$
$$\Box D \equiv \{a \in S : \mathsf{ext}\, a \subseteq D\} \quad \forall\text{-simulation of } D.$$

Since a basic pair assumes no condition on $\Vdash$, all which applies to X will apply symmetrically to S too. So, if $\Diamond x \equiv \{a \in S : x \Vdash a\}$ is what S understands of an element $x \in X$, then a region $U \subseteq S$ will be simulated by X in two different ways:

$$\mathsf{ext}\, U \equiv \{x \in X : \Diamond x \between U\} \quad \exists\text{-simulation of } U,$$
$$\mathsf{rest}\, U \equiv \{x \in X : \Diamond x \subseteq U\} \quad \forall\text{-simulation of } U.$$

Since X, S and $\Vdash$ are arbitrary, a region in one domain can be totally different from its simulations in the other. What is most interesting here, however, is that X and S still can communicate about their regions with full agreement, if they are a bit careful and follow some rules.

Given a region D, X can talk about it with S with mutual understanding, providing they make clear whether they are considering its $\exists$-simulation $\Diamond D$ or its $\forall$-simulation $\Box D$. In the first case, X has to consider her own $\forall$-simulation of $\Diamond D$, that is $\mathsf{rest}\,\Diamond D$. This is because one can prove that[4]:

$$\Diamond D = \Diamond\, \mathsf{rest}\, \Diamond D \quad \text{for every } D.$$

So the $\exists$-simulation by S of D coincides with that of $\mathsf{rest}\,\Diamond D$, which means that nothing changes to the eyes of S when X passes from D to $\mathsf{rest}\,\Diamond D$. This means that they can agree on regions of X, if S considers $\exists$-simulations, that is subsets of the form $\Diamond D$ for some D, and X replaces a region D with its $\forall\exists$-simulation $\mathsf{rest}\,\Diamond D$.

They can extend this mode of communication also to regions of S, with analogous precautions. One can prove the equation:

$$\mathsf{rest}\, U = \mathsf{rest}\, \Diamond\, \mathsf{rest}\, U \quad \text{for every } U.$$

This means that, when S wants to talk about his region U, X has to consider its $\forall$-simulation $\mathsf{rest}\, U$, and S has to consider his own $\exists$-simulation of $\mathsf{rest}\, U$, that is $\Diamond\, \mathsf{rest}\, U$. By the above equation, the $\forall$-simulation by X of U and of $\Diamond\, \mathsf{rest}\, U$ coincide.

One can prove that $\mathsf{rest}\,\Diamond$ is *saturation* on X, that is, it preserves inclusion of subsets ($D \subseteq E \rightarrow \mathsf{rest}\,\Diamond D \subseteq \mathsf{rest}\,\Diamond E$), it is expansive ($D \subseteq \mathsf{rest}\,\Diamond D$) and idempotent ($\mathsf{rest}\,\Diamond D = \mathsf{rest}\,\Diamond\,\mathsf{rest}\,\Diamond D$). A region D is called *saturated* if $D = \mathsf{rest}\,\Diamond D$. By the equation $\mathsf{rest} = \mathsf{rest}\,\Diamond\,\mathsf{rest}$, D is saturated iff it is the $\forall$-simulation $\mathsf{rest}\, U$ of some region U of S. On these regions X and S find immediate agreement; that is, when $\Diamond$ is restricted to saturated regions, it has an inverse rest.

Similarly, $\Diamond\,\mathsf{rest}$ is a *reduction* on S, that is it preserves inclusion, it is contractive ($\Diamond\,\mathsf{rest}\, U \subseteq U$) and idempotent. A region U is called *reduced* if $U = \Diamond\,\mathsf{rest}\, U$. Because of $\Diamond = \Diamond\,\mathsf{rest}\,\Diamond$, U is reduced iff it is the $\exists$-simulation by S of some region in X. Also on these regions X and S agree, because when rest is restricted to them, $\Diamond$ is its inverse.

[4] This and the next equation are an immediate consequence of the fact that $\Diamond$ is left adjoint to rest, which means that the equivalence $\Diamond D \subseteq U$ iff $D \subseteq \mathsf{rest}\, U$ holds for every D and U.

In conclusion, $\Diamond$ is a bijection between saturated subsets of X and reduced subsets of S, and rest is its inverse. With a little more work, one can also prove that the family of all saturated subsets of X is actually a complete lattice, the same for reduced subsets of S, and that $\Diamond$ is actually a complete lattice isomorphism. So, in this precise sense, X and S can communicate perfectly well, provided they keep in mind that $\exists$-simulations in one direction correspond to $\forall$-simulations in the other.

Everything that has been said up to now works equally well if the roles of X and S are interchanged. Or, which amounts to the same effect, if X and S are kept fixed, but $\forall$-simulations and $\exists$-simulations, and more generally $\forall$ and $\exists$, are interchanged. So, for instance, if one starts from a region D of X and its $\forall$-simulation $\Box D$ by S, then X must pass from D to her $\exists$-simulation of $\Box D$, that is $\mathsf{ext}\Box D$. This has no effect on S, because the following equations holds:

$$\mathsf{ext}\, U = \mathsf{ext}\Box\,\mathsf{ext}\, U \quad \text{for every } U,$$
$$\Box D = \Box\,\mathsf{ext}\Box D \quad \text{for every } D.$$

Without repeating details, one has that $\mathsf{ext}\Box$ is a reduction on X and that $\Box\,\mathsf{ext}$ is a saturation on S; reduced subsets of X and saturated subsets of S form isomorphic lattices, via the isomorphism $\Box$ with inverse ext.

What is the relevance of all this for topology? Think of X as a set of points, and $\mathsf{ext}\, a \subseteq X$ $(a \in S)$ as a family of neighbourhoods. Then for every subset $D \subseteq X$, the common definition says that x is in the interior of D, written $x \epsilon \mathsf{int}\, D$, if there is a neighbourhood $\mathsf{ext}\, a$ of x such that $\mathsf{ext}\, a \subseteq D$. In the present notation, this is

$$x \epsilon \mathsf{int}\, D \equiv \exists a(x \Vdash a \;\&\; \mathsf{ext}\, a \subseteq D).$$

Since by definition $x \Vdash a \equiv a \epsilon \Diamond x$ and $\mathsf{ext}\, a \subseteq D \equiv a \epsilon \Box D$, it is also

$$x \epsilon \mathsf{int}\, D \equiv \exists a(a \epsilon \Diamond x \;\&\; a \epsilon \Box D) \equiv \Diamond x \between \Box D \equiv x \epsilon \mathsf{ext}\Box D,$$

that is, $\mathsf{int} = \mathsf{ext}\Box$. So the previous discussion says that X and S can agree on a region D of X if this region is *open*, that is $D = \mathsf{int}\, D$ holds.

The common definition says that x is in the closure of D, written $x \epsilon \mathsf{cl}\, D$, if every neighbourhood of x has some point in common with D. In the present notation, this is

$$x \epsilon \mathsf{cl}\, D \equiv \forall a(x \Vdash a \rightarrow \mathsf{ext}\, a \between D)$$

and one can easily find out that $\mathsf{cl} = \mathsf{rest}\Diamond$. As usual, D is said to be *closed* if $D = \mathsf{cl}\, D$.

It is now obvious that the definition of cl is logically dual to that of int, in the sense that one is obtained from the other by interchanging $\exists$ with $\forall$ and & with $\rightarrow$. Although obvious, this fact was apparently not explicitly noticed before; certainly, some of its immediate consequences had been ignored.

In all the previous discussion it has also been obvious that, since the ingredients of a basic pair are totally arbitrary, one can argue by symmetry and save half of the work. In particular, $\Box\,\mathsf{ext}$ is a saturation on S, here denoted by $\mathcal{A}$, and $\Diamond\,\mathsf{rest}$ is a reduction on S, denoted by $\mathcal{J}$. What is their topological meaning? An element a of the set S is an index, or name, for a basic neighbourhood $\mathsf{ext}\,a \subseteq X$. Sometimes a is called a *formal basic neighbourhood*, or simply an *observable*. Spelling out their definitions, we see that:

$$a \in \mathcal{A}U \equiv \mathsf{ext}\,a \subseteq \mathsf{ext}\,U \equiv \forall x(x \Vdash a \rightarrow \exists b(x \Vdash b \;\&\; b \in U)),$$
$$a \in \mathcal{J}U \equiv \mathsf{ext}\,a \between \mathsf{rest}\,U \equiv \exists x(x \Vdash a \;\&\; \forall b(x \Vdash b \rightarrow b \in U)).$$

So $a \in \mathcal{A}U$ means that the family of neighbourhoods $\mathsf{ext}\,b \subseteq X$ $(b \in U)$ covers the neighbourhood $\mathsf{ext}\,a$. We say that a is covered by U; note that this directly relates an observable with a subset of observables, and that points occur only in its definition. The meaning of $a \in \mathcal{J}U$ is that $\mathsf{ext}\,a$ is inhabited by a point all of whose neigbourhoods are in U. We say that a is positive with U.

As with any saturation or reduction, one can easily see that the families $Sat(\mathcal{A})$ of $\mathcal{A}$-saturated and $Red(\mathcal{J})$ of $\mathcal{J}$-reduced subsets, that is subsets U for which $U = \mathcal{A}U$ and $U = \mathcal{J}U$ holds, respectively, can be given the structure of a complete lattice. By symmetry, $Sat(\mathsf{cl})$ and $Red(\mathsf{int})$ are also complete lattices; they are the familiar lattices of closed and of open subsets of X, respectively. Due to the links between the operators $\Diamond$, rest and between ext, $\Box$ through which int, $\mathcal{A}$ and cl, $\mathcal{J}$ are defined, one can prove that one has the isomorphisms $Red(\mathsf{int}) \cong Sat(\mathcal{A})$ and $Sat(\mathsf{cl}) \cong Red(\mathcal{J})$.

As we have just seen, it is very natural to introduce the reduction $\mathcal{J}$, either by symmetry, starting from int, or by duality, starting from $\mathcal{A}$. And yet the notion of $\mathcal{J}$ had not appeared before in topology. It seems evident that the reason for this was the lack of a topological interpretation of the structure of basic pair. In fact, this is what allows one to perceive the presence of symmetry and logical duality in topology. We will see below some foundational reasons which can explain why they had not appeared before.

Besides the notions of open, closed, cover, and positivity, one finds that a clear structure underlies also the notion of continuity. In fact, the basic form of continuity can be seen to coincide with a very clear dynamics between two basic pairs, that is a pair of relations producing a commutative square. These are the arrows between basic pairs.

Basic pairs in which the lattice of open subsets is a topology on X are characterized by the two simple conditions B1 and B2. If one reads $x \Vdash a$ as a is an approximation of x, these express refinability, or convergence of approximations: B2 says that every point has some approximation, and B1 says that for every two approximations of a point, there is always a third improving on both. One can see that continuity does not include preservation of convergence: arrows of basic pairs preserve convergence if and only if they are topologically equivalent to a function. So traditional point-set topology can be seen as the special case in which convergence and its preservation are assumed.

2.2 The Definition of Basic Topology

The notion of concrete space is not sufficient to develop topology constructively. In fact, in many classical examples of topological spaces, points do not form a set in any effective sense; a typical example is the continuum, defined via Dedekind cuts. The aim of formal topology is to obtain such spaces as the collection of "virtual," or formal, points over a suitable point-free structure on the set of observables. In fact, as a rule, even when points do not form a set, the formal structure of open subsets can be given in a fully effective way, usually by an inductive definition. For instance, a base for a topology on the continuum is given by intervals with rational endpoints, which can be indexed by the set of pairs of rational numbers, and this set can be equipped with a convenient pointfree structure.

Points of a classical topological space, such as the continuum, become again what they had been originally (for Dedekind himself), that is, idealizations. It is a task of the minimalist foundation to distinguish the collection of formal points, which is an ideal entity called formal space, from the structure on observables, or formal opens, which is effective. And also to distinguish the cases in which it happens that points can be given concretely as a set, so that one has a basic pair or a concrete space, from those in which points can be only ideal, and one has a formal space.

The point-free structure to be defined on the set of observables is usually called a *formal topology*. It is crucially important to find the most convenient and convincing definition. To explain the method by which this is done, it seems helpful to resume the metaphor of communication between X and S through $\Vdash$. In fact, for S to find the right definition is the same as to internalize his dialogue with a hypothetical X. That is, S should first equip himself with a saturation $\mathcal{A}$ and a reduction $\mathcal{J}$. Then he should think of $\mathcal{A}U$ as his way of communicating with the virtual X about his region U, when X takes its $\exists$-simulation. To emphasize his "dialogue" with X, he may call $\mathcal{A}U$ "formal open," since he knows that it corresponds to an open subset for her. Analogously, S thinks of $\mathcal{J}U$ as a way to communicate with X about her $\forall$-simulation of U. Since reduced subsets correspond to closed subsets of X, he may call them "formal closed."

A delicate question concerns how S can manifest the fact that he is having a relation with just one person (he could talk with X about her open subsets, through $\mathcal{A}$, and with X' about her closed subsets, through $\mathcal{J}$). One can easily prove that, when $\mathcal{A}$ and $\mathcal{J}$ are defined using concrete points of the same set X and through the same relation $\Vdash$, they satisfy the condition called *compatibility*:

$$\mathcal{A}U \between \mathcal{J}V \leftrightarrow U \between \mathcal{J}V$$

for every $U, V \subseteq S$.

The internalization of the dialogue is, metaphorically speaking, an axiomatization of the structure induced by a basic pair on its set S. Thus one calls *basic topology* any structure $\mathcal{S} = (S, \mathcal{A}, \mathcal{J})$ where S is any set, $\mathcal{A}$ is a saturation and $\mathcal{J}$ a reduction on S, and $\mathcal{A}$ is linked to $\mathcal{J}$ by compatibility.

Now finally one can see how points can be simulated. The only way is to say that a formal point is a subset $\alpha \subseteq S$ which behaves, with respect to the structure of $\mathcal{S}$, exactly as if it were the subset $\Diamond x$ determined by some concrete point x. In this way the relation $x \Vdash a$ is simulated by $a \epsilon \alpha$, and this is why one writes $\alpha \Vdash a$ instead. If $\mathcal{A}$ were defined through points of X, one would obviously have that $x \Vdash a$ and $a \epsilon \mathcal{A}U \equiv \mathsf{ext}\, a \subseteq \mathsf{ext}\, U$ imply $x \epsilon \mathsf{ext}\, U$. This means that α is required to split covers:

$$\frac{\alpha \Vdash a \quad a \epsilon \mathcal{A}U}{(\exists b \epsilon U)\alpha \Vdash b}$$

Similarly, if $\mathcal{J}$ were defined pointwise, one would have that $\alpha \Vdash a$ and $\Diamond x \subseteq U$ imply $a \epsilon \mathcal{J}U \equiv \mathsf{ext}\, a \between \mathsf{rest}\, U$. So α must enter positivity:

$$\frac{\alpha \Vdash a \quad \alpha \subseteq U}{a \epsilon \mathcal{J}U}$$

This is not yet enough. If a concrete point x satisfies $x \Vdash a$ and $x \Vdash b$, then $x \epsilon \mathsf{ext}\, a \cap \mathsf{ext}\, b$. In a concrete space, $x \epsilon \mathsf{ext}\, a \cap \mathsf{ext}\, b$ holds if and only if there is an observable c such that $x \Vdash c$ and $\mathsf{ext}\, c \subseteq \mathsf{ext}\, a \cap \mathsf{ext}\, b$.

As a concrete space is just a basic pair satisfying convergence, so now a *formal topology* is defined as a basic topology with an additional condition induced on S by convergence, or B1 and B2. One can easily see that this is:

$$\mathcal{A}U \cap \mathcal{A}V = \mathcal{A}(U \downarrow V)$$

(where now $\downarrow$ is defined in terms of $\mathcal{A}$ by putting $a \downarrow b \equiv \mathcal{A}\{a\} \cap \mathcal{A}\{b\}$ and then by defining $U \downarrow V \equiv \cup_{a \epsilon U} \cup_{b \epsilon V}\, a \downarrow b$ as before).

Finally, a *formal point* is a subset α of observables of a formal topology $\mathcal{S}$ which splits covers, enters positivity, is inhabited and is convergent, that is, if $\alpha \Vdash a$ and $\alpha \Vdash b$ then $\alpha \Vdash c$ for some $c \epsilon a \downarrow b$.

The collection $\mathcal{P}t(\mathcal{S})$ of formal points of $\mathcal{S}$ is called a *formal space*. If the powerset axiom PSA is assumed, then $\mathcal{P}t(\mathcal{S})$ is a set, since it is a subset of $\mathcal{P}S$; then $\mathcal{P}t(\mathcal{S})$ is a possible conversation partner for $\mathcal{S}$, through the relation $\alpha \Vdash a \equiv a \epsilon \alpha$, even if her points might have no effective value. Predicatively, $\mathcal{P}t(\mathcal{S})$ remains what it is, after all, that is an idealization from $\mathcal{S}$'s mind, and definitely not a set.

The discovery of symmetry and logical duality, together with the method by which definitions have been introduced and justified, have brought point-free structures which are richer than one would have expected previously. The saturation $\mathcal{A}$ is not a novelty. It was part of the original definition of formal topology in (Sambin, 1987). Moreover, pairs $(S, \mathcal{A})$ with $\mathcal{A}$ satisfying convergence $\mathcal{A}U \cap \mathcal{A}V = \mathcal{A}(U \downarrow V)$ can be seen as a predicative presentation of locales (Sambin, 1987; Battilotti and Sambin, 2006). If they are equipped with a predicate Pos as required in (Sambin, 1987), they correspond to open locales (for a proof, see (Vickers, 2006)).

The real novelty is the presence of the reduction $\mathcal{J}$ besides $\mathcal{A}$, that is a positivity relation besides the cover, and of the condition connecting them, namely their compatibility. The reduction $\mathcal{J}$ allows one to speak about closed subsets in point-free terms. Although it is new, from the point of view of the underlying logical structure its introduction is as natural as that of $\mathcal{A}$: $\mathcal{A}$ is a way to express quantifications of the form $\forall\exists$, that is $\Box$ ext, and $\mathcal{J}$ of the form $\exists\forall$, that is $\Diamond$ rest.

A similar remark applies also to all other definitions, including those of arrows: they seem to flow almost by themselves out of purely structural and logical considerations. Actually, the name "basic picture" has been chosen precisely because it can be described as the study of pure dynamics between sets, when they are connected by relations. In this sense it is a sort of applied logic. At the same time, this study is carried on by means of some mathematical definitions which have a strong topological meaning. So the basic picture is also a generalization of topology. Topology in the stricter sense, either with points as in concrete spaces or point-free as in formal topologies, is obtained as a special case simply by adding treatment of convergence.

Besides revealing an elegant, pure structure beneath topology, a fact which should appeal to a logician's taste, the basic picture offers a new, firm foundation for topology, bringing a deeper understanding of familiar notions, and thus should satisfy mathematicians also. Yet, the development of the basic picture is very recent. It is thus natural and interesting to investigate the reason for this, and we do this in the next section.

3 A Different Topology in Different Foundations

The definition of topological space that is commonly given in a foundation in which the powerset axioms PSA is assumed makes no mention of a second set S alongside the set of points X. The reason for this is that the information given by the presence of S is supposed to be redundant. In fact, assume one starts from a basic pair $\mathcal{X} = (X, \Vdash, S)$; the validity of properties turning it into a concrete space does not alter the substance of our discussion and we assume it whenever useful. If PSA is assumed, then $\mathcal{P}S$ is as good a set as S itself. So

$$X \xrightarrow{\Vdash^+} \mathcal{P}S,$$

where $x \Vdash^+ U \equiv \exists b(x \Vdash b \ \&\ b \in U) \equiv x \in \mathsf{ext}\, U$, is as good a basic pair as the original $\mathcal{X}$. But in the new basic pair the family $\mathcal{O}X$ of all open subsets of X is indexed by a set, and hence $\mathcal{O}X$ itself is a set (it is $\{D \in \mathcal{P}X : \exists U \subseteq S(D = \mathsf{ext}\, U)\}$). Moreover, one can easily see that the family of open subsets induced on X by $(X, \Vdash^+, \mathcal{P}S)$ coincides with $\mathcal{O}X$. Hence, as long as one is interested in open subsets of X, there is no difference between the original $\mathcal{X}$ and the basic pair $(X, \epsilon, \mathcal{O}X)$. Then it is clear why there is no need to keep the set S, and hence also why the general notion of basic pair had never been seriously considered: using PSA one can always reduce to the case in which the set S is $\mathcal{O}X$ and the relation $\Vdash$ is membership.

In other words, assuming PSA the information that open subsets can be obtained from a set-indexed base is contained in the common definition of topological space.

From the minimalist perspective, the same fact acquires a quite different interpretation: the information about a set-indexed base is left out not because it is redundant, but rather because an impredicative foundation is unable to perceive it. In fact, the property which distinguishes S from $\mathcal{P}S$, namely inductive generation, is irreparably lost by the impredicative notion of set.

In a predicative approach to topology, as in (Martin-Löf, 1970; Sambin, 1987), the situation seems to be dual: a set of points X does not appear and the treatment starts from the structure of opens, which is given on a set like S. In fact, in the most important examples of spaces, such as those treated in (Martin-Löf, 1970), points can be obtained predicatively only as ideal objects, that is formal points, and they do not form a set. Thus the set X is not absent because it is redundant, but on the contrary because it is an unavailable piece of information. In this way, however, all cases in which points *do* form a set are neglected.

In both cases, one of the two sets of a basic pair is missing. Thus no dynamics is possible. This explains why the symmetry between pointwise and pointfree notions, the logical duality between closed and open, and hence the notion of positivity relation, had not appeared before.

These remarks suggest that different foundations reach different mathematical notions, even when they start from the same reality. In this case, however, it is highly debatable what such "reality" is precisely, prior to its formalization into a rigourous mathematical definition. So it is perhaps more interesting to analyse a case in which there can be general agreement that the reality one starts from is the same. The definition of basic topology is an excellent test for this. We must take for granted that it is obtained by abstraction from the structure induced on the formal side of a basic pair $\mathcal{X}$; we have seen that in the minimalist foundation this is the basic topology $(S, \mathcal{A}, \mathcal{J})$ concretely presented through the definitions $\mathcal{A} \equiv \Box\,\mathsf{ext}$ and $\mathcal{J} \equiv \Diamond\,\mathsf{rest}$. Then we can realize how, starting from the same mathematical "reality" given by a basic pair, the process of abstraction leads to different results, according to the different foundational "lenses" one is wearing.

Assuming classical logic, it is not necessary to consider closed subsets, in the sense that the intrinsic characterization, saying that a subset D is closed if it contains its closure $\mathsf{cl}\,D$, is immediately proved to be equivalent to the "logical" definition of closed subset as complement of open subset. In fact $\mathsf{cl}\,D \subseteq D$ by definition is

$$\forall a(x \Vdash a \rightarrow \mathsf{ext}\,a \between D) \rightarrow x \in D.$$

By classical logic, this is equivalent to $x \in -D \rightarrow \neg\forall a \neg(x \Vdash a \;\&\; \neg(\mathsf{ext}\,a \between D))$ and hence to $x \in -D \rightarrow \exists a(x \Vdash a \;\&\; \mathsf{ext}\,a \subseteq -D)$, which is the definition of $-D \subseteq \mathsf{int} - D$, that is of $-D$ being open.

The same fact can be expressed in a way which makes it evident that it is due only to logical laws. By the definition $\mathsf{cl} \equiv \mathsf{rest}\,\Diamond$, using relativized quantifiers one can express $x \in \mathsf{cl}\,D$ as $(\forall a \in \Diamond x)(\exists y \in \mathsf{ext}\,a)(y \in D)$. The laws concerning quantifiers, in particular the classical equivalence between $\exists\neg$ and $\neg\forall$ as well as that between $\neg\exists$ and $\forall\neg$ which holds also intuitionistically, apply also to relativized quantifiers. So the combination $\forall\exists$, which by double negation is the same as $\forall\exists\neg\neg$, is equivalent to the combination $\neg\exists\forall\neg$, that is $\neg(\exists a \in \Diamond x)(\forall y \in \mathsf{ext}\, a)\neg(y \in D)$, which by the definition $\mathsf{int} \equiv \mathsf{ext}\Box$ is equivalent to $x \in -\mathsf{int} - D$.

So on basic pairs the equation $\mathsf{cl} = -\mathsf{int}-$ holds classically, and hence by double negation also $\mathsf{int} = -\mathsf{cl}-$, that is closure and interior can be defined one in terms of the other through complementation. This shows that, "wearing the lenses" of classical logic, one cannot avoid reducing a quantification of the form $\forall\exists$, characterizing closure, to one of the form $\neg\exists\forall\neg$. This prevents one from perceiving the fine structure of duality between the intuitive notions of closed and open subsets, since it soon boils down to the much simpler duality between a subset and its classical complement, that is between a classical proposition and its negation.

By symmetry, the equations $\mathcal{A} = -\mathcal{J}-$ and $\mathcal{J} = -\mathcal{A}-$ hold. So the notion of basic topology would reduce classically either to a set with a reduction $(S, \mathcal{J})$ or a set with a saturation $(S, \mathcal{A})$. These structures are too impoverished to be considered worthwhile. As a matter of fact, the only pointfree structure which has been introduced up to now is essentially that of locale; here it can be seen as obtained by abstraction from a concrete space, and thus it is a pair $(S, \mathcal{A})$ where $\mathcal{A}$ satisfies convergence. Then the locale is $Sat(\mathcal{A})$ which, in the case of a concrete space, as we have seen is isomorphic to the lattice of open subsets of X.

Closed subsets can also be defined in terms of open ones on the base of intuitionistic logic, provided one has access to impredicative definitions. In fact, given any basic pair $\mathcal{X}$, if PSA holds one can consider all open subsets as basic neighbourhoods, by passing from $\mathcal{X}$ to $(X, \Vdash^+, \mathcal{P}S)$ as defined previously. Here by definition x is in the closure of D if $\forall U(x \Vdash^+ U \rightarrow \mathsf{ext}^+ U \between D)$; since $\mathsf{ext}^+ U = \mathsf{ext}\, U$ and since all open subsets of X are of the form $\mathsf{ext}\, U$ for some U, we arrive at a definition of closure in terms of interior:

$$x \in \mathsf{CL}_{\,\mathsf{int}} D \equiv \forall E(x \in \mathsf{int}\, E \rightarrow \mathsf{int}\, E \between D).$$

Once can easily prove that $\mathsf{CL}_{\,\mathsf{int}}$ coincides with cl in the original basic pair. So also in an impredicative approach one can focus on interior, and consider the notion of closure as defined in terms of it.

By symmetry, given a reduction $\mathcal{J}$ on a set S one can define (impredicatively) a saturation on S by putting

$$a \in \mathcal{A}_{\mathcal{J}} \equiv \forall Z(a \in \mathcal{J}Z \rightarrow U \between \mathcal{J}Z).$$

One can easily prove that actually, for *every* reduction $\mathcal{J}$ on S (that is, not necessarily induced by a basic pair), the operator $\mathcal{A}_{\mathcal{J}}$ is indeed a saturation, that it is compatible with $\mathcal{J}$, and the greatest such. Moreover, when $(S, \mathcal{A}, \mathcal{J})$ is the basic topology induced on S by $\mathcal{X}$, then $\mathcal{A}_{\mathcal{J}}$ coincides with $\mathcal{A}$.

Then one could define a basic topology to be a set with a reduction $(S, \mathcal{J})$. However, such a definition has never been developed; for some good reasons, what corresponds to $(S, \mathcal{A})$ in the present framework has been chosen instead. In fact, while classically the structure of closed subsets by de Morgan laws is completely symmetric to that of open subsets, this is no longer true intuitionistically. A sign of this is the different behaviour of structures of the form $(S, \mathcal{A})$ or $(S, \mathcal{J})$. It is apparently not possible to find a property characterizing, among all reductions, those which are induced by a concrete space; on the other hand, this is easily done for a saturation, and the solution is the condition of convergence mentioned before. In other terms, it is not known how to present the algebraic structures of closed subsets by means of a structure of the form $(S, \mathcal{J})$; on the other hand, the algebraic structures of open subsets, that is locales, can all be isomorphically presented, reasoning impredicatively, as a lattice $Sat(\mathcal{A})$, for some pair $(S, \mathcal{A})$ where $\mathcal{A}$ is a saturation satisfying convergence.

In conclusion, both in the classical and the impredicative approach one has chosen locales to be the main pointfree structure, which here corresponds to the choice of $(S, \mathcal{A})$. The reason for this is that the lattice $Sat(\mathcal{A})$ is isomorphic to the lattice $Red(\mathsf{int})$ of open subsets of X.

The trouble is that, when the only criterion is isomorphism of lattices, one cannot see that in a basic pair int is the trace of existential quantifications (because $\mathsf{int} \equiv \mathsf{ext}\,\Box$ is of the form $\exists\forall$, and every open subset is of the form $\mathsf{ext}\, U$, that is the $\exists$-simulation of a subset of S), while on the other hand $\mathcal{A}$ is associated with universal quantifications (since $\mathcal{A} \equiv \Box\,\mathsf{ext}$ is $\forall\exists$, and $\mathcal{A}$-saturated subsets are $\forall$-simulations of subsets of X).

In other words, even if $Red(\mathsf{int})$ and $Sat(\mathcal{A})$ are isomorphic lattices, one is given by a reduction int and the other by a saturation $\mathcal{A}$. Since this kind of information is *not* preserved by lattice isomorphisms, it is better to retain the operators, that is keep (X, int), $(S, \mathcal{A})$, $(S, \mathcal{J})$, ... and not only the lattices $Red(\mathsf{int})$, $Sat(\mathcal{A})$, $Red(\mathsf{int})$, ... they give rise to. Then the symmetry between int and $\mathcal{J}$ (and between cl and $\mathcal{A}$) remains visible. And, under the inspiration of symmetry, one can see that, just as with int one can define CL, for every reduction $\mathcal{J}$ one can define a saturation $\mathcal{A}_{\mathcal{J}}$ such that $(S, \mathcal{A}_{\mathcal{J}}, \mathcal{J})$ is a basic topology. If a reduction allows one to define a saturation compatible with it, there is no reason why also the converse should hold. So, even assuming that the aim is to keep only one of the operators of a basic topology, the choice for the saturation $\mathcal{A}$ is the wrong one, since it gives rise to a basic topology only classically.

Finally, if the task of characterizing the structure induced by a basic pair on the set S is carried out in the computational view, then one obtains some further information not about the mutual relationship between $\mathcal{A}$ and $\mathcal{J}$, but about their nature. That is, when the axiom of choice AC is valid one can prove that the saturation $\mathcal{A}$ can be generated effectively through an inductive definition. Let us say that an *axiom-set* I, C is given by a family of sets $I(a)\, set\ (a \in S)$ indexed on the set S and a family of subsets $C(a,i) \subseteq S\ (a \in S, i \in I(a))$. From any axiom-set I, C one can generate a saturation $\mathcal{A}_{I,C}$ by postulating, for every $a \in S$ and $U \subseteq S$, that $a \,\epsilon\, \mathcal{A}_{I,C}U$ holds only either if $a \,\epsilon\, U$ or if inductively $C(a,i) \subseteq \mathcal{A}_{I,C}U$ for some $i \in I(a)$.

There are two relevant mathematical results to consider here. One says that, when the saturation $\mathcal{A}$ is concretely presented by a basic pair, using AC one can find an axiom-set I, C such that $\mathcal{A} = \mathcal{A}_{I,C}$ (see Coquand et al., 2003). Therefore it becomes natural to require, among the conditions defining basic topologies, also that the saturation $\mathcal{A}$ should be generated inductively, from some axiom-set I, C.

But then the second result says that, starting from any axiom-set I, C, one can generate effectively, this time by a coinductive definition, also an operator $\mathcal{J}_{I,C}$ and prove that $\mathcal{J}_{I,C}$ is a reduction, that it is compatible with $\mathcal{A}_{I,C}$, and that actually it is the greatest such (see Martin-Löf and Sambin, in press).[5]

In this way one again finds that the notion of closed subset is uniquely determined by that of open subset, and therefore the abstract definition of basic topology retreats to the background. In fact, it can be reduced to the data of an arbitrary set S and an arbitrary axiom-set I, C on S, understanding that from it one can generate $\mathcal{A}_{I,C}$ and $\mathcal{J}_{I,C}$. This option has the advantage that a basic topology is certainly given effectively. The drawback is that it conflicts with the algebraic/geometric view, since the property of being generated cannot be easily expressed in algebraic terms.

4 Benefits of the Minimalist Foundation

The previous analysis shows how the attitude expressed in the minimalist foundation has been relevant for the development of the basic picture. Though it remains in the end only an example, it seems sufficiently meaningful to suggest some reflections of general nature.

4.1 Effective Computations and Ideal Structures in One Framework

Effective computations and ideal algebraic/geometric structures are two vital modes of mathematical thinking. It is hard to deny this. They are present also in constructive mathematics, but the proposed foundational systems focus on just one of them and leave the other in the background, subject usually to an informal, metalinguistic treatment. However, there are no a priori reasons that all of mathematics should be reducible to either of these two aspects. Thus it seems natural and important to aim, as the minimalist foundation does, at expressing both within a single formal system, although at two different levels of abstraction.

Brouwer was right in rejecting the reduction of mathematics to a purely formal manipulation of signs. With hindsight, we can reach a more balanced view and see that the evil is not formal language in itself, but its overestimation. The expression of some mathematical content with full symbolism makes it easier to share it socially and, at an individual level, helps to increase awareness of it. It is an important stage of the transformation which starts from a subjective intuition and aims at an objective mathematical entity. In a dynamic perspective, formalization is therefore one of

[5] It is still an open problem to prove, under the assumptions that $\mathcal{A}, \mathcal{J}$ are presented by a basic pair and that $\mathcal{A} = \mathcal{A}_{I,C}$, that also $\mathcal{J} = \mathcal{J}_{I,C}$ holds.

the facets of the process of abstraction through which every mathematical concept is obtained.

A general criterion for an abstract concept to be healthy is that it is well-founded. That is, it should remain clear what its contact is with the reality it comes from, at least in the sense of something of a lower degree of abstraction. This gives a theoretical rationale for requiring a formal system always to admit a computational interpretation: it is a discipline to anchor it to something real and thus avoid meaningless speech. A pure consistency proof, as required in Hilbert's program, would not be enough for this purpose.

At the same time, the general criterion of well-foundedness does not say that computations should necessarily be the only reality to which one refers for meaningfulness. Reducing all of mathematics to manipulation of numbers and signs, as a machine does, would mean losing the power of high level human abstractions. A purely algebraic structure can be very helpful for grasping the essence of a mechanism beyond the details of specific examples or its actual implementation. And it remains meaningful as long as it clearly refers to something real, which can be computations again but indirectly (as in high level programming languages) or something non-numerical (as happens for instance with the notion of triangle).

Briefly, mathematics should preserve its computational content without sacrificing the power of its abstractness and becoming too close to programming (which many working mathematicians rightly dislike). The actual development of formal topology and the basic picture shows that this is entirely possible.

4.2 Creation of New Mathematics

As happens with a new postulate, the most exciting aspect of a new foundation is that it leads to an extension of mathematics. This is actually what motivates its introduction and makes it meaningful. Since the minimalist foundation is weaker than others—in the sense that it is obtained by subtracting rather than adding principles—the novelties it brings cannot be the solution of specific problems, which is the typical reason for adding a postulate. So its introduction seems fully justified only if it opens new conceptual perspectives or it suggests the exploration of new landscapes (for the same reason it would be worthless to have a bicycle if one could cycle only along motorways).

The minimalist foundation allows one to develop mathematics in its usual extensional and ideal way, and yet always preserve its computational content. This benefit was to be expected, since it was the inspiring motivation.

It was not expected that the same attitude would gradually bring to light some clear structures which had been overlooked by other foundations. These have been delivered mainly by a methodological principle: to consider and express mathematically in an independent way notions with a different logical structure, with no subjection to the identifications forced by classical logic. A logical analysis of the notions of open and closed subset, or better of the operators interior int and closure

cl, shows that the formula by which they are defined is the same, but for the form of quantifications. Assuming classical logic, the positive information of the form $\forall\exists$ which characterizes cl is identified with, and hence replaced by, the combination $\neg\exists\forall\neg$, that is absence or negation of an information of the form $\exists\forall$, which characterizes int. Thus closed subsets are uniquely determined by open subsets by way of negation, or complement, and closure is explicitly definable by $\mathsf{cl} \equiv -\mathsf{int}-$. Also assuming intuitionistic logic, closure can be uniquely determined by interior through an impredicative definition. As a consequence, closed subsets and the operator cl are anyway by and large ignored.

In an intuitionistic and predicative approach, closure and interior are mathematical notions representing two combinations of quantifiers, $\forall\exists$ and $\exists\forall$ respectively, which cannot be reduced to each other by negation. So they must be treated independently. This fact becomes particularly visible and clear when it comes to expressing their properties in algebraic terms. Note that this is a task which *must* be faced and cannot be left out as an optional extra. Actually, we have accomplished it already with the definition of basic topology, which is the necessary first step of a formal approach to topology. In fact, the notion of basic topology was obtained as an axiomatic description, which is to say as an algebraic characterization, of the structure induced by a basic pair on its formal side. But then, by the symmetry of basic pairs, of the operators int and cl all one can say algebraically is that they give rise to a basic topology. That is, int is a reduction, cl is a saturation, and their only link is compatibility:

$$\mathsf{cl}\, D \between \mathsf{int}\, E \leftrightarrow D \between \mathsf{int}\, E$$

for all subsets D, E (the role of compatibility is perhaps made clearer by noting that it is equivalent to $\mathsf{cl}\, D \subseteq \mathsf{CL}_{\mathsf{int}}\, D$ for every D). This is a very simple algebraic structure, whose reverse shows that an intuitionistic algebraic treatment of int and cl was missing. In fact, the only link between int and cl which can be expressed algebraically is compatibility, but apparently no mention of this condition appears in the literature. One can only guess that compatibility went unobserved because to express it perspicuously one needs overlap and, incredibly, a specific notation for it, such as the sign $\between$ adopted here is, has not been systematically used before.

In each of the extensions of the minimalist foundation we have considered, closed subsets are uniquely determined by open subsets. This happens in each foundation in a different way, and in the case of the computational view open and closed are to be meant in the formal sense. As we have seen, this fact explains why one is very unlikely to reach the notion of basic topology in its general form if some strong principles are assumed *in advance*, that is in the process of its definition. Still, a sign of the soundness of the structure of a basic topology is that it is not reduced to something simpler even if the same strong principles are assumed *after* the definition has been given.

In fact, in each case it is possible to start from one of the operators and by using a strong assumption define the other one, so as to obtain a basic topology; but still, such construction will not cover the general case. When the law of excluded middle

LEM is assumed, given a saturation $\mathcal{A}$ one can define a reduction compatible with it by putting $\mathcal{J} \equiv -\mathcal{A}-$. But this does not give all basic topologies: even assuming LEM, all one can say is that a basic topology $(S, \mathcal{A}, \mathcal{J})$ is the same as a structure $(S, \mathcal{A}, \mathcal{A}')$ where $\mathcal{A}'$ is a second saturation with $\mathcal{A}U \subseteq \mathcal{A}'U$ for every subset U (which is classically equivalent to the compatibility of $\mathcal{J} \equiv -\mathcal{A}'-$ with $\mathcal{A}$).

Given a reduction $\mathcal{J}$, one can define impredicatively, as we have seen, a saturation $\mathcal{A}_{\mathcal{J}}$ so that $(S, \mathcal{A}_{\mathcal{J}}, \mathcal{J})$ is a basic topology. In the computational view any saturation $\mathcal{A}$ is to be given by way of an axiom-set I, C which generates it inductively, so that with the same I, C one can define coinductively also a reduction $\mathcal{J}_{I,C}$ compatible with $\mathcal{A}$. But in both cases one cannot rule out basic topologies in which one of the operators is *not* uniquely determined by the other. For example, given any axiom-set I, C and any extension J, D of it, one can easily see that, since $(S, \mathcal{A}_{J,D}, \mathcal{J}_{J,D})$ is a basic topology and $\mathcal{A}_{I,C}U \subseteq \mathcal{A}_{J,D}U$ for every U, then $(S, \mathcal{A}_{I,C}, \mathcal{J}_{J,D})$ is a basic topology too.

So, from the minimalist perspective, only the assumption of strong principles can allow one to make precisely that choice, out of the many equally possible, in which closed subsets can be uniquely determined by open ones. But even strong principles cannot hide the fact that the algebraic properties of closed subsets, and their link with open subsets, cannot be obtained from those of open subsets alone. So when no strong principles are available *or* when an intuitionistic algebraic treatment is required, the only option is to deal with closed and with open subsets independently. In particular, one has to keep a primitive expression both for interior and for closure.

After basic pairs and basic topologies, the independent treatment of the notions of open and closed is systematically kept in all subsequent definitions, beginning with that of their morphisms. Concerning this, by analogy one would expect two distinct and independent conditions, one saying that open subsets and the other that closed subsets are preserved. It is interesting to observe that this actually happens only if one defines morphisms to be continuous relations, rather than functions as usual, because with functions the existential and universal images of subsets coincide, and hence so would do also the two conditions. As concrete spaces and formal topologies are obtained from basic pairs and basic topologies, respectively, by adding a single assumptions of convergence, their morphisms are obtained by adding only its preservation. So topology in the proper sense, both pointwise and pointfree, becomes a special case.

The algebraic character of the operators $\mathcal{A}$ and $\mathcal{J}$ in a basic topology is clear, but still the definition is not purely algebraic since their domain is the collections of subsets of a set, which depends on logic. An abstract algebraic treatment is obtained by replacing the collection of subsets with its algebraic axiomatization. To be able to express compatibility between $\mathcal{A}$ and $\mathcal{J}$, this must count also a primitive $\asymp$ which corresponds to overlap $\between$ between subsets, at the same way as the common notion of partial order $\leq$ corresponds to inclusion $\subseteq$. The axiomatization of properties of operation of subsets with respect to inclusion, which is a universal notion, is the standard one and gives the structure known as Heyting algebra, or locale; by adding an axiomatization of the properties of overlap, which is an existential notion, one obtains a new structure, which has been called an *overlap algebra* (see (Sambin, 201x, Ch. 9) for details).

The algebraic soundness of definitions is confirmed by the fact that they extend smoothly to the general framework of overlap algebras. The result is a purely algebraic formulation of the basic picture, and hence also of topology (Sambin, 201x, Ch. 10). Here the expressive power provided by the extra primitives (for overlap and for closure) is essential. In particular, to be able to put basic pairs and continuous relations in algebraic terms, one must pass through a characterization of relations as pairs of adjunctions linked by a property, called symmetry, which can be expressed algebraically only if overlap is available.

Owing to the methodological choice of compatibility of the minimalist foundation with all others, no ideological commitment is required to appreciate the novelties introduced by the basic picture. In practice the new structures, reached by means of the minimalist attitude, remain new and intelligible, and can be of interest, also for mathematicians with different principles or without active interest in foundational questions.

So the choice of weaker foundational principles is compensated for by richer, purely mathematical structures, which are otherwise difficult to discern. Specifically, one can observe that when the independent treatment of the notions of open and closed is carried out systematically, some notions emerge (overlap, closed, positivity, coinduction) which are somehow dual to more standard ones (inclusion, open, cover, induction, respectively) and equally important. We have seen that one is forced to introduce them somehow when no strong principles are available by which the information they carry is usually reconstructed. In other words, they are the mathematical substitutes for the foundational principles by which they are usually defined or determined.

These appear as the first steps of an investigation searching for a positive and explicit mathematical treatment of that kind of information, mostly of existential form, which otherwise is usually dealt with by negation, or surreptitiously reconstructed using strong principles, or simply left in the dark. The basic picture shows that a systematic exploration of this dark side of the mathematical planet is possible, and provides us with the first tools for undertaking it. This could become its most lasting conceptual contribution.

One can already see some applications confirming this idea outside topology. In computer science, the positivity relation offers a mathematical modelling of the concept of liveness, as cover does with that of safety (Hancock and Hyvernat, 2006. In logic, the positivity relation provides with a complete semantics for a primitive constructive notion of satisfiability which is introduced coinductively (Ciraulo, 2007) and goes along with the well-known fact that the cover gives a complete semantics for derivability (Sambin, 1995).

4.3 Pluralism as a Source of Richness

Since the minimalist foundation is weaker than all others, one could not hope to find any real novelty while remaining inside existing mathematics or by adhering to aseptic metamathematical considerations. Something new has been found only by adopting the minimalist attitude with heart and soul and by putting it to work directly on "reality," that is on such basic ingredients as sets, relations, etc.

It is fidelity to an effective conception of sets which has forced us to keep on stage the information on the base, given by the set S, which a posteriori can be seen as the only road leading to the discovery of duality and symmetry in topology, from which all the rest follows. Forgetting the base and defining closed subsets impredicatively in terms of open subsets makes it is very hard to conceive that closure is not uniquely determined by interior. A sign of this is that the link between closure and interior expressed by their compatibility was not studied. And only by relaxing the requirement of an explicit computational meaning for all statements is one free to see and play with pure algebraic structures.

This explains in particular why an aim has not been that of recasting the given definition of topological space in the most constructive form possible, nor of developing only a predicative version of locales.

In my view, the aim is to create new territories of thought, and the challenge is to do it while keeping the customary rigour of mathematics. It is the lack of means which compels one to be creative. Whatever novelty there is in the basic picture, it has been possible to conceive it *only* because I was forced to it, since no other choice was possible. In fact, the restrictions of the minimalist foundation correspond faithfully to my own limits: I really believe that only what is valid in the minimalist foundation is unquestionable and fully reliable, that is objectively true, and only with that do I feel safely at home. I hope I have shown that it is coherent, and that one can do mathematics with it.

But I observe daily that others feel at home with other principles, which for them are objective truths. Hence the mathematics they develop, which looks to me to be hypothetical reasoning based on specific assumptions, is felt by them to be objectively true, and sometimes also as the only possible reasoning.

This is rightly so. A foundation is a choice of values and hence a number of principles by which one can give a structure to one's mathematical perception of the world. There is here a striking similarity with the embracing of a religion (or even with membership to a specific culture): in both cases, in order to play its role, it must be felt as an unquestionable truth, that is, at a certain level of conscience one must "forget" that its principles are assumptions. One must not only believe in them, but actually let oneself go into them completely to be able to use them effectively, quickly and deeply, enough for instance to create new mathematics. They must simply become a part of one's way of automatic thinking, and thus at the same level as objective reality.

At the same time, a correct perception of facts should bring one to acknowledge a bewildering triviality, which is often overlooked or put aside: different people follow different foundations (or religions) and hence have different notions of objective truth. Assuming a static view, this is a plain paradox, and out of absolute truth one is inclined to see only a dangerous relativism (which Frege called psychologism). From a dynamic perspective, objective truth is a process; so one is free to believe in something as objectively true and still, at a different level of consciousness, recognize that other people have a different creed, without having to fight them for being wrong. Since there is no absolute principle on which to base an agreement, pluralism is unavoidable and tolerance acquires crucial importance.

To reach tolerance, one has to become aware of the assumptions underlying one's "objective truth" and compare them with those of others. It is easier, or even possible in the first place, to realize that a certain principle (e.g. Euclid's parallel postulate) is actually an assumption if one can see that one is left with something meaningful also in its absence (e.g. non-Euclidean geometry). Here the minimalist foundation may serve the purpose, by showing that some principles are not indispensable and that one can also develop reasonable mathematics without them.

Different foundations are possible, because of different aims and conceptions about what is relevant; in each of them one has some specific mathematics, and hence pluralism, when backed by tolerance, is a source of richness. The most interesting and vital aspect of constructivism is the deep interaction it shows between mathematical practice and reflections on foundations, which therefore cannot be left only to a purely theoretical investigation. Since various options are possible, every mathematician contributes to determining the course of events by making a choice. Though the minimalist foundation rests on fewer assumptions (and thus one could say that it is "more objective"), its aim is not to reach universal agreement by a void compromise. The only "objective truth" one can hope for is that of a higher level of awareness, obtained by adding a further step to abstraction. In my opinion, this is the deepest spirit of constructivism. Following it confers several benefits, for mathematics and for contemporary culture in general.

Bibliography

Battilotti, G. and Sambin, G. (2006). Pretopologies and a uniform presentation of sup-lattices, quantales and frames. *Annals of Pure and Applied Logic*, 137:30–61.

Bell, J. L. (1988). *Toposes and Local Set Theories: An introduction*, volume 14 of *Oxford Logic Guides*. Clarendon, Oxford.

Bishop, E. (1967). *Foundations of Constructive Analysis*. McGraw-Hill, New York, NY.

Bishop, E. (1970). Mathematics as a numerical language. In Kino, A., Myhill, J., and Vesley, R., editors, *Intuitionism and Proof Theory*, pages 53–71. North-Holland, Amsterdam.

Bishop, E. and Bridges, D. (1985). *Constructive Analysis*. Springer, Berlin.

Ciraulo, F. (2007). *Constructive Satisfiability*. PhD thesis, Università di Palermo.

Coquand, T., Sambin, G., Smith, J., and Valentini, S. (2003). Inductively generated formal topologies. *Annals of Pure and Applied Logic*, 104:71–106.

Dummett, M. (1977). *Elements of Intuitionism*, volume 2 of *Oxford Logic Guides*. Clarendon, Oxford.

Hancock, P. and Hyvernat, P. (2006). Programming interfaces and basic topology. *Annals of Pure and Applied Logic*, 137:189–239.

Hyland, J. M. E. (1982). The effective topos. In Troesltra, A. S. and van Dalen, D., editors, *The L.E.J. Brouwer Centenary Symposium*, Studies in Logic and the Foundations of Mathematics, pages 165–216. North-Holland, Amsterdam.

Maietti, M. E. (1999). About effective quotients in constructive type theory. In Altenkirch, T., Naraschewski, W., and Reus, B., editors, *Types for Proofs and Programs. TYPES '98*, volume 1657 of *Lecture Notes in Computer Science*, pages 164–178. Springer, Berlin.

Maietti, M. E. (2005). Modular correspondence between dependent type theories and categories including pretopoi and topoi. *Mathematical Structures in Computer Science*, 15:1089–1149.

Maietti, M. E. (2009). A minimalist two-level foundation for constructive mathematics. *Annals of Pure and Applied Logic*, 160:319–354.

Maietti, M. E. and Sambin, G. (2005). Toward a minimalist foundation for constructive mathematics. In Crosilla, L. and Schuster, P., editors, *From Sets and Types to Topology and Analysis: Towards Practicable Foundations for Constructive Mathematics*, volume 48 of *Oxford Logic Guides*, pages 91–114. Clarendon, Oxford.

Maietti, M. E. and Sambin, G. (201x). Choice sequences in a minimalist foundation. In preparation.

Maietti, M. E. and Valentini, S. (1999). Can you add power-sets to Martin-Löf's intuitionistic set theory? *Mathematical Logic Quarterly*, 45:521–532.

Martin-Löf, P. (1970). *Notes on Constructive Mathematics*. Almqvist & Wiksell, Stockholm.

Martin-Löf, P. (1984). *Intuitionistic Type Theory: Notes by G. Sambin of a Series of Lectures Given in Padua, June 1980*. Bibliopolis, Napoli.

Martin-Löf, P. (2006). 100 years of Zermelo's axiom of choice: what was the problem with it? *The Computer Journal*, 49:10–37.

Martin-Löf, P. and Sambin, G. (in press). Generating positivity by coinduction. In *The Basic Picture: Structures for Constructive Topology*. Oxford University Press, Oxford, to appear.

Nordström, B. and Petersson, K. (1990). *Programming in Martin-Löf's Type Theory: An introduction*. Oxford University Press, Oxford.

Sambin, G. (1987). Intuitionistic formal spaces—a first communication. In Skordev, D., editor, *Mathematical Logic and Its Applications*, pages 187–204. Plenum, New York, NY.

Sambin, G. (1995). Pretopologies and completeness proofs. *Journal of Symbolic Logic*, 60:861–878.

Sambin, G. (2002). Steps towards a dynamic constructivism. In Gärdenfors, P., Kijania-Placek, K., and Wolenski, J., editors, *In the Scope of Logic: Methodology and Philosophy of Science, vol. 1*, number 315 in Synthese Library, pages 263–286. Kluwer, Dordrecht.

Sambin, G. (2003). Some points in formal topology. *Theoretical Computer Science*, 305:347–408.

Sambin, G. (201x). *The Basic Picture: Structures for Constructive Topology*. Oxford University Press, Oxford. to appear.

Sambin, G. and Valentini, S. (1998). Building up a toolbox for Martin-Löf's type theory: subset theory. In Sambin, G. and Smith, J., editors, *Twenty-five Years of Constructive Type Theory*, number 36 in Oxford Logic Guides, pages 221–244. Clarendon. Proceedings of a Congress held in Venice, October 1995.

Troelstra, A. and van Dalen, D. (1988). *Constructivism in Mathematics: An Introduction*, volume 2 of *Studies in Logic and the Foundations of Mathematics*. North-Holland, Amsterdam.

Vickers, S. (2006). Compactness in locales and in formal topology. *Annals of Pure and Applied Logic*, 137:413–438.

Chapter 5
The Municipal By-Laws of Thought

David DeVidi

1 JLB

John Bell's arrival at the University of Western Ontario was a turning point in my life. At the time I was a PhD student, a mere few overdue course credits away from beginning a dissertation on Carnap's *Logical Syntax of Language*. What better way to make up a couple of those credits than by taking the set theory course to be taught by the hotshot new professor the philosophy department had recruited from England, thanks in part to Margaret Thatcher's campaign to drive intellectual talent out of the UK? This turned out to be a doubly lucky choice for me. First, being new to the department, John was on the look-out for PhD students who seemed able to do their sums, so the class was a chance to catch his eye. More significantly, the class was a revelation to me. I didn't just learn a lot of set theory but was exposed to a whole raft of (for me) new philosophical issues, and approaches to traditional philosophical problems, that struck me as deeper, subtler and more interesting than anything I'd seen before—no small statement, since I was fascinated by the Carnap issues I was working on until then. The class also gave me my first contact with whole fields that had only been names to me (if even that) before John's arrival, e.g., category theory, type theory, and topos theory. I quickly signed on as John's student even though it meant spending another couple of years in grad school, essentially learning the things I would have learned doing course work had I intended to write on logic from the start.

John and I discovered many respects in which we are kindred spirits, though quite different people. An example that amused us both: one late evening while sharing refreshments we discovered that at about age 15 each of us had undergone a certain crisis. Figuring out that while our interests and talents suggested we might be best suited to physics, neither of us at 15 could imagine a career in physics that didn't somehow serve the interests of the Pentagon, so we steered into something that seemed less harmful. Which makes us seem alike and perhaps suggests similar

D. DeVidi (✉)
Professor of Philosophy, University of Waterloo, Waterloo, ON, Canada
e-mail: ddevidi@uwaterloo.ca

D. DeVidi et al. (eds.), *Logic, Mathematics, Philosophy: Vintage Enthusiasms*,
The Western Ontario Series in Philosophy of Science 75,
DOI 10.1007/978-94-007-0214-1_5,

life experiences, until one accounts for the fact that John had these thoughts as an Oxford undergraduate and steered towards pure maths, while I had them as a grade 10 student in Saskatchewan and steered temporarily towards journalism.

During my years under John's supervision I didn't come to share all his enthusiasms, but the one that I hope rubbed off on me, at least a bit, is his enthusiasm for enthusiasms. Having a teacher and mentor who inspires interest in the work he cares about—his own, and that by those he admires—by his willingness to wear his own enthusiasm for it on his sleeve, and for whom seeming cool matters so little, was liberating for someone who found the sometimes stuffy atmosphere of professional philosophy difficult. But what I have most hoped to incorporate in my own teaching is John's willingness to share the enthusiasms of others. It is no accident that many of John's students have gone on to successful careers chasing things distant from John's own interests or advocating views John probably thinks wrong, and perhaps nutty.

For this collection, I hope John will find it apt that I'm offering something on logical pluralism, a topic that marries my longstanding interest in matters Carnapian to themes learned from him.

2 The Problem

Whether logical pluralism is correct is a philosophical rather than mathematical question, most straightforwardly stated "Is there more than one correct logic?" But philosophers have in recent times been rather cautious about telling practitioners in other fields that they don't know what they're doing. This caution was learned the hard way, of course. And there is a whole range of systems *called* logics investigated by the strange logician down the hall in the philosophy department, or by the higher-paid computer scientists across campus, or even by engineers or physicists. For most philosophers, these probably fall on a scale of increasing obscurity, beginning with familiar tools of the philosophical trade like Aristotelean syllogism, classical propositional and predicate logic, and normal modal logics, running perhaps through relevance logics, intuitionistic logic, non-normal modal logics, fuzzy logics, and belief revision logics, on into things many professional philosophers will never have heard of (dynamic logic? linear logic? dialethic logics? multi-agent epistemic logics? *other* sorts of constructive logic?). However unfamiliar one is with the specifics, and even if one harbours suspicions about some of these systems, it's professional good manners to think that if people are getting grants via the usual peer review processes to investigate such systems they are mostly at least highly promising and at least some of them must be, in some good sense, correct.

But the answer to the question of logical pluralism is at least not *obviously* yes, so the straightforward wording doesn't get to the heart of the issue. The philosophical question is better stated as "Is there an *interesting* and *non-trivial* sense in which there is more than one correct logic?" While the meaning of these adjectives in this context is by no means transparent, and I am unable to nail them down, much of the debate about pluralism will turn on the question of whether a sort of pluralism that is probably true also meets the tests of non-triviality or of being interesting. I will be content for present purposes with giving a brief gloss for each.

A trivial sense of pluralism would be one, for instance, that is based on an equivocation. We are not tempted to say that "there's more than one logic" on the ground that in non-philosophical contexts it is common to describe any sort of bad thinking as "illogical"—e.g., parents driving their kids to school to prevent their being abducted by strangers when stuffing them into cars and hurtling them along roadways at high speed creates greater risks of harm to them and their schoolmates than is posed by stranger abduction. Other trivial senses of pluralism would arise from cases where philosophers might legitimately dispute the use of the word "logic." Some object to the subject matter taught in critical thinking courses being called "informal logic" on the grounds that there's no such thing. Some philosophers dislike the practice in many logic textbooks of first distinguishing between *deductive* and *inductive* logic, usually as a prelude to remarking that inductive logic will not be covered in the book, on the grounds, once again, that there's no such thing—this time, no such thing as inductive logic. I do not intend to suggest that these latter two are trivial disputes. However, I do say that even if there are such things as informal logic or inductive logic these would give us pluralism about logic only in a trivial sense. The interesting question about logical pluralism is more specific than just whether there's more than one sort of good thinking.

It's even trickier to specify what is supposed to be ruled out by the restriction to *interesting* sorts of pluralism, but it's also clear that this is the condition that rules out many of the most familiar examples of distinct logics from giving an easy answer to the question of pluralism. Classical propositional and predicate logic are distinct systems, after all, but nobody wants to say that they show that pluralism is true. Why not? The first being essentially a sub-system of the second, it essentially differs from the first only in being more restricted. For the present inquiry, this cannot be an interesting difference.

Why, then, do we consider propositional logic at all? Among the reasons is that it is a system with some interesting properties—expressive completeness and decidability, for instance—that is also strong enough to represents a significant fragment of the valid reasoning we do. The reason we go on to study the more powerful system of classical predicate logic is that there is valid reasoning that is not captured in propositional logic that the stronger system allows us to capture. This suggests a more general line of explanation for why this example is not an interesting pluralism, one that suggests a defence of logical monism: what makes the propositional/predicate plurality uninteresting is that one is a subsystem of the other, and one "logic" being a subsystem of another means that we are not dealing with an genuine plurality. But, a monist might argue, each of these "logics" is merely a subsystem of a single larger system. Indeed, all the correct logical systems are correct precisely in being sub-systems of the single, correct system of logic, namely the system that includes all and only deductively correct inferences. Each of the systems we actually work with is limited in some ways to make it useful for some practical purposes: when dealing with truth-functionally valid arguments, the extra machinery of predicate logic merely makes it harder to see what is essential to the case at hand; for many purposes, the machinery of temporal or modal reasoning is unnecessary, so we work with standard predicate logic rather than modal or temporal versions, but similarly for much modal or temporal thinking the quantifiers are not needed, so we work with propositional temporal or modal logics. But all these systems—propositional, modal, temporal, predicate, etc.—insofar as they are correct, are merely fragments of the single correct logic. So there are many correct logics, but only in the uninteresting sense that for each we limit attention to a distinct fragment of the too-complex-to-work-with, and perhaps even too-complex-to-state, system of correct inferences.

The goal in this paper, then, will be to investigate the question of whether there is an interesting and non-trivial sense of pluralism that can be plausibly maintained. I shall begin by lodging some complaints against what I call the obvious approach (intending no disparagement by this terminology, since it is one I have at least implicitly advocated elsewhere since it is presumed in much of the discussion in (Bell et al., 2001)). In effect, I suggest that there is a tension between the jobs of showing how different systems of logic call all be correct and explaining why they are interestingly distinct. The obvious approach, at least in the most important presentation of it to date (Beall and Restall, 2006), explains mutual correctness in a way that robs the distinction of interest. Predictably, this is a prelude to my going on to suggest a different line of defense of pluralism. That defense begins with a detour through the question "What makes something logic?" and finds different logics, at least arguably, arise from distinct, legitimate answers to that question—while not making the distinctness a matter of equivocation on the meaning of the word "logic," and so running afoul of the condition that the plurality not be trivial. I do not pretend to adequately defend pluralism here, but do hope to suggest a line of defence that establishes the *plausibility* of logical pluralism.

3 The Obvious Approach?

The title of (Bell et al., 2001) is *Logical Options*, and the goal of the book is to introduce a range of distinct, philosophically significant logical systems. The book takes a "models first" approach. That is, we began with characterizations of certain inter-related and central logical concepts. An argument is *valid* precisely when every case in which its premises are all true is also a case in which its conclusion is true; a set of sentences is *consistent* precisely when there is a case in which all the sentences in the set are true; and so on. The different logical systems can then be seen as arising from different answers to the question "What constitutes a case?"

We claimed no special novelty in introducing the various logical systems this way. It is, indeed, the obvious way to do so. JC Beall and Greg Restall have gone further than merely assuming that one can elucidate a variety of logical systems in this way to using it as a strategy for defending the claim that logical pluralism is correct (Beall and Restall, 2000, 2006). They introduce the

Generalized Tarski Thesis (GTT): An argument is *valid*$_x$ iff in every case$_x$ in which the premises are true, so is the conclusion.

Logical pluralism then becomes the claim that there is more than one way to specify "the cases$_x$."

Of course, this can't be the whole story. The wording of the Generalized Tarski Thesis requires that cases be the sorts of things in which statements can be true or not, but by itself doesn't tell us any more. The difficulty for such a view is going to come when it is time to specify what makes a suitable class of cases for determining a logic. Presumably, for instance, we don't want to allow singleton classes, for instance the class with just the actual world in it. For then every true sentence "follows in the real world" from every set of true sentences, and all sentences follow from any set that includes a false sentence. The room for mischief here is limited only by the one's ingenuity. It might be valid-in-no-umbrella-thinking to infer that one's head will get wet if one walks in the rain, but surely that doesn't mean that thinking about life without umbrellas is a sort of logic, at least in any sense relevant to debates about logical pluralism.

In the book-length version of their argument (Beall and Restall, 2006), this problem is addressed by distinguishing the admissible from the inadmissible specifications of classes of cases. To be admissible, a class of cases$_x$ must meet three conditions [pp. 14–23].

1. The class of cases must determine a suitable *necessary link* between reasons and conclusions in correct inferences, i.e., a sense in which it *cannot* be the case that all the premises be true while the conclusion is false.
2. The class of cases must make explicable how logic is *normative*. "In an important sense, if an argument is valid, then you somehow go *wrong* if you accept the premises but reject the conclusion" (Beall and Restall, 2006, p. 16). They offer little in the way of positive characterization of this important condition, contenting themselves with noting that the paradox of the preface implies that it's too strong to say that one is *irrational* if one accepts the premises and not the conclusion.
3. The resulting notion of validity must be *formal*, i.e., having to do with the form rather than the content of arguments, in some suitable sense. They distinguish three options for this notion of formality: 1-Formality means not depending on propositional content of any particular type; 2-Formality means not depending on the terms involved referring to any particular category of objects; 3-Formality means not depending on the semantic content of the claims involved.

The guiding idea is that these conditions describe the *core* of our concept of logic. Our language is settled enough to determine that logic has certain features (i.e., the ones described by this list), but, as with many of our concepts, this core does not settle all questions we may, with changing circumstances, be led to ask; it's an ordinary language concept, and so should not be expected to correspond exactly to any precisely specified notion. Mathematical investigation of such notions involves their *precisification* when we suggest a technical replacement for the ordinary concept that can be subjected to rigorous investigation. Sometimes, at least arguably, there is only one reasonable precisification (at least, up to equivalence), as for example with the notion of *computable function*; for other concepts, there can be more than one such precisification, for instance with the notion of *necessary truth* (which can be precisified to metaphysical necessity, physical necessity, historical necessity, and perhaps others). Beall and Restall's argument comes down to a claim that in this respect the concept of logical consequence is more like the concept of necessary truth than that of computability. A proper precisification must respect the core of the ordinary concept, but beyond that the question is not which precisification is *correct*, but *which ones work*. And what ultimately makes pluralism correct is that different precisifications work for different purposes.

> Each such precisification purports to incorporate the core features involved in the use of "follows from" or "logical consequence" ... however, none do such a good job that it renders the others useless or otherwise *un*deserving of the role. (Beall and Restall, 2006, p. 28)

Beall and Restall are careful to distinguish their pluralism from that of the most famous logical pluralist of them all. To do so, they give a rather unsympathetic reading to Carnap, saddling him with a pluralism according to which plurality can only exist "between" rather than "within" languages (equating, as he does in *Logical Syntax*, for instance, "building a logic" with giving the rules for a formalized language). They then argue that by taking distinct formal languages not as distinct languages but instead to be differing "account[s] of the *form* of claims expressed in a natural language such as English" one can explain the possibility of a plurality within a language such as English (Beall and Restall, 2006, pp. 78–79).[1] The requirement that any acceptable logic must be a precisification that respects the core of the ordinary concept contrasts, at least, with the official version of tolerance announced in *Logical Syntax*, according to which, famously, "*In logic there are no morals*. Everyone is at liberty to build his own logic ... as he wishes" (Carnap, 1937, § 17). But they share with Carnap the view where the question of correctness gives out, what matters is the practical value of the proposed system—eventually, candidacy for the status of "a logic" depends on usefulness for some purpose.

4 An Uninteresting Plurality?

Each of the conditions Beall and Restall include in "the core" of logic is important. Indeed, ideas akin to some of these will play a role in my own attempt to sketch a route to an interesting sort of pluralism below. However, the use to which they put the conditions seems to me misdirected. In short, I think this approach produces an uninteresting pluralism, or at least, a pluralism less interesting than might be

---

[1] Beall and Restall (2006) provide only a summary of a view defended in (Restall, 2002), so a critique would involve a significant digression. I will simply note that I do not intend to endorse this as an accurate reading of Carnap's pluralism, only to report what Beall and Restall say.

available with a slightly different approach. For only a very limited sort of disagreement is allowed for in Beall and Restall's version of pluralism. In their laudable attempt to make clear the sense in which various systems can all be correct, they have sold short the degree to which they can be genuinely distinct.

I will illustrate the problem by considering the way Beall and Restall defend the claim that both classical and constructive consequence relations are correct.[2] This will seem problematic to many because, famously, in various constructive mathematical systems it is possible to prove things that can be refuted in classical mathematics, or so it seems; for instance, Brouwer offered a proof that all functions are continuous, a claim that is obviously refutable in classical mathematics. The response of Beall and Restall to such cases is to rule out certain sorts of constructive mathematics, opting for "deference to a certain important tradition in constructive mathematics .... This brand of constructive mathematics is explicitly designed to be consistent with classical mathematics" (Beall and Restall, 2006, pp. 117–118). This is the tradition of Errett Bishop, Douglas Bridges, Fred Richman and others. They quote Richman as follows:

> It is a common misconception that intuitionistic mathematics deals with a special class of mathematical objects that are, in some sense, constructive .... But when an intuitionist does group theory, he is developing a constructive theory of groups, not a theory of constructive groups. (Beall and Restall, 2006, p. 26)

As pluralism, this strikes me as weak tea. In this sense, almost any mathematician will count as a pluralist, because all that is required is that one prefer constructive proofs when they are available. Even hardened classical mathematicians are likely to at least pay lip service to this idea, allowing that constructive proofs often contain useful information not supplied by a non-constructive proof. But for Beall and Restall,

> The constructive mathematician who utilises classical reasoning, but who also notes when she departs the strictures of the constructive high road to the broad and comfortable classical low road, is a perfect example of a logical pluralist at work. (Beall and Restall, 2006, p. 126)

It is not clear that this is an *interesting* sort of pluralism, in the sense indicated above. By restricting the range of appropriate application of constructive reasoning in mathematics as they do, Beall and Restall make the relationship between the two systems rather akin to that between propositional and predicate logic—one is merely a subsystem of the other that will never conflict with it. So is the view just that the constructivist refuses, for what are ultimately misguided reasons, to apply indirect proofs, one of DeMorgan's laws, and so on, but since those are misguided reasons

---

[2] A similar argument can be made, I think, for the other star example from the book, the simultaneous correctness of classical and relevant consequence relations. "Disjunctive syllogism is valid when GTT's cases$_x$ are taken to be possible worlds, and it is *in*valid when those cases$_x$ are taken to be situations. That is that. To whether Disjunctive Syllogism is valid, the classical and relevant accounts give different but not rival answers" (Beall and Restall, 2006, p. 56). While there are differences of detail between what I would say about the examples, these differences do not warrant covering essentially the same ground twice here.

he might well have decided not to use some other selection of classically correct principles?

This is not quite the whole story. What originally motivated the constructivist to regard certain principles as suspect isn't the issue; what matters is that they landed on a useful distinction. Not every rejection of a subset of classical principles would result in a *useful* distinction. Beall and Restall argue that this sort of pluralism is, for instance, "richer than" the scruples reflected in the practice of classical set theorist who carefully note when their proofs depend on the axiom of choice in some form. It's not mere scruples about possibly dodgy principles that are in evidence, but an attitude of finding "the distinction between constructively valid and invalid arguments important, that is, to take the constructive counter-examples as marking an important distinction" (Beall and Restall, 2006, p. 126). A better analogy to classical set theory is one where we suppose two universes of sets, only one governed by choice—restricted proofs establish truths that hold in both universes, while proofs using AC establish only that the truth holds in one universe. The constructivists' attitude is one of finding the counterexamples in the non-choice universe interesting, even if incapable of showing that some claim does not hold for the choice universe.

I confess to not finding the difference between the two-universes analogy and the scrupulous documenting of use of AC completely clear. But given the "models first" approach employed in their presentation, putting the key point in terms of regarding constructive counter-examples as marking an important distinction is apt. For regarding constructive logic as a sub-system of classical, and supposing (to keep this discussion simple) that things are logically possible when not refutable, there are *more logical possibilities* from a constructive than from a classical point of view, and the extra possibilities can serve as counter-examples and so be of interest.

But this allows us to put the worry about this approach in stark terms. For what Beall and Restall do is exactly to rule out the most interesting counter-examples available to a constructivist on the grounds of their incompatibility with classical mathematics—they rule them out because they are *classically* refutable, and so allow classical (hence, bivalent) reasoning to arbitrate the possible. Many constructivists would emphatically disagree with Richman's characterization of intuitionistic mathematics: it is, for instance, the development of a constructive continuum that they have in mind, rather than the same continuum the classical mathematician works with, but reasoned about constructively. Hence the pleasure some of them take in the discovery of smooth models in which there are non-zero numbers ε small enough that $\varepsilon^2 = 0$ (Bell, 1998). Or spaces in which all functions are continuous. Or, to take a more purely logical example, cases where $\neg\forall x(Ax \vee \neg Ax)$ is true.

What is wrong with the position Beall and Restall wind up defending is now easily stated. What makes constructive reasoning interestingly distinct from classical is, as they note, that constructive counter-examples are interesting and important, but are not available to a purely classically-minded mathematician. But their way of accounting for how classical and intuitionistic logic can both be correct rules out precisely the most interesting and important cases that count by constructive but not classical lights as logical possibilities. What makes it interesting to claim that classical and constructive logic are both correct is that these are systems that *disagree*.[3] Of course, for the pluralist claim to be correct, there is some explaining to do about how these seeming rivals can both be correct. But to de-fang the disagreement entirely is to remove the interest of the claim. Is there a version of the pluralist claim that leaves the disagreement intact?

5 What Makes It Logic?

Let us begin anew, looking again at the GTT, but this time as an answer to a different question: what makes a principle or operator *logical* rather than something else (e.g., mathematical or physical)? For the GTT encodes what is today the standard answer to that question.

Logic is, among other things, supposed to be the machine that hums in the background when mathematics is done. We define a *semi-group* to be a structure that includes a set together with an associative binary operation (a "multiplication"), a *monoid* to be a semi-group with a neutral element (call it 1), and a *group* as a monoid where every element has an inverse. *Logic* is what you use to prove things like "in a monoid, neutral elements are unique," and you do that by showing that (a suitable formalized version of) the claim "neutral elements are unique" *follows from* the postulates for being a monoid, i.e., suitable formalized versions of the claims involved in the definition just stated in English.

This is a very simple example of a familiar practice. To define a *kind* of mathematical structures, we need only specify the relevant *non-logical* stuff—the non-logical vocabulary and the non-logical axioms. Logic then helps us discover what is true in *all the structures of that kind*. And, supposing we've got a suitable model theory knocking around, we can find out what doesn't follow, too, by finding counter-examples.

Very nice and very familiar. But what distinguishes the *logical* from the *non-logical* here? What makes 1 non-logical while & is logical? Why is the law of non-contradiction a logical principle, while the associativity of binary operators is not? The answer is supposed to be straightforward. Not all binary operations are associative, so you need an axiom to guarantee that your designated operation is associative if it's semi-groups you want to talk about, but the law of non-contradiction *always* holds. Similarly, the rules governing conjunction hold in every structure, while not every structure has an element suitable for naming by 1 in it. So, the short answer

[3] And that various flavours of constructivism disagree with each other. In this connection, it is interesting that Beall and Restall choose the Axiom of Choice as their example. For that axiom, and indeed other choice principles, have a disputed and complicated status within constructive logic, one that John Bell has done as much to illuminate as anyone. See (Bell, 1993, 2006, 2009; Maietti and Valentini, 1999; DeVidi, 2004, 2006). By advocates of some constructive systems AC is regarded as a principle of logic that follows from the constructivist account of the existential quantifier; but in other systems of constructive reasoning it implies the law of excluded middle, and so all of classical logic.

is: non-logical principles hold in some structures but not others, while *logical* principles hold in *every* structure.

Which of course is not really an answer yet, since it raises an obvious question: what counts as a *structure*? There are various ways to describe structures in which $\neg\forall x(Px \vee \neg Px)$ is *true*, for instance either Kripke frames or topological structures for intuitionistic logic; other sorts of structures show disjunctive syllogism, or even the law of non-contradiction, fail to be true. Which is probably a long way from what we want to be discussing when introducing simple axiom systems in a logic class, but this line of reasoning seems to be pushing us in the direction of having to say that logic is a pretty thin system indeed—if everything anyone wants to call a structure is a structure, are there *any* principles that survive?

Of course, the correct response is: "those are not structures in the relevant sense; that is, those are not what *Tarski* meant when he defined mathematical structures." Which is true. And the relevant fact here is that in a Tarski structure conjunctions and universal quantifiers are interpreted by suitable intersections, disjunctions by unions, and negations by complements, which is manifestly not what is done these other so-called models.

But we mustn't be under any illusion about the order of definition here. One can get classical logic out of a definition of Tarski structures because one is taking structures to be defined in *classical* set theory (which guarantees, for instance, the existence of the relevant complements to interpret the negations). Since, at least close enough for government work, *classical set theory* is the theory that results by employing the axioms of set theory in a classical logic rather than some other, if we try to define logical principles as "those true in any structure" and by this understand "true in the Tarski structures" we're begging the question in favour of classical logic.

To summarize: One standard view is that logical principles are principles which hold in every structure. But for this to be any good as an answer to our question, we'd need some non-question begging way of specifying what counts as a structure—that is, a specification independent of a choice of underlying logic. Add the suggestion that there might be more than one principled specification and you have GTT.

But why does the idea of "correct in all cases" strike us as a reasonable way to capture the notion of logicality in the first place? It connects closely to some ideas that are clearly closely related to those associated with logical correctness.

First, there is the view that logic is *topic neutral*, a view naturally associated with Frege.[4] While there are some statements and patterns of reasoning that are correct in certain domains but not in others, this fact is itself enough to show that these are not *logical* truths or inferences—it is these grounds Frege uses to relegate geometry to its merely synthetic a priori (and so non-logical) status because of its reliance of spatial reasoning. On the other hand, his belief that arithmetic truths hold "everywhere" is what motivates his arithmetic logicism. And it is on the same grounds that most nowadays would relegate, for instance, mathematical induction to

---

[4] This is a version of the intuition behind the "formality" condition identified as part of the "core" of the concept of logical concept by (2006, p. 21).

non-logical status, recognizing as we do models in which induction fails. A logically correct principle is one that does not depend for its correctness on any special feature of the domain we are considering.

As Beall and Restall point out in their argument for the centrality of GTT, it also connects the notion of logical consequence to necessity. It is a familiar move from "necessarily true" to "true in all possible cases"; logical truths are necessarily true if anything is, and in a valid argument the truth of premises "necessitates" the truth of the conclusion in some sense. One only needs some explanation for why it is legitimate to smear together the notions of *all structures* and *all possible cases*, and a few obvious further steps will lead us to the Tarski Thesis.

But this is not the *only* plausible characterization of what makes a principle *logical*. It will be useful to gather a few into a list.

1. Logical is *topic neutral*: a logical truth is metaphysically neutral, and does not depend for its truth on any presuppositions about what the world is like.
2. Logic has to do with when the truth of some statements necessitate the truth of others: logically correct inferences are *necessarily truth preserving*.
3. Logic is *the science of correct inference*. (This is probably as close to a standard characterization as one is likely to find, in fact.)
4. (Deductive) inference is supposed to be *non-productive*. The point of the claim is that when you apply logical rules to a set of premises, the conclusion can't be something which wasn't already "implicitly contained" in the premises. Sometimes people use the image of "unpacking" the premises to convey this idea.

This list could be extended, but this brief list will serve for present purposes. It is easy enough to suppose that these different characterizations somehow amount to the same thing, or perhaps that some entail others. We have seen an example of how some such arguments can go already: if we assume that the notion of necessity is adequately captured by quantification over "ways things might have been," then we can bring together the ideas of necessary truth preservation and topic neutrality, for an argument that depends on restriction to a particular topic for its correctness will then have some counterexample to its correctness among the "ways" to do with other topics. But such arguments are not usually given, and more rarely still are they scrutinized. And grounds for logical pluralism are to be found in the fact that these characterizations can be pried apart.

6 The Systems Disagree

The goal, recall, is a pluralism which allows for more than one system of logic to be correct while being *interestingly* distinct. The complaint against the sort of pluralism defended by Beall and Restall was that in order to account for mutual correctness, they have eliminated the really interesting differences between the systems they consider. As a next step to avoiding this problem, let us briefly consider the fact that those who first and most emphatically advocated many alternatives to classical logic were strongly of the view that their systems were *incompatible* with classical logic, and not merely another tool in the logician's toolbox. Classical logic, they argue, is *wrongheaded* in some important way. The complaints are profitably regarded as arguing that classical logic fails to satisfy one of the criteria of logicality recently listed.

We will turn presently to some examples. But first, as a useful contrast, let us look at some less fundamental complaints about the classical logic that philosophers teach students in their first formal logic class.

We might complain about a misguided choice in the inevitable trade-off between simplicity and comprehensiveness. Of course, when you're building a system of logic you want it to be correct—if the system says an inference is correct, you'd like to be able to rely on it. But it had better be useful, too, and this requires that it be *simpler* than, for instance, reasoning in ordinary language. Teaching classical predicate logic in a first course in logic reflects a choice to leave out of account certain features of ordinary language argumentation because the simpler logic is useful for many purposes even without extra devices such as temporal, modal, deontic, or metamathematical (e.g., "is provable in Peano arithmetic") operators. But it's always open to us to add in this extra machinery when we need it. Here we have our familiar, not very interesting "plurality" of logics, all intended to capture the notion of "correct inference," but *agreeing* about what the correct inferences are wherever they overlap—they are distinct logics only in the sense of being distinct fragments of the same correct logic. The single correct logic is too complex to be useful for many purposes, hence the proliferation of special-purpose fragments. If this is all there is to logical pluralism, there are presumably a lot of pluralists around.

A different sort of objection to standard presentation of classical logic notes its failure to satisfy the neutrality condition. According to the standard semantics for classical logic, for any name c in our language $\exists x.x = c$ is valid, as is $\exists x.x = x$. So, one might badly reason, "anything we can name exists," and "it's a necessary truth that something exists ... let's call that necessarily existing thing *God*." While I did suffer through one talk some years ago that went along those lines, most everybody recognizes these as false consequences in spite of their validity in classical logic. Most are happy to say that this falsity is the price one pays for the simplification gained by having to consider only non-empty domains, and by having each term in the language refer to an individual which is in the domain in question. This is not a *real* violation of neutrality, but another aspect of the trade-off between usefulness and correctness. It differs from the previous example where *correctness* was not traded away, only comprehensiveness. But it's a trade-off that is well-understood. When necessary, i.e., when it becomes important to take into account the existence of names which don't refer to anything that exists (e.g., "Pegasus") or empty domains ("the class of thoughtful members of the Canadian federal cabinet in 2009"), we can use a free logic. But free logics are something of a pain to work with, so we don't bother with them for most purposes. What sort of pluralism does this suggest? One in which standard logic is not merely overly simple for some purposes, but in which it is, strictly speaking, incorrect. But it *is* correct, in the special case where certain simplifying assumptions are warranted, as they very often are.

The more interesting systems a pluralist might regard as additional to classical logic were first formulated as its *rivals*. They arise out of arguments that standard logic ought to be rejected—it's somehow misguided as an account of logic.

Probably the most famous critique of classical logic is the one due to constructivists. In their original form, due to Brouwer, these arguments were presented with some peculiar solipsistic accoutrements, but the nature of his fundamental complaint, that classical logic depends on unwarranted metaphysical assumptions, is clear enough. Moreover, plausibly, later formulations of essentially the same arguments have divorced them from the more alarming parts of Brouwer's originals. For instance, in his discussions of realism and anti-realism, Michael Dummett has argued that constructive reasoning is a metaphysically neutral system, specifying patterns of reasoning acceptable whether or not realism is true; the principle of bivalence, and hence all of classical logic, is correct for application in a particular domain when and only when realism is the correct view for that domain. Hence, the law of excluded middle, and the other classically but not intuitionistically correct principles, are metaphysical and not logical. (See, among many places, (Dummett, 1993).)

But there is a different route to the view that intuitionistic rather than classical logic is correct to be found in Dummett's work. Dummett (1991) puts the intuition that logic must be *non-productive* at the center of his argument. Roughly, the idea is that the meaning of any proposed logical operator will be determined by the rules governing its correct use; you'll have rules which allow you to *introduce* that operator as the principal operator of a logically complex sentence (i.e., which specify what is needed, canonically, to *prove* the sentence), and rules which allow you to *eliminate* it (i.e., which specify the *immediate inferences* you can make from such a sentence). The requirement Dummett proposes is that the basic introduction and elimination rules must be in *harmony*: that is, if you introduce then eliminate, you ought to be able to get out of the process no claims that were not already provable without that maneuver, but you ought to be able to get out anything inferrable from what was required for the introduction. From this basis, Dummett argues that several classical principles are not, properly speaking, logical ones, whatever we teach in our Introduction to Logic courses. For instance, $\forall x(Px \vee \neg Px)$, $P \vee \neg P$, $(P \to Q) \vee (Q \to P)$, and $\neg(P \,\&\, Q) \to \neg P \vee \neg Q$ are all rejected because they depend for their justification on inference rules of classical logic which, he argues, are not in harmony—in particular, they depend on the classical interpretation of negation.

Another familiar critique is plausibly taken as beginning from the idea that logic is the science of correct inference, and so is in the business of giving a proper account of "follows from." One of the things over which you have to grit your teeth and say "just learn it" when teaching Intro courses in logic is, of course, the fact that in classical logic Q "follows from" $P \,\&\, \neg P$, *regardless of what* Q *is*. Another is the fact that a tautology like $Q \to Q$ "follows from" any premises at all. Certainly those arguments are classically valid, and the corresponding conditional statements are classical tautologies. But is it really right that from *The Russell set both is and is not a member of itself* it follows that *everything is permitted*, or *my dog is named Flipper*? The starting point of relevance logics is in this sort of consideration, in the view that classical logic gets this wrong. Some advocates of relevance logic may nowadays be happy enough with the idea that classical logic gets something else right,[5] but all of them will agree that the notion of *following from*, and so the proper notion of *consequence*, requires a connection of *relevance* between premises and conclusion. And this involves significant revisions of classical logic.

I do not intend necessarily to endorse any or all of these lines of reasoning, only to note two things: first, and obviously, the genesis of these *rival* systems of logic is in critiques of perceived deficiencies of existing systems, and so in what are at least considered by those creating the rival system to be important differences; and secondly, that that it is the *disagreements* that are interesting about the systems.

7 The Systems Are Correct

Of course, the emphasis on the disagreements among various logical systems raises the other difficulty for pluralists, namely how the various systems nevertheless can make a claim to *correctness*.

What I want to suggest is that Beall and Restall have latched onto the wrong model of the relationship between precise concepts and their ordinary language kin. While the concepts of classical consequence and the various constructive and relevant alternatives are no doubt more precise than our ordinary language concept, it is a mistake to think of them as all extending a fixed, consistent core characteristic of our ordinary concept. For if the arguments of the original advocates of the alternative systems are correct, there is no such fixed, consistent core.

The route to pluralism, then, begins with the view that each of the various answers to the question "What makes it logic?" has a plausible claim to being central to our understanding of logic. In the ordinary run of business to which we turn our logical concepts, the various answers are related closely enough that they seem merely alternative ways of phrasing the same answer—one needn't draw these distinctions in a critical thinking class aimed at large groups of undergraduates with little interest in going further in philosophy, for instance. But that impression is *mistaken*, at least if there is substance to the critiques of classical logic described in the previous section. Focus on different aspects of our ordinary concept of logic leads us to different, i.e., *disagreeing* systems; that is, different answers to the question "What makes it logic?" yield different answers to the question "Which principles are logical?" This implies pluralism, provided each of the answers to the question has a legitimate claim to being part of the ordinary concept of logic.

I think there is a further bonus that comes with this version of pluralism. An advocate of this sort of pluralism can acknowledge the importance of Beall and

[5] But not all. See (Read, 2006) for an emphatic example.

Restall's examples of how different logical systems are useful for different purposes. But they can do one better; rather than merely offering this as reason for allowing that all the various systems should count as logic, a pluralist of the suggested sort can use their status as different sorts of logic to *explain* the variety of uses. For one can expect that at least very often there will be a conceptual link between the *job to be done* and a particular answer to the question "What makes it logic?"

For instance, the value of, and need for, relevance logic is easily made clear to students by noting that we regularly reason successfully in the presence of inconsistent information without drawing disastrously unrelated conclusions, and that reasoning in the presence of inconsistency is unavoidable as is made clear by attention to the history of law or scientific theorizing. But if logic is "the science of correct inference," then attention to the practice of inference, and so to the relations between premises and conclusions that warrant inference, can be expected to include constraints of relevance. On the other hand, if the job at hand is metaphysics then logic must encode only principles that are free of metaphysical baggage. If Dummett is right, then avoiding explosion is no longer required but we must turn a jaundiced eye towards the law of excluded middle.

8 Conclusion

I do not pretend that this is an air-tight argument for pluralism. However, I think that pluralism is a more viable view than it is sometimes given credit for being. Beall and Restall have provided us with a stimulating and helpful book-length defence of logical pluralism, but one that ultimately defends a disappointingly thin sort of pluralism, as I hope to have successfully indicated. I hope to have described a route to a more robust sort of pluralism that is both plausible and worth defending. Whether it is ultimately defensible is, of course, a different and more difficult matter.[6]

Bibliography

Beall, J. C. and Restall, G. (2000). Logical pluralism. *Australasian Journal of Philosophy*, 78: 475–493.

Beall, J. C. and Restall, G. (2006). *Logical Pluralism*. Oxford University Press, Oxford.

---

[6] This paper has been knocking around for a long time in one form or another. I first presented it in 2003 to an audience that included John Bell at the University of Western Ontario, and in the intervening years at the Western Canadian Philosophical Association and to the philosophy departments at Concordia University and the University of Waterloo. I thank members of those audiences, and in particular Bill Demopoulos, Bill Harper, Tim Kenyon, Greg Lavers, Jeff Pelletier and Jim Van Evra, for useful comments and questions. The financial support of the Social Sciences and Humanities Research Council of Canada is gratefully acknowledged. The best thing about the paper is surely the title, so I am sorry to have to confess that I owe it to my hilarious colleague Kenyon.

Bell, J. L. (1993). Hilbert's ε-operator and classical logic. *Mathematical Logic Quarterly*, 39: 323–337.

Bell, J. L. (1998). *A Primer of Infinitesimal Analysis*. Cambridge University Press, Cambridge.

Bell, J. L. (2006). Choice principles in intuitionistic set theory. In DeVidi, D. and Kenyon, T., editors, *A Logical Approach to Philosophy*, Western Ontario Series in Philosophy of Science. Springer, Dordrecht.

Bell, J. L. (2009). *The Axiom of Choice*, volume 22 of *Series in Mathematical Logic and Foundations*. College Publications, London.

Bell, J. L., DeVidi, D., and Solomon, G. (2001). *Logical Options*. Broadview Press, Peterborough, ON.

Carnap, R. (1937). *The Logical Syntax of Language*. Routledge and Kegan Paul, London. Transl. by Amethe Smeaton.

DeVidi, D. (2004). Choice principles in constructive logics. *Philosophia Mathematica*, 12: 222–243.

DeVidi, D. (2006). Assertion, proof and the axiom of choice. In DeVidi, D. and Kenyon, T., editors, *A Logical Approach to Philosophy*, Western Ontario Series in Philosophy of Science, pages 45–76. Springer, Dordrecht.

Dummett, M. (1991). *The Logical Basis of Metaphysics*. Duckworth, London.

Dummett, M. (1993). Realism and anti-realism. In *The Seas of Language*, pages 462–478. Oxford University Press, Oxford.

Maietti, M. E. and Valentini, S. (1999). Can you add power sets to Martin-Löf's intuitionistic set theory? *Mathematical Logic Quarterly*, 45:521–532.

Read, S. (2006). Monism: The one true logic. In DeVidi, D. and Kenyon, T., editors, *A Logical Approach to Philosophy*, Western Ontario Series in Philosophy of Science, pages 193–209. Springer, Dordrecht.

Restall, G. (2002). Carnap's tolerance, meaning, and logical pluralism. *Journal of Philosophy*, 99:426–443.

Part II
Truth, Consistency and Paradox

Chapter 6
Truth and the Liar

Colin Howson

> *"Now, where were we? Read me back that last line".*
> *"'Read me back that last line,'" read back the corporal who could take shorthand.*
>
> Joseph Heller, *Catch 22*

1 Introduction

Frege famously claimed that logic is the science of truth: "To discover truths is the task of all science; it falls to logic to discern the laws of truth" (Frege, 1956, p. 289). But just like the other foundational concept of *set*, truth at that time was intimately associated with paradox; in the case of truth, the Liar paradox. The set-theoretical paradoxes had their teeth drawn by being recognised as *reductio proofs* of assumptions that had seemed too obvious to warrant stating explicitly, but were now seen to be substantive, and more importantly inconsistent. Tarski includes the Liar paradox in his classic discussion of the concept of truth (Tarski, 1956), and developed it, in the form of his famous theorem on the undefinability of truth, as a *reductio* of the assumption that a language could be *semantically closed*, in the sense of being able to contain its own truth-predicate.

This technical result seemed just one more way of seeing a paradox as a source of useful information, just as the set-theoretical paradoxes were used as implicit reminders that sets could not be too extensive in their membership.[1] It inspired the view, which became the orthodoxy until well into the second half of the twentieth century, that a natural language like English could consistently discuss the truth and falsity of any sentence in it only by implicitly assigning it a lower level in an object-language, metalanguage, meta-metalanguage etc. hierarchy. But there was much that was unattractive about this idea, and gradually challenges to it began to appear. One of the most influential, because it used standard model-theoretic techniques in a

C. Howson (✉)
Professor of Philosophy, University of Toronto, Toronto, ON, Canada
e-mail: colin.howson@utoronto.ca

[1] Boolos reports that Henkin showed that Tarski's Convention (T), "'...' is true iff ...," can be used to generate a paradox in a way formally very similar to the way Curry's Paradox arises from the unrestricted Comprehension axiom of naive set theory (Boolos, 1993, pp. 55–56).

D. DeVidi et al. (eds.), *Logic, Mathematics, Philosophy: Vintage Enthusiasms*,
The Western Ontario Series in Philosophy of Science 75,
DOI 10.1007/978-94-007-0214-1_6,

way that made possible a consistent denial of the conclusion of Tarski's Theorem, was published almost simultaneously in (Martin and Woodruff, 1975; Kripke, 1975). The way in question was that of a type of three-valued logic, originally invented for a quite different purpose by Kleene.

Though Kripke's and Martin and Woodruff's papers contained a broadly similar technical result, the latters' was little more than a very brief statement and proof of it, while Kripke presented it as a part of a systematic philosophical theory of truth. It begins with a powerful critique of the orthodox, so-called Tarskian theory, and develops a convincing diagnosis of the causes of the Liar paradox and a theory of truth intended to capture a central idea, that of grounded truth-ascriptions, which can be seen as a plausible and precise way of demarcating genuine propositions from spurious ones. Because it is an outstanding paper philosophically as well as mathematically, his paper captured the philosophical imagination as few other technical advances in logic have.[2] As Sheard observes in the introduction to his survey of modern truth-theories:

> The start of the recent era in the study of self-referential truth is easily dated. Kripke's "An outline of a theory of truth," appearing in 1975, is almost universally cited as the spark for the explosion of interest in the subject. Certainly no major intellectual advance ever arises in a vacuum, and any suggestion that the subject of self-referential truth had lain completely dormant prior to Kripke's paper would be both inaccurate and unfair. Nonetheless, we need only look at how frequently Kripke's paper is cited in prefatory remarks in the literature, and especially how many times it is cited at a personal level as having reawakened an interest in the subject, to recognize the catalytic role of that essay in the recent wave of research and discussion. (Sheard, 1994, p. 1032)

But appearing in that wave of research and discussion were also some strong reservations about what Kripke's theory achieved, and what in principle any such theory can achieve. Perhaps surprisingly, the source of most of these reservations was an objection due to Kripke himself. The problem he identified is a variant of the so-called Strengthened Liar paradox. In Kripke's theory the Liar sentence is not true or false: this is how the theory escapes the conclusion, valid in bivalent logic, that no language can contain its own truth-predicate. But it would seem to follow that the Liar sentence is not true. Since this is just what the Liar sentence asserts it must be true according to that basic, and surely correct, principle known in the literature as Tarski's Convention (T). The paradox can be dissolved by noting that two distinct truth-predicates are involved in it, but one of them, the one appealed to in concluding that the Liar sentence is not true, is a totally-defined predicate and cannot be defined in the supposedly semantically closed language within which the other lives. But the solution shows that even the three-valued approach apparently cannot dispense with the need for a classical metalanguage. This seems to have been Kripke's own conclusion, expressed in the much-quoted remark that "the ghost of the Tarski hierarchy is still with us," and it is the conclusion drawn by nearly every commentator since then.

---

[2] Gödel should probably remain pre-eminent in this respect.

I think that it is a wrong conclusion. There is certainly a problem, but it is not spectral in nature. In fact, it is almost a familiar problem, that of justifying the use of a classical theory to interpret a non-classical one, and famously experienced by no less a thinker than Niels Bohr. His solution was to appeal to a principle he called a *complementarity principle*, and I believe that something like that principle can be invoked here. If any ghost arises from the corpse of the Strengthened Liar, it is Bohr's.

The discussion of Kripke's work brings into relief an important but not-often highlighted role played by the rather artificially structured formal languages of modern logic. By contrast with these, natural language has no precisely defined notion of "well-formed sentence," and it has an intensional, theory-interpreted semantics far from the theory-neutral referential semantics of modern formal logic. This contrast in syntax and semantics presents an obvious problem for anyone wishing to see in these formalised systems a sort of idealisation of, or approximation to, natural language, and hence might be thought to place in doubt the capacity of logical research to say much if anything useful about natural-language representations and reasoning. This would certainly be the wrong conclusion, if for no other reason than that these systems can, and in the present context do, play a quite different but nevertheless extremely informative role: that of precisely articulated *experiments*, testing out claims about what is and is not possible with respect to the functions a language[3] can discharge subject to suitably imposed constraints. For example, it was at one time widely held that self-referential statements are necessarily meaningless, a view which as Kripke pointed out was definitively refuted by Gödel's celebrated paper containing his two incompleteness theorems, theorems which themselves ended centuries of uncertainty about the capacity of deductive theories to capture truths about their fields of enquiry, and about their own consistency. A simple formal language is used to prove that any language capable of self-reference cannot also contain its own truth-predicate; and the same simple formal language was later used to show that this is not true. The Strengthened Liar, bent on revenge, would be difficult if not impossible to analyse correctly without the help of elementary model theory, as we shall see.

2 The Liar Paradox

Let us start where Kripke started, with the Liar paradox. At its simplest, this proceeds as follows. Consider the sentence

1. (1) is false.

Suppose also we require that the following constraint (Convention (T)) must be satisfied by any adequate notion of truth:

---

[3] I am taking "language" here to mean not only a set of descriptive items and rules for their correct combination, but possibly also an associated deductive system.

(T) B is true iff $d(B)$, for every B; $d(B)$ is the denotation (e.g. disquotation where "B" names B by quoting it) of B.

(1) and (T) are clearly inconsistent, reading "false" as "not true," because we infer that (1) is true iff (1) is not true.

One might perhaps be forgiven for thinking that something formally a bit dubious is going on here, particularly in the self-referential nature of (1) which on the face of it looks like an equation whose solution is assumed rather than demonstrated, and where it is far from clear that it has a solution at all (in the domain of meaningful sentences). In fact, this is just one of the questions easily settled with the help of the apparatus of modern formal logic, with the answer that there is certainly a solution in the class of syntactically well-formed sentences[4]: indeed, this can be shown in more than one way. One, the method of (Martin and Woodruff, 1975), is direct and actually just mirrors the simple construction used above. This involves a many-sorted first order language L whose interpretation v is a valuation function over a sequence of sortal domains, one of which can contain the formulas of L. Assume that L contains a monadic predicate Tr. If the interpretation of Tr is fixed by the condition (in effect Convention (T)) that where t is a term whose value $v(t)$ is a sentence S such that $Tr(t)$ and S take the same truth-value under v, then v is said by Martin and Woodruff to *represent truth in* L. Now suppose v assigns the constant a the sentence $S =$ "$\neg Tr(a)$". Clearly, reading "false" as "not true" we have in the identity $v(a) =$ "$\neg Tr(a)$", an exact formal analogue of the informal $d(A) =$ "A is false" above, removing fears that there is anything in principle ill-formed about the latter. And, just as above, we infer that $v(\neg Tr(a)) = t$ iff $v(Tr(a)) = t$ (for Martin and Woodruff this is not a paradox, however, as we shall see later).

A more oblique way of achieving self-reference is of course due to Gödel, famously exploited in his incompleteness theorems (Bell and Machover, 1977, Ch. 7). Here L is some language in which the predicates and functions of elementary arithmetic are definable. The syntax of L can be encoded in a single domain of an ordinary first order language, with the sentences of L defined in that domain (relative to a particular Gödel numbering) by a formula $Sent_L(x)$ of L. $Sent_L$ is a primitive recursive predicate, and theories much weaker than Peano arithmetic (PA) are well-known to be capable of representing all the recursive functions and predicates, and hence this one. PA itself is often taken as the base-theory, which therefore functions as both object-theory and syntax-theory for L. The diagonal function is also primitive recursive and hence representable in PA, and a celebrated consequence, sometimes referred to as Gödel's fixed-point lemma, is that for any

[4] Whether these are the same as the meaningful sentences is difficult to answer without a precise definition of "meaning," though their status as ungrounded in Kripke's theory is a plausible reason for thinking that they are not.

arithmetical predicate $F(x)$ there is a sentence A of L such that $A \leftrightarrow F[A]$[5] is provable in PA and hence true in the standard model N.[6] To avoid a terminological conflict with another and quite different type of fixed point result discussed later, I will follow McGee and call this the self-referential lemma (McGee, 1990, p. 111).

The capacity of L for self-reference can now be combined with the Liar reasoning to produce a celebrated result, called Tarski's Theorem after its eponymous author. Suppose that some formula of L, call it $T(x)$, represents truth in L, i.e. its extension consists of the code numbers of sentences true in N. By the self-referential lemma there is a fixed point in L of $\neg T(x)$, i.e. a formula λ such that $l \leftrightarrow \neg T[\lambda]$ is true in N. But since $T(x)$ represents truth in L the sentence $T[\lambda] \leftrightarrow \lambda$ must be true in N, which is impossible. Thus whatever the method used for achieving self-reference in a language, any attempt to combine it with a representation of its truth-predicate would seem to be impossible. This result was paraphrased informally by Tarski as the statement that no language which permits self-reference is *semantically closed*. He pointed out that an "essentially richer" language than L is needed to define truth for L (this language is usually called the metalanguage for L). A natural language like English attempts to force semantic closure by adopting Convention (T) as an unrestricted axiom schema, and thereby merely generates the Liar paradox as a deductive consequence.

These observations suggested to many (though not actually Tarski himself) a solution of the Liar paradox in terms of a linguistic hierarchy, with the metalanguage containing the truth-definition for some object language, and also all the sentences of that language, or suitable translations of them. That the suggestion works as it should is seen by modelling it using a simple formal language like L above and appropriate expansions of it. To this end add a new monadic predicate Tr to L to obtain $L(Tr)$. Tr represents truth in L, which implies, as we saw above, that all instances, i.e. with sentences A of L, of $A \leftrightarrow Tr[A]$ are true in the expansion of the standard model N of L to (N, T_N), where T_N is the set of Gödel numbers of all sentences of L true in N. To the axioms of PA also add all instances of $A \leftrightarrow Tr[A]$, for sentences A of L, as new "analytic" axioms, together with $\forall x(Tr(x) \rightarrow Sent_L(x))$ (but do not allow any formula containing "Tr" into the induction schema). The strategy has some nice consequences. First, it is easy to see that the Liar paradox cannot be reproduced, since although there is a λ such that $\neg Tr[\lambda] \leftrightarrow l$ is a theorem of PA, by the self-referential lemma, $Tr[\lambda] \leftrightarrow \lambda$ is not among the new axioms: inspection of the proof of the lemma shows that λ will contain an occurrence of Tr (but note that, since $Sent_L[\lambda]$ is provable so are $\neg Tr[\lambda]$ and λ, both of which are therefore *true* in (N, T_N)). In fact, this primitive truth-theory, call it PA$^+$, is demonstrably (relatively) consistent: it is easily shown to

[5] $F[A]$ is the formula $F(t)$ where t is the numeral for the Gödel number $A^\#$ of A. This notation, used by Reinhardt (1986), is simpler than the usual curved bracket-and-corner one. I will use it to denote any acceptable way of referring to object-language sentences within that language.

[6] If, as is usually assumed, the metatheory is standard set theory, the existence of N or a structure isomorphic to it is of course provable.

be a conservative extension of PA.[7] Secondly, each provable arithmetical statement clearly becomes provably true, though it quickly follows from conservativeness that the reflection principle stating that all theorems of the PA are true is not a theorem of PA^+, assuming PA itself is consistent (these points are made variously in (Halbach, 2000; Ketland, 1999)).[8]

Nevertheless, though demonstrably consistent relative to PA, the result, as a model of informal reasoning about truth, is still badly incomplete. PA^+ is bound to be incomplete in the technical sense, since the true arithmetical sentences are not recursively enumerable. But the weakness descends to more mundane levels. In particular, the requirement that as much as possible of informal valid reasoning should be modelled demands that for any statement which we can informally prove true we should be able to show that it is true formally, and as things stand we clearly cannot. This deficit is, however, easily remedied: renaming Tr as Tr_1, T_N as T_0, L as L_0 and $L(Tr_1)$ as L_1, all we need do to achieve this is repeat the procedure by which we expanded L_0 and add a new predicate symbol Tr_2 to obtain a new language L_2, and a corresponding version of Convention (T) restricted to sentences of L_1. And so on. Continuing in this way through the natural numbers an infinite hierarchy of metalanguages and metatheories is generated, each with its associated Tr-predicate and (T)-schema. The Tr_n can be continued to a transfinite progression for the ordinals α using a suitable representation of ordinals by code numbers, such that Tr_α can be coded by a natural number in such a way that it is decidable whether a given number codes an index and, for each pair of index-codes, which codes the greater index. These desiderata can be achieved using Kleene's system of ordinal notations, generating a transfinite hierarchy of languages L_α for countable ordinals α less than the smallest non-recursive ordinal, i.e., the smallest ordinal which does not define a recursive well-ordering of natural numbers (Halbach, 1997, pp. 70–71).

Like many theories which appear to do a job which no other seems capable of, the inadequacies of this one tended to get overlooked until shown up in sharp contrast by a plausible competitor. Here is a familiar list of deleterious features:

1. All the ordinally indexed truth-predicates in the hierarchy are, we naturally want to say, instances of some underlying truth concept; otherwise, why call them truth-predicates? Yet we are precluded from saying this in any precise way without resurrecting the Liar paradox.
2. Natural languages do not have such a layered structure, and so the modelling requirement seems rather obviously violated. The usual response to this is, as we know, that a single global truth-predicate is not a feature that any model should seek to reflect since it is demonstrably *pathological*: in pretending to semantic

[7] Expand any model M of Peano arithmetic to a model (M, T_M) of PA^+. Suppose B is an L-sentence provable in PA^+. Then B is true in all (M, T_M) and so is true in all M. By first-order Completeness, B is a theorem of PA.

[8] It is provable within the truth-theory obtained by adding the clauses of a standard truth-definition, where the clause for the universal quantifier says that a universally quantified sentence is true iff all its instances are, and extending the induction axioms to include formulas of $L(Tr)$ (Feferman, 1991, p. 14). This theory is not a conservative extension of Peano arithmetic, however.

closure natural languages, or their associated truth-theory characterised by an unrestricted Convention (T), are inconsistent.

3. Nevertheless, counterintuitive consequences flow from trying to regiment informal reasoning to accommodate a hierarchy of truth-predicates. Consider Kripke's well-known imaginary Nixon-Dean exchange (concerning the Watergate scandal of 1974; Dean was a White House aide and Nixon was of course US President):

   Dean: "All of Nixon's utterances about Watergate are false."
   Nixon: "All of Dean's utterances about Watergate are false."

   According to the levels-of-language account, the predicate "false" here must be understood as false$_\alpha$ for some α. But clearly neither Nixon nor Dean can succeed in including the other's claim in the scope of his own at whatever level it may be. But intuitively this is wrong: both can be in each other's scope *and* indeed have well-defined truth-values. Suppose Dean had said "Watergate = Watergate." Then Nixon's claim is false. Suppose also all Nixon's other statements about Watergate are false. Then Dean's statement is true. There is an intuitive sense in which there are genuine truth-values for both statements grounded in the truth-values of the nonsemantical, object-level assertions each made about Watergate.

4. The Liar sentence at any level α is, as we saw, provably true at that level. A similar argument to that above shows that the Truth-Teller at level α, i.e. the sentence which is a fixed point of the predicate Tr_α, is provably false. But neither it nor the Liar sentence intuitively make any determinate utterance at all, and to that extent ascribing truth-values to them that in some reasonable sense depend on the "real" (i.e., non-semantic) level-0 facts should be out of the question. Yet they both get one, with that assigned the Truth-Teller especially absurd. In the light of these observations, why give them truth-values which in no genuine sense of "true" do they seem to merit?

The answer is of course that they are bound to have truth-values if the truth-predicate is total, i.e. everywhere defined, and the bizarre values themselves merely reflect this distorting assumption. In that case, it might reasonably be asked, why assume it? Why indeed?—For it turns out that if we give it up, the apparently very deep implication of Tarski's Theorem, which has inspired in some a sort of logical mysticism according to which truth is a necessarily transcendent notion, is to others no more than a consequence of a gratuitous assumption. That truth is a total predicate is an assumption of Tarski's Theorem, and—logically speaking—equally a target for the reductio as that truth is definable in the object-language. To see this, assume there is a predicate Tr in some L capable of discussing its own syntax which under some interpretation of L represents the truth-predicate for L with respect to that interpretation; i.e. we have Convention (T) in the form

$$[\![A]\!] = [\![Tr[A]]\!]$$

where the double brackets signify truth value. Assume also there is a fixed point λ for $\neg Tr(x)$ such that

$$[\![l]\!] = [\![\neg Tr[\lambda]]\!]$$

We immediately infer that $[\![Tr[\lambda]]\!] = [\![\neg Tr[\lambda]]\!]$. If the values are t and f subject to the usual valuation rules this is of course impossible. But it becomes entirely consistent if $[\![\lambda]\!]$, $[\![\neg Tr[\lambda]]\!]$ and $[\![Tr[\lambda]]\!]$ are *undefined*. So why not direct the modus tollens of the Liar reasoning to the assumption that truth is a total predicate?

Since bivalence itself is a modelling assumption certainly not uniformly well-fitted to informal reasoning, engendering as it does the Sorites and other paradoxes, one might think it surprising that the truth-value-gap approach was not adopted from the outset. Evading the Liar paradox by declaring λ neither true nor false was in fact an old suggestion but, as Kripke pointed out (Kripke, 1975, p. 62), it had not previously been accompanied by any developed semantic theory (and by this is meant a theory in which the formal languages become evaluated fairly classically in set-theoretical structures) in which the gaps are systematically derived rather than imposed ad hoc as the occasion of paradox-avoidance demands. Martin and Woodruff, and independently Kripke, remedied the deficit in broadly similar ways, both using standard model-theoretic techniques combined with *three-valued* valuational schemes proposed originally by Kleene (1952, pp. 332–336)[9]; Martin and Woodruff used the weak three-valued tables and Kripke the strong. The weak scheme is so-called because if any argument in a compound function is undefined so is the function at that argument. The strong scheme is characterised by the following tables for $\neg$ and &: $\neg t = f$, $\neg f = t$, $\neg u = u$; & is the same as ordinary conjunction on values in $\{t, f\}$, is always symmetric, and & $(u, t) = u$, & $(u, f) = f$, and & $(u, u) = u$. $\vee$ and $\rightarrow$ are defined in the usual way from $\neg$ and &. In both schemes the quantifiers are interpreted conjunctively (universal) and disjunctively (existential), where all instances are assumed to be nameable in the domain of the base-model. Both have another property which is central to Kripke's (and Martin and Woodruff's) construction, as we shall see shortly. An n-ary partial predicate P is one which may be undefined for certain arguments by a valuation v; i.e. v may assign some n-tuples of domain elements the value u. Kripke calls the set of n-tuples $\mathbf{d}$ in the domain D of an interpretation to which $v(P)$ assigns t the *extension* of P in an interpretation v, and the set of n-tuples $v(P)$ assigns f the *antiextension* of P in v. Suppose M and M' are two ordinary classical model-structures expanded to include respectively the partial predicates P and P', with extension/antiextension pairs (P_D, Q_D) and (P'_D, Q'_D). Let $M \leqslant M'$ signify that $P_D \subseteq P'_D$ and $Q_D \subseteq Q'_D$. In what follows the extension/antiextension pair interpreting P in a structure will be called the *partial interpretation* of P in that structure.

[9] Calling the Kleene rules "three-valued" is misleading: as Kripke emphasised, u is not itself a third truth-value: on the contrary, it signifies the lack of a truth-value. Nobody presumably thinks that ∞ is a member of the domain of the variables when seeing "$x/0 = \infty$"; it is just a shorthand for saying that $x/0$ is undefined.

We shall just be concerned with one partial predicate Tr, interpreted as (partial) truth. Both Kripke and Martin and Woodruff showed how to construct a partial interpretation (T, F) of Tr in which truth for the language containing Tr is represented in the language, in Martin and Woodruff's terminology: Kripke used the general theory of induction on abstract structures due to Moschovakis to exhibit a minimal fixed point in a sequence of expansions of an initial partial structure, and Martin and Woodruff used Zorn's lemma to exhibit a maximal one. At this point Martin and Woodruff stopped, having carried out their objective of showing how ordinary model-theoretic techniques, combined with a three valued logic, allow truth to be represented within a semantically closed language. It was Kripke who carried the discussion further, in a way that stimulated a general philosophical interest, by showing that the minimal model had important properties that among others qualified it to be a theory of truth grounded in non-semantic facts. This also seemed to be what the orthodox account based on linguistic stratification also provided, but as we saw only rather spuriously, and at great cost in terms of being able to express what seem informally valid inferences involving truth.

These remarks indicate how big a role informal methodological criteria play even in such an abstract discipline as logic. But that is because this sort of logic has explanatory objectives. That granted, criteria of what is to count as a good explanation start mattering, and foremost among them are the extent to which something ill-explained by one theory is satisfactorily explained in a non-ad hoc way by a rival, and the extent to which a good theory takes over what is already regarded as meritorious in the existing one. According to these criteria, as Kripke pointed out, his theory scores very well, at least up to a point. The point is however universally regarded at least as a very serious one, but before examining it a brief resumé of Kripke's theory will be useful. Since the details of his construction are well-known I shall present just the features that will matter to the subsequent discussion.

Assume that L is a language whose intended interpretation contains the natural numbers, or some structure isomorphic to them, and can code its own syntax by a Gödel numbering or some other coding function: it should be an acceptable structure in the sense of Moschovakis (L might be the language of elementary arithmetic, or even ZFC). Let $\mathcal{L}$ be L plus a monadic predicate Tr, where in any partial interpretation T and F are disjoint sets of code numbers of sentences. Thus if A is a sentence of $\mathcal{L}$ then $[\![Tr[A]]\!] = t$ in (T, F) if $A^{\#} \in T$ and $[\![Tr[A]]\!] = f$ in (T, F) if $\neg A^{\#} \in T$. The heart of Kripke's construction is the so-called *Kripke jump*: for any partial interpretation (T, F) this is a function g from partial interpretations to partial interpretations such that $g(T, F) = (\{A^{\#} | [\![A]\!] = t \text{ in } (T, F)\}, \{A^{\#} | [\![\neg A]\!] = t \text{ in } (T, F)\})$,[10] where the valuation scheme is the strong Kleene one.[11] The following result is fundamental:

Theorem 1 *g is monotonic with respect to $\leqslant$; that is, if $(T, F) \leqslant (T', F')$ then $g(T, F) \leqslant g(T', F')$.*

---

[10] Kripke includes numbers of non-sentences in F.

[11] This is not essential to the construction, as we shall see.

The consequent states that the set of sentences assigned a determinate value by M' extends the set of sentences assigned a determinate value by M, i.e. for all sentences A of $\mathcal{L}$, if $[\![]\!]_M = t[f]$, then $[\![A]\!]'_M = t[f]$. The theorem is proved by induction on the degree of A, assuming as before that all elements of the base model have names in $\mathcal{L}$ (e.g., numerals).

The next step is to define an ordinally indexed sequence $(T, F)_\alpha$, by letting $(T, F)_0 = (T, F)$ and $(T, F)_{\alpha+1} = g((T, F)_\alpha)$, with limits defined in the usual way. Let $(T_a, F_a) = (T, F)_a$. Using the monotonicity of g, a proof by induction shows that for all α, $(T_\alpha, F_\alpha) \leqslant (T_{\alpha+1}, F_{\alpha+1})$, i.e. $(T_\alpha, F_\alpha) \leqslant g(T_\alpha, F_\alpha)$. Cardinality considerations dictate that at some (countable limit) ordinal σ[12] there is a fixed point, i.e. $(T_\sigma, F_\sigma) = (T_\kappa, F_\kappa)$ for all $\kappa \geqslant \sigma$. Hence $T_{\sigma+1} = \{A^\# | A \text{ is true in } (T_\sigma, F_\sigma)\} = \{A^\# | A \text{ is true in } (T_{\sigma+1}, F_{\sigma+1})\}$; i.e. (T_σ, F_σ) and $(T_{\sigma+1}, F_{\sigma+1})$ make exactly the same sentences of $\mathcal{L}$ true. Similarly for "false." It follows that $[\![Tr[A]]\!] = [\![A]\!]$ in (T_σ, F_σ) for every sentence A of $\mathcal{L}$, since A is true in (T_σ, F_σ) iff $A^\#$ is in T_σ iff $Tr[A]$ is true in $(T_{\sigma+1}, F_{\sigma+1})$ iff $Tr[A]$ is true in (T_σ, F_σ); repeat for "A is false". *Hence the formula $Tr(x)$ represents truth in $\mathcal{L}$, in Martin and Woodruff's terminology, and so $\mathcal{L}$ is semantically closed.*

Note that the existence of fixed points depends on the monotonicity of the jump operation, and the monotonicity of the jump operation depends only on the valuation scheme. The Kleene weak and strong tables are not alone in being monotonic: as Kripke noted, so are many others, including supervaluations. Two familiar non-monotonic schemes are the classical bivalent one and—which will be relevant to the later discussion—the one resulting from adjoining so-called exclusion negation $\sim$ to either of the Kleene schemes, which sends $t[f]$ to $f[t]$ and u to t. If we try the ordinal construction that generated the fixed point for the strong (or weak) Kleene schemes we find that, where λ' is the fixed point of the total predicate $\sim Tr(x)$, λ' oscillates between t and f at successive stages. Indeed, it is easily shown to be *essential* for monotonicity that $\neg u = u$. The combination of bivalent semantics with an inductive construction resembling Kripke's is of course the central feature of so-called Naive Semantics (Martin, 1984a, c).

Finally, let $T = \emptyset$ and $F = \emptyset$. $M_\sigma = (\emptyset, \emptyset)_\sigma$ is the least fixed point relative to the $\leqslant$ ordering: for every fixed point c in the set C of fixed points, $(\emptyset, \emptyset)\alpha \leqslant c$ (the proof is by induction on α; the basis is trivial because every partial model extends $(\emptyset, \emptyset)$). It is easy to see that $\lambda^\#$ is not in the extension or anti-extension of Tr in M_σ, or in any fixed point: it always takes the value u (such sentences are called "paradoxical" by Kripke, as opposed to others which while undefined in M_σ have definite truth-values in other fixed points). So what the Tarski-hierarchy solution of the Liar paradox accomplishes by its typed truth-predicates—consistency in the evaluation of λ—is accomplished here in a single non-stratified language containing its own truth-predicate, and one moreover in terms of which, as Kripke showed, all those of the Tarski hierarchy are definable: all the languages in the Tarski hierarchy are sublanguages of $\mathcal{L}$ and all the associated predicates Tr_α are definable in M_σ, in

[12] σ is not less than the smallest non-recursive ordinal; if the base model is N then σ is that ordinal.

the following way. Let $A_1(x)$ be the formula of L defining the syntactic predicate "x is the code number of a sentence of L." Let $T_1(x)$ be the formula $Tr(x) \,\&\, A_1(x)$ of $\mathcal{L}$. Now let $A_2(x)$ define the code numbers of all formulas from the language whose atomic formulas are those of $\mathcal{L}$ together with the formulas $T_1(x)$, for all variables x, and closed under truth-functions and quantification. $T_2(x)$ is defined by the formula $Tr(x) \,\&\, A_2(x)$, and so on through the natural numbers. Let L_{n+1} be the language obtained by adding T_{n+1} to L_n, where $L_0 = L$. Induction on n proves that each of the $T_n(x)$ is a total predicate and that the extension of $T_{n+1}(x)$ is the set of code numbers of all true sentences of L_n. ((Kripke, 1975, p. 75); this procedure can be continued into the transfinite so that all the Tr_α in the Tarskian hierarchy eventually become defined (Halbach, 1997)). Both the Tarskian and M_σ also assign ordinal levels at which sentences become determinately true or false, though in M_σ there is of course no type-subscript to indicate this: for any sentence A the level of A in M_σ is the ordinal α at which A first acquires a determinate value ((Halbach, 1997, p. 74), shows how the Tarskian and Kripkean levels are systematically related).

3 Groundedness

Let μ be the fixed point of $Tr(x)$ (μ is the so-called Truth-Teller). There are fixed points in which μ is true, and there are fixed points in which it is false. But none of this, as it stands, seems to explain why we intuitively feel that *neither* the Liar nor the Truth-Teller expresses a genuine proposition, but merely a grammatically well-formed pseudo-proposition. One, and possibly the main, reason why Kripke's theory attracted so much philosophical attention is because it does, or did, seem to explain why: these statements are ungrounded in the "real facts" holding in the base structure. For this reason M_σ is of central importance for the interpretation of Kripke's theory. The significance of starting with both T and F empty is that M_σ is simply the inductive closure under the jump operation of the empty interpretation of Tr under the jump operation: the only determinate values enter the hierarchy at the ground level of sentences of L, the values of all other sentences being determined at each stage in the inductive process by the valuation rules. Thus these truth values are grounded, in a formally precise sense, in the non-semantical properties of the base structure. Kripke himself actually cited this formal explication of the informal concept of "grounded" as the principal virtue of his theory (Kripke, 1975, p. 57).

The fact that the base-model determines all truth-values in M_σ in this way is a powerful argument for regarding M_σ among all fixed points as the authentic interpretation of the predicate Tr. Not only does this solve the problem of multiple fixed points when the concern is to model a unitary notion of truth, but it does so in a natural, indeed compelling way: it makes groundedness a part of the *meaning* of truth, which is as it should be if we believe that it is truth in the basic, non-semantic model which should determine the truth or falsity of all sentences in which Tr occurs, and it also provides a natural solution to the problem stated above. For neither the Liar sentence nor the Truth-Teller are grounded, even though there are fixed points which the latter is true.

Note that groundedness cannot just mean "having a determinate value in the least fixed point," since what is in the least fixed point will depend on the valuation scheme adopted, assuming it is monotonic. Groundedness means, or should mean, that those determinate values themselves are determined by what is true or false in L about N. This consideration implies that the rules, besides being monotonic, must also obey the compositionality principle: the values of truth-functional compounds depend on the values of their components and the values of quantified sentences depend on the values of their instances in the model (another way of putting it is to say that the truth-values of the compounds are *reductive*). The Kleene weak and strong rules have this property, but supervaluations, though also monotonic, do not: the classical tautology $\lambda \vee \neg\lambda$ in $\mathcal{L}$ is always true in any supervaluational fixed point though neither λ nor $\neg\lambda$ is. Thus by this argument, which seems a reasonable one, supervaluations seem to be ruled out as determiners of authentic, i.e. grounded truth.

This still leaves the question of how to choose between monotonic rules which are reductive. As far as the rules for the standard connectives and quantifiers are concerned the choice, assuming some semblance of their ordinary meaning is to be retained, is essentially between the weak and strong rules. The strong rules are the nearest, in a well-defined sense (called, in (Kleene, 1952, p. 335),"regularity"), to the classical valuation scheme, a fact exemplified in Kripke's Nixon-Dean exchange above: their respective statements are grounded, true or false, according to the strong rules but not the weak, in accord with our intuitions given the factual assumptions. Kripke voiced doubts, surely well-founded, about whether there is a fact of the matter about which monotonic valuation scheme is "really" correct, but states that this is not his main concern [p. 77]. However, he also uses the Nixon-Dean exchange to argue that a satisfactory formal theory should prove the groundedness of the protagonists' statements, and as we see that does, given the other constraints, seem to point to the strong Kleene rules.

In Maudlin's very interesting approach (Maudlin, 2004), also a three-valued one involving appeal to the smallest fixed point, "is true" is explicitly a logical operator with its own semantic clause, that the truth value of $Tr[A]$ is the same as that of A (so Convention (T) gets embodied in the semantic rules of language).[13] Grounded sentences according to this theory are sentences whose truth-value proceeds as in a directed acyclic graph along paths determined by the valuation rules from boundary sentences assigned a value exogenously. Suppose the formal language is like Martin and Woodruff's, in which constants can denote sentences, and suppose b denotes "$Tr(b)$." Then the truth-value of $Tr(b)$ is determined by that of the sentence denoted by b, which is $Tr(b)$; i.e. we are in a semantic cycle, and $Tr(b)$ is intrinsically ungrounded. Similarly, the truth-value of $Tr(a)$ where a denotes "$\neg Tr(a)$" is

[13] In effect this principle is tacitly adopted in nearly all the discussions of truth, since Convention (T) is always imposed, in an appropriate way, as the sole constraint on the extension of the truth-predicate.

determined by that of $\neg Tr(a)$. These sentences, independently of whether they are paradoxical in the classical sense, are ungrounded, according to Maudlin, because their semantic determination involves such cycles [p. 40].

Although the structure of the class of truth-determinate sentences is extensionally the same as given by the Kripke least fixed point, Maudlin claims that his account is conceptually distinct from Kripke's since it is not mere membership in a fixed point that signifies groundedness, but the intrinsic topological character of its semantic structure. It follows that for him u is a genuine semantical value additional to t and f, signifying "ungrounded" in his absolute sense, whereas for Kripke it merely means "undefined," a notion entirely relative to the fixed point in question: "the 'undefined' sentences are merely those left over without truth-values once a fixed point has been chosen" (Maudlin, 2004, p. 57). So, for example, the Truth-Teller necessarily has the value u for Maudlin, but depending on the fixed point chosen, it may have any one of the three values u, t or f in Kripke's theory.

But the difference between the two approaches reduces to little more than a verbal difference once the least fixed point is chosen to be the authentic truth-determiner, for the truth-values in M_σ are, as we saw, determined by the closure conditions in essentially the same way as they are in Maudlin's theory: the ungrounded sentences for both are those where truth fails to be determined by the structure of the ground model M. The cycle in the evaluation of the Truth-Teller manifests itself as a contradiction arising from the assumption that the sentence is first declared true or false at some ordinal level. Thus the Truth-Teller is as determinately ungrounded, so to speak, as it is in Maudlin's account. Of greater moment is the threat common to both accounts posed by a development known as the Strengthened Liar. It is not so much that this fortified individual poses some new paradox analogous to the Liar, but that its solution seems to reveal a violence to informal reasoning in the fixed-point theory of truth that is, if anything, worse than anything the Tarskian theory is guilty of.

4 The Strengthened Liar

We have seen that Convention (T) survives in the form

$$[\![Tr[A]]\!] = [\![A]\!]$$

for every sentence A of $\mathcal{L}$ in any fixed point. The self-referential lemma also survives, as the statement that for any formula $H(x)$ of $\mathcal{L}$ there is a sentence C such that $[\![C]\!] = [\![H[C]]\!]$, holds for fixed points (McGee, 1990, p. 111). It follows that

$$[\![Tr[\lambda]]\!] = [\![\lambda]\!] = [\![\neg Tr[\lambda]]\!] = u \qquad (*)$$

in every fixed point.

These facts about fixed points quickly dissolve the alleged paradox of the Strengthened Liar, which, following (Burge, 1984, p. 87), proceeds thus:

1. λ in the three-valued system is neither true nor false.
2. Hence λ is not true.
3. But λ says that it is not true.
4. Hence λ is true.
   Contradiction.

Burge charges truth-value-gap theorists, including Kripke, with recognising the problem but having "little illuminating to say about it" [p. 88]. But one thing that is surely illuminating to say is that there is technically no paradox here at all. Representing the reasoning formally shows up the fallacy clearly:

1. $[\![\lambda]\!] \neq t, f$
2. $\therefore [\![\lambda]\!] \neq t$
3. $\therefore [\![Tr[\lambda]]\!] \neq t$ by $(*)$
   $\therefore [\![Tr[\lambda]]\!] = t$
4. $\therefore [\![\lambda]\!] = t$ by $(*)$
   Contradiction.

The unnumbered step is implicit in the informal derivation, but it is incorrect since $[\![Tr[\lambda]]\!] = [\![\neg Tr[\lambda]]\!] = u$.

> Feferman, noting the formal fallacy, remarks that nevertheless consideration of the Extended Liar does leave one with a further bit of malaise about truth-gap approaches, since the formal model-theoretic constructions don't match up with informal usage. (Feferman, 1982, p. 266)

What lies behind Feferman's reservation seems to be something like the following argument:

(i) If M_σ gives the correct interpretation of truth for $\mathcal{L}$, then $\neg Tr[\lambda]$ should be true, because λ is not true.
(ii) But $\neg Tr[\lambda]$ is not true in M_σ.
(iii) Therefore M_σ does not give the correct interpretation of truth for $\mathcal{L}$.

Adjoining the exclusion-negation operator $\sim$ to $\mathcal{L}$, apparently authorising the passage from $[\![Tr[\lambda]]\!] = u$ to $[\![\sim Tr[\lambda]]\!] = t$, will not help, since as we saw earlier adjoining $\sim$ allows the construction of a $\sim$-Liar sentence and thereby destroys the monotonicity of the valuation scheme. Indeed, even the property that the truth-value of a sentence is u is not one definable in $\mathcal{L}$, for exclusion negation would then be definable in $\mathcal{L}$ as "A is false or A has the value u." This three-valued theory avoids the Liar and Strengthened Liar by making λ and $Tr[\lambda]$ into singularities, but the problem is that they are rather like black-hole singularities in swallowing all relevant information (i.e. about their truth-values) with them. The levels-of-language theory was criticised for failing to allow intuitively valid reasoning to be reproduced in it; now the same or worse seems be happening here.

One of the familiar problems about invoking "intuitively valid reasoning" is that while it may be intuitive it is often not valid, which is why the apparatus of formal models of inference was invented and why it continues to be indispensable. And the fact is that as it stands the argument in (i)–(iii) also makes the same erroneous inference as did the derivation of the Strengthened Liar. No sound theory of partial truth can allow that $Tr[\lambda]$ is true *because* λ is not true, for the truth-predicate represented by Tr would cease to be a *partial* truth predicate, and become a total one. Ketland (2003) has a variant of the proof that the predicate "A takes the value u" is not definable in L,[14] which he accompanies with the remark that

> No matter how we squeeze and tug the carpet, certain *semantic concepts* which are well-defined in the informal metalanguage description of the fixed-point MV-language [Many-Valued language] are *inexpressible* within the very MV-language in question. (p. 294; italics in original)

But this is again just a failure to appreciate the fact that, far from there being an expressive deficiency signalled by $\mathcal{L}$'s failure to be able to define the predicate "x takes the value u," *there would certainly be an expressive failure if it could*: it would mean that it wasn't doing its job of defining a partial predicate. If one is going to accept a genuine theory of partial truth then one must accept that it is partial, and not demand something that cannot in principle be given. Partiality of some sort is inevitable. Either it is the partiality of not having a truth-predicate that applies via Convention (T) to statements involving it, as in the Tarskian theory, or it is the partiality of the truth-predicate itself.

Perhaps surprisingly, it turns out that while we cannot have our cake we can (to some extent) still eat it. For there *is* a way in which we can legitimately infer in $\mathcal{L}$ that $\neg Tr[\lambda]$ is true from the fact that λ is not true in the least fixed point. Indeed, we shall see shortly that *there is a consistent first order theory within which that inference is valid.* But now we seem to have generated another paradox: how can this not contradict the conclusion that it is entirely appropriate for that inference to be invalid within a sound theory of partial truth? The answer is, by distinguishing between *two distinct interpretations of* $\mathcal{L}$, which is in fact what is going on implicitly in (i)–(iii) above. Where the base model is N and $M = (N, (T, F))$ is a fixed point, $\neg Tr[\lambda]$ is of course true in the classical structure (N, T) interpreting $\mathcal{L}$, called the "closing off" of $(N, (T, F))$ (Kripke, 1975, p. 80). But neither λ nor $\neg Tr[\lambda]$ are true in M. Thus we infer that $\neg Tr[\lambda]$ is true even though $[\![\neg Tr[\lambda]]\!]_M = u$, because $\neg Tr[\lambda]$ is true in the closed-off structure M_c. We now have a nice, and formally unimpeachable, explanation of why we find the inference involved in (i) and (ii) above intuitively compelling: we are implicitly appealing to these two distinct interpretative structures for $\mathcal{L}$: to M_c in (i), and to the corresponding partial structure M in (ii).

Moreover, since M_c is a classical first order structure, we should in principle be able to find a classical first order theory formulated in $\mathcal{L}$ within which the chain of reasoning that ends with the assertion $\neg Tr[\lambda]$ can be reproduced. And we

---

[14] He uses a Liar-type construction to show that no multi-valued language L which is capable of self-reference, which contains constants denoting truth-values in a subdomain of any model M, and in which identity is bivalent can contain a term t which satisfies the condition that $(t[A])_M = [\![A]\!]_M$. However, the bivalence of identity implies that the truth-values t, u and f in a three-valued system are distinguishable within L and hence that exclusion-negation is definable in L.

can. Constructed by Feferman and referred to in the literature as KF (for "Kripke-Feferman"; the terminology is due to (Reinhardt, 1986)[15]) its axioms are those of Peano arithmetic (the object-theory also providing the syntax theory of $\mathcal{L}$), together with axioms corresponding to the ordinary truth definition for the atomic sentences of $\mathcal{L}$ which do not contain Tr, axioms describing the strong Kleene valuation scheme, and an axiom saying that no sentence is true and false. A type of completeness theorem for KF due to Feferman states that (N, T) is a model of KF iff $(N, (T, F))$, is a fixed point. Feferman's theorem is a proof that KF is (relatively) *consistent*: it certainly has one model, by Feferman's theorem. Moreover since (a) the self-referential lemma is a provable biconditional in KF, so that $\lambda \leftrightarrow \neg Tr[\lambda]$ is a theorem, and (b) $Tr[A] \rightarrow A$ is a theorem of KF for all sentences A of $\mathcal{L}$ (McGee, 1990, p. 93), it immediately follows that $\neg Tr[\lambda]$ is a theorem of KF, as desired. Since $\lambda \leftrightarrow \neg Tr[\lambda]$ is a theorem of KF, it follows that l itself is a theorem of KF.[16] The Strengthened Liar "paradox" is blocked since KF does not license the inference from λ to $Tr[\lambda]$. This failure is not at all ad hoc: the rule is clearly unsound since KF is a classical theory and hence every tautology, including $\lambda \vee \neg\lambda$, is a theorem, but the value of both λ and $\neg\lambda$, and hence their disjunction, is undefined in every fixed point and so $Tr[\lambda \vee \neg\lambda]$ is *false* in M_c (and its negation is a theorem of KF).

As we all know, that is not quite the end of the story. To many commentators, including to some extent Kripke himself, this strategy for saving the validity of the inference from "λ is not true" to "$\neg Tr[\lambda]$ is true" is something of a poisoned chalice. For in appealing to M_c to explain how we can consistently say that λ is true, we are appealing to a truth-predicate that by Tarski's Theorem lives irreducibly in the metalanguage of $\mathcal{L}$:

> Since the object language obtained by closing off $T(x)$ [my $Tr(x)$] is a classical language with every predicate totally defined, it is possible to define a truth-predicate [TR] for that language in the usual Tarskian manner. This predicate will *not* coincide in extension with the predicate $T(x)$ of the object language, and it is certainly reasonable to suppose that it really is the metalanguage predicate that expresses the "genuine" concept of truth for the closed-off language; the $T(x)$ of the closed-off language defines truth for the fixed point before it was closed off. So we still cannot avoid the need for a metalanguage. (Kripke, 1975, p. 81)

Hence, Kripke famously concluded, the "ghost of the Tarski hierarchy is still with us."

This deflating conclusion is echoed by those commentators who feel that Kripke's construction has not, as it was intended to do, supplied a semantically-closed language within which all legitimate truth-claims are to be made. Despite its distinguished advocacy I think that it nevertheless rests on the same confusion, or

[15] In some presentations there is an additional monadic predicate symbol for "false," and in Feferman's original presentation the axiom that no sentence is true and false was absent.

[16] KF is a provably reliable device for exhibiting (a recursively enumerable subset of) truths with respect to M_σ: Feferman's theorem shows that if $Tr[A]$ is a theorem of KF then A is true in M_σ, since if A is a theorem of KF then A is true all models of KF, and hence in all structures (N, T) where $(N, (T, F))$ is a fixed point, and hence in M_σ itself.

conflation, that issued in the alleged Strengthened Liar paradox. The truth that, *qua* semantically closed language, what $\mathcal{L}$ represents is partial truth, and all legitimate truth-claims, in the sense of partial truth, can be made within $\mathcal{L}$ (or some stronger language like that of Zermelo-Fraenkel set theory). There is therefore no necessity at all, from the point of view of describing partial truth, to ascend to a metalanguage: $\mathcal{L}$ *is* the metalanguage. The necessity only arises if one wishes to infer that $\neg Tr[\lambda]$ is true *because* λ is not true in any fixed point. While this may be an entirely legitimate claim about truth in the classical structure M_c it is not a legitimate truth-claim with respect to partial truth.

Once we see the interpretation of $\mathcal{L}$ in M_c as motivated by the desire for an illuminating explanation rather than a necessity implicit in a theory of partial truth, the ghost of the Tarski hierarchy ceases to be malign. Indeed, we merely find ourselves in a type of situation familiar from other contexts: using classical concepts and theories to interpret and explain non-classical ones, in the present case to explain why we feel that the Liar sentence is true in some significant sense. The strategy of appealing to classically based ideas and theories to compensate for the fact that our cognitive processes do not seem to be geared to a direct appreciation of the non-classical has of course a famous pedigree in modern physics, under the name Bohr famously gave it, of *Complementarity*. Though Bohr's own characterisation of his idea was notoriously opaque a major component of it nevertheless was, as he saw it, the indispensability of classical concepts for interpreting the quantum theory as it stood at the time (Bohr, 1934, p. 94). The parallel is especially appropriate if one regards quantum logic as the natural logic of quantum mechanics (though many of course do not), since its non-classicality partly resides in a type of choice-negation (represented by an operator projecting onto the orthogonal subspace).[17]

Bohr's theory that quantum theory requires complementary interpretations may not be one often pressed by physicists today, but the idea of gaining some sort of epistemic access to one theory via an interpretation in another seemingly incompatible with it remains an acknowledged and formally unexceptionable procedure: a less formally controversial example is Gödel's interpretation of Heyting's Intuitionistic propositional calculus in a classical theory of provability formally identical to S4 (Gödel, 1933), which arguably does more to render Intuitionistic logic intelligible than all of Brouwer's explanatory observations put together. Similarly, the fact that human beings, and possibly a large part of the mammal kingdom too, seem to possess an inbuilt exclusion-negation operator which automatically makes every predicate in the informal language total, suggests that the properties of partial truth based on a monotonic valuation scheme will be more easily understood and reasoned about when interpreted within a deductive theory situated within classical bivalent logic[18] like KF or a suitably stronger classical theory; as obtained, for

[17] Though $A \vee \neg A$ is always true, or what passes for true, even if neither A nor $\neg A$ is. Also quantum logic, or perhaps "logic," is *radically* non-classical, unlike the Kleene rules.

[18] There are axiomatisations of three-valued validity for the Kleene strong scheme, but they make so much of informally valid reasoning illegitimate that Feferman is moved to remark that "nothing like sustained ordinary reasoning can be carried on" in them (Feferman, 1982, p. 264).

example, by adjoining suitable reflection principles to KF, like the statement that if A is a sentence of $\mathcal{L}$, then if $Tr[A]$ is provable in KF then A is true, which itself can be formulated as a sentence of $\mathcal{L}$ (Reinhardt, 1986, p. 234), as can also the statement that if $\neg Tr[A]$ is provable in KF then A is not true. As $\mathcal{L}$-sentences both are instances of the Uniform Reflection Principle $\forall x(Bew_{KF}[F(\dot{x})] \rightarrow \Phi(x))$, where $Bew_K F(x)$ is the provability predicate for KF expressed in $\mathcal{L}$, and $[\Phi(\dot{n})]$ is the Gödel number of the formula $F(\bar{n})$ where $\bar{n}$ is the numeral for n. Most importantly, *no metalanguage need be appealed to in this interpretative exercise* because, as KF shows, the classical interpreting theory can be formulated entirely within $\mathcal{L}$ itself.

KF's status as a classical theory of partial truth has aroused a good deal of comment, and as much in the way of reservations. Thus McGee objects that "the simple connection between truth and proof" is broken in KF because in it "we can prove things that are, according to [it], untrue" (McGee, 1990, p. 106): for example, KF proves λ and also proves $\neg Tr[\lambda]$. But McGee's objection seems to rest on the same conflation between two distinct truth-predicates that generated the Strengthened Liar. According to KF's own classical truth predicate which it is a partial axiomatisation of, KF is not proving things that are untrue: λ is actually true according to that. According to the fixed-point truth predicate, interpreted within KF by the formula $Tr(x)$ of $\mathcal{L}$, KF is *not* proving that λ is true (recall that the inference from A to $Tr[A]$ is quite properly, in its interpretative role as formaliser of fixed-point truth, disallowed by KF) though it is proving that it is untrue. Disequivocate, and the ground of McGee's objection disappears. Reinhardt, commenting on the fact that in KF the inference from A to $Tr[A]$ is disallowed, goes to another extreme, claiming that in this respect "KF does not pretend, even to itself, to be more than a formal device" (Reinhardt, 1986, p. 242). This seems unjust too: KF is a classical interpretation of a non-classical theory, which is surely more than being just a "formal device." A stronger objection is that without Feferman's metatheorem, proved of course in the classical metalanguage, that KF's classical models are just the closings off of fixed points, there would be no reason to suppose that KF is a *faithful* interpretation of truth in M_σ. But the objection is unsustainable, for it would imply that no deductive reasoning would ever be justified in default of an explicit soundness theorem, which is impossible on pain of an infinite regress.

5 Conclusion

The problem with total truth, notoriously exposed in the Liar paradox, is its vulnerability to diagonalisation. The fact is that there has got to be partiality somewhere, and the simplest acknowledgment of this is to allow the truth-predicate (or function from sentences to truth-values) to be undefined at "singular" arguments. Partial functions are familiar and indispensable objects in mathematics. The whole edifice of recursive function theory rests on them: the enumeration theorem for partial recursive functions (Kleene, 1952, p. 341), which states that a partial recursive function

$f_n(x_1, \ldots, x_n, z)$ enumerates all the partial recursive functions of $x_1, \ldots, x_n$ as z takes its (natural number) values, includes $g(x) = f_1(x, x) + 1$ in the enumeration by allowing it to be properly partial, undefined at the argument k. Partial predicates are also an essential part of the theory, since they are predicates whose characteristic functions are partial recursive functions (Feferman, 1982, p. 252).

Similarly, there is no problem in principle to including Tr within the class of predicates definable within a language which permits self-reference, by allowing it to be undefined at the argument λ. The apparent sticking point, known affectionately as the Liar's Revenge, is that in asserting as true the "obvious" truth that λ is not true in any fixed point the Tarski hierarchy is thereby reintroduced. But this is not, as we saw, a necessity that a theory of partial truth itself has to accommodate nor should even want to accommodate if it is to be a genuine explication of partial truth, but an interpretative gesture towards the entrenched way of thinking which automatically closes off every predicate to make it total. The ghost summoned by the closing-off of the fixed point remains Bohr's, not Tarski's.

Bibliography

Bell, J. and Machover, M. (1977). *A Course in Mathematical Logic*. North Holland, Amsterdam.

Bohr, N. (1934). *Atomic Theory and the Description of Nature*. Cambridge University Press, Cambridge.

Boolos, G. (1993). *The Logic of Provability*. Cambridge University Press, Cambridge.

Burge, T. (1984). *Semantical Paradox*, pages 237–289. In [Martin, 1984b].

Feferman, S. (1982). *Towards Some Useful Type-Free Theories*, pages 237–289. In [Martin, 1984b].

Feferman, S. (1991). Reflecting on incompleteness. *The Journal of Symbolic Logic*, 56:1–49.

Frege, G. (1956). The thought; a logical enquiry. *Mind*, 65:289–311.

Gödel, K. (1933). Eine interpretatione des intuitionistischen aussagenkalküls. *Ergebnisse eines mathematischen Kolloquiums*, 4:39–40. English Translation in J. Hintikka, ed., *The Philosophy of Mathematics*, Oxford University Press, 1969.

Halbach, V. (1997). Tarskian and Kripkean truth. *Journal of Philosophical Logic*, 26:69–80.

Halbach, V. (2000). Truth and reduction. *Erkenntnis*, 53:97–126.

Ketland, J. (1999). Deflationism and Tarski's paradise. *Mind*, 108:69–94.

Ketland, J. (2003). Can a many-valued language functionally represent its own semantics? *Analysis*, 63:292–297.

Kleene, S. (1952). *Introduction to Metamathematics*. North Holland, Amsterdam.

Kripke, S. (1975). Outline of a theory of truth. *Journal of Philosophy*, 72:690–716. Page references are to reprint in [Martin, 1984b].

Martin, R. (1984a). *Notes on Naive Semantics*, pages 133–175. In [Martin, 1984b].

Martin, R. (1984b). *Recent Essays on Truth and the Liar Paradox*. Oxford University Press, Oxford.

Martin, R. (1984c). *Truth and Paradox*, pages 175–237. In [Martin, 1984b].

Martin, R. L. and Woodruff, P. (1975). On representing "true-in-*L*" in *L*. *Philosophia*, 5:213–217.

Maudlin, T. (2004). *Truth and Paradox. Solving the Riddles*. Clarendon Press, Oxford.

McGee, V. (1990). *Truth, Vagueness and Paradox: An Essay on the Logic of Truth*. Hackett, Indianapolis, IN.

Reinhardt, W. (1986). Some remarks on extending and interpreting theories with a partial predicate for truth. *Journal of Philosophical Logic*, 15:219–251.

Sheard, M. (1994). A guide to truth predicates in the modern era. *The Journal of Symbolic Logic*, 59:1032–2054.

Tarski, A. (1956). The concept of truth in formalized languages. In Woodger, J., editor, *Logic, Semantics, Metamathematics*, pages 152–278. Oxford University Press, Oxford. Translation, by J.H. Woodger, of 1935 German version.

Chapter 7
Necessary and Sufficient Conditions for Undecidability of the Gödel Sentence and its Truth

Daniel Isaacson

I am delighted to be able to join the other authors of this Festschrift in honouring John Bell, and to be able to express my gratitude to John for his exuberant and generous friendship ever since we first met, in 1967. Even before then, John's fame had preceded him by way of an article about his plans to study mathematics in Oxford or Cambridge that had appeared in the San Francisco Chronicle 6 years earlier and impressed me enough to keep. It is wonderful to think back to those days of extraordinary promise from the vantage point of John's tremendous accomplishments in the nearly 50 years since, and his promise of yet more.

1 Introduction

For S any recursively axiomatized Σ_0-complete theory sound with respect to truth in the structure of the natural numbers, there is a sentence G such that $S \nvdash G$ and $S \nvdash \sim G$. This is a weak form of Gödel's First Incompleteness Theorem, weak because the hypothesis of soundness is much stronger than necessary. Gödel strengthened the theorem by weakening the hypothesis, from soundness to ω-consistency (see (Gödel, 1931, pp. 151, 173)). In 1957 Kreisel introduced "a refinement of the concept of ω-consistency," see (Kreisel, 1957), which he labelled n-consistency, for each natural number n, and noted that 1-consistency, the minimal case of ω-consistency, is sufficient for Gödel's First Incompleteness Theorem. In this paper I analyze the relationship between soundness, ω-consistency, 1-consistency, and conditions intermediate between them, and expound properties of these notions. I also discuss on what basis we know that the Gödel sentence is true. I observe that a necessary and sufficient condition for $S \nvdash \sim G$, for G the Gödel sentence for formal system S, is the consistency of $S \cup \{Con_S\}$, and establish that this condition is strictly weaker than 1-consistency of S. I conclude with some remarks about the relation between Gödel's Incompleteness Theorems and Rosser's Theorem.

D. Isaacson (✉)
University Lecturer in Philosophy of Mathematics; Fellow of Wolfson College, Oxford University, Oxford, UK
e-mail: daniel.isaacson@philosophy.ox.ac.uk

D. DeVidi et al. (eds.), *Logic, Mathematics, Philosophy: Vintage Enthusiasms*,
The Western Ontario Series in Philosophy of Science 75,
DOI 10.1007/978-94-007-0214-1_7,

2 Soundness and Consistency

There are three key elements to establishing Gödel incompleteness for a formal system S in a language $\mathcal{L}$ such that for each natural number n there is a formal numeral, $\overline{n}$, which is a term in the language that denotes n when the variables of $\mathcal{L}$ range over the natural numbers.

First, to specify an assignment of numbers to the symbols of $\mathcal{L}$ which is extended to an assignment of numbers to the expressions (i.e., strings of symbols) of the language $\mathcal{L}$ of S, where we write $\ulcorner E \urcorner$ to signify the number of the expression E, by a function $x * y = z$ from pairs of natural numbers into the natural numbers expressible in $\mathcal{L}$ such that for $\frown$ the operation of concatenating two expressions, for any two expressions X and Y of $\mathcal{L}$, $\ulcorner X \frown Y \urcorner = \ulcorner X \urcorner * \ulcorner Y \urcorner$. It is required that the relation of n to $\ulcorner \overline{n} \urcorner$ is expressible in $\mathcal{L}$.

Second, to show that for every formula $F(v_1)$ with one free variable in $\mathcal{L}$, there is a sentence D such that $(D \equiv F(\overline{\ulcorner D \urcorner}))$ is true (Diagonal Lemma).

Third, to construct a formula $Pr(v_1)$ in $\mathcal{L}$ with one free variable such that $S \vdash X$ if and only if $Pr(\ulcorner X \urcorner)$.

The Gödel sentence can be obtained by applying the Diagonal Lemma to the one-place formula $\sim Pr(v_1)$, i.e., it is a sentence G such that $(G \equiv \sim Pr(\overline{\ulcorner G \urcorner}))$ is true.

Theorem 1 *If every theorem of S is true (about the natural numbers), then $S \nvdash G$, G is true, and $S \nvdash \sim G$.*

Proof Assume $S \vdash G$. Then by the third condition, $Pr(\ulcorner G \urcorner)$ is true. Then by the diagonal equivalence, G is false. Then by the assumption that S is sound (only proves true sentences), $S \nvdash G$, which contradicts our assumption that $S \vdash G$. So by propositional logic, $S \nvdash G$. Then by the third condition, $\sim Pr(\ulcorner G \urcorner)$ is true. So by the diagonal equivalence G is true. Then $\sim G$ is false, so by soundness of S, $S \nvdash \sim G$. ■

Remark Despite the exceedingly strong hypothesis that every theorem is true, and the formal similarity (self-referential diagonalization) of the proof of the first incompleteness theorem to the derivation of paradoxes such the Richard Paradox and the Liar Paradox (noted by by Gödel in the introductory section of (Gödel, 1931, p. 149), there is no threat of paradox. Diagonalization of provability in the context of sound provability gives a sentence which is neither provable or refutable, a surprise at the time but no paradox. Diagonalization on truth results in a sentence which is both true and false, a paradox, or—in the context of arithmetized syntax—results in the more refined result that truth for the language of arithmetic is not definable in the language of arithmetic.

Soundness implies consistency since no sentence and its negation can both be true. That consistency does not imply soundness, i.e., consistency is a strictly weaker condition than soundness, is an immediate corollary of Theorem 1, i.e.,

Corollary 2 *If S is sound then $S \cup \{\sim G\}$ is consistent but not sound.*

Proof If S is sound, then by Theorem 1, $S \nvdash G$. Then by propositional logic, $S \cup \{\sim G\}$ is consistent, but also $\sim G$ is false. ■

That consistency is strictly weaker than soundness can be proved without the apparatus of arithmetization of syntax required in the proof by Corollary 2, e.g., as follows.

Theorem 3 *Let $Q^* =_{\mathrm{df}} Q \cup \{\exists v_1 1 + v_1 = v_1\}$, where Q has as axioms the axioms of* PA *minus the Induction Axioms. Q^* is consistent and proves a sentence false in the structure of the natural numbers.*

Proof Q^* is consistent since it can be interpreted in the ordinals less than ω^ω with zero the ordinal 0, and successor, plus, and times interpreted as those ordinal operations, and the sentence $\exists v_1 1 + v_1 = v_1$ is false in the structure of the natural numbers. ■

Consistency of a Σ_0-complete system implies a very limited degree of soundness with respect to truth in the structure of the natural numbers, as follows.

Theorem 4 *If a system S is Σ_0-complete and consistent, it is Σ_0-sound, i.e., if sentence X is Σ_0 and $S \vdash X$, then X is true.*

Proof If X is Σ_0 and false, then $\sim X$ is Σ_0 and true, in which case if S is Σ_0-complete, $S \vdash \sim X$. Hence if S is consistent, $S \nvdash X$, which is to say that for X Σ_0, if $S \vdash X$, X is true. ■
By Theorem 3, this result is best possible.

The converse of Theorem 4 also holds, so we have

Theorem 5 *For any Σ_0-complete system, consistency is equivalent to Σ_0-soundness.*

Proof By Theorem 4 and the fact that if S is Σ_0-sound, then for X any false Σ_0-sentence, $S \nvdash X$, i.e., S is consistent. ■

Remark Consistency is necessary for incompleteness, since if S is inconsistent, then by propositional logic, for every F, $S \vdash F$ and $S \vdash \sim F$, so a fortiori $S \vdash F$ or $S \vdash \sim F$, i.e., S is complete.

Consistency of S is not only necessary but also sufficient for $S \nvdash G$. This is the first half of Gödel's First Incompleteness Theorem, as follows.

The proof depends on two results from arithmetization of the syntax of S.

Fact 6 *There is a Σ_0-formula $Prov(v_1, v_2)$ such that $\exists v_2 Prov(v_1, v_2)$ expresses $\{n : S \vdash E_n\}$, where E_n is the expression in the language of S with Gödel number n, i.e., $S \vdash E_n$ if and only if $\exists v_2 Prov(\overline{n}, v_2)$ is true.*

Lemma 7 *There is a Σ_0-formula $A(v_1, v_2)$ such that $\exists v_1 A(v_1, v_2)$ expresses $\{n : S \vdash E_n(n)\}$, where if E_n is a formula in the language of S with single free variable v_1, $E_n(n)$ is the formula that results by substituting n for the variable v_1 in E_n.*

Proof From Fact 6 by diagonal substitution into $\exists v_2 Prov(v_1, v_2)$. ∎

Theorem 8 *Let* $a =_{\text{df}} \ulcorner \forall v_2 \sim A(v_1, v_2) \urcorner$ *and let* $G =_{\text{df}} \forall v_2 \sim A(\overline{a}, v_2)$. *If* S *is consistent and* Σ_0*-complete,* $S \nvdash G$.

Proof Suppose $S \vdash G$, i.e., $S \vdash \forall v_2 \sim A(\overline{a}, v_2)$. Then $a \in \{n : S \vdash E_n(n)\}$. By Lemma 7, $\exists v_2 A(\overline{a}, v_2)$ is true. Since S is Σ_0-complete and hence Σ_1-complete, $S \vdash \exists v_2 A(\overline{a}, v_2)$. But this contradicts the consistency of S. So $S \nvdash G$. ∎

3 Truth of the Gödel Sentence

Theorem 9 *For any* Π_1*-sentence* X *in the language of a* Σ_0*-complete theory* S*, if* $S \nvdash \sim X$*, then* X *is true.*

Proof Since X is Π_1, it is of the form $\forall v_i F(v_i)$, for $F(v_i)$ a Σ_0-formula. If X is false, $\exists v_i \sim F(v_i)$ is true, so for some natural number k, $\sim F(\overline{k})$ is true. Then by Σ_0-completeness of S, $S \vdash \sim F(\overline{k})$. Hence $S \vdash \exists v_i \sim F(v_i)$, so $S \vdash \sim X$, which contradicts the hypothesis that $S \nvdash \sim X$. So X is true. ∎

Since the Gödel sentence G for any system S is Π_1, by Theorem 9, if $S \nvdash \sim G$, G is true. However, $S \nvdash \sim G$ holds only on a stronger condition than the consistency of S, as we shall see at the beginning of the next section, but as we shall now see, the proof of Theorem 8, that if S is consistent, $S \nvdash G$, shows the stronger result that

Theorem 10 *If* S *is consistent, the Gödel sentence for* S *is true.*

Proof By Theorem 8, if S is consistent, $S \nvdash G$, in which case, $a \notin \{n : S \vdash En(n)\}$. Then by Lemma 7, $\exists v_2 A(\overline{a}, v_2)$ is false, so $\forall v_2 \sim A(\overline{a}, v_2)$ is true, i.e., G is true. ∎

The converse of Theorem 10 holds, i.e., the Gödel sentence for S is equivalent to the consistency of S.

Theorem 11 *If the Gödel sentence for system* S *is true,* S *is consistent.*

Proof If G, i.e. $\forall v_2 \sim A(\overline{a}, v_2)$, is true, then $\exists v_2 A(\overline{a}, v_2)$ is false, in which case, by Lemma 7, $S \nvdash G$. But a system is consistent if there is any sentence that it does not prove, so S is consistent. ∎

Remark The question "how can we know the truth of the Gödel sentence for system S?" is tantamount, by Theorems 10 and 11, to the question "how do we know the consistency of S?" Almost everyone who knows about, for example, Peano Arithmetic believes it to be consistent. However, the most likely reason anyone will offer as grounds for confidence in the consistency of PA will be that the theory is true of the natural numbers, but this is a much stronger condition than (mere) consistency, which is all that is required to establish the truth of the Gödel sentence. Hilbert envisaged purely finitary consistency proofs, which Gödel's Second Incompleteness Theorem show to be unattainable. What is attainable, as the subsequent pursuit of Hilbert's programme established, is constructive consistency proofs. One may wonder if the conviction we have that, e.g., Peano Arithmetic is consistent is really attainable by means of formal consistency proofs, or whether infinitary considerations give more convincing grounds to hold that a given system is consistent, and thereby that the Gödel sentence for that system is true.

4 Gödel's Notion of ω-Consistency

We noted at the end of section 1 that consistency of S is sufficient to establish that $S \nvdash G$, for G the Gödel sentence for S. Consistency of S is not sufficient to show that $S \nvdash \sim G$, as seen by the fact that there are consistent systems that prove the negation of their Gödel sentence, e.g., the following.

Theorem 12 (a consistent theory that refutes its Gödel sentence) *For $P(v_1)$ a provability predicate for S, and for G a sentence in the language of S such that $S \vdash (G \equiv \sim P(\overline{\ulcorner G \urcorner}))$, let S^* be the system $S \cup \{\sim G\}$. Let $P^*(v_1)$ be a provability predicate for S^*, and let G^* be a sentence such that $S^* \vdash (G^* \equiv \sim P^*(\overline{\ulcorner G^* \urcorner}))$. Then $S^* \vdash \sim G^*$.*

Proof (1) For any sentence X, $S \vdash (X \supset (\sim G \supset X))$, by propositional logic. Then (2) $S \vdash (P(\overline{\ulcorner X \urcorner}) \supset P(\overline{\ulcorner (\sim G \supset X) \urcorner}))$, by properties of $P(v_1)$ as a provability predicate for S, and so by thinning, (3) $S^* \vdash (P(\overline{\ulcorner X \urcorner}) \supset P(\overline{\ulcorner (\sim G \supset X) \urcorner}))$. Then (4) $S^* \vdash (P(\overline{\ulcorner X \urcorner}) \supset P^*(\overline{\ulcorner X \urcorner}))$. For X such that $S \vdash \sim X$, by proof of the Second Incompleteness Theorem for S, $S \vdash (\sim G \equiv P(\overline{\ulcorner X \urcorner}))$, and so also (5) $S^* \vdash (\sim G \equiv P(\overline{\ulcorner X \urcorner}))$. Since $S \vdash \sim X$ implies, $S^* \vdash \sim X$, by the Second Incompleteness Theorem for S^*, (6) $S^* \vdash (\sim G^* \equiv P^*(\overline{\ulcorner X \urcorner}))$. By (4), (5), (6), $S^* \vdash (\sim G \supset \sim G^*)$. Then, since $S^* \vdash \sim G$, $S^* \vdash \sim G^*$. ■

The preceding theorem shows that proving the second half of the First Incompleteness Theorem, i.e., that $S \nvdash \sim G$, requires a hypothesis stronger than consistency of S. Gödel invoked a hypothesis which he called ω-consistency.

Definition 13 [ω-consistency] A system S in a language that contains a closed term $\overline{n}$, i.e., a numeral, for each natural number, is said to be ω-consistent if and only if there is no formula $F(v_i)$ with one free variable in $\mathcal{L}$ such that $S \vdash \exists v_i F(v_i)$ and for each natural number n, $S \vdash \sim F(\overline{n})$.

Theorem 14 *If a system is ω-consistent, then it is consistent.*

Proof The contrapositive is immediate by *ex falso quodlibet*: if a system S is inconsistent S proves every formula in the language of S; in particular for any formula $F(w)$ with one free variable, $S \vdash \exists w F(w)$, and for each n $S \vdash \sim F(\overline{n})$, i.e., S is ω-inconsistent. ■

The converse of Proposition 14 does not hold, i.e., consistency is strictly weaker than ω-consistency.

Theorem 15 *There are consistent systems that are ω-inconsistent.*

Proof One such system is Q^* in Theorem 3, shown in the proof of that theorem to be consistent. But also it is ω-inconsistent, as follows. For each natural number n, $Q^* \vdash \sim 1+\overline{n} = \overline{n}$ by meta-induction on n, for which the base case is $Q^* \vdash 1+0 = 1$, so $Q^* \vdash \sim 1+0 = 0$, and induction step is $Q^* \vdash (\sim 1+\overline{n} = \overline{n} \supset \sim(1+\overline{n})' = \overline{n}')$, so if $Q^* \vdash \sim 1+\overline{n} = \overline{n}$, $Q^* \vdash \sim(1+\overline{n})' = \overline{n}'$ and since $Q^* \vdash (1+\overline{n})' = (1+\overline{n}')$, $\vdash \sim(1+\overline{n}') = \overline{n}'$. ■

Theorem 16 *If a system is sound with respect to truth in arithmetic, then it is ω-consistent.*

Proof Let S be a system whose language contains numerals for the natural numbers and which is sound with respect to truth in arithmetic. Suppose $S \vdash \exists w F(w)$. Then $\exists w F(w)$ is true, i.e., there is a natural number n such that $F(\overline{n})$ is true, which is to say that $\sim F(\overline{n})$ is false. So $S \nvdash \sim F(\overline{n})$, which is to say that S is ω-consistent. ■

The converse holds only for sentences up to Σ_2 in the arithmetical hierarchy.

Theorem 17 *If S is Σ_0-complete and ω-consistent, then S is Σ_2-sound.*

Proof Suppose $S \vdash \exists v_1 \forall v_2 F(v_1, v_2)$ and $\exists v_1 \forall v_2 F(v_1, v_2)$ is false, i.e., there is no number n such that $\forall v_2 F(\overline{n}, v_2)$ is true, i.e., for each natural number n, $\forall v_2 F(\overline{n}, v_2)$ is false, i.e., $\exists v_2 \sim F(\overline{n}, v_2)$ is true. Then by Σ_1-completeness of S (an immediate consequence of Σ_0-completeness), for each n, $S \vdash \exists v_2 \sim F(\overline{n}, v_2)$, so by logic in S, for each n, $S \vdash \sim \forall v_2 F(\overline{n}, v_2)$. This contradicts the ω-consistency of S. So $\exists v_1 \forall v_2 F(v_1, v_2)$ is true. ■

Corollary 18 *If S is Σ_0-complete and ω-consistent, then S is Σ_1-sound.*

Proof Let $\exists v_1 H(v_1)$ be a Σ_1-sentence such that $S \vdash \exists v_1 H(v_1)$. Then $S \vdash \exists v_1 \forall v_2 (v_2 = v_2 \supset H(v_1))$. By Theorem 17, $\exists v_1 \forall v_2 (v_2 = v_2 \supset H(v_1))$ is true. So $\exists v_1 H(v_1)$ is true. ■

Proposition 19 (Kreisel 1955) *There is an ω-consistent system that proves a false Σ_3-sentence.*

Proof Let $P(v_1)$ be a formula that expresses $\{n : \text{PA} \vdash E_n\}$. Let

$$H(x) =_{\text{df}} \exists y (P(\ulcorner E_x \supset \exists v_1 E_y \urcorner) \wedge \forall z P(\ulcorner E_x \supset \sim E_y[z] \urcorner)).$$

From the arithmetization of the syntax of PA, $H(x)$ can be written out as a formula in the language of PA.

Let K be a diagonal sentence for $H(x)$, i.e., s.t. that $(K \equiv H(\overline{\ulcorner K \urcorner}))$. So K is true if and only if K when added to PA proves an ω-inconsistency. We show that PA ∪ $\{K\}$ is ω-consistent, as follows.

Suppose that PA ∪ $\{K\}$ were ω-inconsistent. Then K would be true, i.e., when added to PA it results in an ω-inconsistent system. But also since PA is sound with respect to truth, PA ∪ $\{K\}$ would be sound with respect to truth. But any theory with numerals for all the natural numbers that is sound with respect to truth in the natural numbers is ω-consistent. This contradicts the supposition that PA $\cup \{K\}$ is ω-inconsistent, so by *reductio ad absurdum* PA $\cup \{K\}$ must be ω-consistent.

Consequently, K is false, i.e., when added to PA it does *not* result in an ω-inconsistent system, which means that PA $\cup \{K\}$ is an ω-consistent false theory, as required.

Finally we need to show that K is Σ_3.

$$K \equiv \exists y(P(\ulcorner K \supset \exists v_1 E_y \urcorner) \wedge \forall z P(\ulcorner K \supset \sim E_y[z]\urcorner))$$

so $K \equiv \exists y \forall z(P(\ulcorner K \supset \exists v_1 E_y \urcorner) \wedge P(\ulcorner K \supset \sim E_y[z]\urcorner))$ by predicate logic (prenexing). The predicate $P(v_1)$ is Σ_1. So the matrix of this formula contains two existential quantifiers. When they are brought into prenex position there are then two adjacent existential quantifiers within the scope of $\exists\forall$, which can be contracted to a single existential quantifier and bounded quantifiers. The result is a quantifier prefix $\exists\forall\exists$, so K is Σ_3. ■

Gödel states in (Gödel, 1931, p. 176) that the notion of ω-consistency is "much weaker" than the assumption that "every formula is true in the interpretation considered" but gives no argument for this claim.

Remark Though by Proposition 19, ω-consistency does not imply soundness, there is a weak sense in which it does, as follows:

Theorem 20 *True arithmetic is the only ω-consistent extension of* PA *that includes each sentence or its negation.*

Proof We have already seen that any theory for which all consequences are true in arithmetic is ω-consistent, so in particular true arithmetic is ω-consistent, and of course includes each sentence or its negation.

We need to show that every ω-consistent extension of PA that includes each sentence or its negation coincides with true arithmetic.

Let S be such an extension of PA. We argue by induction over the logical complexity of sentences that for all sentences X, S correctly decides X, i.e., if X is true $S \vdash X$ and if X is false $S \vdash \sim X$.

(i) Base case: X is atomic. This case follows immediately by Σ_0-completeness of PA.
(ii) Induction steps:
   (α) X is of the form $\sim Y$ or of the form $(Y \supset Z)$: These cases were dealt with in Problem 1.
   (β) X is of the form $\forall v_i F(v_i)$. By induction hypothesis, S correctly decides every sentence of the form $F(\overline{m})$ for natural number m.
      (1) Suppose X is true. Then by induction hypothesis for each number m $S \vdash F(\overline{m})$. By completeness of S, $S \vdash X$ or $S \vdash \sim X$. Suppose $S \vdash \sim X$.

Then $S \vdash \exists v_i \sim F(v_i)$. But this contradicts the ω-consistency of S. So the supposition is wrong, which means that $S \vdash X$, as required.

(2) Suppose X is false. Then there is a number k such that $\sim F(\overline{k})$ is true. Then by induction hypothesis, $S \vdash \sim F(\overline{k})$. Then by logic in S, $S \vdash \sim\forall v_i F(v_i)$, as required. ■

Remark As for consistency, an ω-consistent system can, if it doesn't decide every sentence in its language, be extended ω-consistently, as follows.

Theorem 21 *If S is ω-consistent, then for X any sentence in $\mathcal{L}(S)$, $S \cup \{X\}$ or $S \cup \{\sim X\}$ is ω-consistent.*

Proof Suppose $S \cup \{X\}$ and $S \cup \{\sim X\}$ are both ω-inconsistent, i.e., there are formulas $A(x)$ and $B(x)$ such that

$$S \cup \{X\} \vdash \exists x A(x) \tag{7.1}$$
$$\text{and } S \cup \{X\} \vdash \sim A(\overline{n}) \text{ for each } n \in \omega \tag{7.2}$$
$$S \cup \{\sim X\} \vdash \exists x B(x) \tag{7.3}$$
$$\text{and } S \cup \{\sim X\} \vdash \sim B(\overline{n}) \text{ for each } n \in \omega \tag{7.4}$$

then

$$S \vdash ((X \wedge \exists x A(x)) \vee (\sim X \wedge \exists x B(x))) \text{ from (7.1), (7.3) and prop. logic} \tag{7.5}$$
$$S \vdash \exists x((X \wedge A(x)) \vee (\sim X \wedge B(x))) \text{ (7.5) and predicate logic} \tag{7.6}$$
$$S \vdash ((X \supset \sim A(\overline{n})) \wedge (\sim X \supset \sim B(\overline{n}))) \text{ (7.2), (7.4), DT, } \wedge\text{-I} \tag{7.7}$$
$$S \vdash (\sim (X \wedge A(\overline{n})) \wedge \sim (\sim X \wedge B(\overline{n})) \text{ (7.7) and prop. logic} \tag{7.8}$$
$$S \vdash \sim ((X \wedge A(\overline{n})) \vee (\sim X \wedge B(\overline{n}))) \text{ (7.8) and prop. logic} \tag{7.9}$$

(7.6) and (7.9) imply that S is ω-inconsistent.

It must thus be the case that either $S \cup \{X\}$ or $S \cup \{\sim X\}$ is ω-consistent. ■

Remark The corresponding property to Theorem 21 for consistency gives rise to Lindenbaum's Lemma, the result that every consistent set of sentences can be extended to a consistent set of sentences in which each sentence or its negation occurs: enumerate the sentences of the language, successively add each sentence or its negation, preserving consistency; the union of all stages is a consistent set that contains every sentence or its negation. By Proposition 19, it cannot be that Theorem 21 implies that every ω-consistent set of sentences can be extended to an ω-consistent set that contains each sentence or its negation. Where the proof of Lindenbaum's Lemma breaks down in the case of ω-consistency is that the union of a chain of ω-consistent sets need not be ω-consistent, i.e., the union may contain an ω-inconsistency (which is a infinite set of sentences) not contained in any one element of the chain (whereas an inconsistency in the union must, being finite, occur in an element of the chain).

Theorem 22 *The addition of a true Π_1-formula A to a system S that is Σ_0-complete preserves ω-consistency.*

Proof (letter from Georg Kreisel 4 April 2005) . Suppose $S \cup \{A\}$ is ω-inconsistent, i.e., there is a formula $\exists x B(x)$ such that:

$$S \vdash (A \supset \exists x B(x)) \quad (7.10)$$

$$\text{and } S \vdash (A \supset\sim B(\overline{n})) \text{ for each } n \in \omega \quad (7.11)$$

A is of the form $\forall x A_0(x)$ where $A_0(x)$ is Σ_0 (quantifier-free). So

$$S \vdash (\exists x \sim A_0(x) \vee \exists x B(x)) \quad (7.12)$$

$$S \vdash (A \supset (A \wedge \exists x B(x)) \text{ (7.10) and prop. logic} \quad (7.13)$$

$$S \vdash (\exists x \sim A_0(x) \vee (A \wedge \exists x B(x))) \text{ (7.13) and prop/pred logic} \quad (7.14)$$

$$S \vdash \exists x(\sim A_0(x) \vee (A \wedge B(x))) \text{ (7.14) and pred. logic} \quad (7.15)$$

$$S \vdash A_0(\overline{n}) \text{ for each } n \in \omega\text{, since } \forall x A_0(x) \text{ is true and } S \text{ is } \Sigma_0\text{-complete} \quad (7.16)$$

$$S \vdash (A_0(n) \wedge (A \supset\sim B(\overline{n})) \text{ for each } n\text{, due to (7.11) and (7.16) and logic in } S \quad (7.17)$$

$$S \vdash\sim (\sim A_0(\overline{n}) \vee (A \wedge B(\overline{n}))) \text{ for each } n \text{ by (7.26) and prop. logic} \quad (7.18)$$

But (7.15) and (7.18) imply that S is ω-inconsistent.

Thus $S \cup \{A\}$ must be ω-consistent. ■

Corollary 23 *If S is ω-consistent, then*

$$S \cup \{((Pr(\overline{\ulcorner \sim G \urcorner}) \supset\sim G)\}$$

is ω-consistent.

Proof If S is ω-consistent, then $\sim Pr(\overline{\ulcorner \sim G \urcorner})$ is a true Π_1^0-sentence. Hence, by the theorem, if S is ω-consistent $S \cup \{\sim Pr(\overline{\ulcorner \sim G \urcorner})\}$ is ω-consistent.

$$S \cup \{\sim Pr(\overline{\ulcorner \sim G \urcorner})\} \vdash (Pr(\overline{\ulcorner \sim G \urcorner}) \supset\sim G) \text{ by prop. logic} \quad (7.19)$$

Thus $S \cup \{(Pr(\overline{\ulcorner \sim G \urcorner}) \supset\sim G)\}$ must be ω-consistent if S is ω-consistent. ■

5 Kreisel's Notions of *n*-Consistency

5.1 The Notion of 1-Consistency

The notion of ω-consistency is considerably stronger than needed for proving the second half of the First Incompleteness Theorem. Since the arithmetized proof predicate for a system S with arithmetized syntax is Σ_1, the result can be proved with just the assumption that there is no Σ_1 ω-inconsistency. This weaker notion is called 1-consistency, for which the precise definition is:

Definition 24 (1-consistency) A system S in a language that contains a closed term $\overline{n}$, i.e., a numeral, for each natural number, is said to be *1-consistent* if and only if there is no Σ_1-formula $\exists v_1 F(v_1)$ with one free variable in the language such that $S \vdash \exists v_i F(v_i)$ and for each natural number n, $S \vdash \sim F(\overline{n})$.

Theorem 25 *For a Σ_0-complete system S, 1-consistency of S is equivalent to Σ_1-soundness of S.*

Proof (i) Suppose S is 1-consistent, and let $\exists v_i F(v_i)$ be a Σ_1-sentence such that $S \vdash \exists v_i F(v_i)$, and suppose $\exists v_i F(v_i)$ is false. Then for each number n, $\sim F(\overline{n})$ is a true Σ_0-sentence. Hence by Σ_0-completeness of S, for each natural number n, $S \vdash \sim F(\overline{n})$. But this violates the hypothesized 1-consistency of S. So $\exists v_i F(v_i)$ is true. So S is Σ_1-sound.

(ii) Suppose S is Σ_1-sound, and suppose $S \vdash \exists v_i F(v_i)$. Then by Σ_1-soundness of S, there is a natural number k such that $F(\overline{k})$ is a true Σ_0-sentence. By Σ_0-completeness of S, $S \vdash F(\overline{k})$. Then $S \nvdash \sim F(\overline{k})$. Otherwise S is inconsistent, so proves every sentence, including false Σ_1-sentences, violating the hypothesized Σ_1-soundness of S. So S is 1-consistent.■

Theorem 26 *For a Σ_0-complete system S, Σ_1-soundness of S is equivalent to consistency of S + all true Π_1-sentences.*

Proof (i) Suppose S is Σ_1-sound and suppose that S + all true Π_1-sentences is inconsistent. From the assumption that PA is Σ_1-sound it follow that PA is consistent. Thus if S + all true Π_1-sentences is inconsistent, it must be that there is a true Π_1-sentence $\forall v_1 F(v_1)$ s.t. PA $\vdash \sim\forall v_1 F(v_1)$. But $\sim\forall v_1 F(v_1) \equiv \exists v_1 \sim F(v_1)$ is then a false Σ_1-sentence, which contradicts the Σ_1-soundness of PA. So S + all true Π_1-sentences is consistent.

(ii) Suppose S + all true Π_1-sentences is consistent and that PA is not Σ_1-sound, i.e., there is a false Σ_1-sentence $\exists v_1 H(v_1)$ such that PA $\vdash \exists v_1 H(v_1)$, so then PA $\vdash \sim\forall v_1 \sim H(v_1)$, where $\forall v_1 \sim H(v_1)$ is a true Π_1-sentence. But then S + all true Π_1-sentences is inconsistent.■

Theorem 27 *If S is 1-consistent then for X any sentence in $\mathcal{L}(S)$, $S \cup \{X\}$ or $S \cup \{\sim X\}$ is 1-consistent.*

Proof First note that the proof of the corresponding result for ω-consistency does not establish this result for 1-consistency since the formula $\exists x((X \wedge A(x)) \vee (\sim X \wedge B(x)))$, for $A(x)$ and $B(x)$ Σ_0-formulas, is not Σ_1 for an arbitrary sentence X. Kreisel has given a proof (letter of 31 March 2005) of which the following is a variant.

Suppose $S\cup\{X\}$ and $S\cup\{\sim X\}$ are both 1-inconsistent, i.e., there are Σ_1-formulae $\exists x A(x)$ and $\exists x B(x)$, which is to say $A(x)$ and $B(x)$ are Σ_0 formulas such that:

$$S \cup \{X\} \vdash \exists x A(x) \tag{7.20}$$

$$\text{and } S \cup \{X\} \vdash \sim A(\overline{n}) \text{ for each } n \in \omega \tag{7.21}$$

$$S \cup \{\sim X\} \vdash \exists x B(x) \tag{7.22}$$

$$\text{and } S \cup \{\sim X\} \vdash \sim B(\overline{n}) \text{ for each } n \in \omega \tag{7.23}$$

$$S \vdash (\exists x A(x) \vee \exists x B(x)) \text{ (7.20), (7.22) and } S \vdash (X \vee \sim X) \tag{7.24}$$

$$S \vdash \exists x (A(x) \vee B(x)) \text{ (7.24) and pred. logic} \tag{7.25}$$

I now claim that for all $n \in \omega$, $S \vdash (\sim A(\overline{n}) \wedge \sim B(\overline{n}))$. For if not, then for some $n \in \omega$:

$$S \nvdash (\sim A(\overline{n}) \wedge \sim B(\overline{n})) \tag{7.26}$$

But $(\sim A(\overline{n}) \wedge \sim B(\overline{n}))$ is a Σ_0-sentence and S is Σ_0-complete, so

$$S \vdash (A(\overline{n}) \vee B(\overline{n})) \tag{7.27}$$

Since S is Σ_0-sound $(A(\overline{n}) \vee B(\overline{n}))$ is true, so $A(\overline{n})$ is true or $B(\overline{n})$ is true. We suppose w.l.o.g. that $A(\overline{n})$ is true. Then by Σ_0-completeness

$$S \vdash A(\overline{n}) \tag{7.28}$$

$$\text{From (7.21) we have } S \vdash (X \supset \sim A(\overline{n})) \text{ by DT} \tag{7.29}$$

$$\text{So } S \vdash (A(\overline{n}) \supset \sim X) \text{ by (7.29) and prop. logic} \tag{7.30}$$

$$S \vdash \sim X \text{ by (7.28), (7.30) and MP} \tag{7.31}$$

But then by assumption in (7.22) and (7.23) we have:

$$S \vdash \exists x B(x) \text{ and for each } n \in \omega \; S \vdash \sim B(\overline{n}) \tag{7.32}$$

whence S is 1-inconsistent.

So the claim that for all $n \in \omega$, $S \vdash (\sim A(\overline{n}) \wedge \sim B(\overline{n}))$ must be correct.

But then (7.25) again implies that S is 1-inconsistent.

It must thus be that either $S \cup \{X\}$ or $S \cup \{\sim X\}$ is 1-consistent. ■

Remark The following theorem stands in contrast with Theorem 20 for ω-consistency.

Theorem 28 *There are 1-consistent extensions of* PA *other than true arithmetic that include each sentence or its negation.*

Proof The false ω-consistent extension of PA proved to exist by Proposition 19 is 1-consistent. By Theorem 27 and the usual method of generating an extension of a consistent set of sentences that includes every sentence or its negation (Lindenbaum's Lemma), we obtain a 1-consistent complete extension. The reason this construction works in the case of 1-consistency and does not, as explained in the

Remark following Theorem 21, in the case of ω-consistency, is that by Theorem 25, 1-consistency is equivalent to Σ_1-soundness, so if the union of the chain of 1-consistent sets of sentences if 1-inconsistent, it must contain a false Σ_1-sentence. But that false Σ_1-sentence must occur in one of the sets of sentences in the chain, in which case that set of sentences is 1-inconsistent, contrary to construction. ■

Remark The notion of 1-consistency readily generalizes to n-consistency for each natural number n, but as we shall see, n-consistency is only a natural notion for $n = 1$ and $n = 2$.

5.2 The Notion of 2-Consistency

Definition 29 (2-consistency) A system S in a language that contains a closed term $\overline{n}$, i.e., a numeral, for each natural number, is said to be 2-consistent if and only if there is no Σ_2-formula $\exists v_1 \forall v_2 F(v_1, v_2)$ in the language of S such that $S \vdash \exists v_1 \forall v_2 F(v_1, v_2)$ and for each natural number n, $S \vdash \sim\forall v_2 F(\overline{n}, v_2)$.

Theorem 30 *For a Σ_0-complete system, 2-consistency is equivalent to Σ_2-soundness.*

Proof (i) Suppose S is 2-consistent and suppose $S \vdash \exists v_1 \forall v_2 F(v_1, v_2)$, where $F(v_1, v_2)$ is a Σ_0-formula, and $\exists v_1 \forall v_2 F(v_1, v_2)$ is false, which is to say that for each natural number n, $\exists v_2 \sim F(\overline{n}, v_2)$ is a true Σ_1-sentence. Then by the Σ_1-completeness of every Σ_0-complete theory and predicate logic, for each natural number n, $S \vdash \sim\forall v_2 F(\overline{n}, v_2)$. But then S is 2-inconsistent. So by RAA, $\exists v_1 \forall v_2 F(v_1, v_2)$ is true.

(ii) Suppose S is Σ_2-sound and suppose $S \vdash \exists v_1 \forall v_2 F(v_1, v_2)$. It follows that $\exists v_1 \forall v_2 F(v_1, v_2)$ is true, so for some number k, $\forall v_2 F(\overline{k}, v_2)$ is true. Suppose $S \vdash \sim\forall v_2 F(\overline{k}, v_2)$. Then $S \vdash \exists v_2 \sim F(\overline{k}, v_2)$. Since S is Σ_2-sound, it is Σ_1-sound, so $\exists v_2 \sim F(\overline{k}, v_2)$ is true. But this contradicts the truth of $\forall v_2 F(\overline{k}, v_2)$, so by RAA, $S \nvdash \sim\forall v_2 F(\overline{k}, v_2)$. This means that S is 2-consistent.■

Theorem 31 *For a Σ_0-complete system S, Σ_2-soundness of S is equivalent to consistency of S + all true Π_2-sentences.*

Proof (i) Suppose S is Σ_2-sound and S + all true Π_2-sentences is inconsistent. Since S is Σ_2-sound, S is consistent, so the inconsistency of S + all true Π_2-sentences means that there is a true Π_2-sentence $\forall v_1 \exists v_2 A(v_1, v_2)$ such that $S \vdash \sim\forall v_1 \exists v_2 A(v_1, v_2)$. But then $S \vdash \exists v_1 \forall v_2 \sim A(v_1, v_2)$, and $\exists v_1 \forall v_2 \sim A(v_1, v_2)$ is a false Σ_2-sentence, which contradicts the assumed Σ_2-soundness of S.

(ii) Suppose S + all true Π_2-sentences is consistent, and S is not Σ_2-sound, i.e., there is a false Σ_2-sentence $\exists v_1 \forall v_2 B(v_1, v_2)$ such that $S \vdash \exists v_1 \forall v_2 B(v_1, v_2)$. Since $\exists v_1 \forall v_2 B(v_1, v_2)$ is false, $\forall v_1 \exists v_2 \sim B(v_1, v_2)$ is true, so S refutes a true Π_2-sentence, so S + all true Π_2-sentences is inconsistent, contrary to assumption. So S is Σ_2-sound.■

Theorem 32 *If S is 2-consistent then for X any sentence in $\mathcal{L}(S)$, $S \cup \{X\}$ or $S \cup \{\sim X\}$ is 2-consistent.*

Proof (letter from Georg Kreisel 7 April 2005) Since S is 2-consistent, S is 1-consistent, so by Theorem 27, $S \cup \{X\}$ or $S \cup \{\sim X\}$ is 1-consistent. There are two cases—either one or both of these extensions are 1-consistent.

Case when just one extension is 1-consistent: We may suppose w.l.o.g. that $S \cup \{X\}$ is 1-consistent and $S \cup \{\sim X\}$ is 1-inconsistent. From the 1-inconsistency of $S \cup \{\sim X\}$ we have that there is a quantifier-free (or Σ_0) formula $B_0(y)$ such that $S \vdash (\sim X \supset \exists y B_0(y))$ and for all $n \in \omega$, $S \vdash (\sim X \supset \sim B_0(\overline{n}))$: call this $(*)$.

We distinguish between the cases when $S \cup \{\sim X\}$ is consistent and when it is inconsistent.

First suppose that it is consistent. By the absoluteness of purely numerical formulae $\sim B_0(\overline{n})$ for all $n \in \omega$ are true, and so by Σ_1-completeness for all $n \in \omega$ $S \vdash \sim B_0(\overline{n})$.

We show that $S \cup \{X\}$ is 2-consistent. Suppose not, i.e., for some Σ_0-formula $A_0(y, z)$ $S \cup \{X\} \vdash \exists y \forall z A_0(y, z)$ and for all $n \in \omega$ $S \cup \{X\} \vdash \exists z \sim A_0(\overline{n}, z)$: call this $(**)$. Since $S \cup \{X\}$ is 1-consistent, it is Σ_1-sound, so $\exists z \sim A_0(\overline{n}, z)$ is true. So by Σ_1-completeness of S, $S \vdash \sim \forall z A_0(\overline{n}, z)$. Hence $S \vdash (\sim \forall z A_0(\overline{n}, z) \wedge \sim B_0(\overline{n}))$ and so $S \vdash \sim (\forall z A_0(\overline{n}, z) \vee B_0(\overline{n}))$ and so $S \vdash \sim \forall z (A_0(\overline{n}, z) \vee B_0(\overline{n}))$.

But also from $(*)$ and $(**)$

$$S \vdash (\exists y \forall x A_0(y, x) \vee \exists y B_0(y))$$

and so

$$S \vdash \exists y \forall x (A_0(y, x) \vee B_0(y))$$

which is to say that S is Σ_2-inconsistent, contrary to hypothesis.

Now suppose that $S \cup \{\sim X\}$ is inconsistent, $S \vdash X$ so $S \cup \{X\} = X$. So if S is 2-consistent then, vacuously, $S \cup \{X\}$ is 2-consistent.

Case when $S \cup \{X\}$ and $S \cup \{\sim X\}$ are both 1-consistent: Assume $S \cup \{X\}$ and $S \cup \{\sim X\}$ are both 2-inconsistent, i.e., that there are Σ_0-formulas $A_0(y, z)$ and $B_0(y, z)$ such that

$$S \cup \{X\} \vdash \exists y \forall z A_0(y, z) \tag{7.33}$$
$$\text{and } S \cup \{X\} \vdash \sim \forall z A_0(\overline{n}, z) \text{ for all } n \in \omega \tag{7.34}$$

together with

$$S \cup \{\sim X\} \vdash \exists y \forall z B_0(y, z) \tag{7.35}$$
$$\text{and } S \cup \{\sim X\} \vdash \sim \forall z B_0(\overline{n}, z) \text{ for all } n \in \omega \tag{7.36}$$

By (7.33) and (7.35) and given that $S \vdash (X \vee \sim X)$,

$$S \vdash (\exists y \forall z A_0(y, z) \vee \exists y \forall z B_0(y, z))$$

and so

$$S \vdash \exists y \forall z \forall u (A_0(y, z) \vee B_0(y, z))$$

Since $S \cup \{X\}$ and $S \cup \{\sim X\}$ are both 1-consistent and hence Σ_1-sound, $\sim \forall z A_0(\overline{n}, z)$ and $\sim \forall z B_0(\overline{n}, z)$ are all true, and hence provable S, and so their conjunction also is:

$$S \vdash \sim \forall z A_0(\overline{n}, z) \wedge \sim \forall z B_0(\overline{n}, z)$$

so

$$S \vdash \sim (\forall z A_0(\overline{n}, z) \vee \forall z B_0(\overline{n}, z))$$

so

$$S \vdash \sim \forall z \forall u (A_0(\overline{n}, z) \vee B_0(\overline{n}, u))$$

Then S is 2-inconsistent, contrary to the assumption. So at least one of $S \cup \{X\}$ and $S \cup \{\sim X\}$ is 2-consistent. ■

5.3 Properties of 3-Consistency

Remark We have seen that for S any Σ_0-complete axiomatic theory and for $n = 1$ and 2, 1-consistency of S, Σ_n-soundness of S, and consistency of S + all true Π_n-sentences coincide, and give content to the notion of 1-consistency and 2-consistency. On the other hand, the notions of n-consistency, Σ_n-soundness, and consistency of S + all true Π_n-sentences come apart from each other for $n \geq 3$.

Theorem 33 *The properties of 3-consistency and Σ_3-soundness are not equivalent.*

Proof Since ω-consistency implies 3-consistency, the theory constructed in Proposition 19 is 3-consistent and not Σ_3-sound. ■

Theorem 34 *The properties of 3-consistency of a system S and consistency of S + all true Π_3-sentences are not equivalent.*

Proof Let K be the false Σ_3-sentence constructed in Proposition 19. Then its negation is a true Π_3-sentence, in which case PA $\cup \{K\} \cup$ {all true Π_3-sentences} is inconsistent, while PA $\cup \{K\}$ is 3-consistent. ■

5.4 Expressing Consistency, ω-Consistency, 1-Consistency, and 2-Consistency

When Gödel said that the condition of ω-consistency is "purely formal" in contrast to the assumption that every theorem is *true*, he was strictly correct if that means "expressible in arithmetized syntax." But if it means expressible in *finitary arithmetic*, and finitary arithmetic is identified with quantifier-free or bounded arithmetic (PRA), then ω-consistency cannot be expressed. This subsection looks at how the notions of consistency, ω-consistency, 1-consistency, and 2-consistency can be expressed in arithmetized syntax. (Recall that the role of the function $x * y = z$ in arithmetized syntax is specified in the second paragraph of Section 2.)

Expressing consistency: Consistency is expressed by a Π^0_1-sentence, $\forall v_0 \sim Prov(\ulcorner 0 = 0' \urcorner, v_0)$, for the Σ_0-formula $Prov(v_1, v_0)$.

Expressing ω-consistency: The property of ω-consistency is Π^0_3, when we arithmetize the definition of ω-consistency: "There does not exist a formula with one free variable $F(v_1)$ such that $S \vdash \exists v_1 F(v_1)$ and for all $n \in \omega$ $S \vdash \sim F(\overline{n})$."

The property of being the Gödel number of a formula is Σ_0, though this is a little bit delicate: F is a formula if and only if there exists a formation sequence which generates it. On the face of it, this is a Σ_1-property (like provability). However, unlike provability, there is a computable bound on the length of the formation sequence, computable from the expression being tested as to whether or not it is a formula. Let $Fm_1(v_0)$ be a Σ_0-predicate (i.e., with bounded quantification only) such that $Fm(\overline{n})$ is true if and only if n is the Gödel number of a formula with one free variable. We can then express ω-consistency by:

$$\begin{aligned}
\sim \exists v_0(F_1(v_0) \wedge \exists v_1 Prov(\overline{\ulcorner \exists v_0 \urcorner * v_0}, v_1) \wedge \forall v_2 \exists v_3 Prov(s(v_0, v_2), v_3))\\
\forall v_0(F_1(v_0) \supset \sim (\exists v_1 Prov(\overline{\ulcorner \exists v_0 \urcorner * v_0}, v_1) \wedge \forall v_2 \exists v_3 Prov(s(v_0, v2), v_3)))\\
\forall v_0(Fm_1(v_0) \supset (\forall v_1 \sim Prov(\overline{\ulcorner \exists v_0 \urcorner * v_0}, v_1) \vee \exists v_2 \forall v_3 \sim Prov(s(v_0, v_1), v_3)))\\
\forall v_0(Fm_1(v_0) \supset \exists v_2 \forall v_1 \forall v_3(\sim Prov(\overline{\ulcorner \exists v_0 \urcorner * v_0}, v_1) \vee \sim Prov(s(v_0, v_1), v_3)))\\
\forall v_0 \exists v_2 \forall v_1 \forall v_3(Fm_1(v_0) \supset (\sim Prov(\overline{\ulcorner \exists v_0 \urcorner * v_0}, v_1) \vee \sim Prov(s(v_3, v_1), v_3)))
\end{aligned}$$

Thus the ω-consistency of a theory is a Π^0_3 sentence.

Expressing 1-consistency: To express 1-consistency for a consistent theory S, we need to say: "For every Σ_0-formula $F(v_0)$, if $S \vdash \exists v_0 F(v_0)$, there exists $n \in \omega$ such that $S \nvdash \sim F(\overline{n})$."

But by the property of completeness of S w.r.t. Σ_0-truth and the consistency of S, this is equivalent to: "For every Σ_0-formula $F(v_0)$, if there exists a proof (in S) of $\exists v_0 F(v_0)$, then there exists $n \in \omega$ such that there exists a proof (in S) of $F(\overline{n})$."

Let $\Sigma_0 Fm_1(x)$ be a Σ_0-formula such that $\Sigma_0 Fm_1(x)$ is true if and only if E_x is a Σ_0-formula with one free variable. 1-consistency/Σ_1-soundness can then be expressed as follows:

$$\forall v_0 \forall v_1(\Sigma_0 Fm(v_0) \wedge Prov(\overline{\ulcorner \exists v_0 \urcorner * v_0}, v_1) \supset \exists v_2 \exists v_3 Prov(s(v_0, v_2), v_3))$$

which is equivalent to

$$\forall v_0 \forall v_1 \exists v_2 \exists v_3(\Sigma_0 Fm(v_0) \wedge Prov(\overline{\ulcorner \exists v_0 \urcorner * v_0}, v_1) \supset Prov(s(v_0, v_2), v_3))$$

So 1-consistency/Σ_1-soundness for a Σ_0-complete theory is expressible by a Π^0_2-sentence.

Expressing 2-consistency: To express 2-consistency we need to say, "For every Σ_0-formula $F(v_0, v_1)$, if $S \vdash \exists v_0 \forall v_1 F(v_0, v_1)$, then there exists n such that $S \nvdash\sim \forall v_1 F(\overline{n}, v_1)$."

This is a Π_3^0-condition in the same way that ω-consistency is, since there is no equivalence:

$$S \nvdash\sim \forall v_1 f(\overline{n}, v_1) \text{ iff } S \vdash \forall v_1 F(\overline{n}, v_1)$$

If we are given that S is consistent, then

$$S \vdash \forall v_1 F(\overline{n}, v_1) \text{ implies } S \nvdash\sim \forall v_1 F(\overline{n}, v_1)$$

But no consistency condition, including ω-consistency, would give

$$S \nvdash\sim \forall v_1 F(\overline{n}, v_1) \text{ implies } S \vdash \forall v_1 F(\overline{n}, v_1)$$

which is a Π_1-completeness condition.

By contrast, in the case of 1-consistency, the needed equivalence is

$$S \nvdash\sim F(\overline{n}) \text{ iff } S \vdash F(\overline{n})$$

for $F(v_0)$ a Σ_0-formula, which holds by Σ_0-completeness and consistency (as in the argument justifying the expression of 1-consistency above).

6 Consistency of $S \cup \{Con_S\}$

Though 1-consistency is a much weaker condition than ω-consistency, it is still stronger than required. It is immediate from $S \vdash (G \equiv Con_S)$, which is provable from the proof of the Second Incompleteness Theorem, that the consistency of $S \cup \{Con_S\}$ is a necessary and sufficient condition for $S \nvdash \sim G$, i.e.,

Theorem 35 *Let $P(v_1)$ be a provability predicate for a system S, and let G be a sentence in the language of S such that $S \vdash (G \equiv \sim P(\ulcorner G \urcorner))$. Let X be any sentence such that $S \vdash \sim X$. Let Con_S stand for $\sim P(\ulcorner X \urcorner)$. Then $S \cup \{Con_S\}$ is consistent if and only if $S \nvdash \sim G$.*

Proof By propositional logic, $S \nvdash \sim G$ if and only if $S \cup \{G\}$ is consistent. By the proof of the Second Incompleteness Theorem, $S \vdash (G \equiv Con_S)$. Therefore $S \nvdash \sim G$ if and only if $S \cup \{Con_S\}$ is consistent. ■

It remains to show that consistency of $S \cup \{Con_S\}$ is strictly weaker than 1-consistency of S. Prima facie it is, given that 1-consistency of S is equivalent to the consistency of S + all true Π_1-sentences, by Theorem 26, and Con_S is a true Π_1-sentence for S any consistent theory. But suggestive as this observation is, it doesn't constitute a proof that consistency of $S \cup \{Con_S\}$ is strictly weaker than 1-consistency of S. We establish this result by constructing a theory S such that if $\mathrm{PA} \cup \{Con_{\mathrm{PA}}\}$ is consistent, S is 1-inconsistent and $S \cup \{Con_S\}$ is consistent.

Theorem 36 *Let* $\mathrm{PA}^+ =_{\mathrm{df}} \mathrm{PA} \cup \{Con_{\mathrm{PA}}\}$ *and* $S =_{\mathrm{df}} \mathrm{PA} \cup \{\sim Con_{\mathrm{PA}^+}\}$. *If* PA^+ *is consistent, then* S *is 1-inconsistent and* $S \cup \{Con_S\}$ *is consistent.*

Proof (i) If PA^+ is consistent, then Con_{PA^+} is a true Π_1-sentence, so $\sim Con_{\mathrm{PA}^+}$ is a false Σ_1-sentence. Then by Theorem 25, S is 1-inconsistent.

(ii) (1) Suppose $S \cup \{Con_S\}$ is inconsistent.
(2) Then by propositional logic, $S \vdash \sim Con_S$.
(3) Similarly, S is inconsistent if and only if $\mathrm{PA} \vdash Con_{\mathrm{PA}^+}$, so
(4) S proves the inconsistency of S iff $S \vdash Pr_{\mathrm{PA}}(\ulcorner \overline{Con_{\mathrm{PA}^+}} \urcorner)$, that is, iff $\mathrm{PA} \cup \{\sim Con_{\mathrm{PA}^+}\} \vdash Pr_{\mathrm{PA}}(\ulcorner \overline{Con_{\mathrm{PA}^+}} \urcorner)$.
(5) Then by the Deduction Theorem, $\mathrm{PA} \vdash (\sim Con_{\mathrm{PA}^+} \supset Pr_{\mathrm{PA}}(\ulcorner \overline{Con_{\mathrm{PA}^+}} \urcorner))$, and so by the contrapositive of (5),
(6) $\mathrm{PA} \vdash (\sim Pr_{\mathrm{PA}}(\ulcorner \overline{Con_{\mathrm{PA}^+}} \urcorner) \supset Con_{\mathrm{PA}^+})$.
(7) By arithmetized Second Incompleteness Theorem, $\mathrm{PA} \vdash (Con_{\mathrm{PA}} \supset \sim Pr_{\mathrm{PA}}(\ulcorner \overline{Con_{\mathrm{PA}}} \urcorner))$.
(8) By arithmetization of the logical fact that if PA (or any other theory) does not prove the consistency of a theory then *a fortiori* it does not prove the consistency of an extension of that theory, $\mathrm{PA} \vdash (\sim Pr_{\mathrm{PA}}(\ulcorner \overline{Con_{\mathrm{PA}}} \urcorner)) \supset \sim Pr_{\mathrm{PA}}(\ulcorner \overline{Con_{(\mathrm{PA} \cup \{Con_{\mathrm{PA}}\})}} \urcorner))$.
(9) By (7) and (8), $PA \vdash (Con_{\mathrm{PA}} \supset \sim Pr_{\mathrm{PA}}(\ulcorner \overline{Con_{(\mathrm{PA} \cup \{Con_{\mathrm{PA}}\})}} \urcorner))$.
(10) By (6) and (9), $\mathrm{PA} \vdash (Con_{\mathrm{PA}} \supset Con_{\mathrm{PA}^+})$.
(11) Then by *Modus Ponens*, $\mathrm{PA} \cup \{Con_{\mathrm{PA}}\} \vdash Con_{\mathrm{PA}^+}$, i.e., $\mathrm{PA}^+ \vdash Con_{\mathrm{PA}^+}$.
(12) From the assumption that $\mathrm{PA} \cup \{Con_{\mathrm{PA}}\} =_{\mathrm{df}} \mathrm{PA}^+$ is consistent, we have $\mathrm{PA}^+ \nvdash Con_{\mathrm{PA}^+}$, i.e., Second Incompleteness Theorem for PA^+.
(13) Then by (1), (11), (12) and RAA, $S \cup \{Con_S\}$ is consistent. ■

Corollary 37 *For systems* S *for which Gödel's Second Incompleteness Theorem holds, the condition that* $S \cup \{Con_S\}$ *is consistent is strictly weaker than the condition that* S *is 1-consistent.*

Proof Since the 1-consistency of S is equivalent to the consistency of S + all true Π_1-sentences (Theorems 25 and 26), and if S is 1-consistent then Con_S is a true Π_1-sentence, 1-consistency of S implies consistency of $S \cup \{Con_S\}$. On the other hand, Theorem 36 shows that for systems S for which the Second Incompleteness Theorem holds, consistency of $S \cup \{Con_S\}$ does not imply 1-consistency of S. ■

Remark Though consistency of $S \cup \{Con_S\}$ is sufficient for $S \nvdash \sim G$, the proof that it is sufficient goes via the Second Incompleteness Theorem. For a proof of the second half of the First Incompleteness Theorem without first proving the Second Incompleteness Theorem, the proof from 1-consistency is best possible.

7 Comparing Gödel's Incompleteness Theorems with Rosser's Theorem

From Theorem 12 we saw that consistency is not sufficient to show that the Gödel sentence for a system S is not refutable in S. This shows that Rosser's Theorem, that for a particular sentence R constructed for a given system S, if S is consistent then $S \nvdash R$ and $S \nvdash \sim R$, is incomparable in strength with and not strictly a stronger theorem than Gödel's incompleteness theorem, contrary to the way is often presented, for example, (Mendelson, 1997, p. 208); (Adamowicz and Zbierski, 1997, p. 178), and more to the point, Kleene when he says, "Rosser in [1936] achieved a noteworthy improvement of the first Gödel incompleteness theorem" (Kleene, 1986, p. 140). Of course what people who talk this way have in mind is that Rosser's theorem shows that any consistent Σ_0-complete system will be Π_1-incomplete, while Gödel's Theorem shows this on an assumption stronger than consistency. However Rosser's result does not show that the Gödel sentence is irrefutable just on the assumption of consistency (which, as we have seen, cannot be done). We also know that Rosser's theorem is no strengthening of Gödel incompleteness in that it cannot give rise to the Second Incompleteness Theorem, and it is the Second Incompleteness Theorem that is the heart of the matter.[1]

Bibliography

Adamowicz, Z. and Zbierski, P. (1997). *Logic of Mathematics: A Modern Course of Classical Logic*. Wiley, New York, NY.

Feferman, S. et al., editors (1986). *Kurt Gödel Collected Works Volume I: Publications 1929–1936*. Oxford University Press, Oxford.

Gödel, K. (1931). Über formal untentscheidbare Sätze der *Principia mathematica* und verwandter Systeme I. *Monatshefte für Mathematik und Physik*, 38:173–198. English translation by Jean van Heijenoort, "On formally undecidable propositions of *Principia Mathematica* and related systems I," (Feferman et al., 1986, 144-195).

Kleene, S. C. (1986). Introduction to [Gödel, 1931]. In [Feferman et al., 1986], pages 126–141.

Kreisel, G. (1957). A refinement of ω-consistency (abstract). *The Journal of Symbolic Logic*, 22:108–109.

Mendelson, E. (1997). *Introduction to Mathematical Logic*. Chapman and Hall, London, fourth edition.

[1] I am grateful to Georg Kreisel and Alex Wilkie for generously answering my many questions on the topic of this paper. I cite sources for various results in this paper but am not claiming originality for any of the results for which I do not cite sources, including those I am aware I figured out for myself, such as Theorems 28, 35, 36 and Corollary 37. I am also grateful to the Arts and Humanities Research Council for a Research Leave Award which gave me time on which to work on the subject of this paper.

Chapter 8
Paraconsistent Set Theory

Graham Priest

1 Introduction: For JLB

John Bell played a pivotal role in my intellectual life. I was first shown the beauty of mathematics at school, by a particularly gifted teacher. Inspired by this, I went up to Cambridge to read the subject. The process was disillusioning. With only few exceptions, the lectures were dull and boring. It was not uncommon for a lecturer to spend an hour writing on the blackboard with his or her back to the audience. Supervisors obviously derived pleasure from solving the problems that I could not solve, but this was with all the emotional engagement of cross-word puzzle solving. The sense of intellectual excitement that I had experienced at school evaporated.

By the end of my undergraduate career I had decided that philosophy was probably more interesting than mathematics, and that my education might best be served by studying mathematical logic; I went off to London to take an MSc in the subject. It was then my great good fortune to meet John, who was lecturing on model theory in the MSc programme. I am sure that the topic could have been made just as dull and boring as the topics I had studied in Cambridge, but John was inspirational. The sense of intellectual excitement in his lectures was palpable. The lectures were crystal clear, delivered with enormous enthusiasm, and the hand-written lecture notes that he handed out were a model of elegance. Many of us would go off with him for coffee and a cigarette after the lecture; John would hold forth on the things he loved, which of course included mathematics. I was reinspired.

At the end of the MSc, I decided to do a PhD, and asked John to be my supervisor. The area I wanted to work on was somewhat peripheral to his interests, but, with magnanimity, he agreed. I was a pretty self-motivated student, and John realised that he could leave me largely to my own devices. I would go and see him at the LSE once a month or so. We would have a coffee, and talk for hours; John, it must be said, would do most of the talking. For some reason, I seem to remember, many

G. Priest (✉)
Boyce Gibson Professor of Philosophy, University of Melbourne, Melbourne, Australia;
Distinguished Professor of Philosophy, Graduate Center, City University of New York,
New York, NY, USA
e-mail: g.priest@unimelb.edu.au

D. DeVidi et al. (eds.), *Logic, Mathematics, Philosophy: Vintage Enthusiasms*,
The Western Ontario Series in Philosophy of Science 75,
DOI 10.1007/978-94-007-0214-1_8,

of the topics started with "m": music, Mozart, morality, masochism, miscegenation, Marxism, Mao Zedong ... and, just occasionally, mathematics. When I explained what I was up to, he would listen, ponder, make helpful suggestions, and wonder, no doubt, where it was all going. He was the ideal supervisor for me: enthusiastic, patient, and—though he was only a few years older than I—with a perspective of the richness of life of which I was in awe.

Paraconsistency and dialetheism were only a twinkling in my eye in those days. John persuaded his research students to go to an alternative logic conference in Uldum, Denmark, in 1971—not alternative in the sense of alternative logic, but politically alternative: it was organised as a protest against the acceptance of NATO funding by the organisers of the usual UK logic conference. It was listening to a talk by Moshé Machover on the philosophy of mathematics at that conference that started me thinking about these matters. (Moshé was another of my lecturers for the London MSc, and in his own way, just as inspirational as John.) Virtually nothing about paraconsistency made its way into my thesis.[1] Nonetheless, it was clear that my interests—driven by a mathematical nominalism—were fairly unorthodox even in those days. John's mathematical interests were in mainstream areas of mathematical logic, but he never tried to push me into those. He was always happy to engage with and encourage things less orthodox. I suspect it appealed to the subversive in him. At one stage of my candidature, Imre Lakatos learned that there was someone with interests in the philosophy of mathematics working in the Mathematics Department, and insisted that I immediately transfer to the Philosophy Department. John protected me from the Lakatosian imperialism.

That was all many years ago. After I finished the doctorate, I left London. But John and I have remained good friends over the years, though living on different continents. My interests have since became even more heterodox. Exactly what he thinks of paraconsistency now, I am not entirely sure. I suspect that he views it as bizarre, though always with a deep chuckle in his voice and glint in his eye. Anyway, the rest of this essay is dedicated to John. Serve him right.

What I will discuss here is paraconsistent set theory. Set theory, at least, is a topic close to John's heart .... Specifically, I will discuss the shape of an acceptable paraconsistent set theory. I will review what is currently known about the matter, and suggest some new ideas. There are, it must be confessed, as many questions as answers. At the end of the essay I will apply the discussion to another important issue for paraconsistency: that concerning its metatheory—and especially the model-theoretic definition of validity. The connection is, of course, that such a metatheory is formulated within set theory.[2]

---

[1] Though I remember discussing some of the material that would become "The Logic of Paradox" (Priest, 1979), with John and my other examiner, Michael Dummett, at the PhD viva in Oxford. Neither, I think, saw much in it.

[2] What follows is essentially Chapter 18 of the second edition of (Priest, 1987). I am grateful for the permission of Oxford University Press to republish the material.

2 Paraconsistent Set Theory: Background

The problem posed by Russell's paradox and its set-theoretic cousins may be thought of as generated by two factors. First, an unrestricted abstraction—or comprehension—principle of set existence, which allows an arbitrary condition to specify a set. Second, various principles of logic which allow certain instances of this (or their conjunction) to entail everything. Since the discovery of these paradoxes, the orthodox reaction has been to maintain the principles of logic in question, but reject the unrestricted comprehension principle. This strategy gives type theory, Zermelo-Fraenkel set theory, and so on.

There is, however, another possible strategy: maintain the comprehension principle and reject, instead, some of the principles of logic in question. There are various ways one may do this, but the one which will be concern us here is the paraconsistent way. Allow for the set theory to entail contradictions, but reject the principle *ex contradictione quodlibet*, or to give it its more colourful name, Explosion, $\{\alpha, \neg\alpha\} \vdash \beta$, and hence obtain a theory that is inconsistent but non-trivial.

How should one do this? Part of the answer is easy. A paraconsistent set theory can naturally be thought of as a theory that endorses the two axioms (or one axiom and one axiom schema):

$$\forall x(x \in y \leftrightarrow x \in z) \rightarrow y = z \qquad Ext$$

$$\exists x \forall y(y \in x \leftrightarrow \alpha) \qquad Abs$$

where x does not occur free in α.[3] The rest of the answer is not easy, however. What is the appropriate underlying logic? In particular, what notion of conditional is being employed in Ext and Abs?

Paraconsistency gives us several choices in answering this question. In making the appropriate choice, there are two constraints that need to be borne in mind. First, the resulting theory should not allow us to prove too much; second, it should not allow us to prove too little.

For the first: although using a paraconsistent logic allows isolated contradictions to be accepted, we still do not want wholesale contradiction. In particular, if *everything* were provable, the theory would be quite useless. And even though contradictions do not imply everything, there may still be arguments delivering triviality. A notorious one is Curry's paradox. Suppose that the conditional of the logic

---

[3] One might also want to add an appropriate version of the Axiom of Choice to these. There are, however, ways of obtaining the Axiom from unrestricted comprehension. One way is to use the machinery of Hilbert's ε calculus. (See, e.g., (Leisenring, 1969, pp. 105–107).) Another, much more radical, way is to take Abs in an absolutely unrestricted form which allows α to contain "x" free. This delivers the Axioms of Choice (see Routley, 1977, p. 924f of reprint) whilst, surprisingly enough, maintaining non-triviality (see Brady, 1989).

satisfies both *modus ponens* and Contraction (or Absorption): $\{\alpha \rightarrow (\alpha \rightarrow \beta)\} \vdash \alpha \rightarrow \beta$. Triviality then ensues.[4]

This fact puts fairly severe constraints on an appropriate underlying logic. In fact, it rules out very many paraconsistent logics. For example, it rules out da Costa's well known C systems. It also rules out many of the best known systems of relevant logic, such as R.[5] Not everything is ruled out, though, as we shall see.

But before we turn to this, let us consider the other constraint: not too little. It is easy enough to choose an underlying logic for paraconsistent set theory that does not give triviality. Choose the null logic (in which nothing follows from anything)! This is obviously not very interesting. A minimal condition of adequacy on a paraconsistent set theory would seem to be that we can get at least a decent part of standard, orthodox, set theory out of it. We might not require everything; we might be prepared to write off various results concerning large cardinality, or peculiar consequences of the Axiom of Choice. But if we lose too much, set theory is voided of both its use and its interest.

It should be remembered, here, that paraconsistency, unlike intuitionism, has never been a consciously revisionist philosophy. The picture has always been that classical mathematics, and the reasoning that this embodies, is perfectly acceptable as long as it does not stray into the transconsistent. It is only there that it goes awry. So the unproblematically consistent bits of orthodox set theory, at least, ought to be delivered by a paraconsistent set theory.

The results of this second constraint are in some tension with the results of the first. Put crudely, the matter is this. If we weaken our logic in a way that is sufficient to avoid triviality, we weaken it so much that it fails to deliver much set theory that we want to keep. We will see how this tension plays out in the following discussion.

3 The Material Strategy

As we have just seen, an underlying logic for a paraconsistent set theory must invalidate either *modus ponens* or Contraction. Both are live options. Let us start with the rejection of *modus ponens*. There are various ways that one can arrange for *modus ponens* to fail in a paraconsistent logic, but undoubtedly the most natural is to take the conditionals (and biconditionals) in Ext and Abs to be material. That is, $\alpha \rightarrow \beta$ is simply *defined* as $\neg\alpha \vee \beta$. ($\alpha \leftrightarrow \beta$ is defined in the usual way as $(\alpha \rightarrow \beta)$ & $(\beta \rightarrow \alpha)$.) In nearly every paraconsistent logic, material detachment fails: $\{\alpha, \neg\alpha \vee \beta\} \nvdash \beta$. I will call this the *material strategy*.[6] (The strategy does

---

[4] See (Priest, 1987, 6.2).

[5] For a survey of paraconsistent logics, see (Priest, 2002).

[6] Adopting the material strategy in some form goes *half* way towards meeting (Goodship, 1996), who advocates taking the main conditional of both the Comprehension Principle and the T-schema to be material. Would treating the conditionals in the two schemas show the paradoxes of self-reference to be of different kinds? No. They still all fit the Inclosure Schema (Priest, 1995, part 3), and so have the same essential structure.

not, of course, mean that the language employed does not contain different kinds of conditional. For example, it may contain a relevant and detachable conditional as well—though it need not.)

A simple and natural choice here is the logic LP (Priest, 1987, Ch. 5). A sound and complete tableau system for this is as follows. (See Priest, 2001, 8.3.) Lines are of the form $\alpha, +$ or $\alpha, -$. A tableau for the inference $\{\alpha_1, \ldots, \alpha_n\} \vdash \beta$ starts with the lines:

$$\begin{array}{c} \alpha_1, + \\ \vdots \\ \alpha_n, + \\ \beta, - \end{array}$$

The rules are as follows:

$$\begin{array}{ccc} \alpha \wedge \beta, + & \alpha \wedge \beta, - & \neg(\alpha \wedge \beta), \pm \\ \downarrow & \swarrow \quad \searrow & \downarrow \\ \alpha, + & \alpha, - \quad \beta, - & \neg\alpha \vee \neg\beta, \pm \\ \beta, + & & \end{array}$$

$$\begin{array}{ccc} \alpha \vee \beta, - & \alpha \vee \beta, + & \neg(\alpha \vee \beta), \pm \\ \downarrow & \swarrow \quad \searrow & \downarrow \\ \alpha, - & \alpha, + \quad \beta, + & \neg\alpha \wedge \neg\beta, \pm \\ \beta, - & & \end{array}$$

$$\begin{array}{c} \neg\neg\, \alpha, \pm \\ \downarrow \\ \alpha, \pm \end{array}$$

$$\begin{array}{ccc} \forall x\alpha, + & \forall x\alpha, - & \neg\forall x\alpha, \pm \\ \downarrow & \downarrow & \downarrow \\ \alpha(x/b), + & \alpha(x/a), - & \exists\neg x\alpha, \pm \end{array}$$

$$\begin{array}{ccc} \exists x\alpha, + & \exists x\alpha, - & \neg\exists x\alpha, \pm \\ \downarrow & \downarrow & \downarrow \\ \alpha(x/a), + & \alpha(x/b), - & \forall\neg x\alpha, \pm \end{array}$$

$$\begin{array}{c} . \\ \downarrow \\ b = b, + \end{array}$$

$$\begin{array}{c} b = c, + \\ \alpha(x/b), \pm \\ \downarrow \\ \alpha(x/c), \pm \end{array}$$

Here, b and c are any terms on the branch, a is a constant new to the branch, and "$\pm$" can be disambiguated consistently either way.[7] The closure rules for a branch are two:

$$\begin{array}{cc} \alpha, + & \alpha, - \\ \alpha, - & \neg\alpha, - \\ \times & \times \end{array}$$

(The second of these enshrines the Law of Excluded Middle.)

The paraconsistent set theory that this logic produces has a number of interesting features. It is provably non-trivial.[8] It validates all those axioms of ZF that are instances of *Abs* (of course). It validates the Axiom of Infinity, but not the Axiom of Foundation. It can also (unlike ZF) demonstrate the existence of a universal set.[9] What theorems of ZF—beyond the axioms—it can (or cannot) establish, is as yet a largely unanswered question. But the failure of material detachment means that most of the natural arguments fail. Whilst this does not meant that there are no unnatural arguments for the same conclusions, the prospects look rather bleak. The failure of detachment is a singular handicap. For the same reason, any other way of pursuing the material strategy does not look promising.

A perhaps more promising variation is to look at the consequences of the axioms, not in LP, but in the non-monotonic extension LPm. (See Priest, 1987, 2nd ed., Ch. 16.) The results of this approach are presently unknown.

4 The Relevant Strategy

A second, and arguably more plausible, strategy is to use a conditional in a logic which validates *modus ponens*, but not Contraction. The most plausible candidate for this is a relevant logic weaker than R, one of the *depth relevant* logics, as they are sometimes called. The following is a tableau system for such a logic.[10] Lines are now of one of two forms. One is $\alpha, +i$ or $\alpha, -i$, where i is a natural number (thought of as representing a world). Premises and conclusion take the number 0. The other is $rijk$, where i, j and k are natural numbers ("r" representing a ternary accessibility relation, as is standard in the semantics for relevant logics). The rules for LP are all present, except that a natural-number world-parameter, i, is added

[7] The rule for $b = b$ means that this can be introduced at any time.

[8] It might be thought that without detachment the axioms cannot be shown to be inconsistent. This is false, though. An instance of *Abs* is $\forall x(x \in r \leftrightarrow \neg x \in x)$. Whence we have $r \in r \leftrightarrow \neg r \in r$; and cashing out the conditional in terms of negation and disjunction gives $r \in r \wedge \neg r \in r$. More generally, whenever α is a classical consequence of Σ, there is a β such that $\alpha \vee (\beta \wedge \neg\beta)$ follows from Σ. (See Priest 1987, Ch. 6.) Hence, any classically inconsistent theory is inconsistent in this logic also.

[9] For details of all this, see (Restall, 1992). Note that he defines "$x = y$" as "$\forall z(z \in x \leftrightarrow z \in y)$."

[10] A semantics with respect to which it is sound can be found in (Priest, 1987, 2nd ed, 19.8).

uniformly. Thus, for example, the rule for $\wedge +$ is[11]:

$$\begin{array}{c} \alpha \wedge \beta, +i \\ \downarrow \\ \alpha, +i \\ \beta, +i \end{array}$$

It is easiest to define the conditional, $\rightarrow$, in terms of a non-contraposing conditional, $\Rightarrow$. Thus, $\alpha \rightarrow \beta$ is $(\alpha \Rightarrow \beta) \wedge (\neg\beta \Rightarrow \neg\alpha)$. The rules for $\Rightarrow$ are as follows. When $i > 0$ (i is an impossible world):

$$\begin{array}{ccc} & \alpha \Rightarrow \beta, +i & \\ & rijk & \\ \swarrow & & \searrow \\ \alpha, -j & & \beta, +k \end{array} \qquad \begin{array}{c} \alpha \Rightarrow \beta, -i \\ \downarrow \\ rijk \\ \alpha, +j \\ \beta, -k \end{array}$$

$$\begin{array}{ccc} & \neg(\alpha \Rightarrow \beta), -i & \\ & rijk & \\ \swarrow & & \searrow \\ \alpha, -j & & \neg\beta, -k \end{array} \qquad \begin{array}{c} \neg(\alpha \Rightarrow \beta), +i \\ \downarrow \\ rijk \\ \alpha, +j \\ \neg\beta, +k \end{array}$$

In the left hand rules, j and k are any numbers on the branch. In the right hand rules, j and k are new to the branch.

When $i = 0$ (i is a possible world), the rules simplify to:

$$\begin{array}{ccc} & \alpha \Rightarrow \beta, +0 & \\ \swarrow & & \searrow \\ \alpha, -j & & \beta, +j \end{array} \qquad \begin{array}{c} \alpha \Rightarrow \beta, -0 \\ \downarrow \\ \alpha, +j \\ \beta, -j \end{array}$$

$$\begin{array}{ccc} & \neg(\alpha \Rightarrow \beta), -0 & \\ \swarrow & & \searrow \\ \alpha, -j & & \neg\beta, -j \end{array} \qquad \begin{array}{c} \neg(\alpha \Rightarrow \beta), +0 \\ \downarrow \\ \alpha, +j \\ \neg\beta, +j \end{array}$$

[11] The rules for identity are an exception. The rules for this are:

$$\begin{array}{c} . \\ \downarrow \\ b = b, +0 \end{array} \qquad \begin{array}{c} b = c, +i \\ \alpha(x/b), \pm j \\ \downarrow \\ \alpha(x/c), \pm j \end{array}$$

In the left hand rules, j is any number on the branch. In the right hand rules, j is new to the branch.

The closure rules are now:

$$\begin{array}{cc} \alpha, +i & \alpha, -0 \\ \alpha, -i & \neg\alpha, -0 \\ \times & \times \end{array}$$

(So the Law of Excluded Middle is guaranteed only at the base world.)

Naive set theories based on relevant logics such as this are known to be inconsistent but non-trivial. Indeed, the logic may be strengthened in various ways, and this is still true—though not, of course, with Contraction. (See Brady, 1989, Priest, 2002, § 8.) Thus, this relevant set theory satisfies the first constraint. What of the second?

To answer this question (at least to the extent that the answer is known), it is useful to divide set theory into two parts. The first comprises that basic set theory which all branches of mathematics use as a tool. The second is the more elaborate development of this, which includes transfinite set theory, as it can be established in ZF, "higher" set theory.

The theory is able to provide for virtually all of basic set theory—Boolean operations on sets, power sets, products, functions, operations on functions, etc. (I will return to the reason for the qualification "virtually" in a moment.) Thus, for convenience, let the language be augmented with set-abstract terms. We may define the Boolean operators, $x \cap y$, $x \cup y$ and $\bar{x}$ as $\{z : z \in x \wedge z \in y\}$, $\{z : z \in x \vee z \in y\}$ and $\{z : z \notin x\}$, respectively, and $x \subseteq y$ as $\forall z(z \in x \rightarrow z \in y)$. We can then establish the usual facts concerning these notions.[12]

How much of the more elaborate development of set theory can be proved is not currently known.[13] What can be said is that the *standard* proofs of a number of results break down. One thing we obviously lose is that kind of argument which appeals to vacuous satisfaction. Thus, for example, suppose that we wish to establish $\forall x(x < \xi \rightarrow A(x))$ by transfinite induction on the ordinal ξ. We can no longer argue in the basis case that since $\neg x < 0$, $x < 0 \rightarrow A(x)$; but we can make the zero case explicit, and perform the induction on $\forall x(\xi = 0 \vee (x < \xi \rightarrow A(x)))$. The first disjunct must then always be considered as a special case. Things not so easy to reconstruct are arguments employing *reductio*, such as Cantor's Theorem. Where α is an assumption made for the purpose of *reductio*, we may well be able to establish that $(\alpha \wedge \beta) \rightarrow (\gamma \wedge \neg\gamma)$, for some γ, where β is the conjunction of other facts appealed to in deducing the contradiction (such as instances of *Abs*). But contraposing and detaching will give us only $\neg\alpha \vee \neg\beta$, and we can get no further.

---

[12] Much of this is spelled out in (Routley, 1977, § 8).

[13] *Added in Press*: In fact, it has recently been shown that standard results about ordinals and cardinals can be proved in this framework, as well as a number of results concerning large cardinals. See (Weber, 2009, Chs. 5 and 6, 2010b).

Even given β, the failure of the disjunctive syllogism prevents us from obtaining $\neg\alpha$.[14] Much remains to be done in investigating higher set theory in this context.

Let me now return to the qualification "virtually." Problems arise with the empty set. There can be no set, φ, such that for every a and b:

(1) $a \cap \overline{a} \subseteq \varphi$
(2) $\varphi \subseteq b$

For let a be $\{x : \alpha\}$ and b be $\{x : \beta\}$. Then, (1) and (2) together give us: $(x \in \{x : \alpha\} \wedge x \in \overline{\{x : \alpha\}}) \rightarrow x \in \{x : \beta\}$. *Abs* then gives $(\alpha \wedge \neg\alpha) \rightarrow \beta$, and the theory is not paraconsistent.

If we define φ_1 as $a \cap \overline{a}$, then this clearly satisfies (1), but it does not satisfy (2). Alternatively, if we define φ_2 as $\{x : \forall y\ x \in y\}$ then it is easy enough to show that this satisfies (2), but not (1). It is provably the case that both φ_1 and φ_2 have no members. One cannot, though, show that they are identical. For $\{\neg x \in y \wedge \neg x \in z\} \nvdash x \in y \leftrightarrow x \in z$. Generally speaking, one cannot expect the global structure of the universe of sets to be a Boolean algebra, as it is classically (albeit the case that, classically, the maximum element of the algebra and some set-theoretic complements are proper classes). What one will have, instead, is a De Morgan algebra.[15]

This might, perhaps, be something that can be accepted. Boolean algebras are, after all, just special cases of De Morgan algebras. But we are not finished yet. It is not only the empty set that has multiple doppelgangers; so does the universal set. In fact, all sets do. For let α be an arbitrary truth; then $x \in y \leftrightarrow (x \in y \wedge \alpha)$ is not *relevantly* valid (from left to right). Thus, even though y and $\{x : x \in y \wedge \alpha\}$ have the same members, we will not have $y = \{x : x \in y \wedge \alpha\}$. What has gone wrong at this point is clear. Ext notwithstanding, the entities in question are not extensional. Nor is this an accident; the identity conditions of the entities in question are given in terms of $\rightarrow$, and this is an intensional functor, more at home in giving the identity conditions for properties than sets.

This suggests changing the biconditional in Ext. A natural thought is to replace it with the material biconditional, $\equiv$. Natural as this thought is, the strategy does not work. For $\{\alpha \wedge \neg\alpha\} \models \beta \equiv \alpha$. Now let α be any provable contradiction. Then for any z, $x \in z \equiv \alpha$. By Ext, it now follows that $z = \{x : x \in \alpha\}$; there is only one set. (Note that this argument does not go through in the material strategy because the material conditional does not detach to give the identity.)

There is another possibility. To see this, consider restricted quantification for a moment. It is natural to express "all As are Bs" using a conditional, thus: $\forall x(Ax \rightarrow Bx)$. If $\rightarrow$ is a standard relevant conditional, then the inference:

---

[14] This is not the only sort of problem. Various natural arguments require the use of principles that involve nested $\rightarrow$s, such as Permutation, $\{\alpha \rightarrow (\beta \rightarrow \gamma)\} \vdash \beta \rightarrow (\alpha \rightarrow \gamma)$. The logic just described does not contain this principle. Whether it can be added whilst maintaining non-triviality is not known. There is certainly triviality in the area. See (Slaney, 1989).

[15] For a more systematic discussion of the issue, see (Dunn, 1988).

1. Everything is B; hence all As are Bs

fails, since it depends on the validity of the inference $B(a) \vdash A(a) \rightarrow B(a)$. Yet inferences of this form are frequently appealed to when employing restricted quantifiers of the kind in question. If we interpret $\rightarrow$ as $\supset$, the material conditional, the inference is valid enough. But now the inference:

2. All As are Bs; a is an A; hence a is a B

fails, since it employs the Disjunctive Syllogism, $A(a), \neg A(a) \vee B(a) \vdash B(a)$. This is even worse.

A solution to this problem is to use another sort of conditional. In many formulations of relevant logics, there is a logical constant, t, which may be thought of as the conjunction of all truths.[16] (So t is true at the base world, 0, and any other world at which all the things true at the base world are true.) The appropriate tableau rules for t are:

$$\begin{array}{cc} . & \alpha, +0 \\ \downarrow & t, +i \\ t, +0 & \alpha, -i \\ & \times \end{array}$$

It is not difficult to check that these validate the following inferences:

$$\vdash t$$

$$\alpha \vdash t \rightarrow \alpha$$

We may now define an enthymematic conditional, $\rightharpoonup$, in terms of t:

$$\alpha \rightharpoonup \beta \text{ is } (\alpha \wedge t) \rightarrow \beta$$

and use this as the conditional involved in restricted universal quantification. Thus, "All As are Bs" is to be understood as $\forall x(A(x) \rightharpoonup B(x))$. We now have:

$$\begin{array}{rl} B(a) \vdash & t \rightarrow B(a) \\ B(a) \vdash & A(a) \rightharpoonup B(a) \end{array}$$

Hence $\forall x B(x) \vdash \forall x(A(x) \rightharpoonup B(x))$. And $\forall x(A(x) \rightharpoonup B(x)) \vdash (t \wedge A(a)) \rightarrow B(a)$. Hence $A(a), \forall x(A(x) \rightharpoonup B(x)) \vdash B(a)$. So both the inferences 1 and 2 are valid.[17]

---

[16] See, e.g., (Dunn and Restall, 2002, p. 10). Sometimes, depending on the context, t gets interpreted as the conjunction of all logical truths.

[17] For a general discussion of restricted quantification in relevant logic, see (Beall et al., 2006), which suggests the use of a different, but closely related, kind of enthymematic conditional.

Now return to set theory. It is natural to hear "y is a subset of z" as "all members of y are members of z," that is, on the present account, $\forall x(x \in y \rightharpoonup x \in z)$. Let us define $y \subseteq z$ in this way. We may now take Ext to be $\forall x(x \in y \rightleftharpoons x \in z) \rightarrow y = z$, where $\rightleftharpoons$ is the biconditional corresponding to $\rightharpoonup$. This is equivalent to $(y \subseteq z \wedge z \subseteq y) \rightarrow y = z$.

Using $\rightleftharpoons$ instead of $\leftrightarrow$ in Ext overcomes many of the problems we noted. Thus, for example, there is only one set that contains everything. $\forall x\, x \in y \vdash \forall x(x \in z \rightharpoonup x \in y)$. So $\forall x\, x \in y, \forall x\, x \in z \vdash y = z$. Moreover, let α be any truth. Then we have $t \rightarrow \alpha$, so $x \in y \rightharpoonup (\alpha \wedge x \in y)$. Since $(\alpha \wedge x \in y) \rightharpoonup x \in y$, we have $y = \{x : x \in y \wedge \alpha\}$. The structure of sets is still not a Boolean algebra since the empty set is still not unique.[18] We do not have $x \notin y \vdash x \in y \rightharpoonup x \in z$. Hence, we do not have $\neg\exists x\, x \in y \vdash y \subseteq z$ or, therefore, $\neg\exists x\, x \in y, \neg\exists x\, x \in z \vdash y = z$. But the empty set is enough of an oddity that this may not matter too much. Reconstructing the reasoning of set theory using $\rightharpoonup$ in Ext and the definition of $\subseteq$ therefore looks much more promising.[19]

5 The Model-Theoretic Strategy

I have discussed the material strategy and the relevant strategy for naive set theory. These do not exhaust the possibilities. Let us return to the axiomatisation that employs a material conditional uniformly. Call this M. (And suppose that the language contains just the standard extensional connectives and quantifiers, as in the usual formulations of ZF—and no set abstracts.) This time, we will consider, not what is provable in M, but what the models of M are. M has many models, many of which are clearly pathological. For example, there is the model with but a single element, which both is and is not a member of itself. (This verifies the trivial theory.)

But M has many other models. We can construct some of these with the Collapsing Lemma.[20] Let $\mathcal{M} = \langle D, I \rangle$ be any model of ZF. Let ξ be any ordinal in $\mathcal{M}$, and a be the initial section of the cumulative hierarchy, V_ξ, in $\mathcal{M}$. (That is, the pair $\langle \xi, a \rangle$ satisfies the formula "x is an ordinal and $y = V_x$" in $\mathcal{M}$.) Define a relation, $\sim$, on D as follows:

$$(x \text{ and } y \text{ are in } a \text{ (in } \mathcal{M}) \text{ and } x = y) \text{ or } (x \text{ and } y \text{ are not in } a \text{ (in } \mathcal{M}))$$

This is obviously an equivalence relation. (Since there are no function symbols, it is vacuously a congruence relation too.) It leaves all the members of V_ξ alone,

---

[18] Note, in particular, that $\rightharpoonup$ does not contrapose. So from the fact that $x = y$ we cannot infer that $\overline{x} = \overline{y}$.

[19] *Added in Press*: Unfortunately not. It is shown that this strategy does not work in (Weber, 2009, Ch. 4, 2010a).

[20] For the Collapsing Lemma, see (Priest, 1991) or (Priest, 1987, 2nd ed., Ch. 16). For the use of the Lemma to construct models of set theory with other properties, see (Priest, 1995, Ch. 11), technical appendix.

but identifies all other members of D. Construct the collapsed interpretation, $\mathcal{M}^{\sim} = \langle D^{\sim}, I^{\sim}\rangle$, with respect to this equivalence relation. The Collapsing Lemma tells us that $\mathcal{M}^{\sim}$ is a model of ZF.

But something else also happens. Let me use boldfacing for names. Then "**a**" refers to a in $\mathcal{M}$, and $[a]$ in $\mathcal{M}^{\sim}$. For all $b \in D^{\sim}$, the sentence $\mathbf{b} \in \mathbf{a}$ is both true and false in $\mathcal{M}^{\sim}$. For if (in $\mathcal{M}$) b is of rank less than ξ, $\mathbf{b} \in \mathbf{a}$ is true in $\mathcal{M}$, and so in $\mathcal{M}^{\sim}$; and if not, there is some x which is also not of rank less than ξ (e.g., $\{b\}$) such that b is in x. (I am not, here, assuming the Axiom of Foundation.) Since x has been identified with a in $\mathcal{M}^{\sim}$, $\mathbf{b} \in \mathbf{a}$ is true in $\mathcal{M}^{\sim}$. Whatever b is, there are elements which do not have rank less than ξ such that b is not a member of them (e.g., $\{c\}$, where c is distinct from b and has rank greater than ξ). Since these have been identified with a in $\mathcal{M}^{\sim}$, $\mathbf{b} \in \mathbf{a}$ is also false in $\mathcal{M}^{\sim}$. Now consider any sentence of the form $\mathbf{b} \in \mathbf{a} \equiv \alpha(x/\mathbf{b})$. The left side is both true and false. Hence the biconditional is true in $\mathcal{M}^{\sim}$ ($\{\beta \wedge \neg\beta\} \models \beta \equiv \alpha$). So $\forall x(x \in \mathbf{a} \equiv \alpha)$ is true, as is $\exists x \forall x(x \in y \equiv \alpha)$. So $\mathcal{M}^{\sim}$ is a model of Abs. It is a model of Ext as well, of course, since this is in ZF. Hence, $\mathcal{M}^{\sim}$ is a model of naive set theory (materially construed).[21]

In fact, we can obtain more than this. Suppose that in $\mathcal{M}$ there are inaccessible cardinals. Let ϑ_1 be the least such, and ϑ_2 be a greater one. Take ξ to be ϑ_2. Since the sets of rank less than ϑ_2, and a fortiori than ϑ_1, remain unaffected in the collapse, both of these are consistent substructures of $\mathcal{M}^{\sim}$ which are models of ZF. Moreover, any theorem of ZF with its quantifiers relativised to V_{ϑ_1} (so that $\exists x\alpha$ becomes $\exists x \in \mathbf{c}\,\alpha$, where "**c**" refers to V_{ϑ_1}; and similarly for $\forall$) holds consistently in $\mathcal{M}^{\sim}$. (This is not true of V_{ϑ_2}, since this set itself behaves inconsistently.[22]) That is, V_{ϑ_1} is a consistent inner model of ZF (which shows that the theory of $\mathcal{M}^{\sim}$ is highly non-trivial).

To take stock, what we have established is that there are interpretations that:

- are models of Ext and Abs
- are models of ZF
- contain the cumulative hierarchy (at least up to V_{ϑ_1}) as a consistent inner model.

We may therefore suppose that the true interpretation of the language of set theory has these properties. This is an appealing picture. The cumulative hierarchy (up to ϑ_1) is a perfectly good, consistent, set-theoretic structure; but it does not exhaust the universe of sets. There may be non-well-founded sets (such as the set of all sets) and inconsistent sets, such as the set of all sets that are not members of themselves. The universe of sets is just much richer than orthodox set theory takes it to be.

Of course, the model $\mathcal{M}^{\sim}$ that we actually constructed using the Collapsing Lemma is still pathological from this perspective. It contains only one inconsistent set, $[a]$, which has to do duty for all inconsistent and non-well-founded sets.

---

[21] The fact that $\mathcal{M}^{\sim}$ is a model of Abs is a special case of a more general lemma, to be found in (Restall, 1992).

[22] In fact, V_{ϑ_2} behaves just like the set of all non-well-founded sets, given Mirimanoff's paradox. It is well-founded, but it is also a member of itself, so is not well-founded.

There are undoubtedly other models (the details of whose natures require further investigation).[23] It should be remembered that, even in classical logic, set theory—and every other theory with an infinite model, but an "intended interpretation"—has an absolute infinity of pathological models. Specifying the correct interpretation is always a further issue. The model $\mathcal{M}^{\sim}$ at least suffices to demonstrate the possibility of interpretations of naive set-theory which have the above properties.[24]

And to return, at last, to the question of what to make of the theorems of orthodox set theory, ZF, on this approach. The answer is obvious. Since the universe of sets is a model of ZF (as well as naive set theory), these hold in it. We may therefore establish things in ZF in the standard classical way, knowing that they are perfectly acceptable from a paraconsistent perspective.[25] We cannot, of course, require the theorems of ZF to be consistently true in that universe; but if, on an occasion, we do require a consistent interpretation of ZF, we know how to obtain this too. The universe of sets has a consistent substructure that is a model of ZF.

6 Metatheory of Paraconsistent Logic

Let us turn, finally, to the issue of paraconsistent model theory. If the paraconsistent strategy for set theory is to be anything more than an intellectual exercise, the underlying logic used must, in some sense, be the right one for reasoning about sets. Hence arise familiar debates about which logic is correct, and why. A frequent objection made against paraconsistency in this debate goes as follows. Paraconsistent logics have metatheories. In particular, they have appropriate semantics, proof systems, and corresponding soundness and (hopefully) completeness results. Now the logic in which such proofs are carried out must be classical, non-paraconsistent, logic.[26] This shows that paraconsistent logic cannot be maintained as the correct logic.

---

[23] Some of these can be obtained by other applications of the Collapsing Lemma. Different methods of constucting models of inconsistent set theory, some of which also model ZF, are discussed in (Libert, 2003).

[24] Criticising the strategy under discussion here, (Weir, 2005, p. 398), says: "It will not do to say ... that the models which ... [do not have the desired properties] are 'pathological' or 'unintended.' All the dialetheist's ZFC models are unintended in the sense that they do not capture anything like the full structure of the naive universe of sets. This compares unfavourably with the unintended models of first-order number-theory: they at least contain the 'real' structure of numbers." This is simply question-begging. The thesis is precisely that one of these models does capture the full structure of the universe of sets. (Or, if there are many equally good models, then each captures the structure of an equally good universe.) From the dialetheic perspective, it is precisely the cumulative hierarchy that is an incomplete fragment of the universe of sets. And the models in question do contain the cumulative hierarchy as a fragment (at least up to an inaccessible cardinal).

[25] In particular, the argument constructing the interpretation $\mathcal{M}^{\sim}$ above can be carried out in ZF, and so is perfectly acceptable.

[26] Rescher (1969, p. 229) documents this claim, though he does not endorse it.

The argument is far too swift. For a start, the logic of the metatheory of a theory need not be classical. For example, an intuitionist metatheory for intuitionist logic is well known.[27] Is there a metatheory for paraconsistent logics that is acceptable on paraconsistent terms? The answer to this question is not at all obvious. First, the standard proofs in the metatheories of paraconsistent logics are usually given, as are most mathematical proofs, in an informal way. The question, then, is how to interpret the proofs formally. A normal assumption is that the proofs are carried out using classical logic. And indeed, this would seem to be sufficient for the purpose. This point is not definitive, however. Most paraconsistent logics are generalisations of classical logic in one way or another. In particular, they coincide with classical logic in those cases (models) which are consistent (i.e., in which all formulas behave consistently). Hence, if an informal argument concerns a consistent situation, and can be regimented using classical logic, it is perfectly acceptable for a paraconsistent logician.[28] Can a paraconsistent logician, or at least, one who subscribes to paraconsistent set theory, look at the metatheoretic arguments concerning paraconsistent logic in this way? The answer, unfortunately, is "no". For metatheoretic constructions are carried out in set theory; and paraconsistent set theory is not consistent.

In the model theory of paraconsistent logic, we must therefore use paraconsistent set theory, however that is best construed. To what extent model theory can be developed on the relevant strategy for naive set theory is still an open question. But the model-theoretic strategy for naive set theory provides a simple way of accommodating paraconsistent model theory. One may think of the metatheory of the logic, including the appropriate soundness and completeness proofs, as being carried out (as we know it can be) in ZF. According to the model-theoretic strategy, the results established in this way can perfectly well be taken to hold of the universe of sets, paraconsistently construed. The paraconsistent logician can, therefore, simply appropriate the results.

It might be thought that this approach to the metatheory of paraconsistent logic suffers from a problem. In the material and model-theoretic strategies for paraconsistent set-theory, the relationship between the premises and the conclusion of a valid inference is expressed by a material conditional. Thus, simplifying to the one-premise case for perspicuity, and writing the relation "α holds in $\mathcal{I}$" as "$\mathcal{I} \Vdash \alpha$", an inference from α to β is valid iff:

Val for every interpretation $\mathcal{I}$, $(\mathcal{I} \Vdash \alpha \supset \mathcal{I} \Vdash \beta)$

Now, the material conditional does not support detachment. Hence an inference can be valid, yet this does not licence the detachment of the conclusion from the premise. Surely this deprives the notion of validity of its punch?

No. The disjunctive syllogism is perfectly acceptable provided that the situation is consistent. (See, again, (Priest, 1987, Ch. 8).) Provided that we do not have $\mathcal{I} \Vdash \alpha$ and $\mathcal{I} \nVdash \alpha$, we can get from $\mathcal{I} \Vdash \alpha$ to $\mathcal{I} \Vdash \beta$. In particular, then, provided that $\mathcal{I}$ is in part of the universe of sets that is consistent (the cumulative hierarchy, or a

---

[27] See, e.g., (Dummett, 1977, Ch. 5), esp. p. 197.

[28] For further discussion, see (Priest, 1987, Ch. 8).

sufficiently generous part thereof), we have business as usual. (Note: this does not mean that the set of things made true by $\mathcal{I}$ is consistent. "$\mathcal{I} \Vdash \alpha$ and $\mathcal{I} \nVdash \alpha$" is quite different from "$\mathcal{I} \Vdash \alpha$ and $\mathcal{I} \Vdash \neg\alpha$.") If $\mathcal{I}$ is a set outside this part of the universe, matters are different. Thus, we may expect that there is an interpretation, $\mathcal{M}$, that is in accord with the actual, in the sense that for any γ, γ iff $\mathcal{M} \Vdash \gamma$. One should not expect this interpretation to be in the hierarchy. Appropriate techniques of diagonalisation will give us sentences, α, such that $\mathcal{M} \Vdash \alpha$ and $\mathcal{M} \nVdash \alpha$. In such cases, even though **Val** holds, the fact that α (i.e., $\mathcal{M} \Vdash \alpha$) will not allow us to detach β (i.e., $\mathcal{M} \Vdash \beta$). However, such αs will be unusual. In standard cases **Val** will provide a licence to get from α to β.

It might still be thought odd to have the validity of a deductive inference grounded in a defeasible inference such as the disjunctive syllogism. But a little thought should assuage this worry. The difference between a material $\mathcal{I} \Vdash \alpha \supset \mathcal{I} \Vdash \beta$ and a relevant $\mathcal{I} \Vdash \alpha \rightarrow \mathcal{I} \Vdash \beta$ is not as great as might be thought in this context. Both are simply true (or false) *statements*. Inference, by contrast, is an *action*. Given the premises of an argument an inference is a *jump* to a new state. No number of truths is the same thing as a jump. (This is the moral of Lewis Carroll's celebrated dialogue between Achilles and the Tortoise, (Carroll, 1895).) None the less, truths of a certain kind may *ground* the jump, in the sense of making it a reasonable action. There is no reason why a sentence of the form $\gamma \supset \delta$ may not do this just as much as one of the form $\gamma \rightarrow \delta$. It is just that one of the latter kind always does, whilst one of the former kind does so only sometimes.

If it is still not clear how a sentence can function in this way, consider sentences of the form:

($*$) You promised to do x

The truth of ($*$) normally grounds doing x, in the sense of making it reasonable to do it. But, to use a celebrated example, suppose that ($*$) is true, where the x in question is the returning of a weapon to a certain person. And suppose that that person comes requesting the weapon, but you know that they intend to use it to commit suicide. Then the truth of ($*$) does not, in this context, ground the action. Just as with validity and the material conditional, the truth of a sentence of a certain kind may ground an appropriate action in normal circumstances, but fail to do so in unusual circumstances.

This objection dealt with, there would seem nothing to prevent the paraconsistent logician from simply appropriating all the classical metatheoretic results in the way explained. The appropriation might be thought to have all the charms of theft over honest toil (as Russell said in another context); on the other hand, why reinvent the wheel?

7 Conclusion: The Shock of the New

At various times in its history, mathematics has been shocked by the discovery of new kind of entity: irrational numbers, infinitesimals, transfinite sets, and so on. The reception by the mathematical community of these entities has often been controversial and contentious; and the discovery has always been followed by a process of rethinking mathematical reasoning in the light of these entities and their properties. The discovery of inconsistent objects, such as the Russell set—of all those sets that do not contain themselves—is the most recent, and perhaps the most contentious, episode of this kind; and we are still in the process of thinking through its ramification for mathematical reasoning. In mathematical revolutions of this kind, it is always important to preserve the central parts of previous mathematical thought. What I have been engaged in here is a contribution to this project.

Bibliography

Beall, J., Brady, R., Hazen, A., Priest, G., and Restall, G. (2006). Restricted quantification in relevant logics. *Journal of Philosophical Logic*, 35:587–598.

Brady, R. (1989). The non-triviality of dialectical set theory. In Priest, G., Routley, R., and Norman, J., editors, *Paraconsistent Logic: Essays on the Inconsistent*, chapter 14. Philosophia Verlag, Munich.

Carroll, L. (1895). What the tortoise said to Achilles. *Mind*, 4:278–280.

Dummett, M. (1977). *Elements of Intuitonism*. Oxford University Press, Oxford.

Dunn, J. (1988). The impossibility of certain second-order non-classical logics with extensionality. In Austin, D., editor, *Philosophical Analysis*. Kluwer Academic Publishers, Dordrecht.

Dunn, J. and Restall, G. (2002). Relevance logic. In Gabbay, D. and Guenther, F., editors, *Handbook of Philosophical Logic*, volume 6, pages 1–128. Kluwer Academic Publishers, Dordrecht, second edition.

Goodship, L. (1996). On dialethism. *Australasian Journal of Philosophy*, 74:153–161.

Leisenring, A. (1969). *Mathematical Logic and Hilbert's ε–Symbol*. Gordon and Breach, New York, NY.

Libert, T. (2003). ZF and the axiom of choice in some paraconsistent set theories. *Logic and Logical Philosophy*, 11:91–114.

Priest, G. (1979). Logic of paradox. *Journal of Philosophical Logic*, 8:219–241.

Priest, G. (1987). *In Contradiction*. Martinus Nijhoff, Dordrecht. Second, extended edition, Oxford University Press, 2006.

Priest, G. (1991). Minimally inconsistent LP. *Studia Logica*, 50:321–331. Reprinted with minor changes as Ch. 16 of the second edition of Priest (2006).

Priest, G. (1995). *Beyond the Limits of Thought*. Cambridge University Press, Cambridge. Second, extended edition, Oxford University Press, 2002.

Priest, G. (2001). *Introduction to Non-Classical Logic*. Cambridge University Press, Cambridge. Second, extended edition, 2008.

Priest, G. (2002). Paraconsistent logic. In Gabbay, D. and Guenther, F., editors, *Handbook of Philosophical Logic*, volume 6, pages 287–293. Kluwer Academic Publishers, Dordrecht, second edition.

Rescher, N. (1969). *Many-Valued Logic*. McGraw-Hill, New York, NY.

Restall, G. (1992). A note on naive set theory in LP. *Notre Dame Journal of Formal Logic*, 33:422–433.

Routley, R. (1977). Ultralogic as universal. *Relevant Logic Newsletter*, pages 50–89 and 138–175. Reprinted as an appendix in *Exploring Meinong's Jungle and Beyond*, Canberra: Research School of Social Sciences, 1980.

Slaney, J. (1989). RWX is not Curry paraconsistent. In Priest, G., Routley, R., and Norman, J., editors, *Paraconsistent Logic: Essays on the Inconsistent*, chapter 17. Philosophia Verlag, Munich.

Weber, Z. (2009). *Paradox and Foundation.* PhD thesis, University of Melbourne.
Weber, Z. (2010a). Extensionality and restriction in naive set theory. *Studia Logica*, 94:87–104.
Weber, Z. (2010b). Transfinite numbers in paraconsistent set theory. *Review of Symbolic Logic*, 3:71–92.
Weir, A. (2005). There are no true contradictions. In Priest, G., Beall, J., and Armour-Garb, B., editors, *The Law of Non-Contradiction: New Philosophical Essays*, chapter 22. Oxford University Press, Oxford.

Chapter 9
Paradox, ZF, and the Axiom of Foundation

Adam Rieger

It is a great pleasure to contribute to this *Festschrift* for John Bell. No-one has done more than he has to demonstrate the fruitfulness of the interplay between technical mathematics and philosophical issues, and he is an inspiration to all of us who work somewhere in the borderland between mathematics and philosophy.

I also owe him a great personal debt. I arrived at the LSE dejected and disillusioned by my experiences of the Mathematical Tripos at Cambridge, but it is impossible to be downhearted for long in the company of John. His enthusiasm, humour and warmth were the perfect antidote to the stuffiness and inhumanity of Cambridge and helped hugely to rebuild my interest and self-confidence. John's energy levels must be seen to be believed, and an evening with him is an unforgettable experience. It generally starts about 4 p.m. and ends around 5 in the morning, when the last of his companions (never John, who always gives the impression that he could go on talking indefinitely) finally succumbs to sleep.

At John's suggestion, I wrote my M.Sc. dissertation at the L.S.E. on truth, which led on eventually to an Oxford D.Phil. which concerned both the semantic and set-theoretical paradoxes. It is the concept of set—example *par excellence* of one that straddles philosophy and mathematics—that is the subject of this essay.

1

At the beginning of the twentieth century there was a crisis in the foundations of mathematics. The crisis centred around the concept of *set*, which suddenly achieved prominence in two different ways. Firstly, Cantor's theory of the transfinite showed that sets were of great intrinsic mathematical interest. And secondly through the work of Frege and Russell it emerged that sets were central to the philosophical project—logicism—of reducing mathematics to logic.

The discovery by Russell and others that, if handled carelessly, sets give rise to contradictions, threatened not only the logicist programme, but mathematics itself;

A. Rieger (✉)
Senior Lecturer in the Department of Philosophy, University of Glasgow, Glasgow, UK
e-mail: A.Rieger@Philosophy.arts.gla.ac.uk

D. DeVidi et al. (eds.), *Logic, Mathematics, Philosophy: Vintage Enthusiasms*, The Western Ontario Series in Philosophy of Science 75,
DOI 10.1007/978-94-007-0214-1_9,

for large parts of mathematics, in particular analysis, make essential use of the completed infinite, and the paradoxes seemed to show that this was a risky practice.

A hundred years later there is no longer a foundational crisis. Why is this? By 1930, mathematicians had found a way of coping with the paradoxes. A system of axiomatic set theory, *Zermelo-Fraenkel* set theory (ZF for short) had been developed, which allowed mathematicians to do all they wanted to do with sets, whilst maintaining consistency. ZF is not the only axiomatic set theory, but at the present time it has a completely dominant position amongst such theories. For the purposes of university mathematics courses, for example, set theory just *is* ZF.

Does ZF really deserve its elevated position? Below I examine three sorts of argument which can be adduced in support of ZF:

1. Argument from the paradoxes
2. Argument from the iterative conception
3. Argument from pragmatic mathematical considerations

and conclude that none of them are convincing.

2

There is a widespread view that one needs some kind of hierarchy, and hence ZF or something like it, to avoid the paradoxes.[1] Let us go back to basics to examine the merits of this claim.

According to the *naive conception* of set, any arbitrary collection forms a set.[2] This entails the truth of the naive comprehension schema

$$\exists x \forall y (y \in x \leftrightarrow \varphi(y)).$$

But this schema is in fact logically false: for as Russell noticed (Russell, 1902), on letting φ be $y \notin y$ we obtain a contradiction.

Now whilst this shows that there is a fatal defect with the naive conception, it does not yield an illuminating explanation of what exactly is wrong with it, an explanation that might be of some use in the reform of the concept of set which must inevitably follow.

Such an explanation is, however, available.

Suppose a is a set, and consider the set

$$b = \{x \in a : x \notin x\}.$$

[1] Logical *cognoscenti* know that this is false, but the view is common amongst those who are mathematically, but not logically, well-informed.

[2] As far as I know, nobody has ever explicitly put forward the naive conception, though it is implicit in (Frege, 1893).

It is easy to see that $b \notin a$. So we have a recipe which, for any set a, gives us a set which is not a member of a. The set b "diagonalizes out" of a.

If a already has all sets as members, trouble arrives in the shape of the Russell paradox. And naively, there must be such a set a, for example the set of everything whatever. (Just take $\varphi(y)$ to be $y = y$ in the naive schema above.) On this way of looking at it, contradiction arises because, under the naive conception, we have both *extensibility* (the ability to extend any given set by finding something that is not one of its members) and *universality* (the existence of a set which contains everything). Clearly these cannot co-exist consistently.

Essentially the same diagnosis can be given for the other standard set-theoretic paradoxes. In the so-called Cantor paradox, extensibility arises from the power set operation—since the power set $P(a)$ of a is always strictly larger than a, there must be elements of $P(a)$ which are not in a. So again the existence of a universal set leads to contradiction.

In some of the paradoxes the extensibility and universality are relativised to particular sub-collections of the universe. For example, the Burali-Forti paradox hinges on the tension between (i) the principle that, for any initial segment of ordinals, we can, by considering the ordinal number of the segment itself, find an ordinal not in that segment, and (ii) the principle that there is a well-ordered set of *all* ordinals. The slightly less well-known Mirimanoff paradox concerns the set W of all well-founded sets.[3] Since all the members of W are well-founded, so is W, but then it should be a member of itself, and so, after all, not well-founded. Here extensibility is obtained by considering, for any set a of well-founded sets, the set $b = a \cup \{a\}$; this is another set of well-founded sets (since a itself must be well-founded), yet it has a member (namely a itself) which is not a member of a (else we have $a \in a$, so a is not well-founded).[4]

3

A consequence of this diagnosis is that there is a neat and rather natural way to solve the paradoxes: ban universality by not allowing very large collections (e.g. the universe and the collection of all ordinals) to be sets. Remarkably, this solution was hit upon by the originator of set theory, Georg Cantor, before the "paradox industry" had even got under way. In a letter he wrote to Dedekind (Cantor, 1899) we find the following passage:

> ... it is necessary, as I discovered, to distinguish two kinds of multiplicity .... For a multiplicity can be such that the assumption that *all* of its elements "are together" leads to a contradiction, so that it is impossible to conceive of the multiplicity as a unity, as "one

---

[3] For the definition of well-founded, see below.

[4] For more details, including attempts to apply the same idea to the semantic paradoxes, see (Rieger, 1996, Ch. 1) or (Priest, 1994) (Priest uses the terms *transcendence* and *closure*). Dummett's idea of an indefinitely extensible concept (Dummett, 1991, p. 316) and the book (Grim, 1991) are also relevant. The basic idea can be found in (Russell, 1906c), which I discuss below.

finished thing." Such multiplicities I call *absolutely infinite* or *inconsistent multiplicities* .... If on the other hand the totality of the elements of a multiplicity can be thought of without contradiction as "being together" ... I call it a *consistent multiplicity* or *set*.[5]

Cantor has discovered what in modern parlance is called the set/class distinction, usually attributed to (von Neumann, 1925). The key idea is that some infinite collections are all right, and they form the sets; others are just too big, and are either abolished altogether, or allowed in as some other kind of entity (proper classes).[6] The idea is, in the light of our diagnosis, thoroughly motivated and not at all ad hoc.

And, indeed, this is exactly what happens with ZF. This is sometimes expressed, somewhat misleadingly, by saying that ZF incorporates the doctrine of *limitation of size*—misleading because this phrase suggests that there is some cardinal magnitude below which collections are safe but above which they are paradoxical, whereas the point is not that sets must be below some particular size but that they must not be as big as the universe.

Nothing we have said so far, however, requires sets to be arranged in a *hierarchy*. But the ZF axioms embody such a requirement. In particular, the *axiom of foundation* states that every (non-empty) set x has a member y which is minimal, in the sense that no member of x belongs to y.[7] Another way of putting this, equivalent in the presence of the axiom of choice, is that there is no infinite descending membership chain $x \ni x_1 \ni x_2 \ni \ldots$. So there cannot be, for example, a set which is a member of itself, or a member of a member of itself.

A suspicion therefore arises that ZF restricts the notion of set more than is necessary to avoid the paradoxes, and therefore offends against the following methodological principle: *when forced by paradox to reform a naive concept, preserve as much of it as possible*. The naive concept of set does not obey the axiom of foundation: it allows such self-membered sets as the set of absolutely everything, and the set of all things discussed in this paper. ZF rules out both of these, but the principle of restricting universality seems to deny sethood only to the first.[8] Can there be a consistent theory which allows the second?

Indeed there can. One sort of theory with this property is discussed by (Aczel, 1988). Briefly, the idea is to take the axioms of ZF *except* foundation, and add to

---

[5] Though it is not quite as explicit, the distinction between the transfinite and the absolute infinite can be found much earlier in Cantor's writings (e.g., Cantor, 1883, p. 205). It seems likely that, having realized that any set can be enlarged by the power set operation, Cantor drew immediately the conclusion that there can be no universal set. Cantor is sometimes accused of believing in naive set theory (e.g., (Körner, 1960, p. 44): "Cantor's theory of classes, by admitting as a class any collection, however formed, leads to contradictions"). This is quite unjustified: rather "his conception of set ... was one in which the paradoxes cannot arise" (Menzel, 1984, p. 92); see also (Hallett, 1984, p. 38 and *passim*).

[6] More precisely, the principle that a collection is too big to form a set iff it can be put into 1-1 correspondence with the universe can be taken as the basis for an axiomatization of set theory, as is done in (von Neumann, 1925).

[7] A set x satisfying this condition is said to be *well-founded*.

[8] To make the example work, interpret "discussed in this paper" so that it applies to only a small (e.g. finite) number of things.

them some version of an *anti-foundation* axiom. This is best understood in terms of membership graphs. There is a natural sense in which (directed) graphs can be regarded as pictures of sets: for example, Fig. 9.1 is a picture of the von Neumann ordinal $2 = \{\emptyset, \{\emptyset\}\}$.

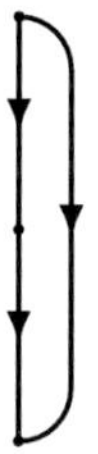

Fig. 9.1 An exact picture of 2

Only well-founded graphs (graphs without infinite paths) can be pictures of sets in the ZF universe; to obtain the richer non-well-founded universes we allow *any* graph to be a picture of a set. Thus Fig. 9.2 is the picture of a set $a = \{a, \emptyset\}$.

Fig. 9.2 A non-well-founded graph

By constructing a graph model from a model of ZF, Aczel proves that these systems are consistent if ZF is.[9]

Summary: it can be rigorously proved that ZF restricts the notion of set more than the paradoxes demand.

4

How did the idea take hold that a hierarchy is necessary to solve the paradoxes? To answer this it will be necessary to take a short historical detour.

In December 1905 Russell read a remarkable paper to the London Mathematical Society, later published as (Russell, 1906c). In it he states clearly that the lesson of the paradoxes is that naive comprehension must be rejected:

> What is demonstrated by the contradictions we have considered is broadly this: "A propositional function of one variable does not always determine a class." (Russell, 1906c, pp. 144–145)

[9] For more details see (Aczel, 1988). I discuss the merits of the various anti-foundation axioms in (Rieger, 2000).

And he gives essentially the diagnosis above:

> ... there are what we may call *self-reproductive* processes and classes. That is, there are some properties such that, given any class of terms[10] all having such a property, we can always define a new term also having the property in question. Hence we can never collect *all* the terms having the said property into a whole; because, whenever we hope we have them all, the collection which we have immediately proceeds to generate a new term also having the said property. (Russell, 1906c, p. 144)

This insight would seem to lead naturally to the conclusion outlined above, that a solution to the paradoxes may be obtained by ensuring that there is no universal class, no class of all ordinals, etc. However, Russell does not simply draw this inference; rather he considers three different responses, all of which would indirectly ban the offending classes: the "zigzag" theory, the "limitation of size" theory, and the "no-classes" theory, tentatively suggesting that the last of these offers the most promising route for a solution.[11]

Less than a year later, Russell had changed his mind about the paradoxes. In a paper published in September 1906, he wrote this:

> I recognise, however, that the clue to the paradoxes is to be found in the vicious-circle suggestion. (Russell, 1906b, p. 198)

The "vicious-circle suggestion" is

> ... that whatever in any way concerns *all* o[r][12] *any* or *some* (undetermined) of the members of a class must not itself be one of the members of a class.[13] (Russell, 1906b, p. 198)

A new concept, *circularity*, has now entered the discussion. Russell believes that all the paradoxes result from the (allegedly) circular practice of allowing totalities containing members which, in some appropriate sense to be discussed below, "concern" that very totality.

It might perhaps seem at first sight that this is just another way of stating the original diagnosis. For is not this circularity the key feature of the "self-reproductive" classes identified in the earlier paper? But in fact the difference is dramatic. According to the previous diagnosis, there cannot be a class of *all* ordinals or *all* things (for this would lead to the contradictory consequence that there is a new ordinal (thing) which both must and must not be in the class). But in the later paper, Russell advocates the very much stronger principle that there can be *no class whatever*, large or small, which has members concerning that very class. This inevitably imposes a hierarchical structure on the universe of classes. *At this point an alien constructivism*

---

[10] "Term" here just means "object."

[11] It might be thought that "limitation of size" embodies exactly the idea of restricting universality, but it is clear that Russell does not think of it in this way: rather he sees the theory as posing the question "how far up the series of ordinals it is legitimate to go" (p. 53), a question which he cannot see any prospect of answering.

[12] The original has "of," which seems to be a misprint.

[13] The occurrence of "any" (and "some") may seem puzzling here: since anything presumably concerns itself, the principle seems to rule out anything ever being a member of a class. But Russell should be read as forbidding any member of a class concerning quantification over the class.

was imported into classical mathematics, vestiges of which are still visible today. For the vicious circle principle is inextricably linked with a constructivist view of the metaphysics of mathematics.[14]

5

What had happened to Russell in the few months between the two papers? He had been reading Poincaré. The second paper was, in fact, written in reply to [Poincaré, 1906]. In that work Poincaré blames the paradoxes on circularity. His treatment is sketchy, and he discusses only Richard's paradox[15] in any detail, claiming that "the same explanation serves for the other antinomies, as may be easily verified" (p. 190). Thus applied to classes,

> ... the definitions that must be regarded as non-predicative are those which contain a vicious circle.[16] (p. 190)

Though he does not explicitly formulate the VCP, it is clear from a later passage in the paper that he has the same conception of it as Russell:

> ... if the definition of a notion N depends on *all* the objects A, it may be tainted with the vicious circle, if among the objects A there is one that cannot be defined without bringing in the notion N itself. (p. 194)

Now Poincaré's views on the VCP arise completely naturally from his wider views on the philosophy of mathematics, in particular, his view on mathematical existence. He is explicit in his constructivism. In a paper written in 1912, he declares that a mathematical object "exists only when it is conceived by the mind" (Poincaré, 1963, p. 72). He considers a "genus" (set) G with a member X, and writes of the members of G

> ... they will exist only after they have been constructed; that is, after they have been defined; *X exists only by virtue of its definition,*[17] which has meaning only if all the members of G are known beforehand, and X in particular.

This conception of existence of course provides a motivation for the VCP. If mathematical objects are brought into existence by their definitions, then it seems that no totality can possibly contain members defined in terms of that very totality.

---

[14] To avoid confusion, I should perhaps make it clear that here and throughout the paper I use "constructivism" as a name for a metaphysical view about mathematics, roughly that mathematical objects are brought into existence by some activity of human minds. The term is sometimes now used for mathematics without the law of excluded middle, but I shall use it in its earlier sense.

[15] This is a semantic paradox, introduced in (Richard, 1905), which concerns the collection E of all reals definable in a finite number of words; by a diagonal argument we can obtain a new real, not in E yet definable in a finite number of words.

[16] The original italicises this sentence. At this point "non-predicative" means simply "not defining a class"; confusingly, Russell, having accepted the diagnosis, started using "impredicative" to *mean* "violating the vicious circle principle."

[17] My italics.

However, Russell adopted Poincaré's views on impredicativity without accepting the constructivist outlook. By doing so he landed the classical mathematical community in a philosophical confusion from which it has yet to emerge.[18]

6

Armed with his diagnosis of the paradoxes and aided by Whitehead, Russell embarked on reworking mathematics whilst obeying the VCP: the result was the ramified type theory of (Whitehead and Russell, 1910). This is not the place for a detailed discussion of that work, but for present purposes it is enough to note that the system obtained is a (rather complex) hierarchy of propositional functions; the position of a function in the hierarchy depends not only on (i) its arguments but also on (ii) the ranges of any quantifications in its definition (this latter refinement making the hierarchy "ramified"), both (i) and (ii) being required to be lower down in the hierarchy.

But despite Whitehead and Russell's efforts, their system has never been accepted as a foundation for mathematics. Instead, the system of axiomatic set theory developed in continental Europe, mostly by Zermelo, proved much easier to work with. Most of the axioms appear first in (Zermelo, 1908), which contains versions of: extensionality, empty set, pairs, separation, power set, union, choice and infinity, that is, all the axioms in the now-standard theory except for replacement and foundation. Interestingly Zermelo's motivation at this point seems only partly to have been the paradoxes; primarily he was concerned to analyse exactly which principles concerning sets he had used in his proof that every set can be well-ordered (Zermelo, 1904). The central idea is to replace naive comprehension by separation: that is, we cannot in general form the set of absolutely all y such that $\varphi(y)$, but only the set of all members of some set a such that $\varphi(y)$. Paradox is avoided because there is no way to prove that the universe is a set; indeed the Russell paradox becomes a proof that it is *not* a set.[19]

The replacement axiom was added later by Fraenkel (1922) and Skolem (1922). As for the axiom of foundation: the issue first seems to have been considered by (Mirimanoff, 1917a, b), who distinguishes "ordinary" sets which do not have infinite descending membership chains from "extraordinary" ones which do. He does

[18] Goldfarb (1989) attempts to reconcile Russell's predicativism with his lack of constructivism, arguing that his views on variables and their ranges of significance can lead to ramification of intensional entities (in particular propositions and propositional functions) even on a realist conception. But even if this is right—and Goldfarb says he is only making a "first step" (p. 27) towards a full treatment of the issue—the fact remains that Russell advocates the VCP in full generality. As Goldfarb admits (pp. 30–31), it is hard to see how the ramification of *sets* can be justified except on a constructivist view.

[19] The axiom of foundation immediately rules out a universal set, for such a set would be a member of itself. But the point is that such a set is ruled out anyway by the other axioms. Foundation plays no role in solving the paradoxes.

not assert, however, that there is anything wrong with the extraordinary sets. von Neumann (1925) describes non-well-founded sets as "superfluous" [p. 404] and gives an axiom [p. 412] which excludes some, but not all, of them. Three years later, in (von Neumann, 1928), he formulates the axiom of foundation in the form $\forall x(x \neq \emptyset \rightarrow \exists y \in x(y \cap x = \emptyset))$. However, it is not until (Zermelo, 1930) that the axiom of foundation is explicitly adopted as a postulate. With this paper all the axioms of standard modern set theory are in place.

7

In comparing type theory with ZF, it is useful to try to get clearer about what the VCP is saying. Russell never provided a single clear statement of it. Here are two attempts:

> I. No totality may contain members defined in terms of itself.
>
> (Russell, 1908, p. 75)
>
> II. Whatever involves *all* of a collection must not be one of the collection.
>
> (Russell, 1908, p. 63)

Now it seems that Russell took these, and other, statements to be different formulations of the same principle.[20] However, (Gödel, 1944) pointed out that it looks like there is more than one principle here, in particular one to do with *definitions*, the other to do with the notion of *involving*.[21] Let us call these VCP I and VCP II, and try to see more precisely what implications they have for sets.

VCP I seems closest to the constructivist spirit of Poincaré. It rules out *impredicative definitions*: for example definitions of x which quantify over a collection of which x is a member.[22]

Since sets presumably "involve" their members, VCP II rules out sets which are members of themselves. It also seems reasonable that "involving" is transitive, so that a set also involves the members of its members, and so on. Hence a reasonable explication of "x involves y" in the case of sets is "y is a member of the transitive closure of x,"[23] in which case VCP II will be obeyed if sets satisfy the axiom of foundation.[24]

---

[20] Some other statements from Russell's writings are to be found at the pages cited above, and also (Russell, 1906b, p. 204); (Whitehead and Russell, 1910, p. 37).

[21] Gödel claims to discern a third principle, concerning "pre-supposing," which I shall not discuss here.

[22] This will not do as it stands as a *characterization* of impredicative definitions. For example it will be equally objectionable if, instead of x itself being a member of the totality, some second object y, defined using x, is a member. Presumably to make this rigorous we would require some notion of well-foundedness for definitions; I shall not attempt to supply details here.

[23] The transitive closure of x is the set whose members are the members of x, the members of the members of x, and so on.

[24] Though he does not state it explicitly, this seems to be what Gödel has in mind in his paper (see p. 131 with its footnote reference to Mirimanoff). It is a little too strong to say that VCP II *entails*

Now the system of *Principia Mathematica* obeys both VCP I and VCP II. Impredicative definitions are rigorously avoided, and the universe has a hierarchical structure. That ZF obeys VCP II is, as I have said, guaranteed by the axiom of foundation. But ZF violates VCP I. The axiom of separation allows us, for any set a, to define a new set b by admitting only those members of a satisfying some formula $\varphi(x)$. But there is no restriction on the quantifiers that may occur in $\varphi(x)$: they may range over the whole universe. Impredicative definitions are perfectly allowable in ZF.

8

As I recounted above, most of the axioms of ZF resulted from Zermelo's attempt to defend his proof that every set could be well-ordered. This does not apply, however, to the axiom of foundation. I conjecture that it was inspired by type theory, but I do not know of anything explicit in the early literature which supports this. In any case, somewhat later a new way justifying the axioms developed. This is the second of the three ways I mentioned of arguing in favour of ZF.

The idea is as follows. We all have an intuitive grasp of the concept of natural number, that is, we grasp a structure which we refer to as "the natural numbers." If someone wanted to justify the Peano axioms for number theory, they would appeal to the evident truth of the axioms in this intuitively understood structure. The claim is that something analogous can be done for set theory. There is an intuitive conception of set, the *iterative* conception, which gives rise to an intuitively understood model, the *cumulative hierarchy*. The axioms (or at least, a number of them)[25] are then justified by appealing to the fact that they are true in this model.

How does the model work? Start with the empty set.[26] Call this V_0. V_1 is the power set of V_0, and in general, we obtain the next level after V_n by taking the power set. V_ω is just the union of all the V_n for finite n, and $V_{\omega+1}$ is the power set of V_ω. Continue through the ordinals, forming power sets at each stage and taking unions at limit ordinals. The result is a hierarchy in which sets only have members from lower down the hierarchy. As (Lavine, 1994, p. 144) points out, "the iterative conception gives the Axiom of Foundation center stage."

---

the axiom of foundation, for an infinite descending membership chain $x_1 \ni x_2 \ni \ldots$ in which all the x_i are *different* violates foundation without circularity. Such a chain seems equally offensive to the constructivist intuitions underpinning the VCPs, and suggests that they do not fully capture those intuitions.

[25] There is disagreement, for example, on whether the axiom of replacement is derivable from the iterative conception.

[26] A variation is possible in which instead we start with some *atoms* or *urelements*, that is, some non-sets. Though this is probably more natural from a naive point of view, mathematicians standardly work with a universe of *pure* sets, where everything is a set, since this is technically smoother (for example the quantifiers can simply be taken to range over all sets) and does not result in any limitation in structure. For present purposes the difference in the two approaches is not important.

The cumulative hierarchy was hinted at in (Mirimanoff, 1917a) and introduced explicitly by von Neumann (1929) for the purposes of a consistency proof, but the idea of using it as an intuitive model justifying the axioms only came later. It is suggested by Gödel (1947, pp. 474–475), but only becomes explicit around 1970, when a number of papers appeared roughly simultaneously. Shoenfield (1967, 1977), Boolos (1971) and Wang (1974) are representative. Let us examine some of the passages in which they justify that part of their conception of set which gives rise to the axiom of foundation.[27] First Shoenfield:

> Sets are formed in *stages*. For each stage S, there are certain stages which are before S. At each stage S, each collection consisting of sets formed at stages before S is formed into a set.... When we are forming a set z by choosing its members, we do not yet have the object z, and hence cannot use it as a member of z. The same reasoning shows that certain other sets cannot be members of z. For example, suppose that $z \in y$. Then we cannot form y until we have formed z. Hence y is not available as an object when z is formed, and therefore cannot be a member of z. (Shoenfield, 1977, p. 323)

Boolos actually claims that the iterative conception of set has an intuitive plausibility independent of the paradoxes, and that one might have come to see it as superior to naive set theory (as embodied in the naive comprehension axiom) even if the paradoxes had never been discovered. That is (though this is not the way Boolos expresses it), there are really two versions of naive set theory, one captured by naive comprehension, the other by the iterative conception, and the latter has at least as great an intuitive appeal as the former:

> ZF ... is not only a consistent (apparently) but also an independently motivated theory of sets: there is, so to speak, a "thought behind it" about the nature of sets which might have been put forth even if, impossibly, naive set theory had been consistent. (Boolos, 1971, p. 490)

Boolos observes that naive comprehension implies that there is a set of all sets, and that this set is then a member of itself. He continues

> It is important to realise how odd the idea of something's containing itself is. Of course a set can and must *include* itself (as a subset). But *contain* itself? Whatever tenuous hold on the concepts of *set* and *member* were given one by Cantor's definitions of "set" and one's ordinary understanding of "element," "set," "collection," etc. is altogether lost if one

---

[27] More detailed marshallings of evidence against the iterative conception may be found in (Lavine, 1994, Ch. V; Hallett, 1984, Chs. 5–6). The overall conclusion of Hallett's book, however, that "we have no satisfactory simple heuristic explanation of *why* it [ZF] works," seems to me to be too strong. It is not mysterious that ZF avoids the paradoxes, since it is apparent from the axioms that the paradoxical collections are denied sethood. Hallett also makes much (in Chapter 5) of the technical result that we have very little idea of the size of the power set of ω, arguing that this refutes ZF's claim to embody a "limitation of size" conception. This, however, seems to depend on thinking of "limitation of size" in the style of Russell, as "no sets allowed that are bigger than such-and-such a cardinal"; rather, as I have been trying to convey, the point is that however big it is, $P(\omega)$ is still a set, and therefore not as large as the universe. There is, however, another sense of "why ZF works" considered by Hallett: why it (or indeed any set theory) is adequate as a foundation for mathematics. I agree that this is genuinely mysterious, and I shall not try to solve the mystery here.

> is to suppose that some sets are members of themselves. The idea is paradoxical not in the sense that it is contradictory to suppose that some set is a member of itself, for, after all, "$(\exists x)(Sx \ \& \ x \in x)$"[28] is obviously consistent, but that if one understands "$\in$" as meaning "is a member of," it is very, very peculiar to suppose it true. For when one is told that a set is a collection into a whole of definite elements of our thought, one thinks: Here are some things. Now we bind them up into a whole (footnote: We put a "lasso" around them, in a figure of Kripke's.). *Now* we have a set. We don't suppose that what we come up with after combining some elements into a whole could have been one of the very things we combined (not, at least, if we are combining two or more elements). (pp. 490–491)

Wang says simply:

> A set is a collection of previously given objects. (Wang, 1974, p. 530)

What I want to emphasize here is the constant appeal, in these passages, to constructivist images and terminology. All three authors use temporal words: "before," "yet," "until," "when," "now," "previously." The question is, in what sense are we to take this? Clearly all agree that it is not to be taken literally: there is not actually a time t_0 at which only the empty set exists, another (later) time by which the singleton of the empty set has been formed, and so on. The constructivist language is supposed only to be metaphorical. Boolos for example, having presented the intuitive idea in constructivist language, then back-pedals: "From the rough description it sounds as if sets were continually being created, which is not the case" (p. 491).

It is clear, then, that the conception of set advanced is not supposed to be literally constructivist, but apparently only constructivist "in principle," under some liberal interpretation. The trouble is, I shall argue, that the sense has to be so liberal that it is no longer entitled to be called constructivist at all.

Wang admits (p. 531) that there is an element of "idealization" in supposing that we can "run through" an infinite number of objects in the way required in his description of the cumulative hierarchy. But all the authors are silent on what *exactly* this means. If the talk of "formation," "collection" and so on are to have any force, there must surely be envisaged an *agent* who is doing the forming and collecting. What properties do we take this agent to have? (Parsons, 1977, p. 507) raises some problems concerning this:

> It is hard to see what the conception of an idealized mind is that would fit here: it would differ not only from finite minds but also from the divine mind as conceived in philosophical theology, for the latter is thought of either as in time, and therefore as doing things in an order with the same structure as that in which finite beings operate, or its eternity is interpreted as complete liberation from succession.

To elaborate: if the agent is conceived of as working in ordinary time, there is just not enough of it to generate the whole hierarchy (at least if time consists of continuum-many instants). The agent needs to occupy a "super-time" with perhaps a class of instants isomorphic to the ordinals. On the other hand, we must not let the agent be too powerful; if he could move backwards and forwards in time at will then it is mysterious why the sets need to be constructed in order at all.

---

[28] Boolos is using "Sx" for "x is a set".

Even if the notion of the ideal agent could be satisfactorily clarified there remains the problem of the status of the ordinals. The cumulative hierarchy is obtained by iterating the power set operation up the entire collection of ordinals. If these are assumed as given from the start this seems a platonistic rather than constructive foundation for the whole enterprise. Wang (p. 532) suggests that the conception of what ordinals there are can develop as the hierarchy is generated. But only countably many ordinals can ever be defined, so it seems that some kind of platonistic conception is inevitable.

Worse than this, however, is the issue of impredicativity. A *sine qua non* of constructivism is that objects are conceived of as occurring in an order, such that at any point in the construction process, only those objects occurring earlier in the order are available. It seems therefore that no theory allowing impredicative definitions can rightly claim to be constructive: one simply cannot quantify over objects which, if the constructivism is taken seriously, do not exist ("at the time"). But the ZF axioms of which the hierarchy is an intuitive model involve impredicative quantifications. Most striking is the axiom of power set in tandem with the axiom of separation. From the power set axiom we know that for any set x the power set $P(x)$ is also a set; the axiom of separation can then be used to pick out individual subsets by means of a formula φ. But this formula can contain quantifications over anywhere in the universe. To put it informally, what subsets there are of a particular set depends not only on what happens at the level of the set, and the next higher level, but also on what happens in the whole hierarchy—as (Bell and Machover, 1977, p. 509) put it, "the size of a power set Pu of a given set u is proportional not only to the size of u but also to the 'richness' of the entire universe."[29] This seems incompatible with any constructive interpretation.

It is not that the authors are at all unaware of this; it is just that they are silent on the conflict between it and the constructivist heuristic which they give for the iterative conception of set. Wang for example (p. 532) says explicitly "we do not concern ourselves over how a set is defined, e.g. whether by an impredicative definition" and admits (p. 560) that "if we adopt a constructive approach, then we do have a problem in allowing unlimited quantifiers to define other sets," but he seems to see no conflict between his own use of constructivist terminology and his advocacy of impredicativity.

The justification of ZF as constructivist in principle is an attempt to have the best of two incompatible worlds, and results in a hybrid position which is philosophically bankrupt and ought to satisfy nobody. A symptom of the philosophical confusion upon which ZF rests is the status of the axiom of choice. This is accepted by most mathematicians, but is not usually regarded as just another of the axioms of set theory—it has a more dubious status. It is customary to state carefully whether or not any theorem requires it, and to do without it if possible. It is almost as though people feel a little guilty in using it. Why is this? I suggest that the explanation is that

[29] In technical terms, the power set operation is not *absolute*. The issue is discussed by Hallett (1984, pp. 206–207, 221) and Hallett (1994, pp. 83–92).

the strongly non-constructive feel of the axiom conflicts with the (false) idea that the rest of ZF is constructive. But in fact the axiom of choice is fully in the spirit of the rest of set theory—the damage its absence does to the theory of cardinal arithmetic is one demonstration of this. If it were clearly realised that ZF is not constructive at all, the axiom of choice would cease to be regarded as a second-class citizen and take up its rightful position as just another of the axioms of set theory.

The conclusion of this section, then, is that ZF does not embody a philosophically coherent notion of set. There is a coherent constructivist position, which entails repudiating impredicative definitions, obeying VCP I, and ending up perhaps with something like ramified type theory. It seems, however, that such a position will not lead to a foundation for classical mathematics. (Whitehead and Russell famously had to postulate the axiom of reducibility to make possible the derivation of mathematics in their system, but this axiom is unmotivated in the light of the VCP. And alternative versions of constructivism, for example intuitionism, are more damaging yet to classical mathematics.) There is also a coherent anti-constructivist position, which rejects the metaphysics of constructivism and its resultant inability to justify classical mathematics. This position rejects the VCP in all its forms. But ZF is an uneasy compromise between these two: it pays lip-service to constructivism without really meaning it, and in doing so forfeits its claim to philosophical justification.

9

Suppose it is admitted that ZF cannot be given a coherent philosophical justification. It seems there is still a third and final argument a defender of it might use: we might call it the argument from mathematical pragmatics.

ZF has proved adequate as a foundation for mathematics, in the sense that all known mathematics can be carried out in ZF. It is convenient to work with: for example, the well-foundedness of the sets allows inductive definitions to be handled smoothly. So whether or not it can be thought of as the axiomatization of a coherent notion of set, it is reasonable—so the argument goes—for it to occupy the position it does as the dominant theory of sets.

One reply to this is that it is no longer clear that ignoring non-well-founded sets gives a theory which is optimal for applications. In recent years uses have started to be found for non-well-founded theories—indeed the current revival of interest started with Aczel's realization that the modelling work he was doing in computer science (on parallel processing) was much simpler if one abandoned foundation (Aczel, 1988, Ch. 8). Rather than attempt to describe this application in detail, I will try to give the general flavour with some simpler examples.

It is common in mathematical modelling to use (ordered) n-tuples $\langle x_1, \ldots, x_n \rangle$. There is a standard way of handling n-tuples in set theory: for example for the pair $\langle a, b \rangle$ we use the set $\{\{a\}, \{a, b\}\}$. It can happen that we want an entire n-tuple to be equal to one of its elements, and this will be forbidden by the axiom of foundation.

Thus in the treatment of the liar in (Barwise and Etchemendy, 1987) (so far the best-known application of non-well-founded sets) the aim is to model a proposition which asserts its own falsehood. Propositions are modelled by pairs,[30] so what we need is a proposition p which satisfies $p = \langle \mathbf{F}, p \rangle$ (where **F** is an atom representing falsehood). This is possible only if we abandon foundation.

A similar example, this time from computer science: a *stream* is a sequence of data items, and can neatly be defined as an ordered pair where the first element is an item of data and the second element is a stream. Then an infinite sequence of zeroes is a stream s satisfying $s = \langle 0, s \rangle$. Once again the axiom of foundation prevents this from being modelled in a natural way. This example is from Barwise and Moss (1996, p. 34). The book explores applications in other areas, for example the theory of games and the model theory of modal logic.

It is true that a hard-line supporter of ZF cannot be *forced* to repudiate foundation. We can always carry out these modellings by choosing appropriate objects in the well-founded universe. But such an approach is analogous to a hard-line disbeliever in complex numbers insisting on them as mere pairs of reals. As more and more applications are discovered, it becomes clearer that there is no good reason for not accepting non-well-founded sets as genuine sets.

10

There is a second and deeper reply to the "pragmatic" argument for ZF. A theory of sets should, I think, be answerable to our informal concept of set as completely arbitrary collection, as well as to the needs of mathematicians. Thus, even if mathematicians can get by using only some special class of sets, it does not follow that we should rest content with a theory which says that these are all the sets there are. Only a non-well-founded theory can convincingly be shown to modify the naive conception as much as, but no more than, is required by the paradoxes; and only in adopting such a theory can we obtain a truly satisfactory solution.

Bibliography

Aczel, P. (1988). *Non-well-founded Sets*. CSLI, Stanford.

Barwise, J. and Etchemendy, J. (1987). *The Liar*. Oxford University Press, Oxford.

Barwise, J. and Moss, L. (1996). *Vicious Circles*. CSLI, Standford.

Bell, J. L. and Machover, M. (1977). *A Course in Mathematical Logic*. North-Holland, Amsterdam.

Benacerraf, P. and Putnam, H., editors (1983). *Philosophy of Mathematics: Selected Readings*. Cambridge University Press, Cambridge, second edition.

Boolos, G. (1971). The iterative conception of set. *Journal of Philosophy*, 68:215–232. Reprinted in [Benacerraf and Putnam, 1983].

Cantor, G. (1883). *Grundlagen einer allgemeinen Mannigfaltigkeitslehre*. B.G. Teubner, Leipzig.

Cantor, G. (1899). Letter to Dedekind. Translation in [van Heijenoort, 1967] pp. 113–117.

[30] I am simplifying the details of the theory to bring out the essential point.

Dummett, M. (1991). *Frege: Philosophy of Mathematics.* Harvard University Press, Cambridge, MA.
Fraenkel, A. (1922). Zu den Grundlagen der Cantor-Zermeloschen Mengenlehre. *Mathematische Annalen*, 86:230–237.
Frege, G. (1893). *Grundgesetze der Arithmetik, begriffsschriftlich abgeleitet, I.* Pohle, Jena.
Gödel, K. (1944). Russell's mathematical logic. In Schilpp, P. A., editor, *The Philosophy of Bertrand Russell.* Tudor, New York, NY. Reprinted in [Benacerraf and Putnam, 1983].
Gödel, K. (1947). What is Cantor's continuum problem? *American Mathematical Monthly*, 54:515–525. Reprinted with revisions in [Benacerraf and Putnam, 1983].
Goldfarb, W. (1989). Russell's reasons for ramification. In Savage, C. W. and Anderson, C. A., editors, *Rereading Russell: Essays in Bertrand Russell's Metaphysics and Epistemology*, volume XII of *Minnesota Studies in the Philosophy of Science*, pages 24–40. University of Minnesota Press Minneapolis, MN.
Grim, P. (1991). *The Incomplete Universe.* MIT Press, Cambridge, MA.
Hallett, M. (1984). *Cantorian Set Theory and Limitation of Size.* Oxford University Press, Oxford.
Hallett, M. (1994). Putnam and the Skolem pardox. In Clark, P. and Hale, B., editors, *Reading Putnam*, pages 66–97. Blackwells, Oxford.
Körner, S. (1960). *The Philosophy of Mathematics.* Dover, New York, NY.
Lavine, S. (1994). *Understanding the Infinite.* Harvard University Press, Cambridge, MA.
Menzel, C. (1984). Cantor and the Burali-Forti paradox. *Monist*, 67:92–107.
Mirimanoff, D. (1917a). Les antinomies de Russell et de Burali-Forti et le problème fondamental de la théorie des ensembles. *L'Enseignement Mathématique*, 19:37–52.
Mirimanoff, D. (1917b). Remarques sur la théorie des ensembles et les antinomies Cantoriennes (I). *L'Enseignement Mathématique*, 19:208–217.
Parsons, C. (1977). What is the iterative collection of set? In Butts, R. E. and Hintikka, J., editors, *Proceedings of the 5th International Congress of Logic, Methodology and Philosophy of Science 1975, Part I: Logic, Foundations of Mathematics, and Computability Theory*, pages 335–367. D. Reidel Dordrecht. Reprinted in [Benacerraf and Putnam, 1983] and in [Parsons, 1983].
Parsons, C. (1983). *Mathematics in Philosophy: Selected Essays.* Cornell University Press, Ithaca, NY.
Poincaré, J. H. (1906). Les mathématiques et la logique. *Revue de metaphysique et de morale.* Translation in [Poincaré, 1952].
Poincaré, J. H. (1952). *Science and Method.* Dover New York, NY.
Poincaré, J. H. (1963). *Mathematics and Science: Last Essays.* Dover, New York, NY.
Priest, G. (1994). The structure of the paradoxes of self-reference. *Mind*, 103:25–34.
Richard, J. (1905). Les principes de mathématiques et le problème des ensembles. *Revue générale des sciences pures et appliqués*, 16:541. Translation in [van Heijenoort, 1967].
Rieger, A. (1996). *Circularity and Universality.* D.Phil. thesis, University of Oxford.
Rieger, A. (2000). An argument for Finsler-Aczel set theory. *Mind*, 109(434):241–253.
Russell, B. (1902). Letter to Frege, 16th June. In [van Heijenoort, 1967].
Russell, B. (1906a). Les paradoxes de la logique. *Revue de Métaphysique et de Morale*, 14: 627–650.
Russell, B. (1906b). On insolubilia and their solution by symbolic logic. Translation of [Russell, 1906a]; in [Russell, 1973].
Russell, B. (1906c). On some difficulties in the theory of transfinite numbers and order types. *Proceedings of the London Mathematical Society, series 2*, 4:29–53. Reprinted in [Russell, 1973].
Russell, B. (1908). Mathematical logic as based on the theory of types. *American Journal of Mathematics*, 30:222–262. Reprinted in [Russell, 1956].
Russell, B. (1956). *Logic and Knowledge,* ed. Robert C. Marsh. Allen and Unwin, London.
Russell, B. (1973). *Essays in Analysis* ed. Douglas Lackey. George Allen and Unwin, London.
Shoenfield, J. R. (1967). *Mathematical Logic.* Addison-Wesley, Reading, MA.

Shoenfield, J. R. (1977). Axioms of set theory. In Barwise, J., editor, *Handbook of Mathematical Logic*. North-Holland, Amsterdam.

Skolem, T. (1922). Einige Bemerkungen zur axiomatischen Begründung der Mengenlehre. Matematikerkongressen i Helsingfors den 4–7 Juli, Den femte skandinaviska matematikerkongressen, Redogörelse. Translation in [van Heijenoort, 1967].

van Heijenoort, J., editor (1967). *From Frege to Gödel: A Source Book in Mathematical Logic, 1879-1931*. Harvard University Press, Cambridge, MA.

von Neumann, J. (1925). Eine Axiomatisierung der Mengenlehre. *Journal für die reine und angewandte Mathematik*, 154:219–240. Translation in [van Heijenoort, 1967].

von Neumann, J. (1928). Über die Definition durch transfinite Induktion und verwandte Fragen der allgemeinen Mengenlehre. *Mathematische Annalen*, 99:373–391.

von Neumann, J. (1929). Über eine Widerspruchsfreiheitsfrage in der axiomatischen Mengenlehre. *Journal für die reine und angewandte Mathematik*, 160:227–241.

Wang, H. (1974). *From Mathematics to Philosophy*. Routledge and Kegan Paul, London. Pages 181–223 reprinted as *The concept of set* in [Benacerraf and Putnam, 1983].

Whitehead, A. N. and Russell, B. (1910). *Principia Mathematica, Volume 1*. Cambridge University Press, Cambridge.

Zermelo, E. (1904). Beweis, dass jede Menge wohlgeordnet werden kann. *Mathematische Annalen*, 59:514–516. Translation in [van Heijenoort, 1967].

Zermelo, E. (1908). Untersuchungen über die Grundlagen der Mengenlehre, I. *Mathematische Annalen*, 65:261–281.

Zermelo, E. (1930). Über Grenzzahlen und Mengenbereiche: Neue Untersuchungen über die Grundlagen der Mengenlehre. *Fundamenta Mathematicae*, 16:29–47.

Chapter 10
Absoluteness and the Skolem Paradox

Michael Hallett

1 Introduction

When seen in the "correct" light, the contradictions of set theory are by no means disastrous, but instructive and fruitful. For instance, the antinomies of Russell and Burali-Forti live on in the systems of axiomatised set theory in the guise of established theorems. Zermelo used the Russell-Zermelo argument to prove that every set possesses a subset which cannot be an element of that set, and from which it follows that there can be no universal set ((Zermelo, 1908b, pp. 264–265), p. 203 of the English translation), and the essentials of the Burali-Forti argument can be used to prove that there is no ordinary *set* of all (von Neumann) ordinals.[1] The fact that these contradictions reappear as theorems in set theory is not surprising given that the reasoning involved is (or can be turned into) set-theoretic reasoning, and that we always have the choice of treating the derivation of contradictions as arguments in *reductio* form, choosing one premise as responsible for the contradiction. Indeed, much of the early discussion of the arguments was concerned with

M. Hallett (✉)
John Frothingham Chair of Logic and Metaphysics, Department of Philosophy, McGill University, Montreal, QC, Canada
e-mail: michael.hallett@mcgill.ca

[1] The Burali-Forti argument is explicitly used in this way in (von Neumann, 1928, p. 721), though the argument was clearly known to von Neumann much earlier, since all the essentials are present in von Neumann's new account of the ordinals from the early 1920s (see von Neumann, 1923), and this form of the Burali-Forti argument is explicitly mentioned by von Neumann in a letter to Zermelo from August 1923. (See Meschkowski, 1966, pp. 271–273.) Zermelo had given the definition of the "von Neumann" ordinals by 1915, and possibly as early as 1913. In a lecture course in Göttingen in the Summer Semester of 1920 entitled "Probleme der mathematischen Logik," Hilbert shows (pp. 15–16) how the Burali-Forti Paradox can be reproduced in the framework based on Zermelo's definition: if W is the set of all ordinals, then W would itself be an ordinal according to the definition, and must therefore be a member of itself, contradicting one of the central theorems about these ordinals. The lectures can be found in Chapter 2 of Ewald and Sieg (2011). Despite Zermelo's precedence here, von Neumann is still the real discoverer of the von Neumann ordinals, for he was the first to give a complete presentation of the relevant theoretical material, and in particular to recognise the importance of the Axiom of Replacement.

D. DeVidi et al. (eds.), *Logic, Mathematics, Philosophy: Vintage Enthusiasms*, The Western Ontario Series in Philosophy of Science 75,
DOI 10.1007/978-94-007-0214-1_10,

assessing which of the premises used was in fact so reduced. Once absorbed into set-theoretic frameworks designed especially to avoid the known contradictions, the paradoxes give rise to arguments which reveal something deep and interesting about the existence of sets and the structure of the universe of sets itself. These systems *analyse* what is exposed by the antinomies; unsurprisingly, the stronger the system, the more refined the analysis tends to be. The most striking example is von Neumann's system, which allows that there *is* a greatest (von Neumann) ordinal; it is just that this ordinal cannot be an ordinary set, and thus cannot give rise to an even greater ordinal. Moreover, in this setting, the universal set, the Russell set, and the set of all ordinals are all equipollent, all maximally "big," and all equally "too big" to be sets. (See Hallett, 1984, pp. 288–295.)

The situation with the "semantic" antinomies is somewhat different. They, too, give rise to concrete, mathematical results. For instance, the Liar Paradox, developed within a consistent theory which allows for the right representability, yields, instead of a contradiction, Tarski's Theorem on the undefinability of the truth-predicate for that theory. However, the form of the argument is specific neither to set theory nor to set-theoretic reasoning, but applies to a wide range of languages/theories.

Nevertheless, there is a famous semantic paradox which *is* specific to set theory, namely the Skolem Paradox which goes back to (Skolem, 1923). Unlike the antinomies, this does not arise from a pre-axiomatic, set-theoretic contradiction; indeed, it is a "paradox" only made possible by the first-order axiomatic representation of sets, which Skolem (in the paper mentioned) was the first to present.[2] Skolem's argument shows that, while the axioms can prove the existence of sets with increasing infinite cardinality, yet the central concepts concerning the different infinite cardinalities must be "relative inside axiomatic set theory," since one can find an interpretation of the axioms in the natural numbers. Thus, the axiom system, if consistent, has an interpretation which cannot be the one intended, since it is based on only countably many objects. As Skolem himself pointed out, there is nothing directly contradictory about this. The argument can be sharpened (see below, p. 200f.), and crucially then internalised, to produce a contradiction, which can then be used as the basis of *reductio* arguments of great import, as Gödel first showed. What this internalisation demonstrates is that Skolem's relativity is reflected within the theory itself (i.e., without any reference to "exterior" models) in the notion of *absoluteness*. In particular, it can be used to show that, while the natural numbers are absolute, the continuum (and the power set operation generally) is non-absolute, where "absolute" can be given a precise theoretical sense.

The present paper is concerned with this internalisation. There are two things in particular which I wish to bring out. The first is historical. The non-absoluteness of the continuum focuses on a feature of extensions which was first isolated (and

---

[2] For discussions of Skolem's assessment of the argument, and his changing views on its consequences, see (Benacerraf, 1985; George, 1985). For more general discussion of the consequences of the argument, see (Wright, 1985).

objected to) by Poincaré. Poincaré identified this as an instability in the extensions of certain sets, and saw this as problematic, since it conflicts with his generational view of sets. Poincaré's views are set out in the first section of the paper. Axiomatic set theory, as was clear from Zermelo's initial axiomatisation, rejects Poincaré's generational view, and Poincaré's association of set existence with definability. (In large part, Poincaré was reacting to the view of sets put forward by Zermelo.) However, the internalisation of the Skolem Paradox serves to refocus both Poincaré's and Skolem's reservations. According to set theory, there is nothing unstable or relative about the continuum. Nevertheless, I will suggest that the involvement of the crucial notion of absoluteness in both Gödel's and Cohen's arguments for the consistency and independence of the GCH indicates rather a conceptual weakness in the fundamental notion of cardinality. This was something glimpsed by Skolem himself in 1923.

2 "Impredicative" Extensions

The isolation of non-absolute sets is foreshadowed in both Russell's and Poincaré's diagnoses of the antinomies.[3] The term "impredicative" was originally used by Russell in a quite general way to refer to properties whose extension cannot be sets and in this sense the properties used to specify the paradoxical sets are demonstrably "impredicative": see (Russell, 1907, p. 34).[4] This same paper (marked "Received November 24th 1905.—Read December 14th 1905"), written before Russell had fixed on a solution, and before he had stated the VCP, contains the following striking passage:

> ... there are what we may call *self-reproductive* processes and classes. That is, there are some properties such that, given any class of terms all having such a property, we can always define a new term also having the property in question. Hence we can never collect *all* the terms having the said property into a whole; because, whenever we hope we have them all, the collection we have immediately proceeds to generate a new term also having the said property. (Op. cit., 36.)

Russell describes clearly this kind of "self-reproduction" found in the traditional antinomies.

Suppose we call V a temporary universe. Let ψ be some property. It seems that we can define a set u as $\{x : x \in V \wedge \psi(x)\}$. Either $u \in V$ or $u \notin V$. But suppose that ψ is the property involved in the Russell contradiction, and suppose further that $u \in V$. Then we have immediately that $u \in u \leftrightarrow u \notin u$, i.e., the Russell contradiction. Suppose now that ψ is the property $Ord(x)$. We can easily show that

---

[3] That Poincaré was essentially concerned with non-absoluteness is a suggestion I first heard propounded in a lecture given by Wilfrid Hodges in London in 1974 or 1975. Fuller treatments of Russell's and Poincaré's views can be found in, e.g., (Goldfarb, 1988, 1989).

[4] In this paper, and elsewhere, Russell uses the term "proper class" where we would now use the term "set".

$u = \{x : x \in V \wedge \psi(x)\}$ is an ordinal (a von Neumann ordinal), and thus is not a member of itself. Thus, if $u \in V$, we would have, on the contrary, that u *does* belong to itself, again the well-known contradiction. Hence, if we accept the definitions of u as good, the conclusion must be that, with respect to the properties ψ involved in these two paradoxes, V is not the *full* universe; it is indeed only temporary, for there are perfectly good sets, like our u's, which cannot possibly belong to it. We could say that what Russell shows is that there are ψ such that for any temporary universe V it must be the case that $\{x: \psi(x)\}^V \neq \{x: \psi(x)\}$.[5] Put more in Russell's way, whenever we call a halt to the "process" of collecting together the ψ, we find that there is at least one more ψ that is yet to be accounted for. If, in addition, we think of *producing* sets, then these ψ are naturally described, in Russell's phrase, as "self-reproductive" properties.

Russell came to the conclusion, as did Poincaré, that what characterises these collections is a certain circularity in their specification, and both believed that what is needed is some adherence to (a form of) the *vicious circle principle* (VCP) in order to avoid this. Unlike Poincaré, Russell held that what is required is an alteration to "current logical assumptions," and came to believe that these alterations must be guided by the VCP.[6] But what concerns us here is not so much the solution to the problem, but rather its diagnosis. It is clear that the idea of the VCP owes much to the analysis above. Here, for example, is Russell's statement from 1908:

> Thus all our contradictions have in common the assumption of a totality such that, if it were legitimate, it would at once be enlarged by new members defined in terms of itself.
>
> This leads us to the rule: "Whatever involves all of a collection cannot be one of the collection"; or, conversely: "If, provided a certain collection had a total, it would have members only definable in terms of that total, then the said collection has no total."†
>
> †When I say that a collection has no total, I mean that statements about *all* its members are nonsense.[7]

[5] The analogy with Zermelo's argument that the universe is not a set is clear. Russell is close to isolating the notion of non-absoluteness, though his condition is stronger; a property ψ is absolute when it is possible to find at least one V such that $\{x: \psi(x)\}^V \neq \{x: \psi(x)\}$, not that this necessarily holds for all V.

[6] For the remark about "current logical assumptions," see (Russell, 1907, p. 37). Gödel points out (Gödel, 1944, p. 135) that there are actually three distinct formulations of the VCP relied on in Russell's writings. For Gödel's discussion of these, see op. cit., pp. 455ff. Goldfarb suggests in (Goldfarb, 1989) that for Russell these formulations may be more intimately connected than Gödel's discussion allows.

[7] Russell (1908, p. 225). The relation between the "self-reproductive" properties isolated by Russell and the VCP was well summed up by Gödel:

> I mean in particular the vicious circle principle, which forbids a certain kind of "circularity" which is made responsible for the paradoxes. The fallacy in these, so it is contended, consists in the circumstance that one defines (or tacitly assumes) totalities, whose existence would entail the existence of certain new elements of the same totality, namely elements definable only in terms of the whole totality. This led to the formulation of a principle which says that no totality can contain members definable only in terms of this totality [vicious circle principle]. ((Gödel, 1944, p. 133). The square brackets are in the original.)

Russell came to focus more on the direction taken by his footnote, though I want to focus here on the formulation "If, provided a certain collection had a total, it would have members only definable in terms of that total, then the said collection has no total," for this is the closest to Poincaré's view of sets, to which I now turn.

Poincaré takes as his starting point neither Russell's antinomy nor Burali-Forti's, but rather Richard's, an antinomy which we now classify as "semantic," involving essentially a linguistic component. Consider the set E of all real decimals which can be defined in a finite number of words; E is obviously denumerable. The contradiction now goes, quoting Poincaré, as follows:

> Suppose the enumeration [of E] effected, and let us define a number N in the following manner. If the nth decimal of the nth number of the aggregate E is
>
> 0, 1, 2, 3, 4, 5, 6, 7, 8, or 9
>
> the nth decimal of N will be
>
> 1, 2, 3, 4, 5, 6, 7, 8, 9, or 0
>
> As we see, N is not equal to the nth number of E, and since n is any number whatsoever, N does not belong to E; and yet N should belong to this aggregate, since we have defined it in a finite number of words.[8]

Poincaré endorses what he says is Richard's own solution to the antinomy, namely:

> E is the aggregate of *all* the numbers that can be defined in a finite number of words, *without introducing the notion of the aggregate E itself*, otherwise the definition of E would contain a vicious circle; we cannot define E by the aggregate E itself. Now it is true that we have defined N by a finite number of words, but only with the help of the notion of the aggregate E, and that is the reason why N does not form a part of E. ((Poincaré, 1906, p. 307); see also (Poincaré, 1908, pp. 206–207), pp. 480–481 and 190 respectively of the English translations.)

The concentration on "vicious circles" is then taken to be the way to avoid all the antinomies:

> But the same explanation serves for the other antinomies, and in particular for that of Burali-Forti. . . .
>
> Thus *the definitions that must be regarded as non-predicative* [in Russell's sense] *are those which contain a vicious circle*. The above examples show sufficiently clearly what I mean by this. ((Poincaré, 1906, p. 307); see also (Poincaré, 1908, p. 207), pp. 481 and 190 respectively of the English translations.)

The connection to the formulation of Russell's VCP which I picked out (i.e., "If, provided a certain collection had a total, etc.") is clearest in Poincaré's phrase "we cannot define E by the aggregate E itself." From this, Poincaré questions the boundaries of the sets picked out:

---

[8] Poincaré (1906, pp. 304–305); see also (Poincaré, 1908, p. 202), pp. 478 and 185–186 respectively of the English translations. Recall that Poincaré, in this section of his paper ("Les Antinomies Cantoriennes") is explicitly discussing (Russell, 1907).

> A "non-predicative class" [thus, one whose definition suffers from a vicious circle] is not an empty class, but a class with unsettled [*indécise*] boundaries. ((Poincaré, 1906, p. 310); see also (Poincaré, 1908, pp. 206–207), pp. 481–482 and 191 respectively of the English translations.)

This is elaborated in a discussion from 1909:

> There is no actual infinity, and when we speak of an infinite collection, we mean a collection to which we can add new elements continually (similar to a subscription list which will never close, waiting for new subscribers). For the classification could only be closed properly when the list was closed; for every time that one adds new elements to the collection, one modifies it. It is therefore possible that the relation between this collection and the elements already classified is modified; and since it is according to this relation that these elements have been arranged in this or that drawer, it can happen that, once the relation is modified, the elements are no longer in the right drawer, and it will be necessary to move them. While there are new elements to be introduced, it is to be feared that the work [of classification] will have to begin all over again; and it will never happen that there will not be new elements to introduce. The classification will never be finished.
>
> From this there emerges a distinction between two types of classification applicable to the elements of infinite collections, *predicative* classifications, which cannot be disrupted [*bouleversées*] by the introduction of new elements, and *non-predicative* classifications for which the introduction of new elements requires constant reshaping [*remanier*]. ((Poincaré, 1909, p. 463), or (Poincaré, 1913a, pp. 9–10), p. 47 of the English translation.)

Let us examine a little more closely what exactly worries Poincaré.

In the 1909 essay, Poincaré says that the problem concerns collections which are "mutable [*muables*]" (the collections with "unsettled boundaries" from 1906), while the passage quoted above focuses on "disrupted classifications." Though connected, the two are not exactly the same. What does he mean by a "mutable" collection? By this, Poincaré means a collection whose extension does not remain fixed over time, so that in using a term for it, its reference may vary in the course of an argument, thus rendering the conclusion of the argument uncertain. That this is one thing which worries him is made clear at the beginning of the 1909 paper. Suppose we want to use simple syllogistic reasoning to conclude that, since two soldiers are in the same regiment, they are in the same division. The proof only works, Poincaré says, as long as the soldiers stay in the same regiment; the conclusion might well be false if *in the course of the argument* one of the soldiers is transferred to a different regiment. As Poincaré says:

> What is then the condition under which the rules of this logic are valid? It is that the classification adopted is *immutable* [*immuable*]. ((Poincaré, 1909, p. 461), or (Poincaré, 1913a, p. 8), p. 45 of the English translation.)

The relevant classification here is the separation of the collection of all soldiers into regiments, and a soldier shifting regiments changes the classification.

Poincaré is clearly assuming that the argument about the soldiers is not timeless. In most ordinary cases, we rely on general assumptions about short-term stability in the macro-world which more or less guarantees that the conclusion will apply to the situation stated at the beginning of the argument. In any case, with ordinary arguments, mutability will usually depend on contingencies which are outside the scope of the argument; this is enough to suggest caution, for bad luck could dictate that the assignment of soldiers to regiments might change during the course of an argument. But it is different with the antinomies. These (for Poincaré) present cases where the classification *must* change, not because of external contingencies, but by the very nature of the argument itself.

Take for instance the Richard antinomy. We start with a term "E," which refers to the list of all definitions. Then we construct a definition d (of N), which we have reason to think cannot be in the list denoted by E. But d is a definition, so it must be in the list E, since, by assumption, E is the list of all definitions. Poincaré's point, it seems, is that the extension of the term "E" is no longer the same at the *end* of the Richard argument as it was at the beginning, i.e., it is of necessity "muable." At the start, d was not in the extension of E; indeed, the intention on which d rests is predicated on the assumption that it is *not* in E. Yet, once d has been framed, it seems that we are forced to admit that it must belong to E. Hence, the extension of E is no longer the same, and the reference of "E" is uncertain. The extendability of E rests in an essential way on d: it seems that we cannot formulate d *without* incurring a change in the extension of E. In other words, mutability is *intrinsic* to the argument, and this is what forces a contradiction. We are forced to the conclusion that we are dealing with properties which, in Russell's phrase, are "self-reproductive." What seems to underline the ambiguity is that d includes a universal quantifier over E. If the extension of E changes, d will no longer mean the same *after* it is formulated (with the extended E) as it was intended to mean at the moment it was formulated. Consequently, the truth-value of $\forall x \in E\ \psi(x)$ will in general vary as the extension of E varies. Thus, we have a claim similar to Russell's in the passage quoted above (see p. 192), namely that part of the worry concerns the scope of universal quantification.

Poincaré's conclusion is that the whole line of argument is doomed from the beginning, and that it was not just dangerous to employ the term for E, but illegitimate, since it can never fulfill one of his conditions on the correct use of names, that its reference be 'entirely determined,' to use Poincaré's words from a later paper.[9]

Poincaré applies this analysis not only to the antinomies, but to the Cantor diagonal argument as well. Here, we start with a countable list E_R of real numbers, and then define a number N in much the same way as above, i.e., so that it is immediately clear that it cannot be in the list E_R. Although the definition of N contains a reference to the list E_R, there is, it seems, no obvious circularity of the kind that worries Poincaré, since *there is no reason whatever* to think that N should itself be in the list E_R. However, we *can* generate a contradiction by assuming that E_R consists of *all* real numbers, and thus that it must include N. In this case, there *is* an indirect reference in the definition of N to N itself, for the extension of E_R is assumed to include *all* reals. But what does this contradiction show? One way to read it is as showing just that the assumption that E_R contains all real numbers is demonstrably

[9] See (Poincaré, 1912, p. 8), or (Poincaré, 1913a, p. 90), p. 71 of the English translation.

false. In this case, there is not the stretching of the extension of E_R that Poincaré perceives in the argument in Richard's antinomy; nothing in the proof forces us to conclude that the extension of E_R is not fixed. But Poincaré rejects this reading of the argument. Why?

The answer is that Poincaré assumes what might be called a "genetic" or "generational" view of mathematical objects, according to which the stock of mathematical objects varies over time, just as the composition of a regiment will change over time. The clearest confirmation of this is to be found in Poincaré's paper from 1912 where he distinguishes between two points of view, the view that sets are picked out by what he calls "comprehension," and the view that they are created by what he calls "extension." Poincaré introduces the term "Pragmatist," and then says:

> The Pragmatists adopt the point of view of extension, and the Cantorians the point of view of comprehension. For finite sets, the distinction can only be of interest to formal logicians, but it appears to us much more profound where infinite sets are concerned. Adopting the extensional viewpoint, a collection is constituted by the successive addition of new members. By combining old objects, we can construct new objects, and then, with these, newer objects; if the collection is infinite, it is because there is no reason for stopping.
>
> On the other hand, from the point of view of comprehension we start from a collection where there are pre-existent objects, objects which appear to us indistinct at first, but some of which we finally recognise because we attach labels to them and arrange them in drawers. But the objects precede the labelling, and the objects will exist, even though there may not be a curator to classify them.[10]

Poincaré adopts the "Pragmatist" or "extensional" view against the "comprehension view." According to the comprehension view, objects pre-exist, and definitions then only serve the function of selecting certain of them. Indeed, this is the core of Zermelo's defence of impredicative definitions against Poincaré's objections. Poincaré's extensional view, however, is most certainly a "generational" view, according to which objects are created in stages, as the reference to "successive additions" makes clear. But according to what, for Poincaré, does the "successive addition of objects" take place? As the passage above says, this occurs through construction, or by constructive definition. For this view, definition does not serve the same function it does for the comprehension view; correct definition is taken itself to be the criterion of existence, and does not itself rely on a supposition of existence independent of the definitional prescription. What this means is that objects are indissolubly tied to the definitions of them.

This has certain strong consequences for Poincaré's "Pragmatist." First:

> For example, the Pragmatists admit only those objects which can be defined in a finite number of words. Possible definitions, being expressible in sentences, can always be enumerated with the ordinary numbers from one to infinity. ((Poincaré, 1912, p. 5), or (Poincaré, 1913a, p. 88), p. 68 of the English translation.)

---

[10] Poincaré (1912, p. 4, 1913a, pp. 87–88), pp. 67–68 of the English translation. The mention of drawers in this passage recalls what Poincaré has to say in his earlier paper from 3 years before; see p. 194 above. It is very likely that Poincaré distinguishes between the "extensional" and the "comprehension" views precisely because Zermelo hints at such a distinction in the section of (Zermelo, 1908a, pp. 117–118) which replies to Poincaré. See also (Hallett, 2010, pp. 109–112).

A little later in the same paper, Poincaré says:

> And why do the Pragmatists refuse to admit objects which are not capable of a definition in a finite number of words? It is because they consider that an object only exists when it is thought, and that one will not be able to conceive of an object independently of a thinking subject. ((Poincaré, 1912, p. 9), or (Poincaré, 1913a, p. 94), p. 72 of the English translation.)

Here there is an implicit assumption that the thought of the cognising subject is necessarily tied to its linguistic expression. On its own, this is innocuous enough, and certainly would not rule out the construction of a power set $P(\omega)$ from ω in one step. But Poincaré adopts the further view that a set cannot be said to exist *before* its members have been shown to exist, thus that the existence of the members of a set must be logically prior to the existence of the set itself. (This surely must be one reason for the designation "extensional" for the position Poincaré supports.) This leaves it quite vague as to what "logical priority" amounts to, and as to how exactly the stages of existence are to be marked. But the fact that existence must be tied to linguistic specification reveals the central aspect of Poincaré's "logical priority" view. As Goldfarb points out

> Since it is first the specification that legitimizes the entity specified, that specification can in no way depend on the existence of the entity. Therefore, the ranges of the quantifiers in the specification cannot include the entity. (Goldfarb, 1989, p. 25.)

Thus, if we claim that $v \in u$, then we must be able to define v with a specification φ_v which does not involve reference to the set u, either directly or through a quantifier. This, plus the priority thesis, entails that the set itself can be referred to neither in its own specification nor in the specification of any of its members.

This position can be summarised in the following two theses:

> *The Linguistic Specification Thesis* (LST): A mathematical object (in particular, a set) cannot be said to exist until there is a finite linguistic specification (definition) of it.

This is combined with:

> *The Individual Specification Thesis* (IST): Before a set A can be said to exist, we must be in possession of specifications of all of its potential members; needless to say, to avoid circularity, these specifications cannot make reference, either direct or indirect, to the set A.

It follows that a set A cannot be defined by direct or indirect reference to itself, and neither can it contain members which are defined by reference to A.

The following passage from (Poincaré, 1912) shows that what we have just described is a fair reflection of Poincaré's "Pragmatist" position. Poincaré considers the impredicative definition of an object X whereby first X is defined using its relation to all the members of a collection G, and where it is then asserted that X is itself a member of G. He says:

> For the Pragmatists, such a definition implies a vicious circle. One cannot define X without being acquainted with [*sans connaître*] all the objects of the *genre* G, and consequently without being acquainted with X which is one of these individuals. The Cantorians do not admit this. The *genre* G is given to us; consequently we are acquainted with all these

individuals, and definition has only the aim of discerning [*discerner*] among them those which have the relation stated to their fellows.

No, reply their adversaries, the knowledge of the *genre* does not allow you to know all the individuals; it only gives you the possibility of constructing as many of them as you wish. They exist only after they have been constructed, that is to say after they have been *defined*. X only exists by its definition which has sense only if one knows in advance all the individuals of G, and in particular X. ((Poincaré, 1912, p. 7), or (Poincaré, 1913a, p. 91), pp. 70–71 of the English translation.)

Note that Poincaré (the Pragmatist) quite explicitly rejects the position, attributed to the "Cantorians," that specification of a set is enough to give us "acquaintance" with its members.

The use of impredicative definitions is then tied to the worry about the instability of extensions, for Poincaré says the following:

Why do the Pragmatists make this objection [to the impredicative definition of X]? It is because the *genre* G appears to them as a collection which is capable of increasing indefinitely whenever new individuals are constructed which possess the appropriate characteristics. Thus G can never be posited *ne varieteur* as the Cantorians posit it, for we are never sure that G will not become G' in the light of new annexations. ((Poincaré, 1912, p. 9), or (Poincaré, 1913a, p. 93), p. 72 of the English translation.)

And in a later paper, he makes the same explicit connection:

...Richard's law of correspondence lacks a property which, borrowing a term from the English philosophers, one can call "predicative." (With Russell, from whom I borrow the word, a definition of two concepts A and A' is non-predicative when A appears in the definition of A' and conversely.) I understand by this the following: Every law of correspondence assumes a definite classification. I call a correspondence predicative when the classification on which it rests is predicative. However I call a classification predicative when it is not altered by the introduction of new elements. With Richard's classification this, however, is not the case. Rather, the introduction of the law of correspondence alters the division into sentences which have a meaning and those which have none. What is meant here by the word "predicative" is best illustrated by an example. When I arrange a set of objects into various boxes, then two things can happen. Either the objects already arranged are finally in place. Or, when I arrange a new object, the existing ones, or a least a part of them, must be taken out and rearranged. In the first case I call the classification predicative, and in the second non-predicative. ((Poincaré, 1910, p. 47), p. 1073 of the translation.)

This analysis is what is directly applied to the Cantor diagonal argument:

For example, the Pragmatists admit only objects which can be defined in a finite number of words; ... [W]hy then do we say that the power of the continuum is not that of the whole numbers? Yes, being given all the points of space which we know how to define with a finite number of words, we know how to imagine a law, itself expressible in a finite number of words, which makes them correspond to the sequence of whole numbers. But now consider the sentences in which the notion of this law of correspondence figures. A few moments ago, these sentences had no sense since this law had not yet been invented, and they could not serve to define points of space. Now they have acquired a sense, and they will allow us to define new points of space. But these new points of space will not find any place in the classification adopted, and this will compel us to upset it [*la bouleverser*]. And it is this which we wish to say, according to the Pragmatists, when we say that the power of the continuum is not that of the whole numbers. We wish to say that it is impossible to establish between these two sets a law of correspondence which is secured against this sort of disruption [*bouleversement*]; ... ((Poincaré, 1912, p. 5), or (Poincaré, 1913a, pp. 88–89), p. 68 of the English translation. See also (Poincaré, 1909, pp. 463–464), (Poincaré, 1913a, p. 10), p. 47–48 of the English translation.)

When we are given a list of real numbers, a finite string S of words specifies this list, giving a sequence E_R of reals. This brings into existence a new mathematical object, namely this sequence E_R. Sentences which previously contained the string S were strictly speaking meaningless, and therefore could not possibly define real numbers. Once it is recognised that E_R exists, then such sentences have a perfectly good meaning, and some of them may well define real numbers. But no such real numbers (as so defined via E_R) were available for classification when E_R itself was being defined, and so cannot be assumed to figure in E_R. Indeed, the argument shows how to define (with direct reference to E_R) a real number N which *cannot* be in the list. Once E_R is available, the classification of the real numbers currently available thus has to begin anew, and will necessarily lead (so the argument) to a list different from E_R.

This is really the same as the analysis of the Richard antinomy. Unsurprisingly, Poincaré sees the two arguments as simply two aspects of the same process. We can always specify a list of definitions of real numbers, and this specification will then turn certain sentences which were hitherto meaningless into perfectly good definitions of reals. This will necessitate a new list, and so on *ad infinitum*. Indeed, the analysis reconciles an apparent contradiction between the two arguments, a contradiction which can be put starkly by stating that Richard demonstrates that there are only countably many numbers (since there are only countably many possible definitions), while Cantor shows that there are uncountably many:

In this there lies the solution of the apparent contradiction between *Cantor* and *Richard*. Let M_0 be the set of all whole numbers, and M_1 the set of all points of our line definable by finite sentences of our table after a first run through the list, and let G_1 be the law coordinating both sets. Using this law, a new set M_2 is definable and is thus added. For $M_1 + M_2$ there is a new law G_2, using which there arises a new set M_3, and so on. *Richard*'s proof teaches us that, wherever I break off the process, then there's a corresponding law, while *Cantor* proves that the process can always be continued arbitrarily far. There is thus no contradiction between these conclusions. ((Poincaré, 1910, pp. 46–47), p. 1073 of the translation.)

The assimilation of, say, the Burali-Forti antinomy to Richard's may seem a little odd to modern eyes; since (Ramsey, 1926), it has been usual to point to a semantic element behind Richard's antinomy which is not present in Burali-Forti's or Russell's. But the reason for Poincaré's assimilation is clear. Not only does he adopt a generational view of sets, whereby the existence of the elements of a set precede that of the set itself, he makes the existence and the constitution of a set dependent on its mode of definition. Given this, the Burali-Forti antinomy does look somewhat similar, as we saw via Russell's "self-reproductive processes." The set u of all ordinals could not exist without being defined (LST), and the definition uses a predicate $Ord(x)$ which u itself would satisfy, showing that u must be a member of u, thus violating IST. The central principles here, LST and IST, are decisively rejected by modern set theory.[11] Does this mean that all the important aspects of Poincaré's analysis must similarly be rejected? The answer is clearly no, as will be shown in what follows.

3 The Paradox Internalised

Before proceeding to the internalisation of the Skolem argument, let us quickly run through the latter's main steps.

Consider the first-order Zermelo-Fraenkel system (ZFC) with a standard list of axioms, including the Axiom of Choice (AC), and assume that it is consistent. Since the theory is written in a countable language $\mathcal{L}_{ZF}$, with "$\in$" as its only non-logical predicate, then by the Completeness Theorem, it must have a countable model $\mathcal{M} = <M, E>$, where M is the $\mathcal{M}$-domain, and E is $\mathcal{M}$-"membership." M has in it an element m_c, which is the interpretation in $\mathcal{M}$ of ZFC's term "c" for the continuum. Moreover, ZFC can prove the statement "$\neg C(c)$," where "$C(x)$" is the usual $\mathcal{L}_{ZF}$ formula expressing countability. $\mathcal{M}$, being a model of ZF, therefore says that the continuum m_c is uncountable. But it is clear that m_c has only *countably* many $\mathcal{M}$-members, i.e., the collection $\{x : x \in M \ \& \ xEm_c\}$ must be countable, since M has only countably many members. Therefore, it seems, ZFC can have models in which the continuum is both countable and uncountable.

Although this conclusion is somewhat odd (a "paradox"), it is no antinomy. For one thing, the $\mathcal{L}_{ZF}$ statement "$\neg C(c)$" simply says that there is no surjective function whose domain is ω and whose range is c. And since $\mathcal{M}$ is a model of set theory, this must mean that there is no object in M which plays the role of such a function in $\mathcal{M}$, despite the countability of m_c. In other words, m_c is uncountable in just the way that c is.[12]

One might think that what this result shows is simply that first-order set theory has unintended interpretations, if any at all, and this we now regard as nothing unusual. For one thing, so much of the work using interpretations, going back to Hilbert's construction of models for various geometries, and including Henkin's construction of a model for any consistent theory from its syntax, shows us that there is a sense in which the "material" of an interpretation is often quite irrelevant; what matters, rather, is the way it is *arranged.* The non-standard models of first-order arithmetic underline this; it is not the nature, or quantity, of the elements themselves which matters, but they way they are structured. Is this also the case with the Skolem Paradox? For example, it is quite possible that m_c itself is *uncountable*, i.e., has uncountably many *real* members, although contains only countably

[11] Something like the IST is pursued in Martin-Löf's constructive type theory; see e.g., (Nordström et al., 1990, p. 27).

[12] This resolution of the paradox was pointed out by Skolem in his original paper, (Skolem, 1923, p. 223).

many E-elements. Is the oddity just due to the fact that the relation E of the model $\mathfrak{M}$ is itself just non-standard? Short reflection shows that this sanguine reaction is misplaced.

Suppose that ZFC has a model $\mathcal{V}$, with domain V, which we regard as an *intended* model, and thus has the real set membership over it, thus $\in$. By the Downward Löwenheim-Skolem Theorem, there exists a countable sub-structure $\mathcal{V}' = <V', \in'>$ of $\mathcal{V}$, where $V' \subseteq V$, and $\in'$ is just $\in$ itself restricted to V'; this sub-structure satisfies precisely the same $\mathcal{L}_{\text{ZF}}$ sentences as does $\mathcal{V}$, and its membership relation really is the same, intended set-membership relation. Although itself countable, this V' might still contain uncountable members. $\mathcal{V}'$ satisfies the Axiom of Extensionality, so V' must be extensional; if we also insist that it is well-founded, we can apply the Mostowski Collapsing Lemma to get a transitive $\mathcal{V}'' = <V'', \in''>$ isomorphic to $<V', \in'>$. Since it is isomorphic to $\mathcal{V}'$, this model, too, satisfies exactly the same $\mathcal{L}_{\text{ZF}}$ sentences as does $\mathcal{V}$, and its membership relation, too, must be the real one (or behave just like it, which is the same thing). However, in $\mathcal{V}''$ all sets are transitive, and thus are either finite or countable, including the set representing the continuum. Thus, the continuum of this model is a *genuine* set, cut out of the intended model $\mathcal{V}$, and is *really* countable.

This stronger version of the paradox can be internalised. Let us remind ourselves briefly of how this proceeds.

Since the paradox essentially concerns modelling of the language $\mathcal{L}_{\text{ZF}}$, we have to have available some means of talking about the language of ZF inside ZF. This can be done via a standard coding which translates the relevant concepts into set-theoretical ones. Following this Gödelisation, it can then be shown that there is a one-place formula $F(x)$ of $\mathcal{L}_{\text{ZF}}$ which expresses the notion of being a formula of $\mathcal{L}_{\text{ZF}}$, i.e., holds of x just in case x is the code (set) of a formula of $\mathcal{L}_{\text{ZF}}$. This formula determines a set F, and we can show similarly that there are formulae and sets $Sen(x)$, S, $Ax(x)$, A respectively, corresponding to sentences and axioms. With this as a basis, the standard model-theoretic elements can be reproduced fairly straightforwardly inside ZF. The most important of these is the three-place predicate $Sat(v_1, v_2, v_3)$, which says that v_1 satisfies v_2 inside v_3; this allows us to say, inside the theory, that a set v_3 is a model of a formula of $\mathcal{L}_{\text{ZF}}$, coded as v_2, under a satisfaction sequence v_1 (an eventually constant sequence of members of v_3). We can prove two central theorems inside ZF, first that Sat obeys the Tarski conditions on satisfaction, and, following this, that:

$$\text{ZF} \vdash Val(\ulcorner \sigma \urcorner, u) \leftrightarrow \sigma^u \tag{10.1}$$

where "$Val(v_2, v_3)$" is the 2-place predicate obtained by universal quantification from $Sat(v_1, v_2, v_3)$, "σ" refers to a sentence of $\mathcal{L}_{\text{ZF}}$, "$\ulcorner \sigma \urcorner$" its code name, and "$\sigma^u$" refers to the standard relativisation of the sentence to the set u. What $Val(\ulcorner \sigma \urcorner, u)$ in effect says is that all satisfaction sequences drawn from u satisfy σ in u, and thus that "σ is true in the interpretation u." In other words, $Val(\ulcorner \sigma \urcorner, u)$ is precisely the counterpart in ZF of the normal model-theoretic $<u, \in\restriction u> \models \sigma$. (10.1) shows the equivalence of this with the standard relativisation.

Using the notions expressed in "*Val*" and the predicates derived from it, it is now possible to express the notion of being a model of a set of sentences, to prove an internal version of the Completeness Theorem for first-order theories, and to express the notion of being an elementary substructure, written "$ES(x, y)$." With these in hand, we can now demonstrate the Downward Löwenheim-Skolem Theorem inside ZFC:

$$\text{ZFC} \vdash \forall y \subseteq u\big[[\aleph_0 \leq card(u)] \rightarrow \\ \exists z[y \subseteq z \;\&\; card(z) = card(y) \;\&\; ES(z, u)]\big] \qquad \text{(DLST)}$$

(Note the "ZFC"; this means that essential use is made of AC, just as with the conventional version.) On reflection, none of this is a great surprise, since one might argue that model theory is really nothing but informal set theory with a few rhetorical flourishes, and it would be utterly surprising if this conceptually straightforward piece of set-theoretical mathematics could *not* be rendered in our standard axiomatic set theories. But given (DLST), it is now not surprising that we can get some version of the Skolem Paradox, choosing y such that $card(y) = \aleph_0$. But what we get now (under the *same* assumption that ZF is consistent) is neither contradictory nor paradoxical, but rather the connection with *absoluteness*, summed up in the following result concerning $C(x)$:

$$\text{If the system ZF is consistent, then the predicate } C(x) \text{ is not absolute.} \qquad (10.2)$$

By the same token, $\neg C(x)$ fails to be absolute, too. Before explaining this in detail, we must first look at the notion of absoluteness.

Despite differences in the way absoluteness is presented, the notion, as the name suggests, is one of invariance or stability, the singling out of those terms or formulae which keep the *same* value (in the former case, set-value, in the latter, truth-value) as the domain in which they are evaluated varies.[13] Invariance itself, however, is not a fixed notion; more and more things will be absolute as more conditions are put on the domains to be used for the evaluation.

To illustrate this, the first definition of absoluteness (adapted from (Kunen, 1980, p. 117)) is as follows:

Definition 1 (Absoluteness$_1$) The formula ψ with free variables $x_1, x_2, \ldots, x_n$ is said to be *absolute$_1$ for the formula* φ if we can show:

$$\text{ZFC} \vdash \exists x\varphi(x) \;\&\; \forall x_1, x_2, \ldots, x_n\big[[\varphi(x_1) \;\&\; \varphi(x_2) \;\&\; \ldots \;\&\; \varphi(x_n)] \rightarrow \\ [\psi^{\varphi}(x_1, x_2, \ldots, x_n) \leftrightarrow \psi(x_1, x_2, \ldots, x_n)]\big]$$

[13] In what follows, we will slip rather sloppily between talk of formulae, the domains they determine, and the extensions of the formulae, thus sets or proper classes.

This definition covers set terms as well: for the term τ, just take the formula $x = \tau$. Here, φ acts as the universe with respect to which the formula ψ or the term τ is to be evaluated.

In this definition, nothing is said about the demands made by the formula φ, apart from there being something which satisfies it. Hence, it is unlikely that things satisfying a formula ψ looked at from φ's perspective will satisfy ψ invariantly, i.e., be *absolute*$_1$ for φ. It is easy to show that absoluteness in this sense is preserved by the propositional connectives, but it is not difficult to exhibit very elementary notions which are not *absolute*$_1$; for example, the notion of something's being a subset of something else, i.e. "$x \subseteq y$." However, this formula *is* invariant in the sense we are investigating as soon as we demand that the evaluation take place in a *transitive* evaluative class φ; then the things which φ "recognises" to be subsets are precisely the things which ZF can *prove* to be subsets. As soon as we demand transitivity, we can show that formulas built up from atomic formulas using only propositional connectives and *bounded* quantification, the so-called Δ_0 formulas, are absolute as well.[14]

Another thing which may affect what is or is not absolute is the question of which principles of set theory are available in the class φ, that is to say, which principles φ "recognises to be true." This is important because the definition of a term, say, and the proof that the set which the term attempts to define exists, will in general depend on the availability of some assumed background, thus on some of the axioms. Hence, these axioms ought to be available in the classes φ if there is to be a fair assessment of invariance. For this, it is enough to look at models of the finitely many axioms used in the proof that the definition of the term in question is a good one, for the "background knowledge" required for a specific purpose is limited in exactly this way.

There is a perfectly good formal analogue to this informal talk of "availability." Saying that an axiom σ is "available in φ" is really just to say that ZF $\vdash \sigma^\varphi$, for this latter, as (10.1) shows, really means that σ holds in the structure $<\varphi, \in\restriction\varphi>$.[15] From this, there follows an obvious way in which the requirement of background knowledge can be inserted into the definition of absoluteness. Suppose the background knowledge which shows that a term is well-defined, or that a formula expresses what it is intended to express, is summed up in the axioms $\sigma_1, \ldots, \sigma_n$, then in assessing whether the formula or term is absolute for φ, we can demand that ZF $\vdash \sigma_1^\varphi \;\&\; \ldots \;\&\; \sigma_n^\varphi$. In this way, we get the notion of an *absoluteness sequence* for a formula or term. Again, more things will become absolute if we do this.

Based on these observations, we can adopt a second, general definition of absoluteness:

---

[14] See (Kunen, 1980, pp. 118–119).

[15] Whereas it is quite straightforward what σ^φ means in ZF, speaking strictly "$<\varphi, \in\restriction\varphi>$" makes no sense, since the extension of φ might not be a set. However, this abuse of notation is of a piece with the sloppiness pointed out in n. 13.

Definition 2 (Absoluteness$_2$) The formula ψ (or term τ) is said to be *absolute$_2$* if there is an absoluteness sequence $\sigma_1, \ldots, \sigma_n$ for ψ (or for $x = \tau$) for which we can show, for each provably non-empty transitive formula φ, that if $ZF(C) \vdash \sigma_1^{\varphi}\ \&\ \ldots\ \&\ \sigma_n^{\varphi}$, then ψ is *absolute$_1$* with respect to φ.[16]

Note that we can show in ZF(C) the existence of set models of any finite collection of axioms $\sigma_1, \ldots, \sigma_n$, i.e., that ZF(C) $\vdash \exists x \left[\sigma_1^{\varphi}\ \&\ \ldots\ \&\ \sigma_n^{x}\right]$.[17] Hence, to show that a formula ψ is not absolute$_2$, it suffices to show in ZFC that, for any sequence of axioms $\sigma_1, \ldots, \sigma_n$, there is a transitive set u such that $\sigma_1^u\ \&\ \ldots\ \&\ \sigma_n^u$, and such that the (i.e., ZFC's) evaluation of ψ in u differs from the evaluation of ψ in the "universe" (i.e., ZFC's evaluation of ψ).

Note also that these are not the only notions of invariance which we might take as a sign that something is "absolute." For example, Gödel in his original work defined a formula φ as absolute if its value in the constructible universe is the same as its actual value.[18] Nonetheless, *absoluteness$_2$* seems a fairly natural notion, perhaps the most natural which is not tied to a specific predicate like constructibility. The point here is that, even using such a weak notion of absoluteness, while ω is absolute, the theorem (10.2) asserts that $\neg C(x)$, and indeed $P(\omega)$ (ZF's term for the power set of ω), are *not* absolute. The proofs which establish these failures of absoluteness proceed precisely by the reasoning which yields the Skolem Paradox. Let us outline the steps which lead to this.

Assume that ZF (and hence ZFC) is consistent. Now assume that $C(x)$ *is* absolute in the sense of *absoluteness$_2$*, and that $\sigma_1, \ldots, \sigma_n$ is an absoluteness sequence for it, a sequence we abbreviate by σ. ZF can prove Cantor's Theorem, from which it follows that $\exists x \neg C(x)$; let $\tau_1, \ldots, \tau_k$ be the list of axioms used in the proof of this, this time abbreviated by τ. As was mentioned, we can always find set models of finitely many axioms. Hence, there must be an extensional set z for which σ^z and τ^z hold, thus showing that z is a model for σ and τ. The set z might be very large; but we can apply the (DLST) to cut it down to a countable, extensional subset s, an elementary substructure of z. Thus σ^s and τ^s both hold, too. Moreover, since s is extensional, the Mostowski Collapsing Lemma says that there must be a transitive set u which is isomorphic to s; this is likewise countable, and it follows that σ^u and τ^u also both hold. Now, since the axioms τ used in the proof of Cantor's Theorem hold in z, it is easy to see that $\exists x \neg C(x)$ holds relative to z, i.e., $[\exists x \neg C(x)]^z$, and the same must hold for the countable set u, for the axioms τ also hold there. Hence $[\exists x \neg C(x)]^u$, which in effect says that u is a countable model of the statement "There exists an uncountable set," just as in the original Skolem Paradox. There is nothing contradictory about this in itself. But we can now use the absoluteness claim for $C(x)$ to *generate* a contradiction. $[\exists x \neg C(x)]^u$ is the same as $\exists x \in u \neg C^u(x)$; but since u is transitive, and since also σ^u, the absoluteness of $C(x)$ entails that $\forall x \in u[C^u(x) \leftrightarrow C(x)]$. Hence, $\exists x \in u \neg C(x)$. On the other hand, since $x \in u$,

[16] This notion of absoluteness is taken from (Bell and Machover, 1977, p. 502).

[17] See (Kunen, 1980, p. 134ff.).

[18] See (Gödel, 1940, p. 42), or (Gödel, 1990, p. 76).

and u is transitive, we must have $C(x)$, since u is countable, from which it follows that there is an $x \in u$ for which $C(x)$ and $\neg C(x)$, which is a contradiction. This now gives the absurdity from which can conclude that $C(x)$ is not, after all, absolute. The argument that the term $P(\omega)$ is also non-absolute is entirely similar.

It is readily seen that this internal argument follows just the same pattern as the informal argument for Skolem's Paradox, except that, here, the assumption that $C(x)$ is absolute provides the genuine contradiction which the Skolem Paradox does not quite yield. Given this, the Skolem Paradox is transformed into a highly significant (meta-)theorem, parallel to the transformation of the genuine antinomies into theorems in the system.

As we have said, ω is absolute, and so is "being an ordinal number" (the predicate $Ord(x)$), but crucially not the predicate "being a cardinal number." What is the central difference?

The absoluteness of ω means that its value when evaluated in any φ will be just ω itself, provided that φ recognises the truth of the axioms in the absoluteness sequence for ω. On the other hand, that $P(\omega)$ is not absolute means that, for any finite choice of axioms, the value of $P(\omega)$ will in general change as φ varies; i.e., given any finite sequence of axioms, we can find a non-empty, transitive φ satisfying these axioms such that $\varphi \cap P(\omega)$ is not the same as $P(\omega)$. Why does this happen?

The key is the presence of unbounded quantifiers, for these are what in general stand in the way of formulas being absolute.[19] As pointed out, a formula ψ will be absolute for φ (even in the sense of *absoluteness*$_1$) if φ is transitive, and if ψ contains no unbounded quantifiers. But whether or not the definition of a predicate ψ or a set-term t contains unbounded quantifiers depends crucially on the background knowledge available. To illustrate this, let us look at ω.

The set ω is absolute, so the formula $x = \omega$ is an absolute formula. The usual proof of this proceeds by showing that the formula $x = \omega$ can be expressed using atomic formulae ($x = y$, $x \in y$), propositional connectives, and bounded quantifiers. To say that x is ω is to say that x is a limit ordinal while none of its members (predecessors) are, and all of this can be put in the right form providing the formula $Ord(x)$ contains only bounded quantification. But does it? On the face of it, the answer is negative. To say that u is an ordinal is to say that u is well-ordered by the $\in$-relation, and this involves an *unbounded* quantifier, for the special clause governing well-ordering begins by referring to all subsets of u, i.e., begins $\forall x\big[\forall w[w \in x \rightarrow w \in u \ldots$. The same holds if we want to define ordinals, not as certain well-ordered sets, but as certain *well-founded* sets, for the well-foundedness condition begins in exactly the same way. But now suppose the Axiom of Foundation is present; there is then no need to build in either the condition of being well-ordered by $\in$, or of being well-founded. It is enough to demand that u be transitive and totally ordered by $\in$, and these notions can be expressed by formulae

[19] This seems to provide some link between the notions of absoluteness and impredicativity. See (George, 1987) for references to the idea that impredicativity concerns unbounded quantifiers. The link between the two notions is implicit in Poincaré's analyses of the antinomies given in §2. But as George points out, this cannot be all there is to the notion, which seems irredeemably imprecise.

which involve only *bounded* quantification, propositional connectives, and atomic formulae. Thus, it is clear that $x = \omega$ will be absolute for any φ which is transitive and which satisfies the Axiom of Foundation, together with other bits of ZF (minus the power set axiom) which we require to prove that the definition is adequate.

What of $P(\omega)$? Can we "hide," in a similar way, any unbounded quantifiers involved in the specification of members of this? The answer is that we cannot.

Suppose u is a transitive set which satisfies enough of ZF for it to be shown that u contains a set v which is the power set of ω as far as u is concerned. It is important to note that the power set axiom taken on its own is quite weak, for it only says that all the subsets of a given set which the theory recognises can be collected together into a set, and this is quite consistent with its being provable that only very few subsets exist. However, the power set axiom gets its strength by being combined with the Axiom of Separation (Zermelo's *Aussonderungsaxiom*), which (in first-order set theory) is the schema:

$$\forall x \exists y \forall z [z \in y \leftrightarrow (z \in x \ \& \ \psi(z))] \qquad \text{(SEP)}$$

This holds for any ψ in the language, regardless of how formed, thus, in particular, for any ψ containing unbounded quantifiers.[20] Thus, v will contain only those sets which the "model" u can recognise as subsets of ω, and it is clear that this need not be all the genuine subsets of ω. Suppose that $y \subseteq \omega$, and that y is given by an application of separation to a formula $\psi(z)$ which contains unbounded quantifiers. We can think of ψ and the condition "$z \in \omega$" coalesced into the one condition φ, in other words, that y is given by the abstraction term $\{z : \varphi(z)\}$. Allow that u is a model of those axioms needed to prove that $\{z : \varphi(z)\}$ defines a set, in particular the requisite instance of (SEP). The key question now is whether v has y as a member, or, in other words, how u will evaluate the abstraction term $\{z : \varphi(z)\}$. It is immediately obvious that u will evaluate this term as $\{z \in u : \varphi^u(z)\}$, and hence that all the unbounded quantifiers in φ will now be bounded by u. Thus, we now have quite a different abstraction term from $\{z : \varphi(z)\}$, one which *a priori* we have no reason to think corresponds to y itself, although it yields a subset of y. Thus, although v will certainly contain $y^u = \{z \in u : \varphi^u(z)\}$, it is quite possible that it will fail to contain $y = \{z : \varphi(z)\}$, hence, quite possibly will not contain members matching all the subsets of ω.

What this suggests is that u will not "recognise" the existence of a set (or rather, the full extension of the set) corresponding to an abstraction term which contains unbounded quantifiers, for it will automatically read those quantifiers as being bound by u. However, (SEP) tells us that there are subsets of ω given by abstraction terms which have quantifiers which range beyond u. Thus, apparently, any attempt to pin down the extent of $P(\omega)$ which only takes into account formulas ψ restricted

[20] Let us note in passing that, in general, axiom systems are not the simple sum of their axioms, but that the axioms (so to speak) co-operate. Thus, although apparently weak on its own, the power-set axiom is enormously powerful when combined with the Axioms of Separation (or Replacement) and Infinity.

to what is inside u, and not the full range of the ψ which (SEP) permits, will not succeed. In this sense, then, the extent of $P(\omega)$ does seem to depend on that of the "richness of the universe."[21]

To summarise, it seems to me that this "internalisation" both disperses the mystery which is sometimes glimpsed as a result of the Skolem Paradox, and actually to a large extent explains it. It converts the paradox into a genuine contradiction, which is then exploited via an appropriate *reductio*. Moreover, in doing so, Poincaré's existence principles (LST and IST) are firmly rejected, as is clear in the formulation of (SEP); consequently, many sets/extensions which were taken as illegitimate by Poincaré are taken to exist as sets in an unproblematic way with perfectly determinate extensions. Given this, Poincaré's worry about variation in the extension of sets translates directly into the property of absoluteness. In particular, this argument shows that there is a radical difference between the set of natural numbers and the central focus of Poincaré's disquiet, the Cantor-Zermelo continuum, the former being absolute, and the latter not, making it clear that there is a sharp separation between the determinateness of the continuum and its non-absoluteness. This is what, in the end, Zermelo's insistence on the cogency of impredicative definitions amounts to.

These results, of course, are far from paradoxical. But although the internalisation of the Skolem Paradox both dismisses Poincaré's worries about incoherence and takes away any sense of paradox, it does serve to shift attention to a genuine conceptual difficulty within the theory of sets, namely that the cardinal notion of uncountability is insufficiently tied to the ordinal notions which Cantor had adopted to explain it. This is revealed by both Gödel's proof of the consistency of the Generalised Continuum Hypothesis (GCH) and Cohen's proof of the independence of the Continuum Hypothesis (CH). In these consistency and independence proofs, the notion of non-absoluteness (and indeed something close to the Skolem Paradox) plays a central role. A quick inspection of the main lines of argument will show this. It is to this that I turn in the final section.

4 Absoluteness, Consistency and Independence

Let us look first at Gödel's consistency proof, for which the technical notion of absoluteness was first introduced.

The key to the proof is the constructibility predicate $L(x)$ and the notion of constructible subset on which it is based. The constructible subsets of a given set u are, in effect, the subsets of u definable in u, in other words, just the subsets of u which satisfy a defining formula whose quantifiers are restricted to u. This loose account can be rendered inside $\mathcal{L}_{ZF}$ itself, given the formal notion of satisfaction. Let $Sat(x, y, z)$ be the satisfaction predicate described before, and let $ec(u)$ be the set of eventually constant sequences of u, effectively the set of satisfaction sequences

[21] This way of putting the matter is taken from (Bell and Machover, 1977, p. 509).

defined over u. (Note that, if u is infinite, $ec(u)$ has the same cardinality as u.) We can then say that a subset v of u is a constructible subset of u if there is a $z \in F$ and a $w \in ec(u)$ such that

$$v = \{y : y \in u \ \& \ Sat(w(0/y), z, u)\}$$

where $w(0/y)$ denotes the sequence w with the only change (if any) being that w's 0-th choice is replaced by y. Thus, v is the collection of all those y for which there is a formula z and a satisfaction sequence whose first element is y satisfying the selected formula inside u.

The crucial thing about this is that satisfaction has to take place inside u, in other words, given one of the central adequacy theorems concerning the predicate Sat, a definable subset v of u is just the set of all y in u which satisfy some formula φ whose quantifiers are restricted to u. But this just means, in effect, that $v = \{y \in u : \varphi^u(y)\}$, and that v is a collection of members of u picked out in a way that does not make reference, either explicitly or implicitly, to sets which are not already present in u itself. The abstraction term for v is just what the abstraction term $\{y : \varphi(y)\}$ becomes when it is evaluated in u.

These subsets of u can now be collected together into a set, $D(u)$, often called *the predicative power set of* u, meaning that it consists of just those subsets of u which are capable of definition using only quantification restricted to u itself. In usual predicative systems, such as that of Russell, quantification is restricted to some stage or type, and the object being defined is then taken to belong to the next stage or type. Here u should be regarded as the stage or type itself, which is what we get, in effect, if we demand that types be cumulative.

Not unexpectedly, what we now find is that, quite unlike the full power set $P(u)$, if the set u is infinite, then the predicative power set $D(u)$ has the same cardinality as the starting set u, the number of formulas available for use in definitions (thus, the size of F) being only countably infinite, and both $ec(u)$ and the collection of parameters available have the same size as u. Moreover, whereas $P(u)$ is not absolute, for infinite u, $D(u)$ is, which is no surprise, given the bounded nature of the quantification. Hence, for any non-empty transitive φ which satisfies the absoluteness sequence for $D(u)$, $\varphi \cap D(u)$ is equal to $D(u)$.

The predicate $L(x)$ is now defined with the help of the operation $D(x)$. First, sets L_α are defined for all ordinals using a transfinite recursion; for successor ordinals $\alpha = \beta + 1$, $L_\alpha = D(L_\beta)$, and when α is a limit ordinal $L_\alpha = \bigcup\{L_\beta : \beta < \alpha\}$. Hence, the L_α form a cumulative hierarchy, the so-called *constructible hierarchy*. (In sum, for all ordinals α, $L_\alpha = \bigcup\{D(L_\beta) : \beta < \alpha\}$.) Since the operations that go into this recursion (in particular, the $D(x)$ construction and the predicate $Ord(x)$) are all absolute, we can show that both $x = L_\alpha$ and $x \in L_\alpha$ are absolute formulas too. The definition of $L(x)$ is now simply:

$$L(x) \leftrightarrow \exists\alpha[Ord(\alpha) \ \& \ x \in L_\alpha]$$

The formula for $L(x)$ contains an unbounded quantifier, so is not absolute in our sense, although it is invariant for transitive models of ZF which contain all the ordinals. Nevertheless, the absoluteness of $x = L_\alpha$ and $x \in L_\alpha$ is enough to be able to prove the central relative consistency results, as we will see.

It can now be shown that the constructible sets form an "inner model" of set theory in which both the Axiom of Choice (AC) and the GCH hold. To show this requires showing three things[22]:

(1) Each axiom σ of ZF (AC is *not* included) is provable in ZF when relativised to the predicate L, i.e., ZF $\vdash \sigma^L$.
(2) The statement that all sets are constructible, i.e., $\forall x \exists \alpha [x \in L_\alpha]$ (always abbreviated as V = L) is also provable in ZF when relativised to the predicate L, thus ZF $\vdash$ (V = L)^L.

(1) and (2) together are enough to show that V = L is consistent relative to ZF. It is now shown that inside ZF both AC and GCH follow from V = L, i.e.,

(3) ZF + V = L $\vdash$ AC and also ZF + V = L $\vdash$ GCH.

Hence, AC and GCH must also be consistent relative to ZF.

What interests us here is (3), or rather that part of it which concerns the derivation of GCH from V = L, for this very closely mirrors the argument behind the internalised Skolem Paradox. Key use is made of the fact that the formula $x \in L_\alpha$ *is* absolute, which turns on the fact that predicative power-set formation, unlike full power-set formation, is absolute. The heart of the proof is the Lemma, which Gödel calls "an axiom of reducibility for sufficiently high orders" (Gödel, 1938, p. 556), that if all sets are constructible (i.e., if V = L), then any subset of $L_{\aleph_\alpha}$ (i.e., any member of $P(L_{\aleph_\alpha})$) is not just *somewhere* in the L-hierarchy, but must be constructible *before* the stage $L_{\aleph_{\alpha+1}}$. This time we take an absoluteness sequence σ for $x \in L_\beta$, in much the same way as before we started with a supposed absoluteness sequence for $C(x)$, and we assume V = L is in σ. Take any subset u of $L_{\aleph_\alpha}$, and consider the set $L_{\aleph_\alpha} \cup \{u\}$. There must be a transitive set v which includes $L_{\aleph_\alpha} \cup \{u\}$ for which σ^v & [V = L]^v. Since it is easily proved that $card(L_{\aleph_\alpha}) = \aleph_\alpha$, then $card(L_{\aleph_\alpha} \cup \{u\})$ is $\aleph_\alpha$, too. Hence, applying the (DLST), just as before, we can find an elementary substructure w of v which is of size $\aleph_\alpha$ and which also includes $L_{\aleph_\alpha} \cup \{u\}$. (The only difference here is that we have focused on $\aleph_\alpha$ rather than $\aleph_0$.) Hence, σ^w & [V = L]^w. It can be argued easily that, although w need not be transitive, it must be extensional since v is transitive, and thus must satisfy the Axiom of Extensionality. Therefore w also satisfies it, so we can apply the Mostowski Collapsing Lemma to collapse w to a transitive set x isomorphic to it. But since $L_{\aleph_\alpha} \subseteq w$ and $L_{\aleph_\alpha}$ is transitive, the collapsing isomorphism applied to each member of $L_{\aleph_\alpha}$ (thus each member of u) must be the identity. Hence, it follows that u itself must be a subset of x.

22 See (Bell and Machover, 1977, pp. 480–481).

Since now $[V = L]^x$, this must amount to $\forall y \in x \exists \beta \in x[y \in L_\beta]^x$ (appealing to various facts about transitive sets). This is where the absoluteness of the formula $y \in L_\beta$ comes in, for this allows us to replace "$[y \in L_\beta]^x$" with "$y \in L_\beta$" in the formula, yielding $\forall y \in x \exists \beta \in x[y \in L_\beta]$. Since $u \in x$, then $\exists \beta \in x[u \in L_\beta]$. Let $\beta \in x$ with $u \in L_\beta$. But since x is transitive, this must mean that $\beta \subseteq x$. Hence, $card(\beta) \leq card(x) = \aleph_\alpha$, and so $\beta < \aleph_{\alpha+1}$, which means that $L_\beta \subseteq L_{\aleph_{\alpha+1}}$. Hence, u, the subset of $L_{\aleph_\alpha}$ we are considering, must be a member of $L_{\aleph_{\alpha+1}}$, and hence constructible before the stage $L_{\aleph_{\alpha+1}}$. Accordingly, we have proved that

$$\mathrm{ZF} + \mathrm{V} = \mathrm{L} \vdash P(L_{\aleph_\alpha}) \subseteq L_{\aleph_{\alpha+1}},$$

which means that

$$\mathrm{ZF} + \mathrm{V} = \mathrm{L} \vdash 2^{\aleph_\alpha} \leq \aleph_{\alpha+1};$$

and, since it is an elementary theorem of ZFC that $\aleph_{\alpha+1} \leq 2^{\aleph_\alpha}$, it follows that

$$\mathrm{ZF} + \mathrm{V} = \mathrm{L} \vdash 2^{\aleph_\alpha} = \aleph_{\alpha+1};$$

thus the GCH.[23]

Hence, we might say that the availability of the machinery which enables us to "internalise" the Skolem argument is *precisely* that which is deployed in the result central to showing the relative consistency of the GCH. And the non-absoluteness of $C(x)$ (or of the power-set operation), or alternatively, the absoluteness of the predicative power-set operation, is central to the proof.[24]

Skolem's original 1923 paper shows remarkable prescience in matters concerning the consistency and independence phenomena. According to Skolem, Zermelo's axiomatisation rests on a logically prior notion of "domain",[25] and Skolem clearly thinks that there is something odd, if not circular, in trying to found the notion of set

---

[23] For clear accounts of the proof of AC from ZF + V = L, as well as Gödel's main "reducibility lemma," see (Kunen, 1980, pp. 174–175) or (Bell and Machover, 1977, pp. 517–522). Bell and Machover point out the similarity of the proof of the main lemma with the proof of the non-absoluteness of $P(x)$; see p. 522.

[24] It is worth pointing out that it is the iteration of $D(x)$ through the classical ordinals that prevents the constructible hierarchy of the L_α being an appropriate setting for predicative mathematics. See for example (Kreisel, 1960, p. 386). According to what Kreisel calls "the fundamental idea of predicativity" (ibid., p. 387), an ordinal α is predicatively legitimate for use in defining a given level Σ_α of predicative definitions if there is a lower level Σ_β (with $\beta < \alpha$) such that there is a well-ordering of the natural numbers of type α definable by a formula from Σ_β. (See also (Gödel, 1944; Wang, 1954).) This is not in general true of the L-hierarchy.

[25] In his first paper on the axiomatisation of set theory (Zermelo, 1908b, p. 262), Zermelo says the following:

> Set theory is concerned with a "*domain*" $\mathfrak{B}$ of objects, which we will call "*things*", and among which are the sets.

on something set-like, and fully unexplained, like the notion of domain. In particular, he takes it that this rules out the use of the set-theoretic axioms in investigating dependence phenomena for the set-theoretic system itself, useful and fruitful as this might be for investigating these questions for logical systems in general. This is because an assumption of consistency would have to be made, and this would in turn rest on the assumption that "domains" exist, which, properly speaking, would demand axioms for the notion of domain, potentially leading to an infinite regress. Thus:

> If one is not to base oneself again on axioms for domains (and so on, ad infinitum), I see no other choice but to turn to considerations like those which were applied above in the proof of Löwenheim's Theorem, for which the idea [*Vorstellung*] of the finite whole number is assumed as the basis.[26]

Gödel's result has a rather different conceptual basis, in particular because it is consciously a *relative* consistency proof, and one pursued by the method of inner models first used for set theory by von Neumann in (von Neumann, 1929). But crucially it has at its core precisely an application of "Löwenheim's Theorem", as Gödel himself pointed out:

> This ['reducibility'] lemma is proved by a generalization of Skolem's method for constructing enumerable models. (Gödel, 1939, p. 93.)

Skolem's comments are prescient, not just with respect to Gödel's work, but also with respect to Cohen's later proof of the independence of CH. Skolem considers the question of whether the Zermelo axioms isolate a unique domain (up to isomorphism), for clearly the conceptual dependence of set theory on a notion of domain would then be somewhat less serious. He suggests how it might be shown that this is not the case, namely by taking a domain B, and trying to adjoin to it a set $a \notin B$ in something like the way a new object is adjoined to an algebraic structure, though the set theory case will be a good deal more involved. Skolem then says the following:

> Much more interesting would be to be able to prove that one can adjoin a new subset of Z_0 [Zermelo's set of natural numbers] without giving rise to contradictions. This however will be very difficult.[27]

In a footnote, Skolem then goes on:

> Since the Zermelo axioms do not determine the domain B [of all sets] uniquely, it is very unlikely that all problems concerning powers will be decidable using these axioms. For example, it is quite possible that the so-called Continuum Problem, namely whether $2^{\aleph_0} >$ or $= \aleph_1$ is simply not solvable on this basis; nothing need be decided about it. It could be that the situation here is just the same as in the following case: an undetermined field [*Rationalitätsbereich*] is given, and one asks whether there is present in this domain a

[26] Skolem (1923, pp. 229–230) (English translation, p. 299). See also (Wang, 1970, p. 39).

[27] Skolem (1923, p. 229) (English translation, pp. 298–299).

magnitude [*Grösse*] x such that $x^2 = 2$. Because of the ambiguity of the domain this is just not determined.[28]

None of this can be called an anticipation of the Cohen *methods*, but the idea of adjoining a set is a fair general description of Cohen's *goal*, perhaps most easily seen in the adjunction of a non-constructible set to a model which satisfies V = L. The genius of Cohen's work is showing how it is possible to adjoin sets to a model of ZF in such a way that one obtains a new model of ZF. In doing this, the construction of both countably infinite interpretations and of absoluteness again play key roles. In the following pages, I will try to say what this connection is, by giving a very brief sketch of Cohen's proof.

Assume we have a countable, transitive model $<M, E>$ of ZF, the basis of the Skolem argument in the sharpened form. Since $<M, E>$ is countable and transitive, its continuum c^M must be countable. But the cardinal number $\aleph_2$ of this model must also be countable; hence, it seems that there must be a function f which is a bijection between c^M and $\aleph_2^M$ (or ω_2^M), even though M cannot recognise this function. Suppose we can now produce a new structure $<N, E'>$ which preserves as much of $<M, E>$ as possible, enough anyway to remain a model of ZF, but which also *contains* f as an element. Then $<N, E'>$ will be a model of ZF together with the sentence 'The cardinality of the continuum is $\aleph_2$,' that is, providing $<N, E'>$ is constructed properly. This (very crudely put) is the basis of the Cohen independence proofs. Note that the very starting point for the idea of Cohen's proof is rooted in the phenomenon which lies at the basis of the Skolem Paradox, namely the ((DLST)) and the existence of countable models.[29]

Let us try to be a bit less crude. First of all, we can use ZFC itself to carry out the construction, since $<M, E>$ does not have to be a model of the whole system, but only of a finite fragment of it, namely of the axioms which we actually use. If we list these axioms, then ZF can prove that there is a model based on a set M which satisfies them, meaning in particular that we can regard the relation E as just $\in$. Just as before, we can use the (DLST) and the Mostowski Collapsing Lemma in tandem to produce a countable, transitive set which also satisfies these axioms. Thus, in effect, we can regard M as a countable, transitive model of enough of ZF as we need. What the Cohen technique does is to show how to take such an M and create

---

[28] Skolem (1923, p. 229) (299 of the English translation). Skolem's original has "Alef" and not "ℵ." van der Waerden (1937, p. 40) gives "Rationalitätsbereich" as an alternative to "*Körper* [field]." The English translation of Skolem's paper has "commutative field."

[29] As Cohen himself said (Cohen, 2005, p. 2417):

> For example, he [Skolem] pointed out the existence of countable models of set theory. ... But certainly he was aware of the limitations on what can be proved. In a remarkable passage, he even discuses how new models of set theory might be constructed by adding sets having special properties, although he says he has no idea how this might be done. This was exactly the starting point for my own work on set theory, although I was totally unaware that Skolem had considered the same possibility.

For a more detailed discussion, see (Kanamori, 2008).

in ZF an expansion N, still countable and still a model of ZF (or the bits that we need), which contains the same ordinals as M, but also contains a function like f showing that the continuum in the expansion is equinumerous with its second aleph, thus violating the CH.

Suppose that p is an arbitrary partially ordered set in the countable, transitive model M. Let $G \subseteq p$ be a filter in p; G is said to be *generic* if its intersection with any set *dense* for p in M is non-empty. (d is said to be dense for p in M if $d \in M$, and if $\forall x \in p \exists y \in d[y \leqslant x]$.) Cohen showed that for each p there will always be a generic set. Since M is transitive, we have $G \subseteq M$, but in general G will not belong to M as p does. If it does not, then we can define the smallest model N (often denoted by $M[G]$) which contains M and which also has G as a member. (G will be the new subset adjoined to $<M, E>$, or the basis of the new function, which amounts to the same.) Cohen showed that this $M[G]$ exists for generic G, and indeed is also a countable, transitive model of ZF if M is, and indeed contains exactly the same ordinals. The existence of $M[G]$ is exactly what the forcing construction shows; the members of p are at the basis of this, the so-called forcing conditions. Note that since G is in $M[G]$, and $M[G]$ is a model of ZF, then $[\bigcup G]^{M[G]}$ is in $M[G]$. But union is an absolute operation, so $[\bigcup G]^{M[G]} = \bigcup G$, and this set is therefore in $M[G]$.

So far, all of this is quite general. The particular details of what $M[G]$ will satisfy over and above the axioms of ZF will depend on exactly what are chosen as the forcing conditions. Let us go back to our bijection f between ω_2^M and c^M. So far as we know, there is no such f in M; indeed, we can insist that M satisfies the GCH so that we know there cannot be. We can put $[P(\omega)]^M$ instead of c^M, because these two sets have the same cardinality in M, as does the set of all functions (characteristic functions) from ω^M to 2^M, i.e., the set $[2^M]^{\omega^M}$. There will of course be plenty of maps g in M going between $\omega_2^M \times \omega^M$ and 2^M; but since M possesses no injection between ω_2^M and $[P(\omega)]^M$, then none of these maps will be such that the corresponding $g^* : \omega_2^M \to [2^M]^{\omega^M}$ is injective. However, the finite fragments of such a map will be in M, and indeed so will be the set of all these finite fragments. Thus put

$$\begin{aligned}\mathrm{Fn}\left(\omega_2^M \times \omega^M, 2^M\right) = \{p : card(p) < \omega^M \wedge \mathrm{dom}(p) \subseteq \omega_2^M \times \omega^M \wedge \\ \mathrm{ran}(p) \subseteq 2^M\}\end{aligned}$$

This Fn $\in M$. So, in a sense, M recognises all the finite approximations to a function f like the one we want to consider, even though it cannot recognise such a function itself.

Fn is partially ordered by reverse inclusion, i.e., $x \leq y$ iff $x \supseteq y$, so we can take Fn as the set of forcing conditions. Now suppose G is any filter on Fn. Then $\bigcup G$ must be a function from $\mathrm{dom}(\bigcup G) \subseteq \omega_2^M \times \omega^M$ and $\mathrm{ran}(\bigcup G) \subseteq 2^M$. But the sets

$$D_i = \left\{ p \in \mathrm{Fn}\left(\omega_2^M \times \omega^M, 2^M\right) : i \in \mathrm{dom}(p) \right\}$$
$$D_j = \left\{ p \in \mathrm{Fn}\left(\omega_2^M \times \omega^M, 2^M\right) : j \in \mathrm{ran}(p) \right\}$$

are all dense, which must mean that if G is generic, it intersects them all, and we can conclude from this straightforwardly that for a generic G, $\mathrm{dom}(\bigcup G) = \omega_2^M \times \omega^M$ and $\mathrm{ran}(\bigcup G) = 2^M$, which means that $\bigcup G$ is a function from $\omega_2^M \times \omega^M$ onto 2^M.

We can now easily show that this $\bigcup G$ generates a function f of the kind we are seeking. For each α in ω_2^M, we get a function f_α from ω^M to 2^M (given by $f_\alpha(n^M) = \bigcup G(\alpha, n^M)$), and we can show that if $\alpha \neq \beta$, then $f_\alpha \neq f_\beta$. In other words, as the α run through ω_2^M, the f_α run through distinct subsets of ω^M. $\bigcup G$ $(= f)$ is clearly just the amalgamation of the f_α. Hence, in $M[G]$ we can show that there are ω_2^M *different* subsets of ω^M.

It is at this point that we can see the importance of the absoluteness of ω and the non-absoluteness of $P(\omega)$. We know that ω, like 2 (and any other finite ordinal), is absolute, so therefore n^M and ω^M are the same in $M[G]$, i.e., just n and ω respectively. Thus, $M[G]$ knows that there are ω_2^M subsets of ω. But this will only give us a violation of CH if we know that $\omega_2^{M[G]}$ is the same as ω_2^M, in other words, if we know that the shift from M to $M[G]$ preserves this particular uncountable cardinal. We know that cardinality, as opposed to ordinality, is not an absolute notion, although it turns out that in this particular construction ω_2^M *does* equal $\omega_2^{M[G]}$. Thus CH is violated in $M[G]$; $M[G]$ adds subsets of ω to M's $[P(\omega)]^M$.

We can now see the importance of the fact that the continuum is not absolute, for the cardinality of $M[G]$'s continuum is $\aleph_2$ (in $M[G]$), and we have seen that this cardinal is the same in both M and $M[G]$. But the cardinality of the continuum in M is *not* $\aleph_2$ because M satisfies GCH. Hence it must be the case that $c^M \neq c^{M[G]}$, which could not be the case if $P(\omega)$ *were* absolute. Thus, we have traded essentially on the fact ω is absolute but $P(\omega)$ is not.

The method roughly sketched here is extremely flexible. For one thing, by varying the finite partial functions chosen, one can produce models $M[G]$ in which the power of the continuum is almost any uncountable cardinal. For another, even the stability of the uncountable cardinal chosen is something which depends on the forcing conditions. That ω_2^M equals $\omega_2^{M[G]}$ is a consequence of the fact that Fn satisfies the *countable chain condition* (c. c. c.) in M.[30] In fact, if $M[G]$ preserves cofinalities (i.e., if $\mathrm{cf}(\gamma)^M = \mathrm{cf}(\gamma)^{M[G]}$), then any cardinal in M will also be a cardinal in $M[G]$, and if the set of forcing conditions p satisfies c. c. c. in M, then

[30] A *chain* in a partially ordered set p is a subset of p in which the ordering relation is total. Two elements x, y of p are *compatible* if there is an element z of p such that $z \leq x, z \leq y$, and incompatible if there is no such z. An *antichain* in p is a subset q of p such that any two distinct elements of q are incompatible. The c. c. c. for p then says that any antichain of p is countable. See (Kunen, 1980, p. 53).

cofinalities *will* be preserved in the shift to $M[G]$.[31] This is important to realise, since by choosing sets of forcing conditions which do not have c. c. c., one can violate cardinal preservation, or indeed 'collapse' cardinals in M to smaller ordinals in $M[G]$. In short, the forcing models show how radically the size of the continuum is undetermined by the standard axioms of ZF, thereby underlining the "relativity" of the central notions of cardinality and power.

5 Conclusion

The central focus of this paper is the internalisation of the Skolem Paradox, showing how this latter is transformed from a puzzle into an important technical result, a result which, I have argued, is fundamental in the study of modern axiomatic set theory. I have also suggested that the result is a clear reflection, in a very different setting, of what is at the heart of Poincaré's diagnosis of the antinomies, namely that the Cantorian continuum possesses an instability property not possessed by the collection of natural numbers. Poincaré wished to demonstrate, through his analysis, that the Cantorian continuum, indeed "Cantorism" generally, is incoherent. Modern set theory rejects that conclusion. In fact, it might be argued that the non-absoluteness of $P(\omega)$ shows that set theory gives a sufficiently refined account of the continuum to recognise a sharp conceptual distinction between the continuum and the set of natural numbers. Nevertheless, the internalisation of the Skolem Paradox ultimately shows that Poincaré's diagnosis, when reflected in the way I have indicated, points to a central conceptual difficulty, namely set theory's inability to solve the most basic non-trivial questions about exponentiation in the ordinal theory of infinite power.[32]

Bibliography

Aspray, W. and Kitcher, P., editors (1988). *History and Philosophy of Modern Mathematics.* University of Minnesota Press Minneapolis, MN.

Bell, J. and Machover, M. (1977). *A Course in Mathematical Logic.* North-Holland Publishing Company, Amsterdam.

---

[31] See (Kunen, 1980, pp. 204–208).

[32] The material in this paper has been through many incarnations over the last 20 years. The inspiration for it, though, is John Bell, who first made me aware of the internalised version of the Skolem Paradox. For discussions on this and related matters, I am also deeply indebted to the late George Boolos, William Demopoulos, Michael Friedman, Moshé Machover, the late John Macnamara, Mihaly Makkai, Stephen Menn and Gonzalo Reyes. As acknowledged in n. 3, I also owe a substantial intellectual debt to Wilfrid Hodges. I am also very grateful for the generous support of the Social Sciences and Humanities Research Council of Canada over many years, as well as the FQRSC of Québec, formerly FCAR. It is also a pleasure to acknowledge the gracious support of the Alexander von Humboldt Stiftung and the Akademie der Wissenschaften zu Göttingen. Note that the translations which appear in the text are my own, even where published translations are referred to as well as the originals.

Benacerraf, P. and Putnam, H., editors (1964). *Philosophy of Mathematics: Selected Readings.* Basil Blackwell, Oxford.
Benacerraf, P. and Putnam, H., editors (1983). *Philosophy of Mathematics: Selected Readings.* Cambridge University Press, Cambridge, second edition.
Benacerraf, P. (1985). Skolem and the skeptic. *Aristotelian Society, Supplementary Volume*, 59: 85–115. First half of a symposium with Crispin Wright; see (Wright, 1985).
Cohen, P. (2005). Skolem and pessimism about proof in mathematics. *Philosophical Transaction of the Royal Society, Series A: Mathematical, Physical and Engineering Sciences*, 363: 2407–2418.
Ewald, W. and Sieg, W., editors (2011). *David Hilbert's Lectures on the Foundations of Logic and Arithmetic, 1917–1933.* Springer, Heidelberg, Berlin, New York. Hilbert's Lectures on the Foundations of Mathematics and Physics, Volume 3.
Ewald, W., editor (1996). *From Kant to Hilbert. Two Volumes.* Oxford University Press, Oxford.
George, A. (1985). Skolem and the Löwenheim-Skolem theorem: a case study of the philosophical importance of mathematical results. *History and Philosophy of Logic*, 6:75–89.
George, A. (1987). The imprecision of impredicativity. *Mind*, 96:514–518.
Gödel, K. (1938). The consistency of the axiom of choice and of the generalized continuum hypothesis. *Proceedings of the National Academy of Sciences*, 24:556–557. Reprinted in (Gödel, 1990, 26-27).
Gödel, K. (1939). The consistency of the generalized continuum hypothesis. *Bulletin of the American Mathematical Society*, 45:93. Reprinted in (Gödel, 1990, 27).
Gödel, K. (1940). *The Consistency of the Continuum Hypothesis.* Annals of Mathematics Studies. Princeton University Press Princeton, NJ. Reprinted in (Gödel, 1990, 31–101), with additional notes from 1951.
Gödel, K. (1944). Russell's mathematical logic. In Schillp, P. A., editor, *The Philosophy of Bertrand Russell*, pages 125–153. The Open Court Publishing Co., La Salle, IL. Reprinted in (Benacerraf and Putnam, 1964, 211–232), (Benacerraf and Putnam, 1983, 447–469), and in (Gödel, 1990, 119–141).
Gödel, K. (1990). *Kurt Gödel: Collected Works, Volume 2.* Oxford University Press, New York, Oxford. Edited by Soloman Feferman et al.
Goldfarb, W. (1988). Poincaré against the logicists. In *(Aspray and Kitcher, 1988)*, pages 61–81.
Goldfarb, W. (1989). Russell's reasons for ramification. In *(Savage and Anderson, 1989)*, pages 24–40.
Hallett, M. (1984). *Cantorian Set Theory and Limitation of Size.* Clarendon Press, Oxford.
Hallett, M. (2010). Introductory note to Zermelo's two papers on the well-ordering theorem. In Zermelo (2010), pages 80–115.
Heinzmann, G. (1986). *Poincaré, Russell, Zermelo et Peano. Textes de la discusion (1906–1912) sur les fondements des mathématiques: des antinomies à la prédicativité.* Albert Blanchard, Paris.
Kanamori, A. (2008). Cohen and set theory. *The Bulletin of Symbolic Logic*, 14:351–378.
Kreisel, G. (1960). La predicativité. *Bulletin de la société mathématique de France*, 88: 371–391.
Kunen, K. (1980). *Set Theory: An Introduction to Independence Proofs.* Studies in Logic and the Foundations of Mathematics, Volume 102. North-Holland Publishing Company, Amsterdam.
Meschkowski, H. (1966). *Probleme des Unendlichen: Werk und Leben Georg Cantors.* Vieweg und Sohn, Braunschweig.
Nordström, B., Petersson, K., and Smith, J. (1990). *Programming in Martin-Löf's Type Theory : An Introduction.* Clarendon Press, Oxford.
Poincaré, H. (1902). *La science et l'hypothèse.* Ernst Flammarion, Paris. English translation as (Poincaré, 1905b), and retranslated in (Poincaré, 1913b).
Poincaré, H. (1905a). *La valeur de la science.* Ernst Flammarion, Paris. English translation in (Poincaré, 1913b).

Poincaré, H. (1905b). *Science and Hypothesis*. Walter Scott Publishing Company. English translation by W. J. G. of (Poincaré, 1902). Reprinted by Dover Publications, New York, NY, 1952.

Poincaré, H. (1906). Les mathématiques et la logique. *Revue de métaphysique et de morale*, 14:294–317. Reprinted with alterations in (Poincaré, 1908), Part II, Chapter 5, and, with these alterations noted, in (Heinzmann, 1986, 79–104). English translation in (Ewald, 1996), 1052–1171.

Poincaré, H. (1908). *Science et méthode*. Ernst Flammarion, Paris. English translation in (Poincaré, 1913b), and retranslated by Francis Maitland as *Science and Method*, Dover Publications, New York, NY.

Poincaré, H. (1909). La logique de l'infini. *Revue de métaphysique et de morale*, 17:462–482. Reprinted in (Poincaré, 1913a, 7–31).

Poincaré, H. (1910). Über transfinite Zahlen. In *Sechs Vorträge über ausgewählte Gegenstände aus der reinen Mathematik und mathematischen Physik*. B. G. Teubner, Leipzig, Berlin. Partial English translation in (Ewald, 1996, Volume 2, 1071–1074).

Poincaré, H. (1912). La logique de l'infini. *Scientia*, 12:1–11. Reprinted in (Poincaré, 1913a, 84–96).

Poincaré, H. (1913a). *Dernières Pensées*. Ernest Flammarion, Paris. English translation published as *Mathematics and Science: Last Essays*, Dover Publications, New York, NY, 1963.

Poincaré, H. (1913b). *The Foundations of Science*. The Science Press, New York NY. English translation by G. B. Halsted of (Poincaré, 1902), (Poincaré, 1905a) and (Poincaré, 1908), with a Preface by Poincaré, and an Introduction by Josiah Royce.

Ramsey, F. P. (1926). The foundations of mathematics. *Proceedings of the London Mathematical Society*, 25, Second Series:338–384. Reprinted in (Ramsey, 1931, 1–61), and (Ramsey, 1978, 152–212).

Ramsey, F. P. (1931). *The Foundations of Mathematics and Other Logical Essays*. Routledge and Kegan Paul, London. Edited, with an Introduction, by R. B. Braithwaite, xviii, 292.

Ramsey, F. P. (1978). *Foundations: Essays in Philosophy, Logic, Mathematics and Economics*. Routledge and Kegan Paul, London. Edited by D. H. Mellor, with Introductions by D. H. Mellor, T. J. Smiley, L. Mirsky and Richard Stone, viii, 287.

Russell, B. (1907). On some difficulties of the theory of transfinite numbers and order types. *Proceedings of the London Mathematical Society*, 4 (Second Series):29–53. Marked 'Received November 24th, 1905–Read December 14th, 1905'. Reprinted in (Russell, 1973, 135–164) and (Heinzmann, 1986, 54–78).

Russell, B. (1908). Mathematical logic as based on the theory of types. *American Journal of Mathematics*, 30(3):222–262. Reprinted in (Russell, 1956, 59–102), and (van Heijenoort, 1967, 152–182).

Russell, B. (1956). *Logic and Knowledge*. George Allen and Unwin, London. Edited by R. C. Marsh.

Russell, B. (1973). *Essays in Analysis*. George Allen and Unwin, London. Edited by Douglas Lackey.

Savage, C. W. and Anderson, C. A., editors (1989). *Rereading Russell: Essays on Bertrand Russell Metaphysics and Epistemology*. University of Minnesota Press Minneapolis, MN.

Skolem, T. (1923). Einige Bemerkungen zur axiomatischen Begründung der Mengenlehre. *Matimatikerkongressen i Helsingfors den 4–7 Juli 1922, Den femte skandinaiska matematikerkongressen, redogörelse, 1923*, pages 217–232. Reprinted in (Skolem, 1970), 137–152 (which preserves the original page layout). English translation in (van Heijenoort, 1967), 290–301.

Skolem, T. (1970). *Selected Papers in Logic*. Universitetsforlaget, Oslo. Edited by Jens Erik Fenstad.

van der Waerden, B. (1937). *Moderne Algebra, unter Benutzung von Vorlesungen von E. Artin und E. Noether. Zweite verbesserte Auflage*. Julius Springer, Berlin.

van Heijenoort, J., editor (1967). *From Frege to Gödel: A Source Book in Mathematical Logic*. Harvard University Press, Cambridge, MA.

von Neumann, J. (1923). Zur Einführung der transfiniten Zahlen. *Acta Litteratum ac Scientiarum Regiæ Universitatis Hungaricæ Francisco-Josephinæ. Sectio Scientiæ-Mathematicæ*, 1: 199–208. Reprinted in (von Neumann, 1961, 24–33). English translation in (van Heijenoort, 1967, 346–354).

von Neumann, J. (1928). Die Axiomatisierung der Mengenlehre. *Mathematische Zeitschrift*, 27:669–752. Reprinted in (von Neumann, 1961, 339–422).

von Neumann, J. (1929). Über eine Widerspruchsfreiheitfrage in der axiomatischen Mengenlehre. *Journal für die reine und angewandte Mathematik*, 160:227–241. Reprinted in (von Neumann, 1961, 494–508).

von Neumann, J. (1961). *John von Neumann: Collected Works. Volume 1*. Pergamon Press, Oxford.

Wang, H. (1954). The formalization of mathematics. *Journal of Symbolic Logic*, 19:241–246. Reprinted in (Wang, 1962, 559–584).

Wang, H. (1962). *A Survey of Mathematical Logic*. Science Press, Peking. Reprinted as *Logic, Computers and Sets*, 1970, Chelsea Publishing Co., New York NY.

Wang, H. (1970). A survey of Skolem's work in logic. In Fenstad, J. E., editor, *Thoralf Skolem: Selected Papers in Logic*. Universitetsforlaget, Oslo.

Wright, C. (1985). Skolem and the skeptic. *Aristotelian Society, Supplementary Volume*, 59: 117–137. Second half of a symposium with Paul Benacerraf; see (Benacerraf, 1985).

Zermelo, E. (1908a). Neuer Beweis für die Möglichkeit einer Wohlordnung. *Mathematische Annalen*, 65:107–128. Reprinted (with English translation) in (Zermelo, 2010, 120–159). English translation also in (van Heijenoort, 1967, 183–198).

Zermelo, E. (1908b). Untersuchungen über die Grundlagen der Mengenlehre, I. *Mathematische Annalen*, 65:261–281. Reprinted (with English translation) in (Zermelo, 2010, 188–229). English translation also in (van Heijenoort, 1967, 200–215).

Zermelo, E. (2010). *Collected Works, Volume I: Set theory, Miscellanea*. Springer, Berlin, Heidelberg. Edited by Heinz-Dieter Ebbinghaus and Akihiro Kanamori.

Part III
Logic, Set Theory and Category Theory

Chapter 11
Equalisers of Frames in Constructive Set Theory

Peter Aczel

1 Introduction

In a recent note (Palmgren, 2005) Erik Palmgren has shown that, in a sufficiently strong version of Martin-Löf's type theory (Martin-Löf, 1984) the category of set-presented formal topologies has coequalisers. We refer the reader to (Palmgren, 2005) for the background motivation for this result.

Here we want to get a version of Palmgren's result in a sufficiently strong version of constructive set theory (Aczel and Rathjen, 2001). We prefer to work with the (superlarge) category of set-presented class frames, a category that is equivalent to the opposite of the category of set-presented formal topologies. So we want to show that the category of set-presented class frames has equalisers.

Recall that a *(class) frame* is a partially ordered class having a top element $\top$, sups $\bigvee Y$ of arbitrary subsets Y and meets $a \wedge b$ of elements a, b such that the distributive law

$$a \wedge \bigvee Y = \bigvee \{a \wedge y \mid y \in Y\}$$

always holds. The (superlarge) category Frame of frames has maps that preserve the frame structure; i.e. the top, sups and meets.

A set-indexed family $\{\gamma(a)\}_{a \in S}$ of elements $\gamma(a)$ of a class frame A indexed by a set S is a *family of generators of* A if, for every $x \in A$, the class $S_x = \{a \in S \mid \gamma(a) \leq x\}$ is a set and $x = \bigvee\{\gamma(a) \mid a \in S_x\}$. A function $C : S \to Pow(Pow(S))$ is a *set-presentation of A for the family of generators* if, for $a \in S$ and $U \in Pow(S)$,

$$\gamma(a) \leq \hat{\gamma}(U) \iff (\exists V \in C(a))[V \subseteq U],$$

where $\hat{\gamma}(U) = \bigvee\{\gamma(a) \mid a \in U\}$.

P. Aczel (✉)
Former Professor of Mathematical Logic and Computer Science, University of Manchester, Manchester, UK
e-mail: petera@cs.man.ac.uk

D. DeVidi et al. (eds.), *Logic, Mathematics, Philosophy: Vintage Enthusiasms*, The Western Ontario Series in Philosophy of Science 75,
DOI 10.1007/978-94-007-0214-1_11, © Springer Science+Business Media B.V. 2011

A *set-presented frame* is a frame equipped with a set-indexed family of generators and a set-presentation for the family. The set-presented frames form a (superlarge) category sFrame whose maps $A \to A'$ are the frame maps between the underlying frames of A, A' so that the forgetful functor sFrame $\to$ Frame is full.

If A, A' are frames and $F_1, F_2 : A \to A'$ are frame maps then we may form the subframe $\hat{A} = \{a \in A \mid F_1(a) = F_2(a)\}$ and easily show that it gives an equaliser of F_1, F_2 in the category of frames. Our task now is to prove the following result.

Theorem 1 (CZF + $*$-REA) *If $F_1, F_2 : A \to A'$ are frame maps, where A, A' are set presented frames, then their equaliser $\hat{A}$ also has a set-presentation, so that $\hat{A}$ becomes an equaliser of F_1, F_2 in the category* sFrame.

The theorem is an immediate consequence of the following two results, where $F_1, F_2 : A \to A'$ are frame maps between the frames A, A', with equaliser $\hat{A}$, $\{\gamma(a)\}_{a \in S}$ is a set-indexed family of generators for A and $eq(F_1, F_2) = \{U \in Pow(S) \mid \hat{\gamma}(U) \in \hat{A}\}$. A subset $\hat{S}$ of $eq(F_1, F_2)$ is a *set-base* for $eq(F_1, F_2)$ if whenever $U \in eq(F_1, F_2)$ then, for every $b \in U$ there is $V \in \hat{S}$ such that $b \in V \subseteq U$.

The axiom $*$-**REA** will be explained later when we turn to the proof of Lemma 3.

Lemma 2 (CZF) *If the set-indexed family of generators $\{\gamma(a)\}_{a \in S}$ for A has a set-presentation and $eq(F_1, F_2)$ has a set-base $\hat{S}$ then $\{\hat{\gamma}(U)\}_{U \in \hat{S}}$ is a set-indexed family of generators for $\hat{A}$ that also has a set-presentation.*

Lemma 3 (CZF + $*$-REA) *If A' has a set-presentation then the class $eq(F_1, F_2)$ has a set base.*

We assume that the reader has some familiarity with the axiom systems CZF and $CZF + REA$ for constructive set theory. In order to make the paper more self-contained we recall the Strong Collection Scheme and the Fullness Axiom of CZF and the axiom REA that play an important role. The Strong Collection Scheme states, for each formula $\varphi(x, y)$, that if a is a set such that $\forall x \in a\, \exists y\, \varphi(x, y)$ then there is a set b such that

$$\forall x \in a\, \exists y \in b\, \varphi(x, y) \ \ \& \ \ \forall y \in b\, \exists x \in a\, \varphi(x, y).$$

Another collection scheme, the Subset Collection Scheme, was used in the original formulation of CZF. The Fullness Axiom is an axiom that is equivalent to that scheme, given the other axioms and schemes of CZF, and is formulated using the following notion. If A, B are sets then $\mathbf{mv}(B^A)$ is defined to be the class of total relations from A to B; i.e. it is the class of all relations $R \subseteq A \times B$ such that

$$\forall x \in A\, \exists y \in B \ \ (x, y) \in R.$$

Now the Fullness Axiom states that, for all sets A, B there is a subset D of $\mathbf{mv}(B^A)$ such that

$$\forall R' \in \mathbf{mv}(B^A)\, \exists R \in D\ R \subseteq R'.$$

The Regular Extension Axiom (REA) is a useful addition to CZF. A regular set is a transitive set that has the Strong Collection Property, where a set $\mathcal{A}$ is transitive if every element is a subset and it has the Strong Collection Property if, whenever $R \in \mathbf{mv}(\mathcal{A}^a)$, with $a \in \mathcal{A}$ then there is a set $b \in \mathcal{A}$ such that

$$\forall x \in a\, \exists y \in b\ (x, y) \in R \ \ \&\ \ \forall y \in b\, \exists x \in a\ (x, y) \in R.$$

The axiom REA states that every set is a subset of a regular set. A slight strengthening of REA is the axiom $\bigcup REA$, which expresses that every set is a subset of a union-closed regular set; i.e. a regular set closed under the union operation, $x \in \mathcal{A} \mapsto \cup x \in \mathcal{A}$.

It is worth noting that if $\mathcal{A}$ is regular then, if $a \in \mathcal{A}$ and $f : a \to \mathcal{A}$ the $\{f(x) \mid x \in a\} \in \mathcal{A}$, and if $2 = \{\emptyset, \{\emptyset\}\} \in \mathcal{A}$ then $\mathcal{A}$ is closed under unordered pairs $x, y \in \mathcal{A} \mapsto \{x, y\} \in \mathcal{A}$ and also under binary unions $x, y \in \mathcal{A} \mapsto x \cup y \in \mathcal{A}$ when $\mathcal{A}$ is union-closed.

2 Proof of Lemma 2

As $\hat{S}$ is a set-base for $eq(F_1, F_2)$ the family $\{\hat{\gamma}(U)\}_{U \in \hat{S}}$ is easily seen to be a set-indexed family of generators of $\hat{A}$. We focus on the problem of getting a set-presentation $\hat{C} : \hat{S} \to Pow(Pow(S))$ for the family. For $U \in \hat{S}$ and $\mathcal{U} \in Pow(\hat{S})$ let

$$U \mathrel{\hat{\lhd}} \mathcal{U} \iff \hat{\gamma}(U) \leq \bigvee \{\hat{\gamma}(V) \mid V \in \mathcal{U}\}.$$

We must find $\hat{C}$ such that for $U \in \hat{S}$ and $\mathcal{U} \in Pow(\hat{S})$

$$U \mathrel{\hat{\lhd}} \mathcal{U} \iff (\exists \mathcal{V} \in \hat{C}(U))\ \mathcal{V} \subseteq \mathcal{U}.$$

Let $C : S \to Pow(Pow(S))$ be a set-presentation for $\{\gamma(a)\}_{a \in S}$. Observe that

$$\begin{aligned} U \mathrel{\hat{\lhd}} \mathcal{U} &\iff \bigvee \{\gamma(a) \mid a \in U\} \leq \bigvee \{\gamma(b) \mid b \in \cup \mathcal{U}\} \\ &\iff (\forall a \in U)\ \gamma(a) \leq \bigvee \{\gamma(b) \mid b \in \cup \mathcal{U}\} \\ &\iff (\forall a \in U)(\exists V \in C(a))\ V \subseteq \cup \mathcal{U} \end{aligned}$$

The following lemma gives us what we need.

Lemma 4 (CZF) *Let $C : S \to Pow(Pow(S))$, where S is a set, and let $\hat{S} \in Pow(Pow(S))$. Then there is $\hat{C} : \hat{S} \to Pow(Pow(\hat{S}))$ such that, for all $U \in \hat{S}$ and $\mathcal{U} \in Pow(\hat{S})$,*

$$(\forall a \in U)(\exists V \in C(a))\ V \subseteq \cup \mathcal{U} \iff (\exists \mathcal{V} \in \hat{C}(U))\ \mathcal{V} \subseteq \mathcal{U}.$$

Proof For $U \in Pow(S)$ let

$$\mathsf{MV}(U, C) = \{R \in Pow(C^*) \mid (\forall a \in U)\ \exists V\ (a, V) \in R\},$$

where $C^* = \{(a, V) \in S \times Pow(S) \mid V \in C(a)\}$.

Claim *There is* $Q : \hat{S} \to Pow(C^*)$ *such that, for all* $U \in \hat{S}$, $Q(U) \subseteq \mathsf{MV}(U, C)$ *and if* $T \in Pow(C^*)$ *then*

$$(*) \quad T \in \mathsf{MV}(U, C) \iff \exists R \in Q(U)\ R \subseteq T.$$

Proof of Claim By the Fullness Axiom, for each $U \in \hat{S}$ there is a subset D of $\mathbf{mv}(\overline{C}^U)$ such that $\mathcal{T}(U, D)$, where if $\overline{C} = \bigcup_{a \in S} C(a)$,

$$\mathcal{T}(U, D) \equiv \forall R' \in \mathbf{mv}(\overline{C}^U)\ \exists R \in D\ R \subseteq R'.$$

So, using Strong Collection, there is a set $\mathcal{D}$ of subsets of $\mathbf{mv}(\overline{C}^U)$ such that

$$(\forall U \in \hat{S})(\exists D \in \mathcal{D})\mathcal{T}(U, D) \text{ and } (\forall D \in \mathcal{D})(\exists U \in \hat{S})\mathcal{T}(U, D).$$

For each $U \in \hat{S}$ let $Q(U) = \mathsf{MV}(U, C) \cap (\bigcup \mathcal{D})$. By the Union axiom and Restricted Separation $Q(U)$ is always a set so that $Q : \hat{S} \to Pow(C^*)$ and $Q(U) \subseteq \mathsf{MV}(U, C)$. Now let $U \in \hat{S}$ and $T \in Pow(C^*)$. To prove $(*)$, the implication from right to left is immediate. For the converse direction, let $T \in \mathsf{MV}(U, C)$. Choose $D \in \mathcal{D}$ such that $\mathcal{T}(U, D)$. Then, as $T \cap C^* \cap (U \times \overline{C}) \in \mathbf{mv}(\overline{C}^U)$, there is $R \in D$ such that $R \subseteq T \cap C^* \cap (U \times \overline{C})$. As $R \in \bigcup \mathcal{D}$ and $R \in \mathsf{MV}(U, C)$ we get that $R \in Q(U)$. As $R \subseteq T$ we have derived the right hand side of $(*)$. □

Given $\mathcal{U} \in Pow(\hat{S})$ let $T(\mathcal{U}) = \{(a, V) \in C^* \mid V \subseteq \bigcup \mathcal{U}\}$. By the claim there is $Q_1 : \hat{S} \to Pow(C^*)$ such that, for all $U \in \hat{S}$, $Q_1(U) \subseteq \mathsf{MV}(U, C)$ and, as $T(\mathcal{U}) \in Pow(C^*)$,

$$\begin{aligned}(\forall a \in U)(\exists V \in C(a))\ V \subseteq \bigcup \mathcal{U} &\iff T(\mathcal{U}) \in \mathsf{MV}(U, C)\\ &\iff (\exists R \in Q_1(U))\ R \subseteq T(\mathcal{U})\end{aligned}$$

Let $R \in Q_1(U)$ for some $U \in \hat{S}$. Then

$$\begin{aligned}R \subseteq T(\mathcal{U}) &\iff (\forall (a, V) \in R)\ V \subseteq \bigcup \mathcal{U}\\ &\iff (\forall (a, V) \in R)(\forall b \in V)(\exists W \in \mathcal{U})\ b \in W\\ &\iff (\forall b \in \mathcal{G}(R))(\exists W \in \mathcal{U})\ b \in W\\ &\iff (\forall b \in \mathcal{G}(R))(\exists W \in C_1(b))\ W \in \mathcal{U}\\ &\iff T_1(\mathcal{U}) \in \mathsf{MV}(\mathcal{G}(R), C_1),\end{aligned}$$

where $\mathcal{G}(R) = \bigcup_{a \in S}\{V \mid (a, V) \in R\}$, $C_1(b) = \{W \in Pow(S) \mid b \in W\}$ for $b \in S$ and $T_1(\mathcal{U}) = \{(b, W) \in C_1^* \mid W \in \mathcal{U}\}$. Note that each of these are sets.

Let $\hat{S}_1 = \{\mathcal{G}(R) \mid R \in \bigcup_{U \in \hat{S}} Q_1(U)\}$. By the Union axiom and Replacement this is a set. By the claim again there is $Q_2 : \hat{S}_1 \to Pow\left(C_1^*\right)$ such that, for all $U_1 \in \hat{S}_1$, $Q_2(U_1) \subseteq \mathsf{MV}(U_1, C_1)$ and,

$$T_1(\mathcal{U}) \in \mathsf{MV}(U_1, C_1) \iff (\exists R' \in Q_2(U_1))\ R' \subseteq T_1(\mathcal{U}).$$

So, if $\mathcal{V}_{R'} = \{W \mid (\exists b \in S\ (b, W) \in R'\} \in Pow(S)$ for $R' \in Pow\left(C_1^*\right)$,

$$\begin{aligned} R \subseteq T(\mathcal{U}) &\iff (\exists R' \in Q_2(\mathcal{G}(R)))\ R' \subseteq T_1(\mathcal{U}) \\ &\iff (\exists R' \in Q_2(\mathcal{G}(R)))\ \mathcal{V}_{R'} \subseteq \mathcal{U}, \end{aligned}$$

and hence, if $U \in \hat{S}$,

$$\begin{aligned} (\forall a \in U)(\exists V \in C(a))\ V \subseteq \cup\mathcal{U} &\iff (\exists R \in Q_1(U))\ R \subseteq T(\mathcal{U}) \\ &\iff (\exists R \in Q_1(U))(\exists R' \in Q_2(\mathcal{G}(R)))\ \mathcal{V}_{R'} \subseteq \mathcal{U} \\ &\iff \exists \mathcal{V} \in \hat{C}(U)\ \mathcal{V} \subseteq \mathcal{U}, \end{aligned}$$

where $\hat{C}(U) = \{\mathcal{V}_{R'} \mid R' \in \bigcup_{R \in Q_1(U)} Q_2(\mathcal{G}(R))\} \in Pow(Pow(S))$. □

3 Proof of Lemma 3

We first formulate the axiom $*$-**REA**. A union-closed regular set $\mathcal{A}$ is defined to be a $*$-*regular set* if it has the *Relation Reflection Property*; i.e. if $a_0 \in X \subseteq \mathcal{A}$ and $R \subseteq X \times X$ such that

$$(\forall a \in X)(\exists b \in X)[(a, b) \in R]$$

then there is $X_0 \in \mathcal{A}$ such that $a_0 \in X_0 \subseteq X$ and

$$(\forall a \in X_0)(\exists b \in X_0)[(a, b) \in R].$$

The axiom $*$-REA: Every set is a subset of a $*$-regular set.

Note the following result, where **DC** is the axiom of Dependent Choices; i.e. the axiom that states that whenever $R \in \mathbf{mv}(A^A)$, where A is a set, if $a \in A$ then there is $f : \mathbb{N} \to A$ such that $f(0) = a$ and $(f(n), f(n+1)) \in R$ for all $n \in \mathbb{N}$.

Proposition 5 (CZF+DC) *Any union-closed regular set has the Relation Reflection Property and so is $*$-regular.*

Corollary 6 $*$-**REA** *is a theorem of* **CZF** + **DC** + $\bigcup$**REA**.

To prove Lemma 3 we let $\mathcal{A}$ be a large enough $*$-regular set such that $2 \in \mathcal{A}$. We will explain what large enough means as we go on. For any set X let

$Pow_{\mathcal{A}}(X) = \mathcal{A} \cap Pow(X)$. As $\mathcal{A}$ is a set so is each $Pow_{\mathcal{A}}(X)$. Let $\hat{S} = \{U \in \mathcal{A} \mid U \in eq(F_1, F_2)\}$. As $\mathcal{A}$ is a set we can use Replacement and Restricted Separation to see that $\hat{S}$ is a set. We show that $\hat{S}$ is a set-base for $eq(F_1, F_2)$.

H1: We get $S \in \mathcal{R}$ by requiring that $S \in \mathcal{A}$.

H2: Let $U, V \in \mathcal{R}$. We must show that $U \downarrow V \in \mathcal{R}$. But $(U \downarrow V) = \bigcup\{U \downarrow \{c\} \mid c \in V\}$ where $U \downarrow \{c\} = \bigcup\{(b \downarrow c) \mid b \in U\}$ for $c \in V$ with $(b \downarrow c) = \{a \in S \mid a \leq b, c\}$ for $b, c \in S$. It follows that if we assume that the set $\{(b \downarrow c) \mid b, c \in S\}$ is a subset of $\mathcal{A}$ then, as $U, V \in \mathcal{A}$, so is $(U \downarrow V) \in \mathcal{A}$. This can be defined by Restricted Separation in $\mathcal{A}$, provided we assume that $\{(a, b) \in S \times S \mid a \leq b\}$ and $2 = \{0, \{0\}\}$ are elements of $\mathcal{A}$, so that Restricted Separation does indeed hold in $\mathcal{A}$.

H3: For $i = 1, 2$ and $U \in Pow(S)$ let

$$G_i U = F_i(\hat{\gamma}(U)).$$

Note that

$$U \in eq(F_1, F_2) \iff G_1 U = G_2 U.$$

Let $G_1 U = G_2 U$. We must show that if $b \in U$ then there is $V \in \mathcal{A}$ such that $b \in V \subseteq U$ and $G_1 V = G_2 V$. So let $G_1 U = G_2 U$ and let $b \in U$. Let $X = Pow_{\mathcal{A}}(U)$.

We will assume that $S \subseteq \mathcal{A}$

It follows that, for each $b \in S$, $\{b\} \in \mathcal{A}$ so that $\{b\} \in X$. We will use the following lemma.

Lemma 7

1. $(\forall V_1 \in X)(\exists V_2 \in X)\ [G_1 V_1 \leq' G_2 V_2]$.

2. $(\forall V_2 \in X)(\exists V_1 \in X)\ [G_2 V_2 \leq' G_1 V_1]$.

Proof It suffices to prove 1 as we get 2 by interchanging the roles of G_1 and G_2. For $V \in Pow(S)$ and $i = 1, 2$ let $T_i V = \bigcup_{a \in V} S'_{G_i\{a\}}$.

We assume that the $\left\{S'_{G_i\{a\}} \mid i \in \{1, 2\}\ \&\ a \in S\right\} \subseteq \mathcal{A}$.

We get that $T_i V \in \mathcal{A}$ for all $V \in X$. Observe that

$$\begin{aligned} G_i V &= F_i(\hat{\gamma}(V)) \\ &= \bigvee{}'\{G_i\{a\} \mid a \in V\} \\ &= \bigvee{}'\left\{\bigvee{}'\left\{\gamma'(a') \mid a' \in S'_{G_i\{a\}}\right\}\right\} \\ &= \bigvee{}'\{\gamma'(a') \mid a' \in T_i V\} \end{aligned}$$

So, for $V_1, V_2 \in Pow(S)$,

$$\begin{aligned} G_1V_1 \leq' G_2V_2 &\iff (\forall a' \in T_1V_1)[\gamma'(a') \leq' \bigvee'\{\gamma'(b') \mid b' \in T_2V_2\}] \\ &\iff (\forall a' \in T_1V_1)(\exists V' \in C'(a'))[V' \subseteq T_2V_2] \end{aligned}$$

Now let $V_1 \in X$. Then, as $V_1 \subseteq U$, $G_1V_1 \leq' G_2U$ so that

$$(\forall a' \in T_1V_1)(\exists V' \in C'(a'))[V' \subseteq T_2U].$$

We assume that $\bigcup_{a' \in S'} C'(a') \subseteq \mathcal{A}$ So, for $V' \in \bigcup_{a' \in S'} C'(a')$, $V' \in \mathcal{A}$ so that, using the regularity of $\mathcal{A}$, as $U \subseteq S \subseteq \mathcal{A}$,

$$\begin{aligned} V' \subseteq T_2U &\iff (\forall b' \in V')(\exists b \in \mathcal{A})[b \in U \ \&\ b' \in T_2\{b\}] \\ &\iff (\exists W \in \mathcal{A})[W \subseteq U \ \&\ V' \subseteq T_2W] \end{aligned}$$

So

$$(\forall a' \in T_1V_1)(\exists V' \in C'(a'))(\exists W \in \mathcal{A})[W \subseteq U \ \&\ V' \subseteq T_2W]$$

so that

$$(\forall a' \in T_1V_1)(\exists W \in \mathcal{A})[W \subseteq U \ \&\ (\exists V' \in C'(a')[V' \subseteq T_2W].$$

As $T_1V_1 \in \mathcal{A}$ and $\mathcal{A}$ is regular there is a set $\mathcal{W} \in \mathcal{A}$ such that $\mathcal{W} \subseteq Pow(U)$ and

$$(\forall a' \in T_1V_1)(\exists W \in \mathcal{W})(\exists V' \in C'(a')[V' \subseteq T_2W].$$

Now let $V_2 = \cup\mathcal{W}$. Then $V_2 \in \mathcal{A}$ and

$$(\forall a' \in T_1V_1)(\exists V' \in C'(a'))[V' \subseteq T_2V_2].$$

It follows that $G_1V_1 \leq' G_2V_2$. □

Corollary 8 $(\forall V \in X)(\exists V' \in X)[G_1V \leq' G_2V' \ \&\ G_2V \leq' G_1V']$

Proof Given $V \in X$, by the lemma there are $V_1, V_2 \in X$ such that

$$[G_1V \leq' G_2V_2 \ \&\ G_2V \leq' G_1V_1]$$

Let $V' = V_1 \cup V_2$. Then, $V' = \cup\{V_1, V_2\} \in \mathcal{A}$ and hence $V' \in X$. Also $G_iV_i \leq' G_iV'$ for $i = 1, 2$ so that

$$[G_1V \leq' G_2V' \ \&\ G_2V \leq' G_1V'].$$ □

As $\{b\} \in X$ and $X \subseteq \mathcal{A}$ we may apply the Relation Reflection Property to the corollary to get a set $X_0 \in \mathcal{A}$ such that $\{b\} \in X_0 \subset Pow(U)$ and

$$(*) \quad (\forall V \in X_0)(\exists V' \in X_0)[G_1V \leq' G_2V' \ \&\ G_2V \leq' G_1V'].$$

Let $V = \cup X_0$. Then, as $\mathcal{A}$ is union-closed, $V \in \mathcal{A}$ so that $V \in X$ and $b \in V \subseteq U$. Also

$$\hat{\gamma}(V) = \bigvee\{\gamma(a) \mid a \in V\} = \bigvee\{\hat{\gamma}(V_0) \mid V_0 \in X_0\}$$

so that, for $i = 1, 2$, $G_i V = \bigvee'\{G_i V_0 \mid V_0 \in X_0\}$. Using $(*)$ we can now easily show that $G_1 V = G_2 V$ and we are done. □

We now summarise the assumptions we have made about the $*$-regular set $\mathcal{A}$. We have assumed that $\mathcal{A}$ is a superset of the set

$$\{2\} \cup S \cup \left\{S'_{G_i\{a\}} \mid i \in \{1, 2\} \ \& \ a \in S\right\} \cup \bigcup_{a' \in S'} C'(a').$$

Bibliography

Aczel, P. and Rathjen, M. (2001). Notes on constructive set theory. Technical Report 40, The Royal Swedish Academy of Sciences, Institut Mittag-Leffler. Available at http://www.ml.kva.se/preprints/archive2000-2001.php

Martin-Löf, P. (1984). *Intuitionistic Type Theory. Lecture Notes by Giovanni Sambin*. Studies in Proof Theory. Bibliopolis, Napoli.

Palmgren, E. (2005). Quotient spaces and coequalisers in formal topology. *Journal of Universal Computer Science*, 11(12):1996–2007.

Chapter 12
Analogy and Its Surprises: An Eyewitness's Reflections on the Emergence of Real Algebraic Geometry

Max Dickmann

To John Bell on his 60th birthday, in steadfast friendship.

1 Introduction

The years 1978–1980 witnessed the birth of a systematic and organized corpus of knowledge—a theory worth that name—providing the tools required for a structural understanding of the geometric behaviour of algebraic varieties[1] over the field of real numbers.

The obvious interest of such an enterprise for geometry and physics—even for technological applications—raises immediately a question about the reasons for such a late development, especially in comparison to that of its cousin—algebraic geometry over the field of complex numbers—that developed vigorously and without interruption between the second half of the nineteenth century and the present.

A second, obvious, question is that of examining the conceptual process that led to the establishment of the new discipline of *real algebraic geometry*, with special emphasis on its final steps. Nowhere, as far as I know, has this second aspect been treated with revealing—rather than technical—detail by the *dramatis personae*.

In this paper I want to address these two questions from the informal perspective of an eyewitness to the events, rather than that of a rigorous historical investigation, a task quite beyond my abilities.[2] I wish to convey the development of some key ideas

M. Dickmann (✉)
Member of the Équipe de Logique Mathématique, Université Paris VII, Paris, France;
Associate at the Projet: Topologie et Géométrie Algébriques, Institut de Mathématiques de Jussieu, Paris, France
e-mail: dickmann@logique.jussieu.fr

[1] An *algebraic variety* over a field K is the set of solutions $x \in K^n$ of a finite set of polynomial equations $P_1(v_1, \ldots, v_n) = 0, \ldots, P_m(v_1, \ldots, v_n) = 0$, where the P_i's have coefficients in K.
[2] Questions of an historical nature akin (but not identical) to those dealt with here have been the object of several colloquia on "The origins of real geometry" in 2005–2007, in which the author took part. However, the contents of this paper, as well as its omissions, inaccuracies and errors, are entirely my responsibility.

D. DeVidi et al. (eds.), *Logic, Mathematics, Philosophy: Vintage Enthusiasms*, The Western Ontario Series in Philosophy of Science 75,
DOI 10.1007/978-94-007-0214-1_12,

that brought to birth this new mathematical discipline, trying, as much as possible, to spare the reader a tiresome technical exposition that may risk obscuring my goal. In particular, some unavoidable technical definitions will be relegated to footnotes.

2 The Late Development of Real Algebraic Geometry

The following quotations offer, in my opinion, a revealing insight of the state of mind of two leading mathematicians, as late as the beginning of the 1970s, concerning the existing state and the perspectives of development of real algebraic geometry. Dieudonné (1985, p. 74)[3]:

> In principle, in "abstract" algebraic geometry, algebraic equations with coefficients in any field whatsoever can be taken; but the experience of real geometry, where it is possible that *no* point of $\mathbb{R}^n$ satisfies a polynomial equation of degree > 0, *shows that it would not be possible to have workable geometric statements without being over an algebraically closed field* ... (My emphasis).[4]

In a rather interesting reflection, Thom regrets, as late as 1972, the pitiful state of neglect of real algebraic geometry:

> It might be argued that the importance accorded by analysis to the complex field and the theory of analytic functions during the last century has had an unfortunate[5] effect on the orientation of mathematics. By allowing the construction of a beautiful (even too beautiful) theory which was in perfect harmony with the equally successful quantification of physical theories,[6] it has led to a neglect of the real and qualitative nature of things. Now, well past the middle of the twentieth century, it has taken the blossoming of topology to return mathematics to the direct study of geometrical objects, a study which, however, has barely begun; *compare the present neglected state of real algebraic geometry with the degree of sophistication and formal perfection of complex algebraic geometry*. In the case of any natural phenomenon governed by an algebraic equation it is of paramount importance to know whether this equation has solutions, *real* roots, and precisely this question is suppressed when complex scalars are used. As examples of situations in which this idea of reality plays an essential qualitative role we have the following: the characteristic values of a linear differential system, the index of critical points of a function, and the elliptic or hyperbolic character of a differential operator (My emphasis).[7]

It is often a delicate exercise to give a convincing explanation of something that *has not* happened; in the present case, the lack of an *earlier* development of real algebraic geometry. Besides the "harmful" influence of nineteenth century complex analysis suggested by Thom, there is an obvious conjecture—by no means antithet-

---

[3] This and the next quotations are taken from the respective English translations.

[4] However, Dieudonné registers in the same text (p. 74) that topological properties of the sets of points with real coordinates of (complex) varieties given by polynomials with real coefficients were studied by Harnack, Hilbert and others already in the 19th century.

[5] "Harmful" ("néfaste") in the French original.

[6] The French text speaks of the "quantitative character of the physical laws."

[7] Thom (1975, fn. 4, p. 35). Chapter IV of (Dieudonné, 1985) gives an historical panorama of the dominant role played by complex projective geometry during most of the nineteenth century.

ical to his—based simply on the observation of actual historical record. I will try to sketch it here.

Real algebraic varieties have a behaviour that, with hindsight, may have looked "strange" or "irregular" as measured by existing and long acquired mental habits, namely those arising precisely from the theory of analytic functions and (hence) classical algebraic geometry. Here are some instances, far more "perverse" than that mentioned by Dieudonné:

Examples 1 (1) The "spiriting away" of real solutions of algebraic equations into the complex realm in certain parts of real space. Example: the cubic curve in $\mathbb{R}^2$ given by the equation

$$y^2 - x\,(x^2 - 1) = 0$$

obviously has no solutions where both coordinates are real, whenever $0 < x < 1$ or $x < -1$.

(2) *Irreducible* varieties over the reals may be non-connected, contrary to the situation over the complex field (Shafarevich, 1977, Ch. VII, § 2, pp. 318–324). Examples: the polynomial defining the curve above is irreducible, has two connected components (one for $-1 \le x \le 0$ and another for $x \ge 1$), and no "singularities"; the local dimension at each point is 1: in a small neighborhood of any point the curve can continuously (and algebraically) be deformed into a line.

The cubic curve in $\mathbb{R}^2$ defined by

$$y^2 = x^2\,(x - 1)$$

(irreducible as well) has two components, but one of them has "shrunk" into the single point (0, 0) (dimension 0), necessarily a singularity.

Note, in passing, that even as late as 1957 it wasn't even known whether a real algebraic variety has finitely many connected components (proved by Whitney that year).

(3) The (local) dimension may change from one point to another of a *connected*, irreducible algebraic variety over $\mathbb{R}$, contrary, again, to the behaviour known in the complex case, where all points of an irreducible variety have the same local dimension (Atiyah and Macdonald, 1969, pp. 124–125). Example: The *Cartan-Whitney umbrella* in $\mathbb{R}^3$ given by the equation:

$$x^3 - z\,(x^2 + y^2) = 0$$

In this example the points with coordinates $(0, 0, z)$ $(z \neq 0)$—i.e., those lying in the "handle" of the umbrella—have dimension 1, while the rest (those in the "canvas") are of local dimension 2. □

The obvious conjecture I want to state is that *the absence of techniques capable of dealing with phenomena of this kind*—that is, capable of "taming" phenomena genuinely perceived at the time as "wild"—was responsible for the state of neglect of real geometry regretted by Thom. Mathematicians interested in this area *were empty-handed* to prove basic theorems of sufficient generality to explain the behaviour of real algebraic varieties observed in the simple examples above, as well as in many others. Of course, this conjecture does not itself constitute a "cultural" (or "sociological") explanation of the belated appearance of those techniques. Thom's judgement may well be a point of departure for the (necessary) elucidation of this question.

3 The Sparse Pieces of a Puzzle

To be sure, bits and pieces of knowledge of the "real world" slowly began to emerge from the 1830s on. We begin by succinctly recording, with no claim to completeness, the most relevant among these results.[8]

3.1 The (Slow) Progress of Real Geometry

3.1.1 Real Algebra and Logic

1. Sturm's algorithm (1835) for the separation of the real roots of a real polynomial in one variable, the first recorded work on real varieties in a modern sense. This algorithm originated in Sturm's (and Fourier's) work on differential equations. In 1853 Sylvester extended Sturm's algorithm, and in 1852 he established the "inertia law" for quadratic forms over the reals named after him. Hermite also worked on this subject (1854).
2. Hilbert's work (1888) on the representation of non-negative polynomials as sums of squares of polynomials, and the incidence of this question on the possibility of certain geometrical constructions, leading to the statement of Hilbert's 17th problem (1900).
3. Artin and Schreier's solution to Hilbert's 17th problem (1927) that, in particular, led them to introduce the key notion of a *real closed field* (henceforth abbreviated RCF), and to prove the existence and uniqueness of the real closure of any ordered field.
4. Tarski's work—dating back to 1929–1930, though fully published much later—on quantifier elimination for the first-order theory of the field of real numbers in the language $\mathcal{L} = \{+, -, \cdot, 0, 1, <\}$ for ordered (unitary) rings. This result played a central role in the emergence of real algebraic geometry:

[8] In order to avoid an overextended bibliography, we record in the sequel only the publication year of the works mentioned. References to the original articles can be found in the extensive bibliography of (Bochnak et al., 1998), assorted with useful comments in the bibliographic and historical notes at the end of each chapter. See also the bibliographic references in (Becker, 1986; Dickmann, 1985; Knebusch, 1984).

(a) It implies the completeness and the decidability of the first-order theory of $\mathbb{R}$. Completeness shows, amongst many other things, the identity of Th($\mathbb{R}$) with Artin-Schreier's theory of real closed fields.[9] Decidability is at the root of the constructive-algorithmic aspects of real algebraic geometry that which underwent a vigorous development from the 1980s on (cf. Basu et al., 2006).
(b) It has as a consequence the so-called (second) "transfer principle"—"model-completeness" in the logicians' jargon—a tool of constant use in showing that the satisfaction of (first-order) properties by finitely many elements of a RCF is independent of the particular RCF that contains them.
More importantly,
(c) It gives the fundamental identity

$$\text{“Definable (in } \mathcal{L}) = \text{Semi-algebraic”}$$

for real closed fields,[10] that is, the identity within this theory of two central concepts: one originating in logic, the other in geometry. In particular, Tarski announces in (Tarski, 1931) the, by now usual, geometric characterization of the ($\mathcal{L}$–) definable subsets of the real line.
(d) Tarski (1931) also gave an explicit geometric interpretation of logical operations (e.g., existential quantification = projection) and showed, conversely, that some standard topological operations on definable subsets of real Euclidean space (e.g., closure and interior) can be expressed by logical formulas of the first-order language built on $\mathcal{L}$. Quantifier elimination for the first-order theory of RCFs amounts, in fact, to proving that a projection of a semi-algebraic set is semi-algebraic.

5. An extreme case that confirms the scatteredness of information that afflicted real algebraic geometry until rather recently is that of the PhD thesis (Brakhage, 1954), never published, and completely unknown to the practitioners of the subject until a few years ago. He carried out the first systematic topological study of semi-algebraic sets and functions *over arbitrary RCFs*.[11]
6. J.-L. Krivine's work (1964), originating in questions of real analysis, and largely unnoticed by geometers until well into the 1980s, proved first versions of some foundational results in real algebraic geometry.

---

[9] One cannot refrain from noticing the absence in Tarski's work of any reference to that of Artin-Schreier, though they were contemporaries, working 400 km from each other (!); presumably he did not know it.

[10] A *semi-algebraic* subset of K^n, K a RCF, is a finite Boolean combination of sets defined by polynomial equalities and inequalities. The powerful geometric implications of this identity became clear only much later.

[11] I borrow this information from (Bochnak et al., 1998, pp. 57–58).

(a) A weak version of the real *nullstellensatz*.
(b) A precise algebraic characterization of the real radical of an ideal in any ring, yielding, in particular, the exact form of those polynomials vanishing on a given real algebraic variety. [12]
(c) The (later) so-called Kadison-Dubois representation theorem.

Independently, Dubois (1969) and Risler (1970) proved stronger versions of the real *nullstellensatz*, the latter in its definitive form.

7. Stengle's *positivstellensatz* (1974) characterized those polynomials taking non-negative values on a real algebraic variety (and more).
8. Recio (1977)—and, independently, Coste-Roy (1979) and Delzell (1980)—proved the so called "open" quantifier-elimination theorem (a.k.a. "finiteness" theorem), another key result for later developments: an open semi-algebraic subset of K^n (K a RCF) can be represented as a finite union of finite intersections of *strict* polynomial inequalities.
9. Collins (1975) devised an algorithm for quantifier elimination of RCF based on a geometric result: cylindrical algebraic decomposition. Contrary to the (highly inefficient) algorithm derived from Tarski's result, Collins' is as efficient as possible, that is, of doubly exponential complexity on the entry data (Davenport and Heintz later showed that one cannot possibly do better). □

The works of Artin-Schreier (3) and Tarski (4) appear, in retrospect, to be the most important ground stones in the construction of real algebraic geometry.

3.1.2 Topology of Algebraic Varieties over $\mathbb{R}$

1. Harnack's result (1876) on the number of connected components of non-singular algebraic curves in the real projective plane.
2. Hilbert's extension of Harnack's work leading to (the first part of) Hilbert's 16th problem.
3. The work of J. Nash (1952) and H. Whitney (1957) on the topology of algebraic varieties over $\mathbb{R}$. Nash was led to consider analytic functions on open subsets of $\mathbb{R}^n$ which are solutions of algebraic equations with polynomial coefficients, later baptised *Nash functions* (for example, the function $y = \frac{1}{\sqrt{1+x^2}}$); the theory of Nash functions turned out to be of crucial importance in both real algebraic and real analytic geometry. Nash's work was continued by Akbulut, King, Tognoli and other researchers. Whitney proved that the difference of two algebraic varieties over $\mathbb{R}$ has finitely many connected components.
4. Lojasiewicz (1964) was the first organized rendering of the existing, but sparse information on real *analytic* geometry. Amongst a wealth of information, they

[12] His results apply, directly, to algebraic varieties over arbitrary RCFs (rather than only over $\mathbb{R}$). In the sequel we call the former *real algebraic varieties*, while mentioning explicitly those results valid only over $\mathbb{R}$.

contain the first systematic study of (local and global) semi-algebraic and semi-analytic sets in $\mathbb{R}^n$ (and the original proof of the famous *Lojasiewicz inequality*). *Note*. Many results of topological nature concerning varieties over $\mathbb{R}$ were derived from results known to hold in the field of complex numbers.

Each of these results—and many others not mentioned here—contributed in its own way to shape the theory that eventually emerged at the end of the 1970s. However, in spite of their individual significance, the list looked for a long time as a repertoire of scattered "facts." They had various origins and motivations, emerging from disparate areas of mathematics: differential equations, real analysis, topology, geometry, algebra, and even mathematical logic. A unifying thread could hardly be discerned, even with hindsight, until at least the middle 1960s. Further, none of these (and other) contributions could, by themselves or even together, offer a structural explanation of the behaviour of algebraic varieties over $\mathbb{R}$ known from many examples, nor answer certain questions arising from them. For example, does an arbitrary semi-algebraic subset of $\mathbb{R}^n$ have finitely many connected components? Does there exist a notion of connectedness of semi-algebraic sets over an arbitrary RCF reflecting the phenomena observed in examples over RCFs other than $\mathbb{R}$? (Note that the interval topology of any RCF other than $\mathbb{R}$ is totally disconnected.) Is there a "reasonable" theory of dimension for real algebraic varieties and semi-algebraic sets over general RCFs? Do the sets of equidimensional points of a real algebraic variety have a discernible structure (e.g., are they $\mathcal{L}$-definable)? Are there uniform decompositions of semi-algebraic sets into pieces with a simple geometric structure?

4 The Puzzle Solved: Invention of the Real Spectrum

The discovery of the real spectrum of a ring involved a rather complicated detour through a path parallel to that opened by Grothendieck in his abstract formulation of "classical" algebraic geometry in the early 1960s. Though at the end of the journey it was realized that in the real case one could dispense with this abstract machinery, understanding the trend of ideas that led to its discovery requires travelling again along that rugged path.

4.1 Zariski Spectra, Grothendieck Topologies and étale Schemes

Our point of departure is the classical Zariski spectrum of a ring, endowed with a *structure sheaf*:

4.1.1 Data

A (commutative, unitary) ring A.

- Points of $\mathrm{Spec}(A)$: The prime ideals of A.
- Topology on $\mathrm{Spec}(A)$: Given by the family of sets

$$D_a = \{\mathfrak{p} \in \mathrm{Spec}(A) \mid a \notin \mathfrak{p}\} \qquad (a \in A).$$

as a basis of opens.

- Structure sheaf $\mathcal{S}_A$ on Spec(A): The ring of sections over a non-empty basic open D_a (i.e., $a \neq 0$) is $A[a^{-1}]$.

The stalk of $\mathcal{S}_A$ at $\mathfrak{p} \in \mathrm{Spec}(A)$ is, then, the ring $A_{\mathfrak{p}}$, the localization of A at $\mathfrak{p}$. □

As is well-known, Spec(A) is a spectral space. [13] The pair (Spec(A), $\to \mathcal{S}_A$) is an (affine) *ringed space*.

The interesting case for algebraic geometry (over $\mathbb{C}$, say) is $A = \mathbb{C}[V] = \mathbb{C}[X_1, \ldots, X_n]/\mathcal{I}(V)$, where V is an algebraic variety defined by polynomials in $\mathbb{C}[X_1, \ldots, X_n]$, and $\mathcal{I}(V)$ is the ideal of polynomials vanishing on V; $\mathbb{C}[V]$ is called the *coordinate ring* of V. For a field extension $K \supseteq \mathbb{C}$—in fact, any field containing the coefficients of the polynomials defining V—the set of points in K^n satisfying the equations defining V is denoted by $V(K)$. In this case we have:

4.1.2 Basic Properties of the Zariski Spectrum of a Complex Variety

1. The prime ideals of $\mathbb{C}[V]$ correspond bijectively to the irreducible subvarieties of V (e.g., if V is a surface in $\mathbb{C}^3$, to the irreducible components of $V(\mathbb{C})$, the irreducible curves in $V(\mathbb{C})$, the points of $V(\mathbb{C})$, etc.).
2. The points of $V(\mathbb{C})$ correspond bijectively to the maximal ideals of $\mathbb{C}[V]$: for $x \in V(\mathbb{C})$,

$$x \mapsto M_x = \text{ the (maximal) ideal of polynomials vanishing at } x.$$

3. The restriction of $\mathcal{S}_{\mathbb{C}[V]}$ to $V(\mathbb{C})$ (under the identification $x \leftrightarrow M_x$) is the sheaf of germs at x of regular (rational) functions over V, i.e., rational functions whose denominator does not vanish on $V(\mathbb{C})$.
4. The topology induced by Spec$(\mathbb{C}[V])$ on $V(\mathbb{C})$ (under the identification in (2)) is the *Zariski topology*, where the closed subsets are the subvarieties of $V(\mathbb{C})$, rather than its *Euclidean topology* induced from $\mathbb{C}^n$. □

The trouble with this approach is that the open subsets of the Zariski topology on $V(\mathbb{C})$ *are far too big*, precluding the possility of getting an implicit function theorem—an essential tool for the local analysis of the variety (in particular, the study of its singular points).

4.1.3 The étale Site

Grothendieck's remedy to circumvent this (major) stumbling block was to make a fresh start and endow *the set* Spec(A) with:

---

[13] That is, a T_0 space with a basis of quasi-compact open sets (the D_a's in this example) closed under finite intersections, where every irreducible closed set is the closure of a singleton.

(i) A coherent system of coverings, instead of a flesh and bone topology;
(ii) A different structure sheaf on $\mathrm{Spec}(A)$.

The details of Grothendieck's theory are too technical to be expounded here; we shall only describe it in general terms, of necessity vague, but nevertheless needed for later comparison. The interested reader can consult expository texts such as (Johnstone, 1977) or (Goldblatt, 1979).

Ad (i). The systems of coverings considered—made of maps rather than (open) sets—are required to satisfy appropriate axioms, which define a *Grothendieck topology* or a *site*. When $A = K[V]$, V a variety over the field K, a specific site is considered, called the *étale site* of V.

Ad (ii). The stalks of the sheaf on (the site) $\mathrm{Spec}(A)$ to be considered are *henselian* local rings[14] with *separably closed*[15] residue fields. The process is done in such a way that, *grosso modo*, the stalk of the new structure sheaf $\mathcal{S}_A^{et}$ at $\mathfrak{p} \in \mathrm{Spec}(A)$ is the (strict) henselization of $A_\mathfrak{p}$.[16] This is a delicate point, as the sheaf is not constructed stalkwise but over open subsets; one must consider coverings by certain maps called *étale localizations*.

This recipe guarantees, of course

- The validity of (abstract versions of) the implicit function theorem (see Artin et al., 1972),

at the price of

- Not only having to manipulate quite complicated objects but, more importantly, losing much of the geometrical intuition.

In fact, the price of the ransom is irredeemably lost: the étale site of a complex variety V is *never* the site associated to a topology on $\mathrm{Spec}(\mathbb{C}[V])$, except in trivial cases.

We also point out, for later comparison, that although the process just sketched adds no new points—the underlying set is still $\mathrm{Spec}(A)$—the stalks of the structure sheaf $\mathcal{S}_A^{et}$ have non-trivial A-automorphisms: these arise from non-trivial automorphisms of the separable closure of the residue field $k(\mathfrak{p})$ of the localization $A_\mathfrak{p}$ $(\mathfrak{p} \in \mathrm{Spec}(A))$.

[14] A local ring B (with maximal ideal $\mathfrak{m}$) is *henselian* if every simple root of a monic polynomial $F \in B[X]$ in the residue field $B/\mathfrak{m}$ lifts to a root of F in B (necessarily simple as well).

[15] A field is *separably closed* if it contains all roots of every polynomial having no multiple roots. For fields of characteristic zero this just means *algebraically closed*.

[16] It can be proved, though this is by no means trivial, that any local ring B can be extended, in a "minimal" way, to a henselian local ring B^h with separably closed residue field; B^h is called the strict henselization of B. B^h is unique up to B-isomorphism and, in the geometric case, all relevant morphisms defined on B lift to B^h.

4.2 "Real" Variations on Grothendieck's Theme

A process parallel to that sketched in the preceding paragraph was carried out in the "real" context in (Coste and Coste, 1979; Coste-Roy, 1980), inspired by ideas of Gavin Wraith. We shall review their similarities and differences, which eventually led to a quite unexpected outcome.

4.2.1 A Real Analog to the Zariski Spectrum of a Ring

The initial step of constructing an analog to the Zariski spectrum of a ring, see Section 4.1, can be carried out without major obstacles. The idea is to replace the totality of the prime ideals of a ring by the subset of its *real prime ideals*, i.e., those prime ideals $\mathfrak{p}$ of A such that the domain $A/\mathfrak{p}$ (equivalently, its field of fractions $\mathrm{quot}(A/\mathfrak{p}) = k(\mathfrak{p})$) has a total order compatible with the operations with sum and product. These are the prime ideals $\mathfrak{p}$ satisfying, for each $n \geq 1$, the condition

$$\forall x_1 \ldots x_n \in A \left(\sum_{i=1}^{n} {x_i}^2 \in \mathfrak{p} \Rightarrow \quad x_1, \ldots, x_n \in \mathfrak{p} \right)$$

Warning. There are rings not containing any real prime ideal (e.g., the field $\mathbb{C}$). A necessary and sufficient condition for a ring A to contain at least one real prime ideal is that -1 *is not* a sum of squares in A; rings with this property are called *semi-real*. □

The *real Zariski spectrum* of a ring A, $\mathrm{Spec}_{\mathrm{RZar}}(A)$, is defined by the following

4.2.2 Data

- Points: the real prime ideals of A.
- Topology: the topology induced by that of $\mathrm{Spec}(A)$, see Section 4.1.
- Structure sheaf: the restriction to $\mathrm{Spec}_{\mathrm{RZar}}(A)$ of the structure sheaf $\mathcal{S}_A$ of Section 4.1, denoted $\mathcal{S}_A^R$.

Since a prime ideal $\mathfrak{p}$ is real if and only if the maximal ideal $\mathfrak{p}A_\mathfrak{p}$ of the localization $A_\mathfrak{p}$ is real, the stalks of the ringed space $\left(\mathrm{Spec}_{\mathrm{RZar}}(A),\ \mathcal{S}_A^R\right)$ are local rings with a real maximal ideal, called *residually real local rings*; this property is equivalent to: all elements of the form "1 + sum of squares" are invertible.

So far no big news. In fact, the topology of $\mathrm{Spec}_{\mathrm{RZar}}(A)$ has the same shortcomings as that of $\mathrm{Spec}(A)$: its open sets are too big to get an implicit function theorem.

4.2.3 The Real étale Site

The recipe for recovering some kind of implicit function theorem was to follow Grothendieck's steps in Section 4.1 as closely as possible. Step (i), replacing the topology of $\text{Spec}_{\text{RZar}}(A)$ by a suitable site requires a variant on the classical case mentioned in Section 4.1: étale localizations are to be replaced by their real counterparts.

To carry on with step (ii), the natural course of action was to consider sheaves over $\text{Spec}_{\text{RZar}}(A)$ having as stalks henselian local rings with *real closed*—instead of separably closed—residue fields (called *real closed local rings*). Here are some natural examples of rings of this type:

Examples 2 The rings of germs at a *non-singular* point of an algebraic variety over $\mathbb{R}$, of:

- C^∞ functions;
- Analytic functions;
- Nash functions (see Section 3.1.2 (3)).

These are indeed local rings: the maximal ideal is the germ of functions vanishing at the chosen point of the variety; their residue field is $\mathbb{R}$; and the implicit function theorem known from analysis ensures they are henselian. □

The question is, then, how to construct the required sheaves. A process parallel to that of the classical case (see Section 4.1 (ii) above) can be performed via a real analog of the notion of "étale localization." The following remarks are intended to give an inkling into how this is done.

Given a residually real local ring, B, and an order $\leq$ of its residue field $k_B = B/\mathfrak{m}$, a henselisation $B^{h,\leq}$ of B "along $\leq$" can be constructed, having as residue field the real closure $\overline{k_B}$ of $(k_B, \leq)$. This is achieved by lifting successively the simple roots in $\overline{k_B}$ of all polynomials $F \in B[X]$ (and iterating this process as long as necessary).

In particular, given a real prime ideal $\mathfrak{p}$ of A and carrying out this construction on the residually real local ring $A_\mathfrak{p}$, for each order $\leq_\mathfrak{p}$ of the residue field $A_\mathfrak{p}/\mathfrak{p}A_\mathfrak{p}$ —equivalently of $A/\mathfrak{p}$, since $A_\mathfrak{p}/\mathfrak{p}A_\mathfrak{p} \simeq \text{quot}(A/\mathfrak{p})$—, one obtains a real strict henselisation $(A/\mathfrak{p})^{h,\leq_\mathfrak{p}}$ with residue field $\overline{k(\mathfrak{p})}$ = real closure of $(\text{quot}(A/\mathfrak{p}), \leq_\mathfrak{p})$.

This sketch of the construction makes it clear that:

1. The stalks of the sheaf to be constructed are in one-to-one correspondence with the pairs $(\mathfrak{p}, \leq_\mathfrak{p})$ consisting of a real prime ideal $\mathfrak{p}$ of A and an order $\leq_\mathfrak{p}$ of $A/\mathfrak{p}$,[17] and
2. Since the residue field $\overline{k(\mathfrak{p})}$ is real closed—hence has no automorphisms other than the identity—the henselisation $(A/\mathfrak{p})^{h,\leq_\mathfrak{p}}$ *has no proper A-automorphisms.*

This last property guarantees:

Theorem 3 *(Coste and Coste, 1980) The real étale site of a ring is spatial, i.e., it is the site associated with a topological space.*[18] □

---

[17] See also (Roy, 1982, Prop. 3.3, p. 418).

[18] The first to intuit (and conjecture) the validity of this foundational result was Gavin Wraith. An idea of A. Joyal played a crucial role in the proof.

Remark (1) above leads to the determination of the exact nature of this space:

Definition 4 (The real spectrum.) The *real spectrum* of a (commutative, unitary) ring A, $\mathrm{Spec}_{\mathrm{R}}(A)$, consists of:
Points: The pairs $(\mathfrak{p}, \leq_{\mathfrak{p}})$ where $\mathfrak{p}$ is a real prime ideal of A and $\leq_{\mathfrak{p}}$ is a total order of $A/\mathfrak{p}$.
Topology: Generated by the basis of opens consisting of the sets

$$H(a_1, \dots, a_n) = \{(\mathfrak{p}, \leq_{\mathfrak{p}}) \in \mathrm{Spec}_{\mathrm{R}}(\mathrm{A}) \mid a_i/\mathfrak{p} >_{\mathfrak{p}} 0 \text{ for } i = 1, \dots, n\}$$

for all finite sequences $a_1, \dots, a_n \in A$. □

5 The Inception of Real Algebraic Geometry

The invention of the real spectrum opened the way for a rapid development of algebraic geometry over the field of real numbers and, more generally, over arbitrary real closed fields. It was also the catalyst for many important developments arising from the specific ways in which real algebraic geometry came about.

In this final section we shall complete the foregoing account by drawing up a succint "reading guide" of the principal events that followed the discovery of the real spectrum. The book (Bochnak et al., 1998) gives a comprehensive and very well written and treatment of the new discipline of real algebraic geometry. Accounts of its first steps—of a much more limited scope than (Bochnak et al., 1998) but preceding its publication (1988, first French edition)—are contained in the expository surveys (Becker, 1986; Dickmann, 1985; Knebusch, 1984). There is no reason to repeat the story here; it suffices to record the main landmarks.

5.1 Preliminary Remarks

1. First an obvious issue: why bother with algebraic geometry over *arbitrary* RCFs when the main interest is on $\mathbb{R}$? Fascination with pointless generality (typical of mathematicians, some would say)? The pleasure of the purity of method?

   Well . . ., for the first there is the obvious point that many proofs—but not all—cost the same in either case. Secondly, and more importantly, concrete practice shows that establishing a result in $\mathbb{R}$ may require prior knowledge of other results over arbitrary RCFs. For example, rephrasing a certain geometrical statement as a first-order formula—needed to prove a property over $\mathbb{R}$—may require a dimension argument, or the finiteness of the number of connected components, over arbitrary RCFs.
2. Another word of caution in the same spirit: semi-algebraic sets that may not be varieties are indispensable for the study of the latter.
3. Thirdly, a radical difference between complex and real algebraic geometry is the nearly complete absence of projective space in the latter. This stems from the fact

that projective spaces—and, more generally, grassmannians—over a RCF, K, are embedded as *non-singular* affine *algebraic varieties* in affine space of higher dimension over K. For example, the n-dimensional projective space $\mathbb{P}_n(K)$ is naturally (and algebraically) embedded in $K^{(n+1)^2}$, in such a way that its image, and that of all its non-singular algebraic subvarieties, become smooth affine subvarieties of $K^{(n+1)^2}$. This is the case even for the "complex" projective space $\mathbb{P}_n(K[i])$: by separating "real" and "imaginary" coordinates, it is embedded as a non-singular affine variety in $K^{2(n+1)^2}$. For details, see (Bochnak et al., 1998, § 3.4, pp. 70–75).

5.2 A Simplification

To make life easier, the pairs $(\mathfrak{p}, \leq_{\mathfrak{p}}) \in \mathrm{Spec}_{\mathrm{R}}(A)$ in Definition 4 can be subsumed into subsets α of A subject to suitable requirements (and conversely), by the expedient of setting $\alpha = \{x \in A \mid x/\mathfrak{p} \geq_{\mathfrak{p}} 0\}$. Note that $\alpha \cap -\alpha = \mathfrak{p}$.

The real spectrum of a ring, whenever non-empty, is a spectral space (cf. footnote 13). Further, for $\alpha, \beta \in \mathrm{Spec}_{\mathrm{R}}(A)$,

$$\beta \in \overline{\{\alpha\}}(= \text{ closure of } \{\alpha\}) \;\;\Leftrightarrow\;\; \alpha \subseteq \beta$$

It also has the following special property, known as *complete normality*: for $\alpha, \beta, \gamma \in \mathrm{Spec}_{\mathrm{R}}(A)$,

$$\gamma \subseteq \alpha, \beta \;\;\Leftrightarrow\;\; \alpha \subseteq \beta \text{ or } \beta \subseteq \alpha$$

In particular, the set of maximal elements of $\mathrm{Spec}_{\mathrm{R}}(A)$ is a Hausdorff space. For more information on completely normal spectral spaces, see (Carral and Coste, 1983) or (Dickmann et al., 201x).

5.3 The Geometric Case

Henceforth we focus on the case of interest in real algebraic geometry, where $A = K[V] = K[X_1, \ldots, X_n]/\mathcal{I}(V)$ is the coordinate ring of a variety V over a RCF, K, such that $\emptyset \neq V(K) \subseteq K^n$ (this condition ensures that $K[V]$ is a semi-real ring, i.e., $\mathrm{Spec}_{\mathrm{R}}(K[V]) \neq \emptyset$). As in the complex case, cf. Section 4.1, the points of $V(K)$ correspond bijectively to the maximal elements of $\mathrm{Spec}_{\mathrm{R}}(K[V])$: for $x \in V(K)$,

$$\alpha : x \mapsto \alpha_x = (M_x, \leq)$$

where $M_x = \{P/\mathcal{I}(V) \mid P \in K[X_1, \ldots, X_n] \text{ and } P(x) = 0\}$, and $\leq = \leq_{M_x}$ is the (unique) order of K (recall that $A/M_x \simeq K$).

However, the outstanding result in the present case is that, under this embedding, the topology induced on $V(K)$ by the spectral topology of $\mathrm{Spec}_{\mathrm{R}}(K[V])$ *is the Euclidean topology* inherited from the n-fold product of the interval topology on K, *not the Zariski topology*. This fundamental point deserves a closer look: since the isomorphism of the field $A/M_x \simeq K[X_1, \ldots, X_n]/M_x$ onto K is given by $P/M_x \mapsto P(x)$, and the subbasic open sets of $\mathrm{Spec}_{\mathrm{R}}(K[V])$ are by definition of the form

$$H(P/\mathcal{I}(V)) = \{\alpha \in \mathrm{Spec}_{\mathrm{R}}(\mathrm{K[V]}) \mid (P/\mathcal{I}(V))/\alpha \cap -\alpha = P/\alpha \cap -\alpha >_{\alpha} 0\},$$

we have $(*)$:

$$\begin{aligned}\alpha^{-1}[H(P/\mathcal{I}(V))] &= \{x \in V(K) \mid P/M_x >_{M_x} 0\} = \{x \in V(K) \mid K \models P(x) > 0\} \\ &= P^{-1}[(0, +\infty)] \cap V(K)\end{aligned}$$

Since the family $\{P^{-1}[(0, +\infty)] \cap V(K) \mid P \in K[X_1, \ldots, X_n]\}$ is a subbase for the Euclidean topology of $V(K)$, the equality (*) proves our assertion.

Further, (*) implies that the correspondence

$$H(P/\mathcal{I}(V)) \longmapsto P^{-1}[(0, +\infty)] \cap V(K) \qquad (P \in K[X_1, \ldots, X_n]) \qquad (**)$$

is bijective. Note, incidentally, that the equality (*) also shows that $V(K)$ is dense in $\mathrm{Spec}_{\mathrm{R}}(K[V])$. Using quantifier elimination one gets much more out of this bijection.

Indeed, since inverse images commute with set-theoretic operations, taking finite boolean combinations in (*) gives, as in (**), a one-to-one correspondence between arbitrary semi-algebraic subsets of $V(K)$ and finite boolean combinations of (quasi-compact) subbasic opens of $\mathrm{Spec}_{\mathrm{R}}(K[V])$. The latter are called *constructible sets* and play an important role in the theory: they form the canonical basis for the *constructible topology* associated to the spectral topology of $\mathrm{Spec}_{\mathrm{R}}(K[V])$ (in fact, of any spectral space). Identifying $V(K)$ with its image in $\mathrm{Spec}_{\mathrm{R}}(K[V])$ under the map α of 5.3, one gets:

Theorem 5 *Let V be an algebraic variety over a RCF, K, and let S be a semi-algebraic subset of $V(K)$. Then,*

1. *There is a unique constructible set $\widetilde{S} \subseteq \mathrm{Spec}_{\mathrm{R}}(K[V])$ such that $S = \widetilde{S} \cap V(K)$.*
2. *The correspondence $S \mapsto \widetilde{S}$ is an isomorphism from the boolean algebra of semi-algebraic subsets of $V(K)$ onto the boolean algebra of constructible subsets of* $\mathrm{Spec}_{\mathrm{R}}(K[V])$.
3. *It maps bijectively open (resp., closed) semi-algebraic subsets of $V(K)$ onto quasi-compact open (resp., closed constructible) subsets of* $\mathrm{Spec}_{\mathrm{R}}(K[V])$.
   In fact,

4. *If S is an open (resp., closed) semi-algebraic subset of $V(K)$, then $\widetilde{S}$ is the largest (resp., smallest) open (resp., closed) subset of* $\mathrm{Spec}_{\mathrm{R}}(K[V])$ *whose intersection with $V(K)$ is S.*
5. *The tilde operation commutes with the interior and closure operations (for the Euclidean topology of $V(K)$ and the spectral topology in* $\mathrm{Spec}_{\mathrm{R}}(K[V])$*).* □

For a proof of Theorem 5, see (Bochnak et al., 1998, § 7.2, pp. 142–146).

Remarks (a) Items (1) and (2) use in an essential way quantifier elimination for RCFs, and item (3) the open quantifier elimination theorem mentioned in Section 3.1.1 (9).

(b) By quantifier elimination, the constructible subsets of $\mathrm{Spec}_{\mathrm{R}}(K[V])$ are exactly those of the form

$$\{\alpha \in \mathrm{Spec}_{\mathrm{R}}(K[V]) \mid k(\alpha) \models \varphi[\pi_{\alpha}(P_1), \ldots, \pi_{\alpha}(P_m)]\}$$

where $\varphi(v_1, \ldots, v_m)$ is a formula (without parameters) of the language $\mathcal{L}$ for ordered rings and $\pi_{\alpha}(P) = P/\alpha \cap -\alpha$ is the image of $P \in K[X_1, \ldots, X_n]$ in $k(\alpha)$ = the real closure of (quot$(A/\alpha \cap -\alpha)$, $\leq_{\alpha}$).

(c) The tilde operation extends to *semi-algebraic functions*, that is, functions $f : S \longrightarrow T$ where $S \subseteq K^n$, $T \subseteq K^m$, whose graph is a semi-algebraic subset of $K^n \times K^m$. □

5.4 Connectedness

Since $\mathrm{Spec}_{\mathrm{R}}(K[V])$ is a quasi-compact space, it has finitely many connected components, and each of them, being clopen, is in turn a *finite* union of basic opens, hence constructible. By restriction to $V(K)$, the tilde correspondence of Theorem 5 (2) gives a partition of this variety into finitely many semi-algebraic pieces, each of which is maximal for the following notion, defined for semi-algebraic sets $S \subseteq V(K)$:

> S is *semi-algebraically (s.a.-) connected* if it cannot be split into two disjoint, semi-algebraic, non-empty sets, open in the induced topology.

In other words,

> An algebraic variety over a RCF has finitely many s.a.-connected components, each of which is a semi-algebraic set.

This result, combined with another "soft" topological argument and open quantifier elimination, yields a direct proof of

Theorem 6 *Any semi-algebraic subset of K^n, K a RCF, has finitely many s.a.-connected components, each of which is a semi-algebraic set.*[19] □

---

[19] I found this direct argument while writing this paper. The original argument, in (Coste and Roy, 1982, Thm. 5.5), (i), begins by proving the result for $K = \mathbb{R}$ by induction on the dimension of

5.5 Dimension

Another central geometric result is the *cell decomposition theorem*: it shows that any semi-algebraic subset of K^n, K a RCF, can be decomposed in a finite number of (semi-algebraic) pieces with a simple geometric description, called *cells*. Cells are defined by induction, and they come provided with a well defined *dimension*, which agrees with the intuitive meaning of the word.

Though non unique, cell decompositions give a notion of dimension for arbitrary semi-algebraic sets, verifying the traditional requirements, e.g., if $S \subseteq K^n$, $T \subseteq K^m$ are semi-algebraic sets, then:

- $\dim(S \cup T) = \max\{\dim(S), \dim(T)\}$,
- $\dim(S \times T) = \dim(S) + \dim(T)$,
- $\dim(\overline{S}) = \dim(S)$ [$\overline{S}$ = closure of S for the Euclidean topology],
- If S isopen, then $\dim(S) = n$.

Further, this geometrically defined notion of dimension of a semi-algebraic set $S \subseteq K^n$ agrees with an algebraic notion, akin to the Krull–dimension, detectable in the real spectrum, namely, the length of largest specialization (i.e., inclusion) chain of elements of $\mathrm{Spec}_{\mathrm{R}}(K[X_1, \ldots, X_n])$ contained in $\widetilde{S}$ (see Bochnak et al., 1998, Prop. 7.5.6).

5.6 The Implicit Function Theorem

As we have seen in Sections 4.1 and 4.2, ensuring the validity of this classical result from analysis was a driving force, within the abstract, Grothendieck–style setting, that led to the discovery of the real spectrum. However, once the abstract setting became concrete, i.e., the genuine Euclidean topology (as well as its semi-algebraic version over arbitrary RCFs, cf. Section 5.3) was retrieved within this setting via the real spectrum, the validity of a semi-algebraic version of the implicit function theorem ceased being big news (as it was clear earlier). Indeed, a slight adaptation of its classical proof renders it valid for semi-algebraic functions of any class $\mathcal{C}^k (k \in \mathbb{N} \cup \{\infty\})$. (Note: semi-algebraic $\mathcal{C}^\infty$ functions are the same thing as Nash functions.)

---

the ambient space, using cylindrical decomposition (cf. Coste and Roy, 1982, Thm. 3.5). By the compactness theorem of first-order logic, one gets primitive recursive bounds on the number and the degrees of a suitable family of polynomials intervening in the proof; these bounds are then used to validate the transfer to arbitrary RCFs. Admittedly, this argument gives considerably more information, useful elsewhere. See also (Bochnak et al., 1998, § 2.3 and Thm. 2.4.4).

5.7 The Real Spectrum as a Ringed Space

So far we have considered the real spectrum of an algebraic variety over a RCF as a topological space. A fitting question is: which are natural candidates to be sheaves that may turn $\mathrm{Spec}_{\mathrm{R}}(K[V])$ into a (locally) ringed space? As far as I know only two sheaves have been studied with some detail. The most natural, that fits the abstract approach sketched in 4.2 above, is the sheaf of *Nash functions*:

Theorem 7 ((Roy, 1982); see also (Bochnak et al., 1998, Prop. 8.8.2)) *Let V be an algebraic variety over a RCF, K. There is a unique sheaf, $\widetilde{\mathcal{N}}$, on* $\mathrm{Spec}_{\mathrm{R}}(K[V])$ *whose ring of sections $\widetilde{\mathcal{N}}(\widetilde{U})$ over an open constructible set $\widetilde{U}$, where U is an open semi-algebraic subset of $V(K)$, is isomorphic to the ring $\mathcal{N}(U)$ of Nash functions on U.*[20] *The stalk of $\widetilde{\mathcal{N}}$ at a point $\alpha \in$* $\mathrm{Spec}_{\mathrm{R}}(K[V])$ *is a* henselian *local ring whose residue field is $k(\alpha)$, the real closure of the ordered field* $(\mathrm{quot}(K[V]/\alpha \cap -\alpha), \leq_{\alpha})$.
□

A second choice is the (unique) sheaf $\widetilde{\mathcal{C}_{\mathrm{sa}}}$ whose ring of sections over $\widetilde{U}$ is the ring $\mathcal{C}_{\mathrm{sa}}(U)$ of continuous semi-algebraic functions on the open U (cf. Bochnak et al., 1998, Prop. 7.3.2). The stalk of $\widetilde{\mathcal{C}_{\mathrm{sa}}}$ at $\alpha \in \mathrm{Spec}_{\mathrm{R}}(K[V])$ is a local ring (not necessarily henselian) having $k(\alpha)$ as residue field.

Note. In line with the idea of conveying geometrical intuition, we have only accounted for the sheaf of Nash functions in the geometric case. By adding another layer of abstraction (an abstract description of Nash functions due to Artin and Mazur) it is possible to define a sheaf $\mathcal{N}_A$ of "Nash" functions over $\mathrm{Spec}_{\mathrm{R}}(A)$, for *any* ring A, having properties akin to those of its geometrical ancestor.

5.8 Idempotency of the Real Spectrum and the Substitution Lemma

A property of the Zariski spectrum of an arbitrary ring, A —endowed with either the structure sheaf $\mathcal{S}_A$ (4.1) or the étale sheaf $\mathcal{S}_A^{et}$ (4.1)—is that the ring A can be recovered from (or *represented* by) the sheaf ($\mathcal{S}$, say): A is isomorphic to the ring of global sections of $\mathcal{S}$ (i.e., sections over the whole of $\mathrm{Spec}(A)$).

This is not true of the real spectrum of a variety endowed with the sheaf of Nash functions: there are many more Nash functions than polynomials; cf. Section 3.1.2 (3). A weaker, but nevertheless remarkable, property of the real spectrum of any ring A endowed with the Nash sheaf $\mathcal{N}_A$ (see note above) is *idempotency*: iterating the construction gives back the initial ringed space. More precisely:

[20] The collection $\{\widetilde{U} \mid U \subseteq V(K) \text{ open semi} - \text{algebraic}\}$ is a basis for the spectral topology of $\mathrm{Spec}_{\mathrm{R}}(K[V])$.

Theorem 8 (Roy, 1982, Thm. 5.1) *Let A be any ring and let U be a quasi-compact open subset of* $\mathrm{Spec}_R(A)$. *The locally ringed space* $(U, \mathcal{N}_A(U))$ *is isomorphic to* $\left(\mathrm{Spec}_R\left(\mathcal{N}_A(U)\right), \mathcal{N}_{\mathcal{N}_A(U)}\right)$. □

Remarkably enough, in the geometric case this idempotency theorem is equivalent to a result about Nash functions worth mentioning here. To make sense of it, recall that Nash functions on an open semi-algebraic subset U of, say, $\mathbb{R}^n$, are semi-algebraic, i.e., have a (parametrically $\mathcal{L}$-) definable graph; hence the first-order formula of $\mathcal{L}$ defining a Nash function $f : U \to \mathbb{R}$ makes sense over any real closed extension K of $\mathbb{R}$ and, in fact, defines a function $f^K : U^K \to K$ on the open semi-algebraic set $U^K \subseteq K^n$ defined (in K) by the same formula defining U (in $\mathbb{R}$). One has:

Theorem 9 (The Substitution Lemma; Bochnak-Efroymson) *Let U be an open semi-algebraic subset of* $\mathbb{R}^n$ *and let* $\mathcal{N}(U)$ *be the ring of real-valued Nash functions on U. Let* $\varphi : \mathcal{N}(U) \longrightarrow K$ *be a ring homomorphism into a real closed extension K of* $\mathbb{R}$. *Then,*

(1) $(\varphi(X_1), \ldots, \varphi(X_n)) \in U^K$.
(2) *For* $f \in \mathcal{N}(U)$, $\varphi(f) = f^K(\varphi(X_1), \ldots, \varphi(X_n))$. □

5.9 A Beautiful Application: The Bröcker–Scheiderer Theorem

As an illustration of the kind of geometrical statement that can obtained by use of the real spectrum (combined with other techniques), we state the following remarkable result:

Theorem 10 (Bröcker–Scheiderer; cf. (Scheiderer, 1989)) *Let V be an algebraic variety over a RCF, K, such that* $V(K) \subseteq K^n$. *Then,*

1. *Every basic open semi-algebraic subset of* $V(K)$ *can be represented as the intersection of* at most $\dim(V)$ *sets, each defined by one strict polynomial inequality.*
2. *Every basic closed semi-algebraic subset of* $V(K)$ *(i.e. any finite intersection of sets of the form* $\{x \in V(K) \mid P(x) \geq 0\}$ *with* $P \in K[X_1, \ldots, X_n]$*), can be represented as the intersection of* at most $\frac{1}{2}\dim(V)(\dim(V)+1)$ *sets, each defined by a polynomial inequality* (≥ 0).

These bounds are optimal. □

Besides real spectral techniques, the original proof also used techniques from the abstract theory of quadratic forms. A direct proof was later given by Mahé.

For example, *any* basic open semi-algebraic subset of $\mathbb{R}^2$ is the intersection of *at most two* sets, each defined by a strict polynomial inequality! Of course, one can only expect that the degrees of the polynomials required for such an optimal representation will grow very fast; however, I do not know of any precise result of this kind.

Bibliography

Artin, M., Grothendieck, A., and Verdier, J.-L. (1972). *Théorie des topos et cohomologie étale des schémas (SGA 4)*, volume 270 of *Lecture Notes in Mathematics*. Springer, Berlin.

Atiyah, M. and Macdonald, I. (1969). *Introduction to Commutative Algebra*. Addison-Wesley, Reading, MA.

Basu, S., Pollack, R., and Roy, M. (2006). *Algorithms in Real Algebraic Geometry*, volume 10 of *Algorithms and Computation in Mathematics*. Springer, Berlin, second edition.

Becker, E. (1986). On the real spectrum of a ring and its applications to semi-algebraic geometry. *Bulletin of the American Mathematical Society (NS)*, 15:19–60.

Bochnak, J., Coste, M., and Roy, M. (1998). *Real Algebraic Geometry*, volume 36 of *A Series of Modern Surveys in Mathematics*. Springer, Berlin.

Brakhage, H. (1954). *Topologische Eigenschaften algebraischer Gebilde über einem beliebigen reel-abgeschlossenen Konstantenkörper*. PhD thesis, University of Heidelberg.

Carral, M. and Coste, M. (1983). Normal spectral spaces and their dimensions. *Journal of Pure and Applied Algebra*, 30:227–235.

Coste-Roy, M.-F. (1980). *Spectre réel d'un anneau et topos étale réel*. PhD thesis, Université Paris Nord.

Coste, M.-F. and Coste, M. (1979). Topologies for real algebraic geometry. In Koch, A., editor, *Topos Theoretic Methods in Geometry*, volume 30 of *Various Publications Series*, pages 37–100. Matematisk Institut, Aarhus Universitet, Aarhus.

Coste, M.-F. and Coste, M. (1980). Le spectre étale réel d'un anneau est spatial. *Comptes Rendus de l'Académie des Sciences Paris: Sér A*, 290:91–94.

Coste, M. and Roy, M.-F. (1982). La topologie du spectre réel. In Dubois, D. and Recio, T., editors, *Ordered Fields and Real Algebraic Geometry*, volume 8 of *Contemporary Mathematics*, pages 27–59. American Mathematical Society, Providence, RI.

Dickmann, M. (1985). Applications of model theory to real algebraic geometry; a survey. In Di Prisco, C. A., editor, *Methods of Mathematical Logic*, volume 1130 of *Lecture Notes in Mathematics*, pages 76–150. Springer, Berlin.

Dickmann, M., Schwartz, N., and Tressl, M. (201x). Spectral spaces. in preparation.

Dieudonné, J. (1985). *History of Algebraic Geometry. An Outline of the History and Development of Algebraic Geometry*. Wadsworth Advanced Books, Monterey, CA.

Goldblatt, R. (1979). *Topoi: the Categorial Analysis of Logic*. North-Holland, Amsterdam. Revised Edition 1984; reprinted by Dover.

Johnstone, P. (1977). *Topos Theory*, volume 10 of *London Mathematical Society Monographs*. Academic Press, London.

Knebusch, M. (1984). An invitation to real spectra. In Hambleton, I. and Riehm, C., editors, *Quadratic and Hermitian Forms*, volume 4 of *Conference Proceedings, Canadian Mathematical Society*, pages 51–105.

Lojasiewicz, S. (1964). Ensembles semi-analytiques. mimeographed.

Roy, M.-F. (1982). Faisceau structural sur le spectre réel et fonctions de Nash. In *Géométrie Algébrique Réelle et Formes Quadratiques. Proceedings, Rennes, 1981*, volume 959 of *Lecture Notes in Mathematics*, pages 406–432. Springer, Berlin.

Scheiderer, C. (1989). Stability index of real varieties. *Inventiones Mathematicae*, 97:467–483.

Shafarevich, I. (1977). *Basic Algebraic Geometry*. Springer, Berlin.

Tarski, A. (1931). Sur les ensembles définissables de nombres réels I. *Fundamenta Mathematicae*, 17:210–239.

Thom, R. (1975). *Structural Stability and Morphogenesis. An Outline of a General Theory of Models*. W. A. Benjamin, Reading, MA. Translation of *Stabilité structurelle et morphogénèse (Essai d'une théorie générale des modèles)*, W. A. Benjamin, 1972.

Chapter 13
Euler's Continuum Functorially Vindicated

F. William Lawvere

Contrary to common opinion, the question "what is the continuum?" does not have a final answer (Bell, 2005), the immortal work of Dedekind notwithstanding. There is a deeper answer implicit in an observation of Euler. Although it has often been dismissed as naive, we can use the precision of the theory of categories to reveal Euler's observation to be an appropriate foundation for smooth and analytic geometry and analysis.

Euler observed that real numbers are ratios of infinitesimals. This is not only true but an effective definition of the smooth real continuum; the properties of the smooth continuum, including its canonical map to the less-refined Dedekind continuum, can be expressed in terms of such ratios.

Some of the features of categorical precision are the following. Maps (functions, transformations, etc.) have both definite domains and definite codomains, with operations like restriction to subdomains or expansion of codomains being effected by composition with suitable inclusion maps; such operations change the properties of the map. Equality of maps $A \to B$ can be tested by composing with elements $E \to A$ where E belongs to a class of element-types chosen in a way appropriate to the particular category; the limitation to a bare point $E = 1$ is typically not appropriate (except for the special abstract constant sets that Cantor extracted from mathematics for his particular purposes). It is possible and preferable to assume for basic work in geometry and analysis that the ambient categories are cartesian closed, as was already taken for granted by Euler's teachers: for two spaces X, Y there is a well-determined *map space* M with a structural map $X \times M \to Y$. That structural map has a universal property that implies that the punctual elements of M are in bijective correspondence with maps $X \to Y$ in such a way that the structural map effects evaluation. I use the exponential notation $M = Y^X$.

To realize Euler's vision, I will postulate a space T of infinitesimals with certain properties (whose consistency has been shown many times) and then derive properties of the reals R defined as ratios of T.

F.W. Lawvere (✉)
Professor of Mathematics, State University of New York, Buffalo, NY, USA
e-mail: wlawvere@buffalo.edu

D. DeVidi et al. (eds.), *Logic, Mathematics, Philosophy: Vintage Enthusiasms*, The Western Ontario Series in Philosophy of Science 75,
DOI 10.1007/978-94-007-0214-1_13,

But what are ratios? Much confusion has been caused by the notational presumption that division is an operation on the same footing as addition and multiplication. The slogan "you can't divide by zero" by no means disposes of the issues that arise, especially for variable quantities. (In fact, one can even divide by zero, if it is zero that one wants to divide, getting infinitely many answers.) In algebra, if s is a quantity in a ring A, the result $A[s^{-1}]$ of dividing by s is another ring with a homomorphism

$$A \to A[s^{-1}]$$

typically neither injective nor surjective, because the new ring consists of functions defined on an open subspace

$$\mathrm{spec}(A[s^{-1}]) \hookrightarrow \mathrm{spec}(A)$$

of the domain of the original quantities. (For example if A is a ring of polynomials in one variable, then $A[s^{-1}]$ consists of rational functions whose denominator is a power of s, and the open subspace excludes the zeroes of the polynomial s). This and many other constructions should suggest that disaster can result from naive presumptions that symbols like difference quotients necessarily denote anything unique (Lawvere, 1996).

Because division really refers to the possible existence of inversion to multiplication, we define a (possible) *ratio* to be a general map $A \to B$ and we say, in particular, that "b is divisible by a" if and only if a ratio r exists with $b = ra$:

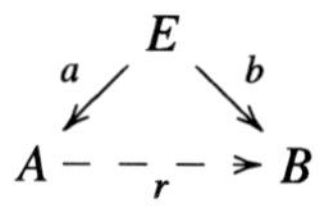

(Thus divisibility is categorically dual to "belonging to" (Lawvere and Rosebrugh, 2003)). For example, these notions apply in a monoid (when $A = B = E$) or in a linear category (where composition is traditionally called multiplication); our special example will be in a nonlinear category, except that the objects involved are so small that all the maps around them are affine-linear.

In the development of Synthetic Differential Geometry (Kock, 2006) over the past 40 years it has usually been assumed that a smooth category contains a line object which actually has the structure of a ring R. Then the subspace D of first-order infinitesimals ($h^2 = 0$) plays a key role. In particular, all the maps $D \to D$ are uniquely represented by elements of R, precisely fulfilling Euler's principle. I show below that reciprocally R can be constructed from a non-coordinatized version T of D, thus achieving a foundation for smooth geometry that is even "radically synthetic" in the sense that all algebraic structure is derived from constructions on the geometric spaces rather than assumed.

There is a direct non-quantitative basis for the necessity of our infinitesimal space T. The most fundamental functor characteristic of smooth and analytic categories is the tangent-bundle functor. Because this functor preserves cartesian products and more, it cries out to be construed as a *representable* functor. By Yoneda's Lemma, a representing object is unique.

The tangent-bundle functor has been (explicitly or implicitly) fundamental in mathematical physics because motion is ubiquitous in the material world. Motion means that a thing is in a certain place and yet elsewhere in the same instant. The generic mathematical model is still the one that involves nilpotent elements; for example, the canonical commutation relation of quantum mechanics is at bottom Leibniz's rule for the derivative. However, other models may be included in the general treatment we give below, in which one might think of T as a generic instant which has a certain point 0 but does not reduce to it.

Therefore we postulate a pointed space T and call the map space X^T the tangent bundle of any space X, with evaluation at the point inducing the bundle map $X^T \to X$. Even the point of T is not really a given structure in the intended examples, for in them $1 \to T$ is unique (but far from being an isomorphism). Thus every actual map $T \to T$ would automatically preserve 0; however, there would still be non-punctual elements $E \to T^T$ of the map space for $E \neq 1$, so we define by pullback

$$\begin{array}{ccc} R & \hookrightarrow & T^T \\ \downarrow & & \downarrow \mathrm{ev}_0 \\ 1 & \xrightarrow{\ 0\ } & T \end{array}$$

the subspace R of *Euler reals*. The object T^T has an intrinsic multiplication arising from composition, and the subspace R is clearly closed under it, so we automatically get "multiplication of reals" as an operation $R \times R \to R$. As often happens, multiplication is more fundamental than addition.

Thus R has the intrinsic structure of a monoid with 0. Moreover, it has always been commutative. To justify that commutativity seems difficult, though intuitively it is related to the tinyness of T, in the sense that even for slightly larger infinitesimal spaces, the (pointed) endomorphism monoid is non-commutative. Note that if we define

$$D_n = \{h \in T^T | h^{n+1} = 0\}$$

then actually $D_n \subset R$, and $D = D_1$ is the object considered in previous Synthetic Differential Geometry work; for that reason a possible isomorphism $D \xrightarrow{\approx} T$ can be thought of as a unit of time (T itself has no intrinsic multiplication). A canonical map $T^T \to R$, retracting the inclusion, is needed to define derivatives and is obtained (together with the commutativity!) by the assumption that the composite of the inclusion followed by the universally commutative quotient monoid is an isomorphism.

Cantor extracted from the cohesive and active world of mathematics a subuniverse of discrete and inert sets (which served not only for cardinal measurements but also as a featureless background in which, dialectically, mathematics could be modeled by Dedekind, Hausdorff, Moore, Frechet, and their twentieth century successors). This contrast can be modeled by a subcategory with a reflector $X \to \pi_0(X)$ satisfying

$$\pi_0(X \times Y) \xrightarrow{\approx} \pi_0(X) \times \pi_0(Y)$$
$$\pi_0(1) = 1$$

(Lawvere, 2007). In particular, the S in the subcategory should satisfy

$$S \xrightarrow{\approx} S^T$$

because these S are spaces of non-Becoming, i.e. spaces in which no motion is possible, not even infinitesimal motion. Typically such a components functor π_0 exists; in particular, any algebraic structure that a space might carry is reflected as a similar structure on its "set" of components.

Proposition 1 $\pi_0(X^T) = \pi_0(X)$ *for all* X *iff* $\pi_0(R) = 1$.

Assume $\pi_0(R) = 1$ (i.e. R is connected). But that leaves many possibilities for $\pi_0(U)$ where $U \subset R$ is the subgroup of invertible elements. The above constructions would also provide a basis for complex-analytic geometry and analysis; in that case we would have $\pi_0(U) = 1$. However the intuition for the real case involves a line which is bi-directional, so that

$$\pi_0(U) = Z_2$$

a multiplicative group of two elements. In all cases, we can consider U_+ defined as the kernel of the natural homomorphism $U \to \pi_0(U)$ (i.e. the component of the identity) as the group of *positive* elements of R.

Further axioms, implying that R has an intrinsic addition, will be discussed below. But first, assuming that R has an addition, we show how to use only a given subgroup P (such as the U_+ defined by using π_0) to derive the structure of an *ordered* rig, that is, the rig R will have moreover a subrig M of non-negative quantities (in general a rig has commutative multiplication and addition satisfying the distributive law, but not necessarily negation). Define

$$A = \{a \in R | a + P \subseteq P\}$$
$$M = \{\lambda \in R | \lambda A \subseteq A\}$$

Proposition 2 *If* P *is a multiplicative subgroup of a rig, then* A *is an additive monoid and hence* M *is a subrig of* R. *If, moreover, 1 belongs to* A, *then* M *is contained in* A *and* P *is contained in* M.

Therefore the relation defined by

$$r \leqslant s \text{ iff } \exists m \in M[r + m = s]$$

has the expected properties of an ordering. The elements h for which

$$0 \leqslant h \,\&\, h \leqslant 0$$

constitute an ideal that contains all nilpotent quantities.

If the ambient category is a topos with generic subobject $1 \to \Omega$ and if we denote by Ω^M the space of order-preserving maps (parameterizing the upclosed parts of M) the natural Dedekind-Yoneda map

$$M \to (\Omega^M)^{\mathrm{op}}$$

is given by the ordering on M itself. This map is neither injective nor surjective; it maps into the inf-completion $V \hookrightarrow (\Omega^M)^{op}$ that plays the role of the nonnegative semi-continuous Dedekind reals and serves as the natural recipient for metrics in the category. (Similarly, R itself can be order-completed, but by using M in a two-sided way.)

There are two ways to insure that R has unique addition: one involves integration and the other differentiation.

1. Integration (or distribution theory) concerns smooth linear functionals. Experience with Taylor series motivates the presumption that for suitable smooth spaces, homogeneity should imply linearity. Thus if we define

$$\Phi(X) = \mathrm{Hom}_R(R^X, R) \hookrightarrow R^{(R^X)}$$

as the space of those functionals φ satisfying just the multiplicative condition

$$\varphi[\lambda f] = \lambda\varphi[f]$$

for λ in R, then the assumed extensivity property of general integration

$$\Phi(X + Y) = \Phi(X) \times \Phi(Y)$$

specializes to

$$\Phi(2) = \Phi(1)^2.$$

From that, $+$ emerges as the unique homogeneous map $R \times R \to R$ which becomes the identity when restricted to both 0-induced axes $R \to R \times R$.

2. The other route from multiplication to addition goes via trivial Lie algebras. The kernel

$$\mathrm{Lie}(R) \hookrightarrow R^T \to R$$

has a binary operation induced from multiplication since the evaluation/projection is a homomorphism. However this operation is usually called addition; the space of endomorphisms of Lie(R) for that operation is a rig that contains (the right action of) R as a multiplicative sub-monoid, so that if we postulate that R exhausts the whole endomorphism space, then R inherits a canonical addition.

The object T is often assumed to be an ATOM; this can be read: "amazingly tiny objectified motion." The objectified motion reflects the intuition that elements of the tangent bundle are arrows obeying differential laws of motion, etc. The qualification "amazingly tiny" refers to the important property (not used in the above discussion) that T is *tiny* with respect to the whole category. This tinyness is partly expressed by (Grothendieck's notation) $f^!$, the surprising additional right adjoint that some topos morphisms f have. The morphism in question is the essential one having $f^*(X) = X^T$, but the additional operation $f^!(Y) = Y^{1/T}$ leads to the representability of differential forms, of laws of motion, and of still unexplored higher infinitesimal constructions. Tinyness is relative to the universe, as crudely measured by the amazing right adjoint:

Proposition 3 *Given any object T in any Grothendieck topos ϵ, then ϵ is a subtopos of a larger one ϵ' where T becomes tiny. But conversely, the smaller ϵ may not be closed under $(\ )^{1/T}$, so that T is typically not tiny with respect to the smaller container ϵ itself.*

Tinyness permits a certain compromise with Robinson's idea that infinitesimal constructions should preserve logic; we must restrict ourselves to geometric logic (also known as coherent, positive, or dynamic logic).

Proposition 4 *If T is tiny and if A is any object equipped with a structure described in geometric logic, then A^T enjoys the same structure with the same geometric properties, as is seen by composing with the morphism to the classifying topos for the structure theory in question.*

Most of the known models for the above discussion are "infinitesimally generated" in the sense that starting from T, all objects of the topos (such as the algebra of operators on Hilbert space) are obtained by the functorial operations of exponentiation, limits, and colimits. Such is the remarkable blossoming of Euler's principle.

Bibliography

Bell, J. L. (2005). *The Continuous and the Infinitesimal—In Mathematics and Philosophy.* Polimetrica, Milan.

Kock, A. (2006). *Synthetic Differential Geometry.* London Math Society Lecture Notes. Cambridge University Press, Cambridge, revised edition.

Lawvere, F. W. (1996). Unity and identity of opposites in calculus and physics. *Applied Categorical Structures*, 4:167–174.

Lawvere, F. W. (2007). Axiomatic cohesion. *Theory and Application of Categories*, 19:41–49.

Lawvere, F. W. and Rosebrugh, R. (2003). *Sets for Mathematics.* Cambridge University Press, Cambridge.

Chapter 14
Natural Numbers and Infinitesimals

J.P. Mayberry

1 Gödel on Nonstandard Analysis

In the introduction to the second edition of his *Non-standard Analysis* (Robinson, 1996) Abraham Robinson included some brief, and, on the face of it, deeply puzzling, remarks by Gödel on the significance of Robinson's theory. In particular, Gödel makes the surprising claim that

> ... there are good reasons to believe that nonstandard analysis, in some version or other, will be the analysis of the future.

The qualifying phrase "in some version or other" is surely necessary here, for as it is conventionally practiced nonstandard analysis is logically parasitic on the standard analysis of Weierstrass, Dedekind, and Cantor, and is perhaps best regarded as merely a special logical *technique* for discovering proofs of standard theorems about the conventional real numbers.

In any case, Gödel goes on to make a very peculiar observation indeed:

> Arithmetic starts with the integers and proceeds by successively enlarging the number system by rational and negative numbers, irrational numbers, etc. But the next quite natural step after the reals, namely the introduction of infinitesimals, has simply been omitted.

But it has been omitted for the very best of reasons, namely, to avoid out-and-out contradiction. For, as is well-known, once the irrational numbers have been introduced in the conventional way—using Dedekind's Cut Axiom, for example—it is a *theorem* that there *are* no infinitesimals. Given any two positive real quantities $0 < a < b$, there is a *finite* multiple of the smaller which exceeds the larger:

$$b < \overbrace{a + a + \cdots + a}^{k \text{ summands}}$$

It is simply inconceivable that Gödel was not aware of all this. Yet he goes on to say

J.P. Mayberry (✉)
Research Fellow in the Department of Philosophy, University of Bristol, Bristol, UK
e-mail: J.P.Mayberry@bristol.ac.uk

D. DeVidi et al. (eds.), *Logic, Mathematics, Philosophy: Vintage Enthusiasms*, The Western Ontario Series in Philosophy of Science 75,
DOI 10.1007/978-94-007-0214-1_14, © Springer Science+Business Media B.V. 2011

> I think, in coming centuries it will be considered a great oddity in the history of mathematics that the first exact theory of infinitesimals was developed 300 years after the invention of differential calculus.

Surely on the face of it this is not odd at all. On the contrary, to most mathematicians the axioms of the theory of complete ordered fields seem self-evidently true of the real numbers, especially if completeness is postulated in the form of Dedekind's Cut Axiom, which is simply an analytic formulation of the familiar fact that two lines intersect in a point.

But these axioms preclude the introduction of infinitesimals in any natural way, and that is why Robinson himself was forced to introduce them by the back door, so to speak, by exploiting the weakness of first order logic as an instrument of definition.[1]

Gödel's final remark, however, is perhaps the most puzzling of all:

> I am inclined to believe that this oddity has something to do with another oddity relating to the same span of time, namely the fact that such problems as Fermat's, which can be written down in ten symbols of elementary arithmetic, are still unsolved 300 years after they have been posed. Perhaps the omission mentioned is largely responsible for the fact that, compared to the enormous development of abstract mathematics, the solution of concrete numerical problems was left far behind.

What could the notorious intractability of simply stated problems in elementary number theory—Fermat's Theorem,[2] Goldbach's conjecture, the twin prime hypothesis—have to do with the supposed "oversight" of not introducing infinitesimals in defiance of a *theorem* that says that infinitesimals don't exist?

These remarks, understood in the conventional way, are, on the face of it, simply absurd. Moreover, since it is, surely, highly unlikely that Gödel is thinking here of the use of Robinson's own brand of nonstandard analysis to attack these problems in natural number arithmetic, we are left with the obvious question: whatever did Gödel have in mind when he made them—or Robinson when he included them in his Preface, come to that?

Surely it is clear that if, as Gödel suggests, infinitesimal analysis is to be the "analysis of the future," it would have to be a *natural* theory rooted in a clear vision of new fundamental principles, not merely a technical device exploiting the definitional weakness of first order logic. In particular, we should need a treatment of infinitesimal analysis that is not parasitic on the traditional, nineteenth century concept of real number, as Robinson's nonstandard analysis unquestionably is.

But when we try to imagine how we might establish Gödel's "analysis of the future," we soon realize that we shall be forced to abandon some our most deeply rooted mathematical convictions. For example, we could no longer assume that we

---

[1] In particular, he makes essential use of the impossibility of giving a first order characterisation of the general concept of *finiteness*.

[2] Of course this been proved by Andrew Wiles. But Wiles' proof (considered along with the other proofs that connect it to the Fermat conjecture) is of a length and complexity that mocks any expectations grounded on the simplicity of the result proved. And the other famous problems remain unsolved.

can always approximate a real number by a strictly increasing ω-sequence of rationals. For if every real r is the limit of such a sequence there is no room for any quantities $r - \epsilon$ strictly smaller than r but separated from it by an infinitesimal distance $\epsilon > 0$.

Indeed, it seems altogether likely that the price we should have to pay for a *natural* theory of infinitesimals, is to abandon our conviction that the natural number system is *absolute*, that is to say, we should have to entertain the possibility of there being inequivalent, but equally "natural," ways of proceeding to infinity. And, of course, if there are inequivalent natural number systems with different properties, then it may well be that certain classical problems in number theory (e.g., the twin prime hypothesis) may not be *well posed* in the sense of having determinate answers: they may have different answers in different number systems.

Now this does suggest how Gödel's brief remarks might be interpreted as making sense. But how could we set about developing his "analysis of the future"? I shall return to this question in Section 7, but first I shall look at issues raised by Gödel's speculations in natural number arithmetic.

2 Natural Numbers or Finite Sets?

Is there a mathematically natural point of view from which it makes sense to call into question the uniqueness of the system of natural numbers? I must emphasize the crucial importance of the word "natural" here, because *of course* it is possible to exhibit non-isomorphic models of formal first order arithmetic.

But from the conventional standpoint, given such an example, at least one of the structures in question must be nonstandard and thus highly *un*natural. And in any case, the Compactness Theorem, on which the construction of nonstandard natural number system depends, itself rests, like so much else in conventional mathematics, on the assumption that the natural number sequence is absolute.

More exactly, what is required for the usual proofs of Compactness is that the notions of *formula*, *proof*, etc. be unequivocal. But the definitions of these notions rest on the notion of *finite sequence* which, in the conventional treatment, depends on the notion of natural number. Thus we need an unequivocal, "absolute" notion of natural number to do formal syntax and formal proof theory in the conventional manner.

On the most naive level there is the conviction, surprisingly widespread among mathematicians, that the natural numbers do not even *require* definition. After all, we all know how to begin the natural number sequence

$$0, 1, 2, 3, 4, 5, 6, 7, 8, 9, 10, 11, 12, \cdots$$

and how to continue it from any point

$$\cdots, 539646, 539647, 539648, 539649, 539650, \cdots$$

What more is there to the natural number sequence other than that it is *generated* from zero by successive *iterations* of the operation of passing from a number to its immediate successor? Moreover, once you have mastered the art of intransitive counting you have acquired all the skills necessary for counting out collections of things. That will enable you to grasp the very meaning of the word "finite": a collection is *finite* if it can be counted out.

But such an *operationalist* account of the concepts of *finite plurality* and *natural number*[3] is manifestly circular: a *finite* plurality is one that can be correlated, one-to-one, with a proper initial segment of the natural number sequence; the natural number sequence, however, consists of those terms that can be reached from zero by iterating the successor operation a *finite* number of times. Anyone familiar with the works of Frege and Dedekind on this question will instantly realise how utterly unilluminating and inadequate such an account is.

The conventional view of theoretical arithmetical rests on the idea that, Frege and Dedekind notwithstanding, the principles of proof by induction and definition by recursion do not, in fact, even *require* justification: they follow immediately from the defining idea that the natural numbers are the successors of zero.

But the idea that the possibility of repeating or iterating an argument or a calculation "forever" justifies *on its own* proof by induction or definition by recursion is simply a fallacy, as both Frege and Dedekind have shown: I propose to call it the *Sorites fallacy*.[4]

It might, at first, appear that the problem here is to explain what natural numbers are, that is, to give an account of what *things* our number words and numerals *name* or *stand for*. But a careful reading of both Frege and Dedekind shows that both of them saw that giving a logically rigorous account of the intuitive idea of finite iteration is the central problem here.

The beginning of wisdom is to realise that there simply are no such *things* as "natural numbers," that natural numbers *as "mathematical objects"* are illusions, non-entities, mere artefacts of our notation, reified and alienated products of our counting and calculating procedures, and that, consequently, to devise a theory of what "they" are *as particular objects* is utterly otiose and, indeed, productive of quite unnecessary confusion.

But what, then, are the facts of arithmetic facts *about*? Not facts about calculating and counting *as activities*, for calculating and counting, as activities, although they may *disclose* the elementary facts of arithmetic, do not actually constitute its subject matter. The facts of arithmetic are facts about finite pluralities (sets), about what our ancestors called "numbers" (*arithmoi* in Greek) until the mathematicians of the seventeenth century changed the meaning of "number," giving it (more or less) its present day meaning.

---

[3] On the conventional view, an account of the natural numbers must of necessity include an account of finiteness, since the natural numbers, after all, are intended, among other things, to name the modes of finite plurality.

[4] The Sorites fallacy is a fallacy even from an infinitary point of view, as Dedekind carefully explains in his "Letter to Kefferstein" (Dedekind, 1967).

Here is the definition of "number" Newton gave in his *Universal Arithmetic*:

> By a Number we understand not so much a Multitude of Unities, as the abstracted Ratio of any Quantity to another Quantity of the same Kind which we take for Unity.

Notice that Newton actually tells us here that his definition is intended to replace the traditional definition of Euclid:

> A number (*arithmos*) is a plurality composed of units

where

> A unit (*monas*) [in a plurality] is whatever it is in comparison with which each of the entities [composing the plurality] is called one.[5]

Moreover, Newton's definition extends the science of arithmetic far beyond its original compass; for it is what allows us to speak of integral, rational, and irrational *numbers*, which form the subject matter of a greatly expanded arithmetic.[6]

Of course the introduction of this altered and expanded version of number was an indispensable step in the development of the calculus and hence of mathematical physics.[7] But the extra power and convenience that this new concept of number provided was purchased at the cost of simplicity, and indeed even intelligibility, in the foundations of mathematics.

In elementary arithmetic it meant replacing finite sets with natural numbers as the subject matter of the science. On the face of it this represents a considerable decrease in clarity. For we have many ordinary, concrete instances of finite sets, but natural numbers are "abstract objects" with individual natures that are opaque to our understanding.

The books now on my desk are the elements of a set (or, in Euclid's terminology, the units of a number) which, as it happens, is a five. But what is the number five *itself*? And, of course, we mustn't allow our facility in employing the familiar rules for counting and calculating with decimal numerals fool us into thinking that we are familiar with the supposed entities that those numerals "name."

In order to understand what is really going on in elementary arithmetic we need to return to the original Greek conception of the subject. We get a modernised version of classical Greek arithmetic by formulating it as the theory of finite sets. This *Euclidean arithmetic* is essentially conventional (i.e., Zermelo–Fraenkel) set theory but with the Axiom of Infinity replaced by the assertion that every set is finite, that is to say, no set can be placed in one-to-one correspondence with any of its proper subsets. This amounts to adopting Euclid's Common Notion Six: *The whole is greater than the part.*

---

[5] *Elements*, Book VII. Thus in a number of horses the unit is a horse, in a number of numbers, a number, etc.

[6] I shall develop a theory of these Newtonian numbers in Sections 6 and 7.

[7] There was no reason to call these abstracted ratios "numbers," however: "abstract quantities" or "real quantities" would have done as well, without altering the traditional meaning of "number."

We can make this identification of Euclidean arithmetic with finitary Zermelo-Fraenkel set theory because the central axioms of the Zermelo-Fraenkel theory—Extensionality, Pair Set, Power Set, Union, Comprehension, and Replacement—all acquire an aura of self-evidence when applied to *finite* sets.[8]

Of course when I say that Euclidean arithmetic just *is* Zermelo-Fraenkel set theory (with the obvious finitary modifications) I am not referring to the familiar first order formal theory that goes by that name, but to the intuitive set-theoretical ideas that underlie and inform it.[9]

3 Euclidean Arithmetic

In Euclidean arithmetic, then, we take the notion of (*finite*) *set*, rather than the notion of *natural number*, as basic. In this theory a set is a finite, extensional plurality composed of definite, well-defined things called its members.[10] A set is itself definite and well-defined and therefore is eligible to occur as a member of other sets. Following Frege I shall call sets, and the things that can be members of sets, *objects*.

To say that sets and their members are "well-defined" is not to say that they have verbal definitions, but rather that they are, so to speak, self contained in the sense of being sharply, indeed, absolutely, distinguished from one another. Given objects a and b either they are identical ($a = b$) or they are distinct ($a \neq b$), *tertium non datur*.

To say that sets are "extensional" pluralities is to say that what a set is—its essence—is determined *only* by what its members are and not by how (or, indeed, whether) we define it. Thus there is nothing to the set whose members are the distinct objects $x, y, \cdots, z$ apart from the fact that its members *are* $x, y, \cdots, z$, and, of course, that, together, $x, y, \cdots, z$ are finite in multitude.

Euclidean arithmetic is a *finitary* theory of finite sets. This means that we must apply constructive principles of reasoning to propositions that require *global* quantification—quantification over the totality composed of all sets and their members.[11]

I shall call this fundamental requirement *Brouwer's Principle*. It incorporates the the idea that the difference between infinite pluralities (*proper classes* or, as I prefer

---

[8] The same is essentially true of Foundation too, but that requires that it be reformulated. See [§§ 4.11 and 8.6](Mayberry, 1994).

[9] Surely it is obvious that no formal, first order theory, finitary or otherwise, can serve as a *foundational* theory for arithmetic.

[10] A plurality that is not finite I shall call an *infinite species*, for example the infinite species of all three element sets.

[11] *Local* quantifiers, whose domains are (finite) sets can be defined using Comprehension

$$(\forall x \in S)\Phi(x) \Leftrightarrow_{\text{def}} \{x \in S : \Phi(x)\} = S$$

These quantifiers satisfy the usual (classical) laws.

to say, *infinite species*) and finite ones (*sets*) is not just a matter of size but actually a matter of *logic*.

In particular, if A is a proposition that can be expressed only with the help of unrestricted quantification over the elements of an infinite species, then the *Law of the Excluded Middle* fails for A and we cannot infer that *either A is true or A is false*. This means that Brouwer's Principle requires us to make a fundamental distinction between *local functions and relations*, which are identified with (finite) sets of ordered pairs in the usual way, and *global functions and relations*, like the power set function $\lambda x \mathrm{P}(x)$ and the membership relation $\lambda xy[x \in y]$, which are defined eveywhere.[12]

For the obvious definition of identity for global functions σ and τ

$$\sigma = \tau \text{ if, and only if, } \textit{for all objects } x, \ \sigma(x) = \tau(x)$$

employs an unbounded universal quantifier over the infinite species of all objects. This means that the Law of the Excluded Middle fails here, that is to say, on this definition we cannot assert, in general, that either $\sigma = \tau$ or $\sigma \neq \tau$, and, in consequence, *global functions cannot be accounted objects*. In contrast to local functions, which are just (*ex*tensional) sets of ordered pairs, they are essentially *in*tensional in character.[13]

Brouwer's Principle also dictates that in using definition by *comprehension* to define a subset $\{x \in S : \Phi(x)\}$ of a set S, the comprehending property Φ cannot require, directly or indirectly, global quantification in its expression. This corresponds to Zermelo's requirement that Φ be *definite*.

But this entails that the principle of induction, *even as restricted to the terms of a finite linear ordering*, can be established in general only for properties Φ which are "definite" in the sense just described.

Thus suppose that

$$L = [\mathrm{First}_L, \cdots, x, \mathrm{Next}_L`x, \cdots, \mathrm{Last}_L]$$

is a (local) linear ordering. Induction along L for the property Φ can be formulated as follows:

From the premises

(i) $\Phi(\mathrm{First}_L)$
(ii) $\Phi(x) \Rightarrow \Phi(\mathrm{Next}_L`x)$, *for all x in the field of L except* $x = \mathrm{Last}_L$.

infer that $\Phi(x)$, for all x in the field of L.

---

[12] A global function, σ (of one argument) is defined by unambiguously specifying, for each object x, the object $\sigma(x)$ that is the *value* of σ at the *argument* x. Similarly, a global relation Φ (of one argument) is defined by unambiguously specifying, for each object x, the truth value, true or false, of the proposition $\Phi(x)$ at the argument x. Of course global functions and relations of two or more arguments can be defined in a similar way.

[13] A similar argument applies to infinite species, so that they cannot be accounted objects either.

This can be proved, but only when Φ is a *definite* property (in Zermelo's sense) which can be formulated without employing unbounded quantifiers. The idea that we can start from (i) and proceed by a number of inferences using (ii) equal to the number of terms in L to infer that $\Phi(x)$, for all x in the field of L, is a clear instance of what I called the Sorites fallacy in Section 2.

This restriction on induction, in turn, entails an analogous restriction on definition by recursion along a linear ordering. Thus we can prove that given a finite linear ordering $L = [\text{First}_L, \cdots, \text{Last}_L]$, a *local* function $g : S \to S$ from a set S to itself, and an aritrarily chosen $a \in S$, there is a unique local function $f : \text{Field}(L) \to S$ defined on the field of L with values in S which satisfies the recursion equations

$$\begin{aligned} f(\text{First}_L) &= a \\ f(\text{Next}_L(x)) &= g(f(x)), \text{ for all x in Field}(L) \text{ except Last}_L. \end{aligned}$$

But for an arbitrarily chosen a and a *global* function γ, we cannot, in general, establish the existence of a local function f defined on the field of L and satisfying analogous recursion equations

$$\begin{aligned} f(\text{First}_L) &= a \\ f(\text{Next}_L(x)) &= \gamma(f(x)), \text{ for all } x \text{ in Field}(L) \text{ except Last}_L. \end{aligned}$$

This discloses a profound difference between *local recursion*—recursion along a finite linear ordering with respect to a *local* function $g : S \to S$—and *global recursion*—recursion along a finite linear ordering with respect to a *global* function γ, even though there is a superficial resemblance between them.

Naively, if f is defined by *local* recursion with respect to $g : S \to S$ along $L = [0_L, 1_L, 2_L, \cdots, k_L]$ starting at $a \in S$ then

$$\begin{aligned} f(0_L) &= a \\ f(1_L) &= g(a) \\ f(2_L) &= g(g(a)) \\ &\vdots \\ f(k_L) &= \overbrace{g(g(\cdots g(a)\cdots))}^{k_L \ gs} \end{aligned}$$

And, by the same token, if f is defined by *global* recursion with respect to γ along $L = [0_L, 1_L, 2_L, \cdots, k_L]$ starting at a then

$$
\begin{aligned}
f(0_L) &= a \\
f(1_L) &= \gamma(a) \\
f(2_L) &= \gamma(\gamma(a)) \\
&\vdots \\
f(k_L) &= \overbrace{\gamma(\gamma(\cdots\gamma(a)\cdots))}^{k_L\ \gamma s}
\end{aligned}
$$

Thus it might appear that recursion corresponds, in each case, to starting at a and *iterating* the function with respect to which the recursion is carried out *as many times as there are terms in the linear ordering* L.

Of course we can iterate both local and global functions *syntactically*, so to speak:

$$\overbrace{g`g`g`\cdots g`a}^{\mathbf{n}}$$

or

$$\overbrace{\gamma(\gamma(\gamma(\cdots\gamma(a)\cdots)))}^{\mathbf{n}}$$

where **n** specifies a number of expressions that we can actually write down.[14]

In any case, this licence to write down longer and longer expressions does not provide a *theoretical* justification for recursion along an arbitrary (finite) linear ordering. On the contrary, to suppose that it does is to commit the Sorites fallacy: it is, in effect, to incorporate into our theory, surreptitiously and indirectly, the assumption that the number system is absolute, thus begging the very question at issue.

In Euclidean arithmetic the idea of the iteration of a function must be given a proper, set-theoretical definition, and cannot be taken as a fundamental datum, as a notion just "given" prior to the rigorous development of mathematics.[15] This

---

[14] Thus $\mathbf{n} = 1729$

$$\overbrace{\gamma(\gamma(\gamma(\cdots\gamma(a)\cdots)))}^{\mathbf{1729}}$$

is a possible value whereas $\mathbf{n} = 10^{1729}$

$$\overbrace{\gamma(\gamma(\gamma(\cdots\gamma(a)\cdots)))}^{\mathbf{10^{1729}}}$$

is not.

[15] Indeed, this is true even in conventional infinitary mathematics, as Dedekind pointed out so eloquently in (Dedekind, 1967).

has profound consequences for the theory of simply infinite systems in Euclidean arithmetic, as I shall explain in Section 5.

4 Arithmetical Functions and Relations

In Euclidean arithmetic, there are two ways of doing conventional "natural number" arithmetic. The first of these is to concentrate theoretical attention on global functions and global relations that are *arithmetical* in the sense I am about to define. The second is to follow Dedekind in the attempt to define "natural numbers" directly using the concept of a simply infinite system. This I shall describe in section 5.

A global function is *arithmetical* if the cardinality of its value depends only on the cardinalities of its arguments. Thus if φ is, say, a binary global function, it is arithmetical if, and only if, for all sets S, T, U and V,

$$S \cong_{\mathrm{c}} U \text{ and } T \cong_{\mathrm{c}} V \Rightarrow \varphi(S, T) \cong_{\mathrm{c}} \varphi(U, V)$$

(where "$\cong_{\mathrm{c}}$" means "equal in cardinality to" and is defined in the usual way).

Similarly, a binary relation is arithmetical if its truth value at given arguments depends only upon the cardinalities of those arguments.

There are global functions and relations arithmetical in this sense corresponding to the familiar basic functions and relations of conventional natural number arithmetic:

(i) Power set:

$$x \mapsto \mathrm{P}(x)$$

(This corresponds to the natural number function $x \mapsto 2^x$.)

(ii) Successor:

$$x \mapsto x \cup \{x\}$$

(This corresponds to the natural number function $x \mapsto x + 1$.)

(iii) Addition (disjoint union):

$$x, y \mapsto (x \times \{\emptyset\}) \cup (y \times \{\{\emptyset\}\})$$

(iv) Multiplication (Cartesian product):

$$x, y \mapsto x \times y$$

(v) Exponentiation:

$$x, y \mapsto \{f \in \mathrm{P}(y \times x) : f \text{ is a function from } y \text{ to } x\}$$

(vi) Bounded sums:

$$\sum_{x=0}^{y} \varphi(x)$$

This is arithmetical if φ is.

(vii) Bounded products:

$$\prod_{x=0}^{y} \varphi(x)$$

This is arithmetical if φ is.[16]

(viii) Cardinal equivalence and ordering:

$$x \cong_{\mathrm{c}} y \text{ and } x <_{\mathrm{c}} y$$

(ix) Bounded arithmetical quantifiers:

$$(\forall x <_{\mathrm{c}} S)\Phi(x) \text{ and } (\exists x <_{\mathrm{c}} S)\Phi(x)$$

These are arithmetical if Φ is.

We can express all the conventional equations and inequalities of natural number arithmetic by using cardinal equality ($\cong_{\mathrm{c}}$), cardinality inequality ($<_{\mathrm{c}}$ and $\leq_{\mathrm{c}}$), and arithmetical functions.

What correspond to "natural numbers" here are not objects but infinite species, in fact, the cardinality classes, that is to say, the equivalence classes under the global relation $\cong_{\mathrm{c}}$ of cardinal equivalence.

Arguments and definitions in Euclidean arithmetic that employ these arithmetical notions are constructive in the strongest conceivable sense: for from the Euclidean standpoint we do not regard a "number theoretic function" $n \mapsto \varphi_{\text{number}}(n)$ as legitimately *defined* unless it is possible to exhibit a well-defined, set-theoretical operation φ_{set} which is arithmetical, and which, when applied to a set S of size n, yields a set, $\varphi_{\text{set}}(S)$, of size $\varphi_{\text{number}}(n)$.

We can use these arithmetical global functions and relations to show that a suitable formalised version of Euclidean arithmetic is equivalent to the theory usually designated $I\Delta_0 + exp$, which is natural number arithmetic based on addition, multiplication and exponentiation, but with the induction schema restricted to formulas all of whose quantifiers are bounded ($(\forall x < t)$ or $(\exists x < t)$, for some term t).[17]

---

16 See (Mayberry, 1994, § 9.1) for the set-theoretical definition of this function and the bounded sum function.

17 This was established by Vincent Homolka in (Homolka, 1983).

In this sense, then, we can regard Euclidean "natural number" arithmetic as something with which we are familiar, even though, strictly speaking, it has no natural numbers. But we must not be too hasty in assimilating it to conventional natural number arithmetic, as it contains surprising and important new features which I am about to disclose.

5 Simply Infinite Systems in Euclidean Arithmetic

The alternative to the theory of arithmetical global functions and relations given in Section 4, is to follow the example of Dedekind (1893) and develop a version of his theory of *simply infinite systems* suitable to Euclidean arithmetic. Dedekind intended his simply infinite systems to be concrete versions of the natural number system.

His theory is an attempt to make mathematically rigorous the naive idea that the natural numbers are generated from zero by successive iterations of the successor function $x \mapsto x + 1 \ (= \sigma(x))$

$$0,\ \sigma(0),\ \sigma(\sigma(0)),\ \sigma(\sigma(\sigma(0))),\ \cdots$$

Every term of the sequence can be obtained by a finite number of iterations of the successor function applied to 0, and, conversely, every term that can be obtained by a finite iteration of the successor function applied to 0 belongs to the sequence.

In fact, Dedekind allowed a simply infinite system to start with an arbitrary element z and an arbitrary successor function σ, provided that σ never assumed the value z nor repeated itself. But, as Dedekind realised, the central problem here is to give a rigorous, mathematical account of the *finite iteration* of a function, that most difficult of mathematical notions whose mysteries we discreetly hide behind the three dots of ellipsis and the Latin phrase *ad infinitum*.

Dedekind's solution to this problem—his theorem on definition by recursion (§126 of (Dedekind, 1893))—makes essential use of a powerful set-theoretical assumption, namely, the existence of a transfinite set S with a power set $\mathrm{P}(S)$.[18] Such infinitary resources are not available to us in Euclidean arithmetic, and therefore, as we shall see, *Dedekind's theorem on the uniqueness up to isomorphism of simply infinite systems*[19] *does not hold in that finitary theory.*

Although in Euclidean arithmetic, where all sets are finite, we cannot carry out Dedekind's original analysis in full, we can nevertheless investigate the problem of finite iteration[20] and thereby obtain a finitary analogue of his notion of simply infinite system. Let me begin by laying down two key definitions.

[18] This means that the set-theoretic machinery required to establish the existence of a unique (up to isomorphism) natural number sequence confronts us with Cantor's intractable *continuum problem.*

[19] §132 of (Dedekind, 1893).

[20] In Euclidean arithmetic it is the finite iteration of *global* functions that is problematic, as I explained in Section 3.

Definition 1 Let σ be a global function of one argument, and a an arbitrary object. Then a linear ordering L is *generated from a by σ* if L is the empty ordering, [ ], or the following conditions obtain:

(i) $\text{First}_L = a$.
(ii) For all $x \in \text{Field}(L)$ except Last_L, $\text{Next}_L(x) = \sigma(x)$.

Notice that we are *defining* the notion of "generation" or "finite iteration" here, not *appealing* to it. This is of crucial foundational significance since it is this "static," purely set-theoretical definition of "generation" that allows us to develop the theory of simply infinite systems (natural number systems) without committing the Sorites fallacy.

The empty ordering [ ] and the one-termed ordering $[a]$ are always generated from a by σ. If $\sigma(a) = a$ these are the only orderings generated. Otherwise σ generates the further orderings

$$[a, \sigma(a)], [a, \sigma(a), \sigma(\sigma(a))], [a, \sigma(a), \sigma(\sigma(a)), \sigma(\sigma(\sigma(a)))], \cdots$$

until an ordering $[a, \sigma(a), \sigma(\sigma(a)), \sigma(\sigma(\sigma(a))), \cdots, k]$ is reached for which

$$\sigma(k) \in \{a, \sigma(a), \sigma(\sigma(a)), \sigma(\sigma(\sigma(a))), \cdots, k\}$$

If this never happens we say that σ generates a *simply infinite system* from a.

But this kind of talk about "generation" as if it were a temporal process of some sort is merely metaphorical, and I employ the metaphor merely to hint at what I am trying to do. In fact, I need to lay down a precise, mathematically acceptable definition.

Definition 2 (Simply Infinite System) Let σ be a global function of one argument, and a an arbitrary object. Then σ *generates a simply infinite system from a*, if for all non-empty linear orderings L,

$$\sigma \text{ generates } L \text{ from } a \Rightarrow \sigma(\text{Last}_L) \notin \text{Field}(L)$$

Notice that the proposition σ *generates a simply infinite system from a* employs a global quantifier in its definition. In particular, it is a Π_1 proposition since its expression requires an initially placed global universal quantifier.

In Dedekind's original definition a simply infinite system is essentially identified with the infinite sequence of *terms*

$$a, \ \sigma(a), \ \sigma(\sigma(a)), \ \sigma(\sigma(\sigma(a))), \ \cdots$$

The corresponding identification in Euclidean arithmetic is with the infinite *species* of *initial segments*

$$[\,],\ [a],\ [a,\sigma(a)],\ [a,\sigma(a),\sigma(\sigma(a))],\ [a,\sigma(a),\sigma(\sigma(a)),\sigma(\sigma(\sigma(a)))],\ \cdots$$

In the latter case we must use segments rather than terms as "numbers" since we need to know the predecessor of a number and it is not always possible to define a global inverse to a global injection σ. Thus each "number" in a simply infinite system is a linear ordering which stands for the cardinality of its own field.

If σ generates a simply infinite system from the initial term a, let us identify the simply infinite system with the species, $\mathcal{N}_{\sigma,a}$ of linear orderings that are generated from a by σ.

Let me give some examples:

(i) *The* ***von Neumann*** *simply infinite system* $\mathcal{VN}$:

(a) $\sigma(x) = x \cup \{x\}$
(b) $a = \emptyset$

(ii) *The* ***Zermelo*** *simply infinite system* $\mathcal{Z}$:

(a) $\sigma(x) = \{x\}$
(b) $a = \emptyset$

(iii) *The* ***Cumulative Hierarchy*** *simply infinite system* $\mathcal{CH}$:

(a) $\sigma(x) = x \cup \mathrm{P}(x)$
(b) $a = \emptyset$

(iv) *The* ***Ackermann*** *simply infinite system* $\mathcal{ACK}$:

(a) $\sigma(x) = y$, where y is the set whose **Ackermann** code is one greater than the Ackermann code of x.[21]
(b) $a = \emptyset$

Of course in Dedekind's original theory we can prove that all simply infinite systems are isomorphic. But it is perhaps the most interesting feature of Euclidean arithmetic that this is no longer the case. In Euclidean arithmetic we have simply infinite systems of differing lengths and with different closure properties with respect arithmetical functions.

This is so strange, and so foreign to the conventional conception of natural number, that I must take care to formulate these claims with mathematical precision.

After all, it seems an outright contradiction to claim that it is possible to have natural number systems of differing lengths. Indeed, given any two simply infinite systems, $\mathcal{N}$ and $\mathcal{M}$, it seems obvious that we can set up a one-to-one correlation between their terms (and hence between the linear orderings which compose them).

---

[21] I include this example because it will be important later. It is, of course, possible to give the successor function of $\mathcal{ACK}$ a set-theoretical definition. For details, see (Mayberry, 1994, Ch. 10, § 10.6).

For let their initial terms, $a_{\mathcal{N}}$ and $a_{\mathcal{M}}$ be correlated, and having correlated n to m, correlate $\sigma_{\mathcal{N}}(n)$ to $\sigma_{\mathcal{M}}(m)$.

The careful reader will perhaps notice that there is a global (and hence illegitimate) recursion being appealed to in this little argument. But, in fact there is something interesting going on here, which leads me to one further example of a simply infinite system.

(v) *Let $\mathcal{N}$ and $\mathcal{M}$ be simply infinite systems. Define a global function τ by*[22]

$$\tau((x, y)) = (\sigma_{\mathcal{N}}(x), \sigma_{\mathcal{M}}(y))$$

Then τ generates a simply infinite system, Inf($\mathcal{N}, \mathcal{M}$) *(the infimum of $\mathcal{N}$ and $\mathcal{M}$), from the initial term $(a_{\mathcal{N}}, a_{\mathcal{M}})$.*

Then it can be shown that Inf($\mathcal{N}, \mathcal{M}$) is shorter than or equal in length to each of $\mathcal{N}$ and $\mathcal{M}$, but is longer than or equal in length to any simply infinite system $\mathcal{N}'$ that is shorter than or equal to each of $\mathcal{N}$ and $\mathcal{M}$ in length – hence its designation as the "infimum" of those two systems.

But we must be careful here, because these key notions of "length" and "closure" must be carefully formulated in a constructive way. Let us start with closure.

Definition 3 (Closure) Let φ be a binary[23] arithmetical global function, let η be a binary global function, and let $\mathcal{N}$ be a simply infinite system. We say that η represents φ in $\mathcal{N}$ if for all $x, y \in \mathcal{N}$

$$\eta(x, y) \in \mathcal{N} \text{ and } \mathrm{Field}(\eta(x, y)) \cong_c \varphi(\mathrm{Field}(x), \mathrm{Field}(y))$$

We say that $\mathcal{N}$ *is closed under the arithmetical global function* φ if we can exhibit a global function η that represents φ in $\mathcal{N}$. This can be expressed symbolically by

$$\mathcal{N} \xrightarrow{\varphi} \mathcal{N} \text{ (via } \eta)$$

To say that $\mathcal{N}$ *is not closed under* φ is to say that it is impossible that η should represent φ in $\mathcal{N}$, no matter what the binary global function η may be.

Similarly, if we want to compare simply infinite systems as to length we must be careful to formulate the concepts to be employed so that they are compatible with constructive reasoning.

Definition 4 (Measures) Let $\mathcal{N}_0$ and $\mathcal{N}_1$ be simply infinite systems, and μ be a unary global function. We say that μ *is a measure for $\mathcal{N}_0$ in $\mathcal{N}_1$* if for all x lying in $\mathcal{N}_0$

---

22 The defining equation defines τ only at arguments that are ordered pairs. At all other arguments we may assign the "don't care" value $\emptyset$.

23 A similar definition can be given for global functions of any number of arguments.

$$\mu(x) \text{ lies in } \mathcal{N}_1 \text{ and } \mathit{Field}(x) \cong_c \mathit{Field}(\mu(x))$$

We say that $\mathcal{N}_0$ *is shorter than or equal to* $\mathcal{N}_1$ *in* length (in symbols: $\mathcal{N}_0 \preceq \mathcal{N}_1$) if we can exhibit a measure μ for $\mathcal{N}_0$ in $\mathcal{N}_1$. To deny that $\mathcal{N}_0 \preceq \mathcal{N}_1$ (in symbols: $\mathcal{N}_0 \npreceq \mathcal{N}_1$) is to say that it is impossible that μ should measure $\mathcal{N}_0$ in $\mathcal{N}_1$, no matter what the unary global function μ may be.

Propositions asserting closure ($\mathcal{N} \xrightarrow{\varphi} \mathcal{N}$) or asserting that one simply infinite system is shorter than another ($\mathcal{N}_0 \preceq \mathcal{N}_1$) are (informally) Σ^1_1 in form, and that raises the question of the "domain of variation" of the informal second-order quantifiers employed in these definitions.

The intention is that the "second-order domain" of global functions be composed of those functions that can be defined from the basic set-theoretical operations (power set, pair set, union, etc.) by composition and the fixing and varying of parameters—we might call the logic employed informally here *predicative second-order logic*. But the notion of "composition" is problematic in the context of Euclidean arithmetic: how many times can we combine and compose previously defined functions to obtain a new one?

However, these predicative second-order definitions allow us to approach questions of closure and length from a conventional, infinitary point of view. Thus we may use facts about the classical standard model, $\mathbf{V}_\omega = (V_\omega, \mathcal{F}, \in)$,[24] of the predicative second-order formal theory of hereditarily finite pure sets to suggest answers to questions concerning the existence of global functions satisfying given conditions.[25]

In particular, if we cannot define a global function with a certain property in $\mathbf{V}_\omega$ classically, we may take that as evidence that no such function can be defined constructively in Euclidean Arithmetic.

For example, it can be shown that in $\mathbf{V}_\omega$ there is no definable global function that measures $\mathcal{Z}$ in $\mathcal{VN}$. This suggests that we may lay down the following *classically witnessed postulate*:

Postulate 1 Given any unary global function μ it is impossible that μ should be a measure for $\mathcal{Z}$ in $\mathcal{VN}$ (in symbols $\mathcal{Z} \npreceq \mathcal{VN}$).

I call this a "postulate" rather than an "axiom," because it is certainly not self-evident, although the fact that it holds classically in $\mathbf{V}_\omega$ strongly suggests that it is valid in Euclidean arithmetic.[26]

Here we encounter a significant difference between Euclidean arithmetic and classical infinitary arithmetic: in Euclidean arithmetic there are non-isomorphic nat-

---

[24] $\mathcal{F}$ consists of all functions $V_\omega^{\mathbf{n}} \to V_\omega$, of all degrees $\mathbf{n}$, definable by terms in the first order theory of V_ω (including terms containing constants for elements from V_ω), and is the domain of variation for the higher order function variables.

[25] This method was invented by Richard Pettigrew, and is to be expounded at length in (Pettigrew, 2008).

[26] This has been established by Richard Pettigrew (2008) using a method inspired by a model-theoretic construction by Steve Popham in (Popham, 1984).

ural number systems. But the difference is even more striking, for it can also be shown that in $\mathbf{V}_\omega$ there is no definable global function that measures $\mathcal{VN}$ in $\mathcal{Z}$, and this justifies our laying down another classically witnessed postulate:

Postulate 2 Given any unary global function μ it is impossible that μ should be a measure for $\mathcal{VN}$ in $\mathcal{Z}$ (in symbols $\mathcal{VN} \not\preceq \mathcal{Z}$).

The fact that both $\mathcal{Z} \not\preceq \mathcal{VN}$ and $\mathcal{VN} \not\preceq \mathcal{Z}$ hold in Euclidean Arithmetic shows that the intuitive picture of a simply infinite system's being laid out sequentially in an extension of infinite length is completely misleading. Since a simply infinite system is the *infinite* species of all linear orderings generated by a successor function σ from a initial term a, it is *intensional* in its very essence, and cannot be said even to have an "extension." In Euclidean arithmetic we can speak of the "extension" only of finite sequences, just as the only extensional collections are finite ones, i.e., *sets*.

Propositions about the comparative lengths of simply infinite systems are, when properly understood, *propositions about the logical and set-theoretical properties of their respective successor functions*. A similar observation applies to the closure properties of simply infinite systems.

Using the method of classically witnessed postulation we can justify the following:

Postulate 3 (i) None of the simply infinite systems $\mathcal{VN}$, $\mathcal{Z}$ and $\mathcal{CH}$ is closed under addition.
(ii) The simply infinite system $\mathcal{CH}$ is strictly shorter in length than both $\mathcal{VN}$ and $\mathcal{Z}$ (in symbols: $\mathcal{CH} \prec \mathcal{VN}$ and $\mathcal{CH} \prec \mathcal{Z}$).

There are, however, simply infinite systems that are closed under addition, indeed under much stronger arithmetical functions. The Ackermann simply infinite system, $\mathcal{ACK}$, can be proved outright (i.e., without making use of classically witnessed postulates) to be closed under addition, multiplication, and exponentiation.

There are various techniques now known for defining new simply infinite systems from previously given ones.[27] I shall now describe one of the simplest.

Let K be a linear ordering of two or more terms which I shall call *K-ary digits*. An *K-ary numeral* is just a non-empty sequence, (L, f), of K-ary digits (so that L is a non-empty linear ordering and f : Field$(L) \to$ Field(K)) whose first term, $f(\text{First}_L)$, is not 0_K (i.e., First_K), unless less L has only one term.

Given a simply infinite system $\mathcal{N}$ define its *K-ary expansion*, $\mathcal{N}[K]$, to be the simply infinite system whose terms are the K-ary numerals (L, f) whose underlying linear ordering L lies in $\mathcal{N}$.[28]

The essential facts about these K-ary extensions are contained in the following:

Theorem 1 *Let $\mathcal{N}$ be a simply infinite system, and let K be a linear ordering, with two or more terms, whose field is measurable by $\mathcal{N}$.*

---

[27] These techniques are presented in (Pettigrew, 2008).

[28] The *elements* of $\mathcal{N}[K]$ are thus linear orderings $[\mathbf{0}, \mathbf{1} \cdots, \mathbf{m}]$ whose *terms*, $\mathbf{0}, \mathbf{1}, \cdots, \mathbf{m}$, are K-ary digits whose lengths lie in $\mathcal{N}$, and which are arranged in their natural order.

(i) $\mathcal{N} \preceq \mathcal{N}[K]$
(ii) $\mathcal{N}[K]$ *is closed under addition.*
(iii) *A necessary and sufficient condition for* $\mathcal{N}$ *to be closed under addition is that* $\mathcal{N}[K]$ *be closed under multiplication.*
(iv) *A necessary and sufficient condition for* $\mathcal{N}$ *to be closed under multiplication is that* $\mathcal{N}[K]$ *be closed under* $\mathrm{Lexp}_K^{\mathbf{1}}$ $(= \lambda x, y(x^{\log_K(y)}))$.
(v) *A necessary and sufficient condition for* $\mathcal{N}$ *to be closed under* $\mathrm{Lexp}_K^{\mathbf{n}}$ *is that* $\mathcal{N}[K]$ *be closed under* $\mathrm{Lexp}_K^{\mathbf{n+1}}$ $\left(= \lambda x, y\left(x^{\mathrm{Lexp}_K^{\mathbf{n}}(\log_K(y), \log_K(y))}\right)\right)$.

Where $\log_K$ *is an arithmetical global function corresponding to the conventional number theoretical function that sends each n to the greatest* $x \leq$ *the base K logarithm of n.*

It is important to realise that, in the (implied) definition of $\mathrm{Lexp}_K^{\mathbf{n}}$, the expression $\mathbf{n}$ is *not* a variable ranging over "natural numbers." The "recursion" in the definition is really a recipe for laying down as many *particular* definitions of these higher "logarithmic exponentials" as one wishes (or as one can). We cannot even "internalize" this recursion to (finite) linear orderings $L = [0_L, \cdots, k_L]$.[29]

But the method of K-ary expansion will not yield closure under exponentiation.

Theorem 2 *Let* $\mathcal{N}$ *be a simply infinite system, and let* K *be a linear ordering, with two or more terms whose field is measurable by* $\mathcal{N}$*. Then the following are equivalent:*

(i) $\mathcal{N}[K]$ *is closed under exponentiation.*
(ii) $\mathcal{N}$ *measures* $\mathcal{N}[K]$.
(iii) $\mathcal{N}$ *is closed under exponentiation.*

So $\mathcal{N}[K]$ is not closed under exponentiation unless $\mathcal{N}$ itself is already closed under exponentiation. Moreover, $\mathcal{N}[K]$ is a *proper* extension of $\mathcal{N}$ unless $\mathcal{N}$ is already closed under exponentiation.

Thus if we start with a simply infinite system, $\mathcal{N}$, not closed under addition (e.g., if $\mathcal{N}$ is $\mathcal{VN}$, $\mathcal{Z}$ or $\mathcal{CH}$), then successive K-ary extensions produce longer and longer expansions

$$\mathcal{N} \prec \mathcal{N}[K] \prec (\mathcal{N}[K])[K] \prec ((\mathcal{N}[K])[K])[K] \prec \cdots$$

closed under stronger and stronger arithmetical functions

$$\lambda x(x+1),\ +,\ \times,\ \lambda x, y(x^{\log_K(y)}),\ \lambda x, y(x^{(\log_K(y)^{\log_K(\log_K(y))})})\ \ldots$$

though all these "logarithmic exponentials" grow more slowly than the exponential function.

[29] Compare the discussion of iterating the application of a global function at the end of § 3.

Notice, however, that if I set $\mathcal{N}_{\mathbf{0}} = \mathcal{N}$ and $\mathcal{N}_{\mathbf{n+1}} = \mathcal{N}_{\mathbf{n}}[K]$, then the ensuing *non-terminating* sequence

$$\mathcal{N}_{\mathbf{0}} \prec \mathcal{N}_{\mathbf{1}} \prec \mathcal{N}_{\mathbf{2}} \prec \mathcal{N}_{\mathbf{3}} \prec \mathcal{N}_{\mathbf{4}} \prec \cdots \prec \mathcal{N}_{\mathbf{n}} \prec \mathcal{N}_{\mathbf{n+1}} \prec \cdots$$

is not "internal" to Euclidean arithmetic, so to speak, and, needless to say, the boldfaced subscripts do not range over "natural numbers."

Rather, the "endlessness" of the "sequence" consists solely in the fact that we can repeat the construction $\mathcal{N} \mapsto \mathcal{N}[K]$ as often as we like—or as we can—so that there is no natural halting place. The subscripts are mere indices or tallies, as it were, recording the number of such expansions that have actually been carried out in a given case.

But even though we cannot use the method of K-ary expansions to construct a simply infinite system closed under exponentiation, nevertheless there are simply infinite systems with this property. I have already called attention to the Ackerman system, ACK, whose successor function generates sets in the order determined by their Ackermann codes, as an example. Richard Pettigrew (Pettigrew, 2008) has exhibited simply infinite systems $\mathcal{N}_1 \prec \mathcal{N}_2$ both of which are closed under exponentiation.[30]

All these facts about simply infinite systems must be seen in the light of the further fact that *no simply infinite system is* "long enough" *to count out every (finite) set*. More precisely, the method of classically witnessed postulation justifies our laying down the following postulate:

Postulate 4 Let $\mathcal{N}$ be a simply infinite system and μ any unary global function. Then it is impossible that for all sets S, $\mu(S)$ lies in $\mathcal{N}$ and $\mu(S) \cong_c S$.

Since no simply infinite system can measure all finite sets, and since there are simply infinite systems with differing lengths and closure properties, we are at a loss to say what it is that corresponds, in Euclidean arithmetic, to *the* natural number sequence of conventional arithmetic. Surely the Euclidean theory of simply infinite systems shows that in Euclidean arithmetic the idea of the *uniqueness* of "the" natural number sequence should be abandoned, and that, from that standpoint, there are many non-equivalent but equally legitimate ways of systematically "proceeding to infinity."

We can regard each simply infinite system as an attempt systematically to exhibit a *unique* member of every cardinality class. In ordinary mathematics, with its infinitary assumptions, this works. But we now know that, in Euclidean arithmetic, any such attempt is bound to fail.

From the standpoint of the philosophy of mathematics, the Euclidean approach to arithmetic relieves us of the temptation to say what those peculiar entities, the natural numbers, "really" are. Mathematically, it allows us to escape from the idea

[30] More precisely, he shows that we can introduce the proposition $\mathcal{N}_1 \prec \mathcal{N}_2$ as a classically witnessed postulate. This will prove to be of significance when we come to discuss the treatment of real numbers in Euclidean arithmetic.

that the essence of arithmetic is to be sought in calculating procedures, an idea that leads directly to what I have called the Sorites fallacy, which manifests itself in the belief that the principles of proof by induction and definition by recursion are self-evidently valid and do not stand in need of justification.[31]

The mathematical facts I have given concerning the lengths and closure properties of simply infinite systems, interesting in themselves from the standpoint of pure arithmetic, allow us to develop a theory of "real numbers" in which infinitesmals occur naturally, rather than as the superadded "ideal elements," that they are in conventional nonstandard analysis. The resulting algebra of real numbers leads, in turn, to a natural development of the infinitesimal calculus.

6 Rational Numbers

A central topic in the ancient version of Euclidean arithmetic dealt with the theory of ratio and proportion in numbers (*arithmoi*). Looking to Euclid for inspiration, let us develop an arithmetic of (positive) *ratios* so that, like the "natural numbers," they correspond to equivalence classes under the appropriate equivalence relation.

We may start by defining *(positive) ratios*, $(S : T)$, to be ordered pairs, (S, T), of sets (where $T \neq \emptyset$), under the equivalence relation

$$(S, T) \cong_{\text{ratio}} (U, V) \textit{ if, and only if, } S \times_{\text{c}} V \cong_{\text{c}} U \times_{\text{c}} T$$

In conventional, contemporary terms the concrete ratio $(S : T)$ corresponds to "abstract" ratio m/n, where S has m elements and T has n. Of course each particular pair (S, T) represents its equivalence class with respect to the equivalence relation $\cong_{\text{ratio}}$. The natural order relation for ratios is

$$(S, T) <_{\text{ratio}} (U, V) \textit{ if, and only if, } S \times_{\text{c}} V <_{\text{c}} U \times_{\text{c}} T$$

In ancient arithmetic the proposition $(S, T) \cong_{\text{ratio}} (U, V)$ was expressed by

$$S : T :: U : V$$

Notice that I have introduced ratios *and the notion of equality* ($\cong_{\text{ratio}}$) *between ratios*, simultaneously, so that ratios are not to be thought of as mere ordered pairs of sets *simpliciter* but rather as ordered pairs of sets *identified or distinguished in accordance with the new notion of* "equality" *defined by the equivalence relation* $\cong_{\text{ratio}}$.

An ordered pair, (S, T), of sets we might call a "concrete" ratio which "illustrates" or "represents" a ratio properly so called in accordance with Euclid's definition of "ratio":

---

[31] In ordinary mathematics we can look on Dedekind's theory of simply infinite systems as rendering the Sorites fallacy harmless, insofar as it bears on mathematical practice, even though it remains a fallacy even there.

A ratio[32] is a relationship in respect of size between two magnitudes of the same kind (*Elements*, Book V, Definition 3).[33]

It is not difficult to define addition, $+_{\text{ratio}}$, and multiplication, $\times_{\text{ratio}}$, for positive ratios, in conformity with the equivalence relation $\cong_{\text{ratio}}$, and this allows us to develop an arithmetic of ratios which parallels the arithmetic of numbers (*arithmoi* – finite sets) sketched in Section 4.

We need, however, not just positive ratios, but *rational numbers*, positive, negative, and zero. Let us therefore define *rational numbers* to be ordered pairs (r, s) of positive ratios, r and s, taken under a new equivalence relation

$$(r, s) \cong_Q (t, w) \textit{ if, and only if, } (r +_{\text{ratio}} w) \cong_{\text{ratio}} (s +_{\text{ratio}} t)$$

The idea here is that the pair (r, s) of positive ratios corresponds to the rational number $r - s$, positive, negative, or zero.

Again we can define addition, $+_Q$, and multiplication, $\times_Q$, of rational numbers in conformity with the equivalence relation $\cong_Q$. Indeed, we can define bounded sums $(\Sigma_{i=0}^{n} a_i)$ and products $(\Pi_{i=0}^{n} a_i)$ of rational numbers, in the obvious way.

It should be clear how to develop the arithmetic and algebra of rational numbers from these definitions. One just needs to bear in mind that the rational numbers actually referred to are really representatives of the $\cong_Q$-equivalence classes to which they belong.

In summary, a rational number q is an ordered pair of the form (r, s), where r and s are positive ratios, so that r and s themselves are ordered pairs of sets, $r = (T : U)\ (= (T, U))$ and $s = (V : W)\ (= (V, W))$, where each of T, U, V and W is a set which is serving as a representative of its cardinality class, and where $U, W \neq \emptyset$. Thus $q = ((T, U), (V, W))\ (= ((T : U), (V : W)))$ is an ordered pair of ordered pairs.[34]

All of this is quite straightforward: after all, everyone knows that the theories of positive ratios, integers, and rational numbers belong, essentially, to arithmetic, and the arithmetic of rational numbers I have described here is, indeed, contained in Euclidean arithmetic.

7 Real Numbers

The conventional reals, which nowadays are usually defined using the axioms for a complete ordered field, were originally introduced as "abstracted ratios" of concrete

---

32 *Logos*, which can mean "reason" in Greek, hence *ratio* which means "reason" in Latin.

33 The reader can verify that the definition of equality for ratios ($\cong_{\text{ratio}}$) I have given is equivalent to Euclid's more general definition (*Elements*, Book V, Definition 5) when the latter is specialized to ratios of *arithmoi*.

34 Needless to say, there are alternatives to the way I have defined (positive) ratios and rational numbers here.

quantities (e.g., "abstracted" ratios of lines to lines or of polygons to polygons, etc.) when they were invented in the seventeenth century.

In Section 2, I quoted Newton's definition of (positive, real) number:

> By a Number we understand not so much a Multitude of Unities, as the abstracted Ratio of any Quantity to another Quantity of the same Kind which we take for Unity.

Of course our positive ratios are essentially abstracted ratios[35] of particular numbers (*arithmoi*) understood in the traditional sense (i.e., as what we call "finite sets"), and are thus genuine "Numbers" in Newton's then novel sense given in the definition I have just quoted.

On the definition Newton gives, the notion of "Number" (i.e., positive real number) is logically dependent on geometrical concepts, so that, for example, to define $\sqrt{2}$ we need to abstract from the ratio of a line to a line, namely, the ratio of the diagonal of a square to its side.[36]

But we no longer believe, as the seventeenth century did, that Euclidean geometry describes the actual space we live in, so this logical dependence on Euclidean geometry is something of an embarrassment. (This no doubt explains why we now define the real numbers axiomatically.) I intend, therefore, to show that we can develop a theory of real number arithmetic which remains Newtonian in spirit, so that real numbers are ratios, but which cuts the notion of real number free of any *logical* dependence on geometrical concepts: a purely *arithmetical* theory of real numbers.

This means that I shall not be required, as Newton was, to define the basic arithmetical operations on real numbers in a geometrical manner—they are already given to us in the corresponding rational-arithmetical operations on the rational numbers.

The theory of real arithmetic I shall develop here *rejects* the conventional idea that the real number field is a proper *extension* of the rational field. On the contrary, *the central idea that informs this approach to real number theory is that the distinction between* ***rational and irrational real numbers*** *rests upon an even more fundamental distinction between* ***large and small sets***. An immediate consequence of this idea is that *the real numbers form a subring of the rational field*.

As we shall see, the distinction between "large" and "small" sets is itself a relative one, depending, as it does, upon which "natural number" system (simply infinite system) $\mathcal{N}$ we use to count out the elements of a set, so, in contrast to the conventional view, *there is no fixed or absolute concept of real number*.

Recall that I have laid down the classically witnessed postulate that no simply infinite system forms a *scale* for all sets, that is to say, given any simply infinite system $\mathcal{N}$ and any candidate universal measure μ with respect to $\mathcal{N}$ (i.e., any unary global function μ), it is impossible that, for every set S, $\mu(S)$ lies in $\mathcal{N}$ and $\mu(S) \cong_{\mathrm{c}}$

---

[35] We can speak of "abstraction" here, because we moderns do our "abstracting" by using equivalence relations and their equivalence classes.

[36] Newton took it for granted that if we "abstract" ratios that are "equal" or "proportional" in the sense of Eudoxus (Euclid's *Elements*, Book V, Definition 5), then we obtain identically the same Newtonian number.

S. In fact this holds even if we confine ourselves to pure sets[37]: there is no *scale* for the species of pure sets. To simplify my exposition I shall confine my attention to pure sets.

Definition 5 Given any simply infinite system $\mathcal{N}$, any set S, and any rational number q we say that

(i) A set S is $\mathcal{N}$-*small* if, and only if, there is an n in $\mathcal{N}$ such that

$$S <_c \text{Field}(n)$$

(ii) A rational number q is $\mathcal{N}$-*small* if, and only if, there is an n in $\mathcal{N}$ such that $|q| <_{\mathcal{Q}} n_{\mathcal{Q}}$, where $n_{\mathcal{Q}} = (\,(\text{Field}(n) : \{\emptyset\}), (\{\emptyset\} : \{\emptyset\})\,)$ is a rational number corresponding to n.

(iii) A set S is $\mathcal{N}$-*large* if, and only if, for all n in $\mathcal{N}$

$$\text{Field}(n) <_c S$$

(iv) A rational number q is $\mathcal{N}$-*large* if, and only if, for all n in $\mathcal{N}$, $n_{\mathcal{Q}} <_{\mathcal{Q}} |q|$, where $n_{\mathcal{Q}}$ is defined as in (ii).[38]

These notions lead to the following key definition.

Definition 6 Let $\mathcal{N}$ be a simply infinite system and let q be any element of $\mathcal{Q}$.

(i) q is said to be $\mathcal{N}$-*real* if it is $\mathcal{N}$-small. $\mathcal{R}_{\mathcal{N}}$ is the species of $\mathcal{N}$-reals.

(ii) q is said to be $\mathcal{N}$-*infinitesimal* if for all $0 \neq n \in \mathcal{N}$

$$q <_{\mathcal{Q}} 1/n_{\mathcal{Q}}$$

$\mathcal{I}_{\mathcal{N}}$ is the species of $\mathcal{N}$-infinitesimals.

(iii) q is said to be $\mathcal{N}$-*rational* if q is $\mathcal{N}$-real and there are m and n in $\mathcal{N}$ such that $|q| \cong_{\mathcal{Q}} m_{\mathcal{Q}} \times_{\mathcal{Q}} n_{\mathcal{Q}}$. $\mathcal{Q}_{\mathcal{N}}$ is the species of $\mathcal{N}$-rationals.

(iv) q is said to be $\mathcal{N}$-*irrational* if q is $\mathcal{N}$-real and for all sets S and $T \neq \emptyset$

$$|q| \cong_{\mathcal{Q}} S_{\mathcal{Q}} \times_{\mathcal{Q}} (T_{\mathcal{Q}})^{-1} \Rightarrow T \textit{ is } \mathcal{N}\textit{-large}$$

where $S_{\mathcal{Q}} = (\,(S : \{\emptyset\}), (\{\emptyset\} : \{\emptyset\})\,)$ and $T_{\mathcal{Q}} = (\,(T : \{\emptyset\}), (\{\emptyset\} : \{\emptyset\})\,)$ represent (the sizes of) S and T in $\mathcal{Q}$.

Note that in accordance with definitions (ii) and (iv) non-zero $\mathcal{N}$-infinitesimals are $\mathcal{N}$-irrational.

---

[37] That is, sets whose transitive closures contain no individuals (non-sets).

[38] Notice that in all these definitions $\mathcal{N}$-*small* is Σ_1 but $\mathcal{N}$-*large* is Π_1. This has important technical consequences.

When it is clear from the context what the counting number sequence $\mathcal{N}$ is, we may drop explicit reference to it and speak simply of "reals," "infinitesimals," "irrationals," etc.

8 The Calculus

I have already remarked that the theory of real numbers sketched here is not absolute, but relative to the choice of a concrete natural number system $\mathcal{N}$. Let us therefore suppose that we are dealing with a fixed simply infinite system $\mathcal{N}$ closed under addition, multiplication, and exponentiation, so that I may drop references to $\mathcal{N}$ and refer to the species, $\mathcal{I}$, of infinitesimals, etc. In the same spirit I shall drop the subscripts from operators etc. when doing so will not lead to confusion.

I shall begin by defining a fundamental equivalence relation on the real numbers:

Definition 7 Let r and s lie in $\mathcal{R}$. Then

$$r \cong_{\mathcal{R}} s \text{ if and only if } r - s \text{ lies in } \mathcal{I}$$

This gives rise to a natural ordering on $\mathcal{R}$ (mod $\cong_{\mathcal{R}}$) defined as follows.

Definition 8 Let r and s lie in $\mathcal{R}$. Then

$$r \preceq_{\mathcal{R}} s \text{ if and only if } \quad [r \cong_{\mathcal{R}} s \text{ or } [r \ncong_{\mathcal{R}} s \text{ and } r <_{\mathcal{Q}} s]]$$

(Note that this definition and the previous one make perfectly good sense if r and s lie in $\mathcal{Q}$.)

It is important to realise that neither $r \cong_{\mathcal{R}} s$ nor $r \preceq_{\mathcal{R}} s$ is *decidable*, e.g., we cannot in general assert

$$r \cong_{\mathcal{R}} s \text{ or } r \ncong_{\mathcal{R}} s$$

$\mathcal{R}$ with $\cong_{\mathcal{R}}$ taken to be the principal notion of equality strongly resembles the classical reals, as we shall see. Indeed, the reals "nearly" form a field in the sense that every real which is *explicitly bounded away from* 0 has a real inverse.[39]

$\mathcal{R}$ under $\cong_{\mathcal{Q}}$ might be thought of as the reals with the infinitesimals added, or, more in the spirit of our enterprise, the reals with the infinitesimals *acknowledged*.[40]

---

[39] A real r is explicitly bounded away from 0 if there is a non-zero $n \in \mathcal{N}$ such that

$$0 <_{\mathcal{Q}} 1/n_{\mathcal{Q}} <_{\mathcal{Q}} |r|$$

Here we are up against the fact that $\neg(\forall n \in \mathcal{N})\Phi(n)$ does not imply $(\exists n \in \mathcal{N})\neg\Phi(n)$.

[40] I shall often drop the subscript $\mathcal{Q}$, e.g., writing "$a = b$" instead of "$a \cong_{\mathcal{Q}} b$" and $a \leq b$ instead of $a \leq_{\mathcal{Q}} b$, since these relations are locally defined and therefore decidable, and all our

Definition 9 Let $\delta >_{\mathcal{Q}} 0_{\mathcal{Q}}$ be a positive infinitesimal. Then

(i) A rational number q lies in $\mathcal{I}_\delta$ if, and only if, for all n in $\mathcal{N}$,

$$|q| <_{\mathcal{Q}} n_{\mathcal{Q}} \times_{\mathcal{Q}} \delta$$

where $n_{\mathcal{Q}}$ is the rational number corresponding to n.

(ii) A rational number q lies in $\mathcal{R}_\delta$ if, and only if, for some n in $\mathcal{N}$,

$$q <_{\mathcal{Q}} n_{\mathcal{Q}} \times_{\mathcal{Q}} \delta$$

where $n_{\mathcal{Q}}$ is the rational number corresponding to n.

(iii) For rational numbers p and q
$p \cong_\delta q$ if and only if $p - q$ lies in $\mathcal{I}_\delta$.
$p \prec_\delta q$ if and only if $p < q$ and $p \ncong_\delta q$.
$p \preceq_\delta p$ if, and only if, $p < q$ or $p \cong_\delta q$

$\mathcal{R}_\delta$ consists of those infinitesimals that are *of the same order of magnitude as* δ, $\mathcal{I}_\delta$ of those infinitesimals that are *infinitesimally small compared to* δ.

We now concentrate our attention on *rational valued functions on the rationals* ($\mathcal{Q}$-functions for short), that is to say, global functions φ whose value at rational number arguments is a rational number. Clearly $+_{\mathcal{Q}}$, $\times_{\mathcal{Q}}$, and $exp_{\mathcal{Q}}$ are functions of this sort.

Our principal concern, however, is with those rational functions that are *real functions of one or more real variables* in accordance with the following definition.

Definition 10 Let φ be a $\mathcal{Q}$-function and J be a subspecies of $\mathcal{R}$. We say that φ is *real function defined on* J if for all x lying in J, $\varphi(x)$ lies in $\mathcal{R}$ (and similarly for functions of more than one argument).

The basic functions $+_{\mathcal{Q}}$, $\times_{\mathcal{Q}}$, and $exp_{\mathcal{Q}}$ are real functions in this sense. In fact, all the "standard" functions used in conventional analysis—the natural logarithm $x \mapsto \ln(x)$, the exponential $x \mapsto e^x$, the trigonometric functions and their inverses, etc.—have analogues in our $\mathcal{R}$.[41]

We are not interested in real functions defined on arbitrary domains $J \subseteq \mathcal{R}$, but those defined on domains $J \subseteq \mathcal{R}$ that are *microstable* in the following sense.

Definition 11 Let J be a subspecies of $\mathcal{R}$.

(i) J is *microstable* if for all x, y lying in $\mathcal{R}$

$$x \cong y \Rightarrow x \text{ lies in } J \text{ if, and only if, } y \text{ lies in } J$$

---

constructions and definitions apply to the rationals. I shall also write "$a \cong b$" and instead of "$a \cong_{\mathcal{R}} b$" and "$a \preceq b$" instead of "$a \preceq_{\mathcal{R}} b$" even though these relations are *not* locally defined. All of this is in the interest of avoiding a proliferation of subscripts.

[41] This can be established by using *polynomials of large degree*, $\sum_{i=0}^{n} a_i x^i$, where n is large.

(ii) Given any infinitesimal δ, J is *δ-microstable* if for all x, y lying in J

$$x \cong_\delta y \Rightarrow x \text{ lies in } J \text{ if, and only if, } y \text{ lies in } J.$$

Of course $\mathcal{R}$ iself is microstable and, indeed, δ-microstable for any infinitesimal δ. We also need microstable versions of the open and closed intervals of classical analysis.

Definition 12 Let a and b lie in $\mathcal{R}$ and let $\delta > 0$ be an infinitesimal.

(i) Let a and b be real numbers, then the *closed interval* $[a, b]$ is the species of all reals x such that

$$a \preceq x \preceq b$$

(ii) Let a and b be real numbers, then the *open interval* (a, b) is the species of all reals x such that

$$a \prec x \prec b$$

(This notational convention means that "(a, b)" can refer to an ordered pair or an open interval. One has to live with this ambiguity when doing conventional analysis too.)

The point of introducing microstability is best understood in conjunction with the notion of continuity.

Definition 13 Let J be a microstable subspecies of $\mathcal{R}$ and φ be a real function defined on J. We say that φ is *continuous on* J if for all x, y lying in J

$$x \cong y \Rightarrow \varphi(x) \cong \varphi(y)$$

This is just the pre-Weierstrassian concept of continuity that prevailed in the eighteenth century. Notice that in accordance with this definition and the definition of real function, *all real functions defined on a microstable domain are continuous on that domain.*

We can relativise the definition of continuity.

Definition 14 Let $\delta > 0$ be a positive infinitesimal, J a δ-microstable subspecies of $\mathcal{R}$, and φ a real function defined on J. We say that φ is *δ-continuous on* J if for all x, y lying in J

$$x \cong_\delta y \Rightarrow \varphi(x) \cong_\delta \varphi(y)$$

The general theory of continuous functions can be developed in a relatively straightforward way: we just have to keep track of when our definitions employ global quantifiers and remember to employ constructive (intuitionistic) logic to propositions involving those quantifiers.

The differential and integral calculus are treated in a manner reminiscent of non-standard analysis. Thus we define the derivative of a real function φ defined on a microstable subspecies J of $\mathcal{R}$ as follows:

Definition 15 Let φ and ψ be real functions defined on a microstable subspecies, J, of $\mathcal{R}$. Then φ is said to be *differentiable on J with derivative ψ* if for all infinitesimals $\epsilon \neq 0$ in $\mathcal{I}$

$$\frac{\varphi(x+\epsilon)-\varphi(x)}{\epsilon} \cong \psi(x)$$

In these circumstances we say that $\psi \cong \varphi'$.

In terms of difference functions this becomes

$$\Delta_{\varphi}(x,\epsilon)\,(=\varphi(x+\epsilon)-\varphi(x)) \cong_{\epsilon} \psi(x)\epsilon$$

so that the difference $\varphi(x+\epsilon)-\varphi(x)$ is ϵ-*infinitesimally* close to the product $\epsilon\psi(x)$ at all arguments x lying in J.

The definite integral also has a definition in terms of infinitesimals:

Definition 16 Let φ be a real function defined on a microstable domain J, let $a, b \in J$ with $a \preceq b$, and let $\epsilon \in I$ be an infinitesimal with $\epsilon > 0$. Then:

(i) The *ϵ-sum of φ from a to b* (in symbols: $\Sigma_a^b \varphi \cdot \epsilon$) is defined by

$$\Sigma_a^b \varphi \cdot \epsilon = \sum_{n=1}^{k} \varphi(a+n\epsilon(b-a))\epsilon(b-a)$$

where k is the $\mathcal{N}$-large natural number such that $k \leq 1/\epsilon < k+1$.

(ii) φ is said to be *integrable from a to b* if for all non-zero infinitesimals ϵ and δ

$$\Sigma_a^b \varphi \cdot \epsilon \cong \Sigma_a^b \varphi \cdot \delta$$

Of course this defines the integral $\int_a^b f$ only up to $\cong$-equivalence.[42] When dealing with an explicitly defined function such as $f = \lambda x(x^2)$, we write $\int_a^b x^2 dx$ in the usual way.

These definitions permit a straightforward development of the Calculus analogous to its development in conventional nonstandard analysis. In particular, an "algebraic" proof of the Fundamental Theorem of the Calculus along Leibnizian lines can be given.

[42] When $\sigma(x) \cong \tau(x)$ for all x in a domain J we write "$\sigma \simeq \tau$ *on* J".

Theorem 3 (The Fundamental Theorem of the Calculus) *Let φ and ψ both be real functions defined on a microstable domain J with $\psi \simeq \varphi'$ on J. Then for all $a, b \in J$ with $a \prec b$ and $[a, b] \subseteq J$, ψ is integrable from a to b and*

$$\int_a^b \psi \cong \varphi(b) - \varphi(a).$$

Proof It will simplify the exposition, though without really sacrificing generality, to consider the case $a = 0$ and $b = 1$. We must show that for all infinitesimal $\epsilon > 0$,

$$\Sigma_0^1 \psi \cdot \epsilon = \sum_{n=1}^{k} \psi(n\epsilon)\epsilon \cong \varphi(1) - \varphi(0)$$

where k is the integer part of $1/\epsilon$ (so that $k\epsilon \cong 1$). Since $\psi \cong \varphi'$ on $[a, b]$

$$\frac{\varphi((n+1)\epsilon) - \varphi(n\epsilon)}{\epsilon} \cong \psi(n\epsilon) \ \ (1 \leq n \leq k)$$

and therefore there are infinitesimals $\delta_1, \cdots, \delta_k$ such that

$$\varphi((n+1)\epsilon) - \varphi(n\epsilon) \cong \psi(n\epsilon)\epsilon + \epsilon\delta_n \ \ (1 \leq n \leq k)$$

Hence

$$\sum_{n=1}^{k} \psi(n\epsilon)\epsilon = \sum_{n=o}^{k-1} [\varphi((n+1)\epsilon) - \varphi(n\epsilon)] + \epsilon \sum_{n=1}^{k} \delta_n$$

But

$$\sum_{n=o}^{k-1} [\varphi((n+1)\epsilon) - \varphi(n\epsilon)] = \varphi(k\epsilon) - \varphi(0)$$

and $\varphi(k\epsilon) \cong \varphi(1)$. Now let $\delta = max\{|\delta_1|, \cdots, |\delta_k|\}$. Then

$$\left| \epsilon \sum_{n=1}^{k} \delta_n \right| \leq \epsilon \sum_{n=1}^{k} |\delta_n| \leq \epsilon k \delta \cong \delta$$

so

$$\int_0^1 \psi \cong \varphi(1) - \varphi(0)$$

as required. □

9 A Finitary Version of the Reals

The theory sketched in Sections 7 and 8 is infinitary in that the real numbers constitute an infinite subspecies of the rationals. I want now to explain how we can develop a *finitary* theory of the reals along essentially the same lines.

Let me begin by observing that we can lay down a classically witnessed postulate

Postulate 5 (i) $\mathcal{N} \prec \mathcal{N}^*$
(ii) Both $\mathcal{N}$ and $\mathcal{N}^*$ are closed under exponentiation.
(iii) $\mathcal{N}^*$ is not a scale for the species of all (*pure*) sets.

Using the fact that $\mathcal{N}^*$ is not a universal scale, we may assume that there is a pure set, S, that is not measurable by $\mathcal{N}^*$.[43]

Under this assumption, the species of sets measurable by $\mathcal{N}^*$ can be proved not to be closed under union, since $\{S\}$ is measurable by $\mathcal{N}^*$ but $\bigcup(\{S\}) = S$ is not. This suggests that we lay down the following definition.

Definition 17 Let $\mathcal{N}$ be a simply infinite system. Then a set S is an $\mathcal{N}$-set if the transitive closure of $S \cup \{S\}$ is measurable by $\mathcal{N}$.

In fact it is more convenient to work with an $\mathcal{N}^*$-large linear ordering

$$L = [\,0_L, 1_L, \cdots, \Omega\,(= \text{Last}_L)\,]$$

in place of an unspecified $\mathcal{N}^*$-large set S. Moreover we may assume that the terms of L consists of L sets[44] representing all the cardinalities strictly less than that of L, arranged in ascending order of cardinality.

We need to work with the *extended Farey series*, Farey(L), based on L which is defined as follows:

Definition 18 (Extended Farey Series) Let L be a linear ordering whose field is composed of L sets whose terms are ordered by increasing cardinality, and for all of whose terms, x, $x <_c Field(L)$.[45] Then the extended Farey series determined by L is the linear ordering, $Farey(L)$, whose terms

$$\{x/y : x, y \in \text{Field}(L), y \neq \emptyset\,(= 0_L)\,\}$$

are ordered linearly in accordance with the global relation $<_Q$.

---

43 Notice that I have said "assume" and not "postulate." From the classically witnessed postulate that $\mathcal{N}^*$ is not a scale for the species of all pure sets we can derive only that the existence of a set too large to measure by $\mathcal{N}^*$ is possible (i.e., not impossible). This makes the stronger assumption that S is actually an *example* of such a set consistent with our theory.

44 That is, sets S for which the transitive closure of $S \cup \{S\}$ is measurable by Field(L). Together the L-sets form a *set*, since the definition of L-set does not require global quantifiers.

45 Note that this condition means that Field(L) contains exactly one representative of each cardinality smaller than that of Field(L) itself, so L "behaves like" a finite initial segment of "the" natural numbers.

Of course, from the standpoint of Euclidean arithmetic Farey(L) is just a (large) linear ordering whose terms form a finite set, since *all* sets are finite in Euclidean arithmetic.

I have now set up the technical machinery that will allow us to recast the account of rational, real, and infinitesimal numbers sketched in Sections 6, 7, and 8 in finitary form. The idea is roughly this: let the species of $\mathcal{N}^*$ sets go proxy for the species of all sets. The species of $\mathcal{N}^*$ sets can function in such a role because $\mathcal{N}^*$ is closed under exponentiation.

Call the species of $\mathcal{N}^*$ sets $\mathcal{V}_*$. Then all the concepts employed in Sections 6 to 8 have their sub-$*$ versions Thus $\mathcal{Q}_* = \mathcal{Q} \cap \mathcal{V}_*$, $\mathcal{R}_* = \mathcal{R} \cap \mathcal{V}_*$, $\mathcal{I}_* = \mathcal{I} \cap \mathcal{V}_*$, etc., so we can rerun the exposition in 6 to 8 replacing all these notions with their sub-$*$ versions, without affecting the mathematical essentials of the exposition.

There is an additional bonus, however. $\mathcal{V}_*$ is a subspecies of the *set* of Farey(L) sets (which we may call $V_{\text{Farey}(L)}$). This means that the finitary theory is more useful when we come to apply it to *minimal parts geometry*, a version of geometry which rejects the infinite divisibility of space and posits instead that any bounded region of space (or even space itself) is composed of a large, though still finite, number of indivisible *minimal parts* which are related by a *nearest neighbour* relation. In this way the minimal parts geometry can be regarded as an undirected graph, the *minimal parts graph.*

By imposing natural symmetry conditions on these graphs we can assign each minimal parts graph a dimension, and can classify all one, two, and three dimensional minimal parts geometries. It is still possible to recover conventional Euclidean geometry in any minimal parts geometry using the analytical machinery I have described.[46]

10 Conclusions

I have described how to carry out a development of the Calculus using a theory of real numbers in which infinitesimals are included among them in a natural way. Moreover, this development arises naturally out of an approach to arithmetic in which non-isomorphic natural number systems also occur in a natural way. This suggests, I believe, a way in which the remarks made by Gödel in the preface to Robinson's book could be vindicated.

There is, however, another way of incorporating infinitesimals naturally into real analysis, a way that begins, not with arithmetic as I have done, but with geometry and kinematics. I am referring to *smooth infinitesimal analysis*, which was invented by F.W. (Lawvere, 1979) and given an elegant, elementary exposition by John Bell in (Bell, 1998).[47]

---

[46] The general theory of minimal parts geometry is developed in (Lush, 2003).

[47] Of course smooth infinitesimal analysis can be expounded naturally in conventional infinitary mathematics.

Despite their radically different motivation, and their different logical and foundational approaches, smooth infinitesimal analysis and the theory I have sketched have much in common.[48] In particular, it is significant that the geometrical/kinematical approach that inspires smooth infinitesimal analysis also calls the uniqueness of the natural number system into question.

There are also strong mathematical connections between the theory I have sketched and conventional nonstandard analysis, though the latter is much more conservative in character, since the role it allots to infinitesimals is essentially that of "ideal elements" (like the points at infinity in projective geometry).

Indeed, it is literally conservative over conventional analysis in the technical sense, so that infinitesimals can always be eliminated from proofs of theorems of nonstandard analysis that do not refer to them explicitly. Thus it is unlikely to have the radical consequences described by Gödel in the remarks I have quoted in section 1.

The general theory of numbers—natural, rational, and real—that I have sketched here, and the theory of minimal parts geometry developed in (Lush, 2003), represent only the first steps in giving a finitary account of mathematics in general.

Of course it is not "finitary" in the conventional sense. But it is not clear that conventional finitism is foundationally coherent. It is, after all, based on the iterative conception of "natural number" in which the natural numbers are *defined* as the successors of zero under *finite iterations* of the successor function, and both Dedekind and Frege called attention to the vicious circle contained in this conception of natural number more than one hundred and 20 years ago.

But even if a finitary account of the core notions of mathematics along the lines I am considering here could be given, it would represent only an *alternative* to the infinitary assumptions and techniques currently in use. And those assumptions are so simple, so convincing, so powerful, and so fruitful, that a finitary rival would be a mere curiosity unless it could match these features.

It is not, after all, *obvious* that the infinitary methods of conventional mathematics lack legitimacy, especially after we abandon the fantasy that the notion of the finite iteration of a function, and with it the notion of "natural number," are somehow just "given."

Perhaps if we arrived at such a finitary alternative to conventional infinitary mathematics, we would see these two approaches, not so much as rivals, but just as different ways of viewing the same underlying phenomena. The root problem, on either approach, is to understand what infinity really means. We have made great advances in understanding this problem, but its final resolution still eludes us.[49]

---

[48] It is not yet clear what the technical, mathematical relation between them is.

[49] I have discussed this issue at considerable length in (Mayberry, 1994).

Bibliography

Bell, J. L. (1998). *A Primer of Infinitesimal Analysis*. Cambridge University Press, Cambridge.

Dedekind, R. (1893). *Was sind und was sollen die Zahlen?* Vieweg, Braunschweig. Translated by Wooster W. Berman as *The Nature and Meaning of Numbers*. Dover, New York, 1963.

Dedekind, R. (1967). Letter to Kefferstein. In van Heijenoort, J., editor, *From Frege to Gödel*, pages 98–103. Harvard University Press, Cambridge, MA. Translated by Hao Wang and Stephan Bauer-Mengleberg.

Euclid (1956). *Elements*. Dover, New York, NY. Translated in three volumes and with introduction and commentary by Sir Thomas Heath.

Grzegorczyk, A. (1953). Some classes of recursive functions. *Rozprawy Matematyczne*, 4:1–45.

Homolka, V. (1983). *A System of Finite Set Theory Equivalent to Elementary Arithmetic*. PhD thesis, University of Bristol.

Klein, J. (1992). *Greek Mathematical Thought and the Origin of Algebra*. Dover, New York, NY.

Lawvere, F. W. (1979). Categorical dynamics. In Kock, A., editor, *Topos Theoretic Methods in Geometry*, Various Publications, Number 30, pages 1–28. Mathematick Institut, Aarhus Universitat, Aarhus.

Lush, D. (2003). *Finitary Geometry in Minimal Parts Graphs*. PhD thesis, University of Bristol.

Mayberry, J. (2000). *The Foundations of Mathematics in the Theory of Sets*. Cambridge University Press, Cambridge.

Newton, I. (1728). *Universal Arithmetic: or a Treatise of Arithmetical Composition and Resolution*. London. Reprinted in *The Mathematical Works of Issac Newton* Vol. 2 (ed. Derek T. Whiteside). Johnson Reprint Corporation, New York, NY, 1966.

Pettigrew, R. (2008). *Natural, Rational and Real Arithmetic in a Theory of Finite Sets*. PhD thesis, University of Bristol. Forthcoming.

Pizer, I. (2003). *On a Natural Construction of Real Closed Subfields of the Reals*. PhD thesis, University of Bristol.

Popham, S. (1984). *Some Results in Finitary Set Theory*. PhD thesis, University of Bristol.

Robinson, A. (1996). Non-standard Analysis. Princeton University Press, Princeton; reprinting of the 1974 revised edition, Amsterdam, North-Holland.

Rose, H. E. (1984). *Subrecursion: Functions and Hierarchies*. Clarendon Press, Oxford.

Tarski, A. (1924). Sur les ensembles finis. *Fundamenta Mathematicae*, 6:45–95.

Zermelo, E. (1908). Untersuchungen über die Grundlagen der Mengenlehre. *Mathematische Annalen*, 65:261–281.

Zermelo, E. (1909). Sur les ensembles finis et le principe de l'induction complète. *Acta Mathematica*, 32:185–193.

Chapter 15
Logic in Category Theory

Alberto Peruzzi

1 Motivations

Logic already spanned a great range of topics before the birth of categorical logic. Some celebrated results achieved in logic during the first half of the twentieth century are milestones in the understanding of mathematical relations between syntactic, semantic and algorithmic aspects of the structure of language and reasoning. Logical tools have been exploited in a variety of applications: from linguistics to computer science, from methodology of science to specific physical theories. The very formulation of questions and answers concerning the foundations of mathematics relies on such tools. Finally, mathematical logic has altered the face of philosophy. In view of such outcomes, it is all the more appropriate to consider the impact of categorical methods on logic, since they affect the study of proofs and models, by forging a stricter relationship between a theory and its models, and by enlarging the range of possible models beyond the universe of sets in a way that leads to a substantial refinement of the status ascribed to logic itself.

In the last 30 years the growth of research on mathematical problems through categorical methods has been fast and wide-ranging. If algebra and topology were the first areas in which such methods became customary practice, by now one can acknowledge a categorical turn in logic too, with effects on landmark results already achieved, along with their applications and indeed the very idea of foundations. What follows will offer a synthetic and selective path across the impact of categories on logic, focusing on the main concepts and the overall sense of the transformation. If it is possible for me to draw such a path, it is thanks to the guidance and support provided 25 years ago by John Bell.

Aside from the personal debt, Bell's contributions to categorical logic deserve mention, both for specific results and clarity of design (see, for instance the explanatory picture of the steps involved in passing from Boolean-valued models to the use of topos-theoretic methods (Bell, 1988, pp. 235–245). His logic-oriented view of topos theory (Bell, 1986), emphasizes the resources of the internal language to

A. Peruzzi (✉)
Professor, Department of Theoretical Philosophy, University of Florence, Florence, Italy
e-mail: alberto.peruzzi@unifi.it

D. DeVidi et al. (eds.), *Logic, Mathematics, Philosophy: Vintage Enthusiasms*, The Western Ontario Series in Philosophy of Science 75,
DOI 10.1007/978-94-007-0214-1_15,

express a *local* notion of set within a type-theoretic setting. These resources frame an analysis of Hilbert's ε–operator (Bell, 1993). Bell also extracted from the categorical form of synthetic differential geometry a crystalline presentation of the basics of the calculus (Bell, 1998), one which sheds light on logical issues concerning the continuum and unlocks applications to both quantum and relativistic physics (Bell, 1996, 2006).

Within the range of ideas, techniques and results of categorical logic, only those of the most general character will be considered here. References will be confined to contributions of two kinds: papers of historical importance and survey papers or books, with no presumption of a complete list in each case. By skipping many details, the exposition will remain at an elementary level. Nowadays this conceptual machinery may perhaps seem obvious to many mathematicians, but it was the outcome of an anything but obvious development, the source of which lies in algebraic geometry, and even when the dependence on this source is deemphasized, the sense of the categorical way to logic still remains cryptic for others as well as for many logicians, computer scientists and philosophers.

Aiming to avoid the obvious and the cryptic, the paper reviews some main steps in the growth of categorical logic. The change in perspective did not occur suddenly and was hardly transparent at the birth of category theory 60 years ago. Since emphasis will be placed on the unifying concepts behind the "categorification" of logic, only occasional attention will be paid to topics external to its core, with only brief mention of modalities, relationships with set theory and the concept of meaning. Talk of "logic in category theory" intends to stress the change in perspective; thus the framework of Section 2 aims to show how much more is at stake than merely taking logic as practiced (in manifold ways) since the beginning of the twentieth century and translating it into categorical terms. The turn rather comes with doing logic from scratch in terms of adjoints and, in fact, using this strategy as an "order parameter." How and to what extent can (and must) this line extend to other areas of logic such as inductive or philosophical logic? Though this question remains implicit in what follows, some results described below suggest no less long-range effects. Section 3 introduces, in chronological order, the notions merged into present-day categorical logic. Section 4 outlines the constructions corresponding to propositional, first order and higher-order logic, with emphasis on the unified picture resulting for model theory, proof theory and computability theory (in λ-calculus form). Finally, Section 5 will refer to the extension of this picture related to constructive type theories and propose some general reflections.

2 Perspective Remarks

The problems at the origin of categorical logic were different from those out of which the two souls of mathematical logic were born: In the Boole-Schröder lineage logic was an autonomous branch of algebra, while in the Frege-Russell tradition it carried the weight of a foundation of mathematics as a whole, see (Mayberry,

1994). Through the Hilbert-style axiomatization of logic it became possible to draw clear-cut boundaries among three, by now familiar, layers of logical structure: propositional, first order and higher order. Set-theoretic assumptions confined to the background re-emerged as soon as Tarski established semantics as a branch of logic.

In 1945 Saunders Mac Lane and Samuel Eilenberg introduced the notion of category (Eilenberg and Mac Lane, 1945), as identifying the formal profile of what a mathematical universe of discourse is in general, in which objects $A, B, \ldots$ and maps $f, g, \ldots$ (arrows, morphisms, but not necessarily functions) between objects are on the same footing. A category **C** can be described as both a generalized, or many-sorted, monoid, and a directed graph with added equations between arcs. Each map f has a unique domain/source $dom(f)$ and a unique codomain/target $cod(f)$. To write $f : A \to B$ means that A is the domain $dom(f)$ and B the codomain $cod(f)$. Significantly, the two axioms for a category deal directly with the maps: (i) the existence, for any object A, of an identity map 1_A, such that for any map g of domain A and any f of codomain A, $g \cdot 1_A = g$ and $1_A \cdot f = f$, (ii) associativity of composition $h \cdot (g \cdot f) = (h \cdot g) \cdot f$, provided each composite is defined, i.e., $dom(h \cdot g) = dom(g) = cod(f)$ and $dom(h) = cod(g \cdot f) = cod(g)$. The notion of category was subsidiary in fact to that of *functor* as a map F from a category **C** to another category **D**, expressing a no less general condition for a minimal preservation process, such that $F(1_A) = 1_{F(A)}$ and $F(g \cdot f) = F(g) \cdot F(f)$, for any A, g, f in **C** (if the order of composition is reversed, the functor is contravariant, or equivalently an ordinary (covariant) functor defined on $\mathbf{C}^{op}$, the opposite category of **C** having all **C**-maps reversed). The notion of functor in turn was needed to define the uniformity of transformations, as for example the basis-independent "isomorphism" between a vector space V and its double dual V^{**}: a *natural transformation* between two functors $F, F' : \mathbf{C} \longrightarrow \mathbf{D}$, is a map τ such that $\tau_A : F(A) \to F'(A)$, for every object A in **C**, and, given $f : A \to B$ in **C**, $\tau_B \cdot F(f) = F'(f) \cdot \tau_A$.[1]

In addition to the idea that mathematics is organized into categories—such as *Grp* of groups and homomorphisms, *Spaces* of spaces and continuous maps, *Set* of sets and functions—which are not themselves necessarily organized into one hierarchy, another basic idea is that of investigating the properties of any object through the maps between it and other objects. This includes investigating the properties of any category **C** by considering functors between **C** and other categories. It is not, however, the land of structuralist holism, where the identity of any entity is given only by reference to a larger collections of entities, and so on indefinitely. Categorical logic uses specific resources to compress the information needed to characterize such properties by means of universal constructions and specific patterns of internal and external parametrization. Characterization is "up to isomorphism" for objects of a category and "up to equivalence" for categories, where a map $f : A \to B$

[1] Henceforth, $\cdot$ will be generally omitted. As for the size of the collection of objects and maps, for simplicity's sake the categories considered will be *small*, i.e., with a *set* of objects (rather than a proper class), or *locally small*, i.e., with a *set* of maps between any two objects.

establishes an isomorphism $A \cong B$ if it has a left and right inverse $g : B \to A$, i.e., $gf = 1_A$ and $fg = 1_B$, and a functor $F : \mathbf{C} \longrightarrow \mathbf{D}$ establishes an equivalence $\mathbf{C} \equiv \mathbf{D}$ if there is a quasi-inverse $G : D \longrightarrow C$ such that, for any $\mathbf{C}$-object A and any $\mathbf{D}$-object B, $GF(A) \cong A$ and $FG(B) \cong B$.

One of the leading ideas Mac Lane advocated as central to a categorical "philosophy" of mathematics was that each mathematical form has many different realizations and category theory aims at an axiomatic description of such forms, which also makes the basic patterns of their mutual relations explicit. What about logic? The answer came with Lawvere's contributions in the sixties, leading to a picture different from both the Boole-Schröder and the Frege-Russell lineages. Lawvere conceives mathematical logic as the *logic of mathematics* which finds formal expression by adjoints, and in this way manifests "general laws about the movement of human thinking," (Lawvere and Rosebrugh, 2003, p. 193), deployed across different mathematical universes of discourse. Emphasis is on objectivity of invariants, and so on syntax-independent structure, in a way equally different from the logicistic approach (however type-theoretically rephrased), a formalistic view of the axiomatic method, and an intuitionistic perspective as well, since the character of logical principles remains objective—and support for this claim comes from the way a categorical approach recovers constructive reasoning, as intrinsically emergent out of the structure of variation and cohesion (Peruzzi, 2000a).

Up to the second half of the seventies the relations between logic and category theory concerned only a minority of category-theorists. Then the situation changed. Whereas in 1977 it could appear merely a novelty that categorical logic was included in Jon Barwise's *Handbook of Mathematical Logic*, with two chapters: one by Anders Kock and Gonzalo Reyes and one by Michael Fourman (Kock and Reyes, 1977; Fourman, 1977), now it occurs even as one of the subjects covered in the 2005 expansion of the *Handbook of Philosophical Logic*, with a chapter by Bell (2005). One might presume the subject is finally recognized as belonging to the intersection of mathematical and philosophical logic. That they have non-empty intersection is clear. That the intersection contains categorical logic is not; and yet, it is useful to make a point: Topics of philosophical logic cover additional aspects of language and reasoning and a fortiori they call for more sophisticated mathematical resources, rather than dispensing with them. If the commitment underlying this point is set aside, the only sense to be made of such an intersection relies on the joint effect of two data: (i) higher-order (free) intuitionistic logic being valid and complete for topos-models; (ii) intuitionistic conditions on the meaning of $\neg$, $\vee$ and $\exists$ being of relevance not only for philosophy of mathematics but (after Michael Dummett, see (Dummett, 1975)) for philosophy of language too.[2]

This may seem a puzzling motivation from what might be called a sociological point of view, given that the standard uses of logic by philosophers in the analysis

[2] Some notions yet to be defined—to start with, that of a topos—are exploited in the present section to pinpoint the different components involved in the impact of categorical thinking on logic, thus affecting its use in extra-mathematical applications.

of language completely ignore categories and functors. Actually, if there is a way to logic most strictly linked to mathematics, it is the categorical way—the presence of $=$ as a primitive notion testifies to that link. Far from being puzzling, the motivation is revealing, once a mathematically substantive framework is required to elucidate the problem of how the logical structure of language is connected to space and change, especially in the way space and change pertain to the commonsense world. An articulated discussion of this point would set the order of matters upside-down, by presupposing tools provided by the categorical analysis of object-patterns and action-patterns (Peruzzi, 2000b), the logical import of which is yet to be introduced. But while the interaction between logic and category theory can be appreciated by itself, this can best be achieved by avoiding two assumptions: that category theory deals with aspects of logic, which, *qua* logical, can be elaborated on presumably autonomous grounds, and that it merely offers an auxiliary language to express results (that can be) achieved with no mention of categories. To counter these (related) assumptions, it is necessary to focus on aspects of logical import which have only emerged through category theory: from refinement of core-concepts as those of variable, substitution, connective and quantifier, to an intrinsic relationships of a theory with its models, up to semantics of λ-calculi and constructive type theories.[3]

One obstacle to grasping the change involved was precisely the received view, present in many logic textbooks, that $=$ is not a logical notion. This view could be adopted only insofar as propositions were treated separately from proofs and proofs were supposed to be unsuited for algebraic description. Category theory puts propositions (as objects) and proofs (as maps) on the same footing, thus propositions and proofs are treated in coordination without, at the same time, merely reducing a proposition to the set of its proofs. Though the categorical analysis of logic makes appeal to stronger assumptions than usually realized, such as that True and False enter logical syntax as 0-ary constants, the resulting refinement of basic notions of logic, together with the view of theories-as-categories and of models-as-functors reveals the actual mathematical content of previous results. Two examples: Deligne's Theorem, to the effect that every coherent topos has enough points is equivalent to completeness for geometric theories; and Stone duality for full first order logic was obtained categorically by adapting the ultraproduct construction (Makkai, 1987), so that completeness is framed as a representation theorem and, more generally, representability of a theory T passes through a syntactic category C_T canonically associated with T.

Let us see how this association is defined. First, a signature Σ is specified, starting from ground sorts, many-sorted function and relation symbols, from which Σ-terms and Σ-formulae are recursively defined in the usual way (with resort to the machinery of a many-sorted first-order language containing $=$, $\wedge$, $\top$ and $\exists$). A Σ-theory T is then a deductively closed set of sequents involving Σ-formulae.

[3] Recent survey papers, such as (Pitts, 2001; Bell, 2005), provide a detailed picture of the impact of categories on also other aspects of logic.

C_T has as objects equivalence classes $\{\mathbf{x}.\varphi\}$—up to clash-free renaming of variables and suitable substitutions—for any $\Sigma(T)$-formula φ, where the context $\mathbf{x}$ lists all the free variables in φ; maps $\{\mathbf{x}.\varphi\} \to \{\mathbf{y}.\psi\}$ are the T-provably equivalent classes $[\theta(\mathbf{x}, \mathbf{y})]$ of formulae $\theta(\mathbf{x}, \mathbf{y})$ in T-language, with $\mathbf{x}$ disjoint from $\mathbf{y}$, which are provably functional, in the sense that the sequents $\theta \vdash \varphi \wedge \psi$, $\varphi(\mathbf{x}) \vdash \exists \mathbf{y}\theta(\mathbf{x}, \mathbf{y})$ and $\theta(\mathbf{x}, y) \wedge \theta(\mathbf{x}, z) \vdash y = z$, with each sequent taken in the same context. Composition is directly defined, by taking care of independence from the representatives of such classes, and $[\varphi \wedge \mathbf{x} = \mathbf{x}'] : \{\mathbf{x}.\varphi\} \to \{\mathbf{x}'.\varphi(\mathbf{x}'/\mathbf{x})\}$ acts as the identity. Thus C_T is a category with finite limits and $\{\top\}$ as terminal.[4]

Another consequence of a categorical approach is that quantifiers enter the language simultaneously with the other connectives, yielding a different hierarchy of logical complexity which assigns "geometric logic" (as a subsystem of infinitary intuitionistic logic) special meaning as the logic of geometric theories, where a geometric theory is one axiomatized by sequents $\varphi \vdash \psi$ such that φ and ψ are positive-existential formulae with possibly infinitary disjunctions, reducing to a coherent theory if disjunctions are finitary; the existence of a "generic" model for any geometric theory has been a central theorem, but one can prove it only after setting the syntax-semantics relationships in functorial form.[5]

The logical import of duality and genericity was unrecognized in early category theory. More, it seemed initially (and did for about 20 years) that the basic concepts and the aims of category theory are remote from logic: categories, functors and natural transformations serve at best to investigate the conditions under which any given mathematical structure, say the structure shared by set-theoretic models of a first-order axiomatizable theory, is preserved or not under suitable maps, provided the language of the theory is purely functional and the form of the theory is equational. Had this remained the case, the interest of categories would have continued to be confined to algebra and related aspects of topology and geometry. But the concept of adjoint functor changed all this. The ubiquity of adjunctions across mathematics is more than a matter of architecture and one place to realize it is in logic. Lawvere highlighted this ubiquity, in contrast with the idea that logic only seeks to identify primitive notions with which to

---

[4] The logical aspects of Stone duality are located in a much wider "phenomenology" of dualities, which find unified formulation in categorical language, as results from (Johnstone, 1982). Representability is actually a constant theme of category theory, which offers the general environment to unify results in the line of Cayley and Stone (but also of Grothendieck). A reference point for this unification is the *Yoneda Lemma*: given the "hom-functor" $h_A : \mathbf{C} \longrightarrow \mathbf{Set}$, with $h_A(-)$ = the set of $\mathbf{C}$-maps from A to –, for a fixed object A of $\mathbf{C}$ (see footnote 1), and given any $F : \mathbf{C} \longrightarrow \mathbf{Set}$, there is a bijection between the natural transformations $h_A \to F$ and the set $F(A)$. The Lemma also extends to contravariant functors, in which case the $\mathbf{C}$-maps from – to A are considered to define h^A. The embedding $\mathbf{C} \longrightarrow \mathbf{Set}^{\mathbf{C}^{op}}$ is full and faithful and thus allows one to investigate $\mathbf{C}$ in a wider context with no loss of information.

[5] That the theory of fields and in particular the theory of local fields can be expressed by coherent first-order axioms is relevant for the use of constructive reasoning in algebra, ... as much as that the subtle distinctions involved in axioms of different forms for the concept of field can be confined to Boolean toposes.

achieve a possibly minimal and adequate axiomatization, when he demanded that foundations transform the architecture revealed by those features of mathematical practice common to all its fields into explicit axioms (Lawvere, 1969). This demand is reminiscent of Eliakim Moore's attitude already expressed in (Moore, 1908), which had influenced Mac Lane (McLarty, 2005). This ideal thread is helpful to realize how Lawvere's demand, consistent with the attention to practical needs underlying the very origins of category theory, allows a different concept of "foundations" to emerge. For, by the notion of adjoint the missing links between Bourbaki's *structures mères* are filled and turned into a subject of logical concern, rather than leaving such *structures mères* and their combinations as contingent highlands in a sea of sets. The same ubiquity of adjoints lends itself to identify mathematically relevant systems of logic, hierarchically organised in correspondence with layers of mathematical structure and its functorial preservation.

If the precise correspondence between topology and logic had already been made clear by the pioneering works by Tarski and Stone, as summed up in the textbook (Rasiowa and Sikorski, 1963), a deeper perspective was opened by the French School of algebraic geometry with the introduction of new and more powerful tools in order to solve cohomology problems. Alexandre Grothendieck was led to use the notion of sheaf over a topological space. A *sheaf* of groups (rings, spaces, sets, ...) over a space B is intuitively a family of locally (fibrewise) defined groups (rings, spaces, sets, ...) continuously varying over B, where sections glued with sections over any open have unique restrictions to any smaller open. Grothendieck investigated *schemes*, as sheaves obtained by gluing together spaces of prime ideals in commutative rings. Beyond the obvious contravariant nature of the restriction of any such structure from V to U, for $U \hookrightarrow V$ opens in X, Grothendieck found how to fully express the sheaf notion as a functor $O(X)^{op} \longrightarrow \mathbf{Set}$, and to generalize this notion to a functorial construction $F : \mathbf{C}^{op} \longrightarrow \mathbf{Set}$ over a *site* of definition $(\mathbf{C}, J)$ as a (small) category $\mathbf{C}$ with a pullback-stable system J of covering families of morphisms $f_i : A_i \to A$ for each object A, so that a topology is given by purely categorical conditions and the gluing axiom becomes the condition that $F(A) \to \prod_i F(A_i) \rightrightarrows \prod_{i,i'} F(A_i \times A_{i'})$ is an equalizer diagram. (In this way, also C_T becomes a site.) To have a suitable resource of "meter sticks" in the study of schemes, in 1963 he introduced the category of such sheaves $\mathbf{Sh}(\mathbf{C})$ as a topos (a "Grothendieck topos") conceived of as a "generalized space" and finally investigated the category of toposes *defined over* a base topos, which in turn could vary. Thus he claimed "the right to transcribe mathematics into any topos whatever," as emphasized by Pierre Cartier (Cartier, 2001, p. 395), so that the very interpretation of any statement is internalised in each given topos, while geometric morphisms between toposes are the appropriate functors to express, in full generality, the change-of-base technique developed within algebraic geometry. Not only will there be different hierarchies of sets within different toposes, but also: if the essential constituents of mathematical thought are those invariant under change-of-topos, no specific cumulative hierarchy of sets can be a proper candidate for foundations; and if such constituents are expressed by universal patterns of construction, $\in$-based set theory is unsuited to the task—its constructive versions add extrinsic logical constraints on $\in$-structure—whereas these features are jointly and directly realized within a categorical perspective.

It was not further topological interpretations of autonomously given logical systems, but rather this generalized and point free notion of space which enlarged the horizon of topology itself and prepared the ground for synthetic differential geometry in *smooth* toposes. A *smooth* topos is one with a real-numbers-object $\mathbf{R}$, any $\mathbf{R} \to \mathbf{R}^{(n)}$ smooth, and a subobject $D = \{d \in \mathbf{R} : d^2 = 0\}$ of first order infinitesimals, satisfying the "Kock-Lawvere" axiom: for each $f : D \to \mathbf{R}$ there is a unique real b such that, for any infinitesimal d, $f(d) = f(0) + d.b$. (Note that D, and a fortiori $\mathbf{R}$, is not decidable.) More to the point here, the topos structure gave the mathematical motivation for geometric logic and for theories with such underlying logic. If the gluing condition on sheaves is dropped, we get a further generalization of the initial case $O(X)^{op} \longrightarrow$ **Set**, namely the notion of a *presheaf* of **E**-objects, which is simply a functor $\mathbf{C}^{op} \longrightarrow \mathbf{E}$.

When **E** is a topos, so also is the category of all presheaves $\mathbf{C}^{\mathbf{op}} \longrightarrow \mathbf{E}$. The study of presheaf toposes turned out to be of great utility in approaching set-theoretic independence problems (for $\mathbf{E} = \mathbf{Set}$) in a context more flexible than Boolean-valued models—where a Boolean-valued model $\mathbf{Set}^{\mathbf{B}}$ is the special case of presheaves of sets on a suitable Boolean algebra **B** other than **2**. Now, a sheaf topos is coherent if its underlying site satisfies a suitable refinement of compactness; and the case of geometric theories is to the point both in explaining how the connection between category theory and logic emerged from a specific area of mathematics and to provide a paradigmatic instance of the way constructive reasoning inheres in categorical constructions. In fact, the logic of such theories can be traced right back to the notion of site, as a "base space" no longer described by open sets of points, and the lack of the law of the excluded middle (LEM) is related to covering conditions in the site and their stability under pullbacks. The full logic of sheaves over a site turns out to be intuitionistic in general, but this did not result from a philosophical bias toward an epistemic (verificationist) or specifically idealistic view of mathematics. Whereas Brouwer the topologist made substantial contributions to the understanding of space (dimension, fix-point theorem, degree of a continuous map of orientable manifolds), Brouwer the philosopher of intuitionism dispensed with space, after the failure of the Kantian foundation of geometry, to leave the inner intuition of time as the only remaining source. The rediscovery of intuitionistic logic within topos theory was rather due to the definition of sheaves, and indeed came from a much more general (and basic) concept of space, namely the site of definition.

Bill Lawvere realized that any Grothendieck topos has a truth-value object Ω and that the existence of Ω in a category **C** with finite limits makes it possible to express the comprehension principle, for formulae interpretable in **C**, as a categorical statement. In 1969–1970 he and Myles Tierney arrived at a more general notion of topos than that of Grothendieck, and one elementarily axiomatizable. Since then, the project of a categorical theory of (continuously variable and cohesive) sets grew up as a framework for the foundations of mathematics, alternative to the standard $\in$-based set theory. It turned out that the logic of a topos is, in general, intuitionistic and is reflected in the properties of Ω, but then the converse was also at hand: a Grothendieck topology (or coverage) can be defined as an operator $j : \Omega \to \Omega$, such that (1) $j \cdot \top = \top$, (2) $j \cdot j = j$, (3) $j \cdot \wedge = \wedge \cdot (j \times j)$, thus j preserves truth, is idempotent and distributes over conjunctions, resulting into a "Lawvere-Tierney topology." Hence a purely categorical construction captures the logical import of such a topology and can logically characterize the very notion of sheaf, as shown in full detail by (Bell, 1988, Ch. 5). In particular, for $j = \neg\neg$, the double negation sheaves allow one to obtain elegant categorical independence proofs.

Already in 1964 Lawvere had axiomatized the category of sets (Lawvere, 1964). For some time, the features of such an $\in$-free system, compared to previous foundational theories, were accessible to logicians only through the last chapter of (Hatcher, 1968). But the resources of an elementary topos made clear that logical notions can be expressed in a topos and that their intrinsic behaviour is intuitionistic, since the algebra of subobjects of the terminal 1 is Heyting and there is an order preserving bijection between $Sub(1)$ and the algebra of truth-values $1 \to \Omega$. The implicit constraint is one of ontological homogeneity: truth-values as well as numbers, referred to by statements about **C**, have to be definable in terms of **C**-structure itself. As this paper does not intend to focus on foundations, it will not broach the many issues involved in the set *vs* category theory debate.[6] Nonetheless, two aspects cannot be omitted: (I) membership, and (II) internalization.

(I) Category theory describes any mathematical structure in terms of morphisms rather than by means of $\in$, but this does not mean that elements are ignored. Rather, the concept of element undergoes refinement and generalization in a way which (once more) originates from algebraic geometry, where a map $X \to A$ may be seen as an element of A defined over X, or, for suitable X, as an X-shaped figure in A, or again as a generalized element of A varying over X, with A and X as objects of the same category. Point-elements are only a particular instance and the existence of enough points to characterize an object is far from trivial. (The assumption that it is so for any object, namely *wellpointedness*, coincides with standard extensionality and the fact that its very form is metatheoretically related to a form of completeness was recognised only by means of topos theory.) An elementary topos is precisely a category of variable sets, with variable elements, over a base B. The collapse of the base into any one-point set, as in classical set theory, freezes variation and restricts $\in$ to *global* or constant elements, namely those that can be factored through the terminal. In the case of **Set** that means elements that factor through any singleton set.

Constancy being associated with globality, global elements (as points) are in general not enough to characterize the objects, and when they are the topos is

---

[6] Together with various contributions by Colin McLarty on the fom- and categories-lists, the debate could have benefited more from Bell's fair as well as suggestive remarks in (Bell, 1986).

called *well-pointed.* Finally, $\in_A$ is definable as the domain of the monomorphism $m : U \rightarrowtail \Omega^A \times A$ uniquely determined (up to iso) by the pullback square $ev_A \cdot m = \top \cdot !_U$, which covers the classical case with $\Omega = 2$.[7] But variable elements do not necessarily collapse to constant ones and, at the same time, the condition for being a subobject is stricter, though given "up to isomorphism," than that for a mere subset.

In many categories, for example in categories of sheaves, the terminal is not punctiform as it is in the category **Set** of classical sets; in such a basic a category of sheaves as the *étale* topos **Sh(B)** of spaces locally homeomorphic to a base space B, constant elements (to be denoted by closed terms) are determined by global sections. Type-theoretically, the upshot of this is that there can be sorts with no closed terms (constant terms), and yet any term $t(-) : X$ can be treated by exponentiation as a closed term $1 \to X^{(-)}$. Variable elements are more than multiple possibilities of constancy and the resulting semantics for type-theories is more general than the classical set-theoretic one. A given map $E \to B$ can have many local sections but no global one, i.e. the only existing E-elements are partially defined on B, while the cohesion of objects is reflected into their algebra of parts, which in general is not Boolean.

For any topological space X, a point $x \in X$ determines a continuous map $\hat{}x$: $\{*\} \to X$, and conversely. The topos $Sh(\{*\})$ is nothing but Set. So, the same point x determines a geometric morphism p_x : **Set** $\longrightarrow$ **Sh(B)**, which by extension is also considered as a *point* of **Sh(B)**. Any topos with a geometric morphism to **Set** is said to be an *S-topos.* Such a functor is unique up to iso and any such topos is characterized by having arbitrary set-indexed copowers of 1 (local smallness was assumed at the beginning). A *point* of an S-topos **E** is accordingly a geometric morphism p : **Set** $\longrightarrow$ **E**. The topos **E** has *enough points* if for any two different maps $f, g : A \to B$ in **E** there is a point that distinguishes them, thus a p such that $p^* \cdot f \neq p^* \cdot g$. In this case we also say that the collection of point-functors to **E** is "jointly conservative."

This property has a definite model-theoretic meaning, as shown below. Thus, in contrast with the persistent contraposition of set theory and category theory, it is more proper to say that a categorical foundation pivots around a theory of cohesive sets with variable and partially defined elements, while classical set theory only deals with the special case of variable sets on a one-point space. Since the arithmetization of analysis, the common attitude is that constancy precedes variability, hence a *variable* is just an indeterminate constant entity. Categorically, in order to refer to a *constant*, an argument is needed to show factorizability through 1.

(II) If all of logic (say all of logic necessary to present-day mathematical practice) can be done, and in a more uniform way, within category theory, then even the claim

[7] A topos can be shown to have power objects $P(-)$ (represented as Ω^{-}), in view of the unique correspondence between relation maps $r : R \rightarrowtail B \times A$ and maps $f_r : B \to P(A)$. Vice versa, by the existence of the pullback of $f_r \times 1_A \cdot r$ along $\in_A$, Ω can be obtained as $P(1)$ up to isomorphism.

that mathematics is founded on ∈-based axioms can benefit from the role categorical concepts play in foundations, as logic does. This may seem to be avoiding the question, but it has the virtue of shifting the focus from formal semantics and ontology to logical syntax.

There is indeed a problem with this shift: any (locally small) topos admits, by means of its internal language, a set-theoretical formulation: it is what Bell defined as "local set theory" (from untyped to typed ∈), and the underlying strategy of internalization can be put to work already for categories (and corresponding theories) with weaker expressive resources. The discovery of internal languages may suggest that, in the end, what counts in categorical logic is just *set* theory. It will not be the "GO-FAST" (Good Old-Fashioned Axiomatic Set Theory), but still set theory, with suitably "localized" $=, \in, \subseteq$. In this set theory, so long as all inferences conform to intuitionistic principles, any substantial reference to categorical "jargon" can be gently pushed aside, apart from now talking about categories, rather than mere collections, of sets. And the need to index families of objects can be satisfied by means of suitably "localized" set-theoretic resources. In other words, as soon as the categorical analysis of logic based on Grothendieck's sheaf theory is internalized, any foundational pretension from categorists becomes unnecessary, for, conceding that between the lines use of category theory may be essential in various branches of mathematics, on the foundational front this turns out to be dispensable, in so far as categorical properties can be translated into a purely type-theoretical language out of which to re-frame ∈-based set theory.[8] So, while it may seem that category theory is needed for *making sense* of the change in the received view of sets as well as for the *architecture* of mathematics; yet, exactly as in previous logicist projects for foundations, these needs leave no trace in the axioms. As "Abstract Nonsense" category theory was more productive than expected but, after proper house-cleaning (read: after type-theoretic paraphrase), it is revealed as just that, only in a new sense: precisely in virtue of its content-plasticity, category theory is dispensable or, if it has any foundational import, this import can be conveyed also by means of a non-categorical language. So the argument could go.

Categorical logic provides specific evidence that such a line of argument is flawed. Section 4 will mention model-theoretic results which show the logical meaning of genericity, which are achieved by a variational method aimed at checking the universal property of a generic model. This method agrees with motivations of a "phenomenological" character, in a sense that differs from Husserlian tenets in many respects and ultimately—by an argument that will not be repeated here (see Peruzzi, 2000b)—amounts to identifying the source of formal notions in content-laden patterns of *activity*, as proposed by Mac Lane especially in (Mac Lane, 1986). It is a sense for which indispensability arguments neither arise by an account of use nor through metaphysical assumptions (even just those of an

[8] This argument is different from relying on mutual equiconsistency relative to sets: for instance, the power of the axioms for an elementary topos being equivalent to that of Z_0, i.e., ZF with only bounded quantifiers, is significant only with reference to a universe of discrete sets of points.

all-embracing formalism), being tied rather to explanatory power, heuristics and conceptual deepening. Since, on the other hand, arguments in support of the indispensability of categories may seem superfluous, indeed, unnecessary precisely in this regard, qualification is needed.

First, emphasis on advances of logic by means of categories is not intended to nullify previous achievements but to refine and locate the practice of logicians within a larger horizon of mathematical significance. Secondly, there is already a labyrinth of non-classical, modal, temporal and quantum logics. So there is a growing labyrinth of systems of categorical logic, related to semantically exotic phenomena with weak mathematical substance and unsuited to transferring the constructions involved to other problems—different from the logic of reflexive graphs and monoid-actions in dealing with dynamical systems relevant to continuum mechanics or cognition. In other words, if there is reason to regret the spread of neo-scholasticism in logical studies (especially those included in philosophical logic) before the use of categorical methods, so also one might deplore the overly formal exercises that have resulted from their application. Yet, the range of uses and abuses of any conceptual apparatus is a measure of its potential; and in fact, as is to be expected when a mathematical theory acts as a powerful and flexible framework with general theorems adapted to many different areas, its very solidification also favours a scholastic attitude. At the same time its growth in complexity calls for diversified competence. That all these concerns apply to present-day categorical logic is a tribute to its methods and results; but it also urges consideration of *what* is characteristic of a categorical approach which allows access to logical problems previously undetected and creates the tools to solve them.

In a nutshell, there are four main aspects of categorical logic. One is the already mentioned refinement of the meanings of *variable*, *constant*, *connective* and *quantifier*, which is directly associated with "universal" properties of mathematical construction-patterns; second, the idea of theories-as-categories enables the use of functorial interpretations and preservation of theories across models, even in the absence of the (classical) prenex normal form theorem; third, an efficient notation system for proofs as maps opens the way to the "geometry" of proofs in terms of coherence and homotopy, well beyond mere provability or its absence, and ties the analysis of constructivity to definite mathematical structures; fourth, such different theories as (untyped, typed, dependent, polymorphic) λ-calculi, synthetic differential geometry, realizability and intuitionistic set theory are merged into a unified framework through the expressive resources of categorical language. The tools needed to make these four aspects explicit will be summarized in Sections 3 and 4: specific layers of logical structure in correspondence with different kinds of categories will serve as a guide-line (following a suggestion implicit throughout (Johnstone, 2002, vol. 2, § D)) to identify the perspective peculiar to categorical logic, which in the end can be seen as an ideal lever putting inferential patterns in direct contact with root components of our understanding of space and motion.

3 The Working Mathematician's Path to Logic

Algebraic logic, in the tradition from Boole to Tarski, hosts the core notions and methods which justify the inclusion of logic as a mathematical subject, but it met serious obstacles in previous attempts, by means of polyadic and cylindric algebras, to deal effectively with nested quantifiers in the presence of n-ary predicates, with $n > 1$. It also had to meet Frege's arguments against the idea that "p is true" can be expressed by an equation (as $p = \top$). The Lindenbaum-Tarski construction of a term algebra defined by equivalence classes with respect to provable biconditionals (or equations) was adequate for what concerns provability and completeness but did not cover the structure of proofs. The Hilbert program showed the need of an explicit analysis of proof-structure. Gentzen's natural deduction and sequent calculus were the first answer to such a need and became a cornerstone in proof theory while remaining separate from algebra and model theory. This separation was replaced, years later, by an intrinsic link through the development of category theory. Together with Gentzen, Mac Lane was a PhD student at Göttingen from 1931 to 1933; there he developed interest in logic. In effect, Mac Lane's first paper (Mac Lane, 1935) which grew out of the dissertation prepared under Bernays' and Weyl's guidance, concerned logic; his initial study of the notion of proof ties in with ideas expressed 53 years later in (Mac Lane, 1997). Recalling this would only be suggestive were it not for the amount of research in categorical logic actually accomplished in between. A bit of chronology is thus helpful in first recognizing how the path from categories to logic could appear as a side-track and then making explicit, step by step, how components of a different character, originating within areas far from the typical concern of logicians, merged into the present picture.

As was anticipated in Section 2, the mathematical tools peculiar to such a path, based on features involved in indexing/parametrizing and in functorial change-of-base, were created during the second half of the fifties by the French school of algebraic geometry, through the *Seminaires* at the IHES and especially the pioneering contributions by Grothendieck.[9] The definition of the concept of adjoint functor by Daniel Kan (1958), added a decisive instrument for organizing mathematical constructions into a single powerful pattern and to stimulate its identification in many contexts beyond the original source in algebraic field theory.

Actually, adjunctions generalize Galois connections, such as the one between subsets of a group H and subsets of a field X on which H acts: for H the automorphism group $Aut(X)$, there is a pair of maps $F : P(X) \longrightarrow P(Aut(X))$ and $G : P(Aut(X)) \longrightarrow P(X)$, where F assigns each subset Y of elements of X the subgroup of actions which keeps the elements in Y fixed, and G assigns each subset K of automorphisms its set of fixed points. Thus $F(Y) \geq K$ iff $Y \leq G(K)$. The general pattern is: a functor $F : \mathbf{C} \longrightarrow \mathbf{D}$ is left adjoint to $G : \mathbf{D} \longrightarrow \mathbf{C}$

[9] Unfortunately, a relevant part of the work on schemes, sheaves and fibred categories was accessible to a large audience only much later (Grothendieck, 1970; Grothendieck and Dieudonné, 1971).

(and G right adjoint to F), written $F \dashv G$, if there is a natural bijection φ between **D**-morphisms $F(C) \to D$ and **C**-morphisms $C \to G(D)$, written $\frac{f:F(C)\to D}{\varphi f:C\to G(D)}$, φ being *natural* in the sense that it does not depend on the specific objects and morphisms considered. In addition to the uniqueness of adjoints when they exist, what is crucial in an adjunction is the unique existence of two natural transformations: a unit η_C: $C \to GF(C)$ and a counit ϵ_D: $FG(D) \to D$, with $\eta_C = \varphi(1_{F(C)})$ and $\epsilon_D = \varphi^{-1}(1_{G(D)})$, since they capture a uniform notion of universality. For, any $f : F(C) \to D$ and any $g : C \to G(D)$ can be uniquely recovered by the factorizations $f = \varphi^{-1}(Gf \cdot \eta_C)$ and $g = \varphi(\epsilon_D \cdot Fg)$. The adjunction pattern is, in effect, the *global* form of a universal map, *free* for a functor over *an* object (or its dual, as in the case of Ω in relation to the functor Sub). Such uniformization is instantiated in such diverse ways as the construction of a free algebra over a set of generators, the evaluation map (as in modus ponens, or the β-rule in terms of computability), but also the notion of polarity and finally the construction of a generic model of a theory.

Intuitively, a diagram D of shape I in **C** is a graph map $D : I \longrightarrow$ **C**. Central to any category **C** are the commutative diagrams of maps and the existence of limits and colimits for diagrams of given shape. Each limit is just the terminal object in the category of cones over diagrams of the same shape, with morphisms all the maps required to allow composites to commute. Colimits are the dual of limits. The notion of *pullback* (also known as fibred product) is given as the limit of diagrams of shape $\cdot \to \cdot \leftarrow \cdot$, i.e., for $f : A \to C \leftarrow B : g$, an object $X = A \times_C B$ and a pair of maps $h : A \leftarrow X \to B : k$ such that $fh = gk$, with the universal property that for any other $h' : A \leftarrow X' \to B : k'$ there is a (unique) $u : X' \to X$ with $h' = hu$ and $k' = ku$. The dual (in the opposite category, with all maps reversed) of a pullback is a *pushout*, which yields coproducts and initial objects as particular cases. The pullback notion is an instance of the gain in generality, as it covers inverse images, meets, intersections, monomorphisms (as left cancellable maps), relativized conjunctions and finally products, the latter being pullbacks over the terminal, which in turn is the limit for an empty diagram. A property is pullback-stable if it is preserved by pullbacks. Existence of limits or colimits in ensured by special functors from-to **C** having left or right adjoints. An *equalizer* is a limit for diagrams of shape $\cdot \rightrightarrows \cdot$ or in other words for parallel pairs of maps $f, g : A \to B$. The notion of monomorphism is a *proper* generalization of that of injective function; the same holds for its dual (epimorphism) with respect to surjective function. The dual notion of an equalizer is that of coequalizer, under which quotients of equivalence relations fall as an instance. A functor is *left exact* (right exact) if it preserves all finite (co-)limits.

In 1963, Lawvere's thesis (Lawvere, 1963) introduced the idea of theories-as-categories and models-as-functors, by dealing with sets of equations expressed by commutative diagrams with no use of $\in$. Whereas universal algebra investigated signatures and corresponding varieties, the concept of a *doctrine* as defined by Lawvere allowed a unified analysis of algebraic theories T and their categories of models (T-algebras), with natural transformations as morphisms. An elementary doctrine is given by a base category with finite products **T** of objects-as-types and maps-as-terms, plus a contravariant assignment of a category of attributes $Att(A)$ for each object A and substitution $Att(f) : Att(B) \to Att(A)$ for $f : A \to B$. A doctrine is existential if Att has a left adjoint. This determines the (full) subcategory of functors $\mathbf{T} \longrightarrow \mathbf{E}$ from a theory $\mathbf{T}$ (and then from its classifying category) to an arbitrary $\mathbf{E}$, with natural transformations carrying the structure. For instance, the axioms for the doctrine of groups give rise to different instantiations of a group-object: an ordinary group when functorially mapped to sets, a topological group in the category of topological spaces, a Lie group in smooth manifolds, a commutative group in the category of groups.[10]

The treatment of theories as categories fits in with Lawvere's general picture, according to which a theory T is the "abstract general," the category of T-models in any given ambient category is the "concrete general," whereas a "concrete particular" is a given category $\mathbf{C}$ to which the process of extraction of a concept (structure) may be applied: by adjointness, each object of $\mathbf{C}$ can be canonically interpreted as a concrete structure of the given abstract kind. The whole suggests approaching logical properties of theories independently of presentation and syntax.

In 1963–1964, the *exposés* at the "Seminaire de Géométrie Algébrique du Bois-Marie" in Paris, later collected as (Artin et al., 1972), displayed the potentialities of categorical language in the description of Grothendieck toposes. Charles Ehresmann's definition of sketches in 1965, as given by an oriented graph with a set of finite diagrams and cones over them, starting from the presentation of a category with finite products, would also prove to be of logical interest later on. In 1968 Jim Lambek described for the first time deductive systems as graphs, which turn into categories when equations between proofs are added (Lambek, 1968).

These seemingly heterogeneous contributions, together with Grothendieck's concept of a fibred category, became in the following decades the axes for a whole area of logical investigation. They gained that role thanks to the seminal ideas and results in Lawvere's research papers of 1966–1969 (Lawvere, 1966, 1968, 1969), which made clear the following points: (1) The logical content of cartesian closed categories, i.e., categories with binary products, provided by the right adjoint to the diagonal functor Δ (with the terminal as the empty product) and a right adjoint to the product functor, so that there is natural bijection $\frac{A\times B\to C}{A\to C^B}$, for arbitrary objects A, B and C, thereby $\times \dashv exp$ (for exponentiation); when 1 is $\top$, $\times$ is $\wedge$, exp is $\Rightarrow$, one directly gets a fragment of propositional logic, with the Deduction Theorem built in. (2) The gain in expressive power as adjoints and indexed structure are taken together, the latter being associated with a variable algebra of predicates, as, after the notion of a doctrine, a hyper-doctrine is a fibred structure given by a functor $H : \mathbf{T}^{op} \longrightarrow \mathbf{S}$ from a category with finite products to a category $\mathbf{S}$ (usually a poset), where each $H(A)$ is a cartesian closed category (usually a Heyting algebra) the structure of which is homomorphically preserved by re-indexing $H(f) : H(B) \to H(A)$, for

---

[10] The idea of a doctrine extends to the consideration of "monads" and 2-categories, as categories with also a further "vertical" dimension of composition. This line will not be pursued here for simplicity's sake, but it is what allows connecting rewriting system for λ-calculi (Seely, 1987) with proof-homotopy as remarked below.

$f : A \to B$, and finally the substitution functor has both a left and a right adjoint (the source of models for polymorphic linear logic is in this notion); also equality can be characterized in hyperdoctrines as left adjoint to the contraction functor $\Delta^* : P(I \times I) \to P(I)$, such that $Eq(X) \subseteq Y$ iff $X \subseteq \Delta^*(Y)$, for a given set I and a *predicate* X defined on I, i.e., $X \in P(I)$ and $Y \in P(I \times I)$. (3) The categorical unification of syntactic and semantic self-reference in the form: an epimorphism $X \to Y^X$ determines a fixed point for any $Y \to Y$, see (Yanofsky, 2003) for a clear introduction. (4) The general role of adjoints in foundations, as supporting the proposal of a "dialectical" view, at odds with the empiricist and analytic legacy spread throughout contemporary philosophy of logic:

> Foundations will mean here the study of what is universal in mathematics. Thus Foundations in this sense cannot be identified with any "starting point" or "justification" for mathematics, though partial results in these directions may be among its fruits. But among the other fruits of Foundations so defined would presumably be guide-lines for passing from one branch of mathematics to another and for gauging to some extent which directions of research are likely to be relevant. (Lawvere, 1969, p. 281)

As already mentioned, in 1969–1970 Lawvere and Tierney succeeded in extracting an elementary formulation of the topos concept: an elementary topos is a cartesian closed category with a subobject classifier, i.e., an object Ω with a map $true : 1 \to \Omega$ such that for any subobject $a : A \rightarrowtail X$, there is characteristic map χ_a such that the equation $true!_A = \chi_a a$ corresponds to a pullback square. Note that a subobject $A \rightarrowtail X$ can be defined, independently of $\subseteq$, as the equivalence class of monomorphisms $a' : A' \rightarrowtail X$ for which there is an iso $f : A \to A'$ with $a'f = a$.)

At the 1970 Nice Colloquium, Lawvere presented a paper which explicitly defines quantifiers as adjoints to substitution (as described below in Section 4) (Lawvere, 1971). 1972 was rich in publications: the collection of essays (Lawvere, 1972) came with an Introduction in which the various ingredients of a categorical approach to logic were presented for the first time in a unified perspective, finally accessible to logicians with no previous acquaintance with categories; Peter Freyd gave a systematic exposition of results about elementary toposes (Freyd, 1972, §§ 2.4 and 4) within which logical features of Ω are neatly described; (Mitchell, 1972) first described in print the internal language of a topos, also identified by André Joyal and Jean Benabou, while Michel Coste gave it detailed formulation; Joyal also introduced the idea that Cohen, Robinson and Kripke forcing are instances of one general semantic pattern in sheaf toposes: it is what was later known as the *Beth-Joyal* or *Kripke-Joyal* semantics; finally, Lawvere began the seminal "Perugia Lectures," the notes of which (Lawvere, 1973) exercised widespread influence.

One of the first uses of sheaves within classical model theory was made by Angus MacIntyre in (MacIntyre, 1973). The following year, Michael Fourman submitted his thesis at Oxford, in which the Heyting algebra structure of Ω is extensively investigated, while Gonzalo Reyes offered an important synthesis, addressed to logicians working in model theory, of the steps leading "from sheaves to logic" (Reyes, 1974). The collection (Lawvere et al., 1975) gave the state of the art in 1975. In particular, the contribution by Gerhard Osius elaborated the comparison, started in (Osius, 1974), between internal and external semantics in a topos context.

At Montréal, which was then becoming the center for categorical studies, André Boileau's thesis *Types vs Topos* refined the link between the type-theoretical features of the internal language and topos-theoretic properties; the successive paper (Boileau and Joyal, 1981), provides a sharp and handy axiom-system. In 1975 Lawvere's paper "Continuously variable sets: algebraic geometry = geometric logic," presented at the Logic Colloquium 1973, was published in the proceedings volume (Lawvere, 1975). The very title of that paper sums up the core-idea behind the logical meaning of geometric morphisms. Finally, Benabou's investigation of small and locally small fibrations proved to be a crucial advance for both foundational issues, as emerges from (Benabou, 1985), and successive applications to theoretical computer science, after the needs of higher-order functional programming led to the acknowledgement of the effectiveness of a categorical semantics. The following year, Corrado Mangione gave the first historical reconstruction of the development of categorical logic (Mangione, 1976).

In 1977 two books appeared which, taken together, cover almost every aspect of the linkage between topos theory and logic at the time. One is the handbook (Makkai and Reyes, 1977), which presents the correspondence between degrees of first order complexity and sheaf categories, with emphasis on model-theoretic aspects, and exemplifies the topological motivation for infinitary logic. The other is Peter Johnstone's advanced *summa* of topos theory (Johnstone, 1977), which for the first time organized the vast amount of information about elementary and Grothendieck toposes into a systematic framework, though logic could only receive limited attention within it. The two aforementioned papers, (Kock and Reyes, 1977; Fourman, 1977), dealt with specific topics. John Zangwill presented the same year a thesis at Bristol in which the internal language of toposes is exploited to introduce local set theory (Zangwill, 1977). Thanks to the clear-cut and full elaboration Bell later accomplished in (Bell, 1988), the potentialities of a local set theory became fully recognisable to a larger audience.

In 1979, *Applications of Sheaves* appeared (Fourman et al., 1979), and it soon became a standard reference for Ω-sets, with Ω a complete Heyting algebra, i.e., satisfying the infinitary distributive law $x \wedge \bigvee y_i = \bigvee (x \wedge y_i)$, also described as a frame or as a locale (depending on the "direction" of morphisms between any two such structures), giving rise to opposite categories, with different properties as concerns duality. Ω-sets are mainly related to independence proofs, so that it is part of categorical set theory rather than categorical logic. In this way, both Boolean-valued models and topological models of intuitionistic analysis given by Dana Scott found a unified treatment, while Scott and Michael Fourman worked out an elegant semantic treatment of intuitionistic *free* first-order logic and axiomatized the associated theory of definite descriptions for partially defined elements (Fourman and Scott, 1979). The same year, Robert Goldblatt's lattice-theoretically oriented introduction (Goldblatt, 1979) allowed logicians to learn of semantics for first-order theories in toposes; unfortunately, adjoints only appear in Ch. 16 which was added to the 2nd (1984) edition, and play no role before this.

Growth was so intense in the 1980s that the short selection of works per year listed so far, however incomplete, should be tripled at least, in order to reach the present state of the art. The ideas did not change but rather found further applications and were more and more articulated. An extremely active area which led to a considerable change in focus for the subject was the recognition of systematic connections with computability theory, notably Martin Hyland's discovery of the *effective* topos *Eff* (Hyland, 1982), where all functions $N \to N$ are recursive. This is also known as the realizability topos as its objects $(X, =_X)$ come with a partial equivalence relation $=_X$ measured by means of natural numbers, so that an arithmetical sentence is true in *Eff* iff it is Kleene-realized, see (McLarty, 1992, Ch. 24).

Once λ-calculus and category theory came to be thought of as two sides of a theory of functions, there was an explosion of research on categories to match the various forms of λ-calculi: from simply typed and untyped λ-calculus to constructive type theory and proof theory, motivated by linear logic and dependent and polymorphic type theory (thanks especially to Jean Yves Girard and Per Martin-Löf), finally pointing at a new unification of logic in terms of fibrations. The relationships of topos theory with both higher order intuitionistic logic and various forms of the λ-calculus is extensively covered in the (by now canonical) reference volume (Lambek and Scott, 1986), while the introduction (Asperti and Longo, 1991) is addressed to computer scientists.

Colin McLarty's introduction (McLarty, 1992, Ch. 13–16) gives a crisp and synthetic account of logical aspects of topos theory; for a survey paper see (Peruzzi, 1991a). Francis Borceux deals with the subject in vol. 3, Ch. 6–7, of his wide-ranging treatise on categorical algebra (Borceux, 1994). Reyes and Macnamara, together with a research group in Montreal, developed the first systematic application of categorical semantics to the analysis of typing in ordinary language (Macnamara and Reyes, 1994). For an up-to-date presentation of logic in category theory, the best references are the relevant part Johnstone's great work (Johnstone, 2002, vol. 2, part D), and Bart Jacobs type-theoretically oriented handbook (Jacobs, 1999), while two recent, mutually complementary primer papers, are the already mentioned (Pitts, 2001; Bell, 2005); several remarks in (Lawvere and Rosebrugh, 2003), especially Appendix A, clarify topics touched in passing here and convey the essentials of the categorical spirit in logic. At present, the most extended source for logic in this spirit remains (Mac Lane and Moerdijk, 1992).

4 Layers of Logical Structure

Subsystems of zero–, first– and higher-order logic have long been investigated for and applied to reduction classes, model-theoretic preservation and definability hierarchies. Categorical logic has re-created some of these layers of structure and identified others with general mathematical content, associated with levels of integration between connectives and quantifiers. In correspondence with a hierarchy of functorial properties, layers of mixed structure (of connectives and quantifiers), some of them previously undetected, have thus become the subject of extensive research.

Full intuitionistic propositional logic IL_0 is obtained just by means of suitable adjoints, as shown in the following diagram for the category **P** of propositions and proofs, where **1** is the terminal category with only one object and one map, the identity.

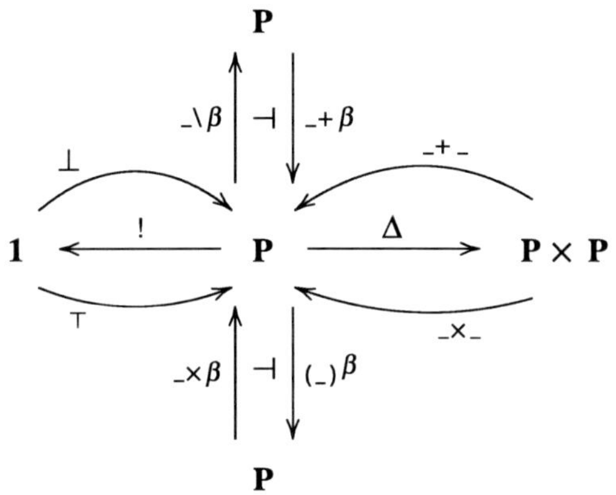

Between **1** and **P** each functor in the picture is right adjoint to the one above it, and the same between **P** and **P**$\times$**P**. The diagonal functor $\Delta : \mathbf{P} \longrightarrow \mathbf{P}\times\mathbf{P}$, is defined so that for any propositional formula φ and any proof f, $\Delta(\varphi) = \langle\varphi, \varphi\rangle$ and $\Delta(f) = \langle f, f\rangle$. This functor is left adjoint to the product functor because $\Delta(\varphi) \to \langle\psi, \gamma\rangle$ iff $\varphi \to \psi \times \gamma$, where $\times$ is just $\wedge$ in **P** and a map $\varphi \to \varphi'$ corresponds to an implication $\varphi \vdash \varphi'$. The adjoints suffice to determine a set of axioms and rules that makes two main kinds of data explicit: (1) the existence of *special* maps (identity, projection to conjuncts, evaluation (namely, *modus ponens*) and injection of disjuncts into their coproduct, with uniqueness of $!_\varphi : \varphi \to \top$ and $0_\varphi : \bot \to \varphi$, (2) the principles for composition of proofs, as well as for product, coproduct and exponential transpose of proofs (matching introduction and elimination rules), which characteristically come with a set of *equations* between proofs, as shown in full detail by (Lambek and Scott, 1986, Part I). Soundness and completeness have now the form: there is a **P**-morphism $x : \top \to \varphi$ (or, in logic notation, $\vdash \varphi$) iff the defining commutative diagrams are preserved by every functor to categories of appropriate kind to interpret **P**-structure. Many results take the form of showing that, for theories of given form, limited classes of functors already suffice.

Moreover, if a left adjoint $\setminus$ is supposed to exist for the coproduct map $\vee$, a co-Heyting algebra structure is obtained, to which a new connective, *subtraction*, corresponds, such that $\varphi \setminus \psi \to \gamma$ iff $\varphi \to \psi \vee \gamma$, which in case $\varphi = \top$ gives a different kind of negation, $-\psi$, dual to $\neg\psi = \bigvee\{\alpha : \alpha \wedge \psi = \bot\}$, and "$-$" generally differs from "$\neg$" in a bi-Heyting algebra. Lawvere suggests a distinction between contradiction and inconsistency which relates to this negation in contrast to the intuitionistic law $\varphi \wedge \neg\varphi \vdash \bot$ which follows directly from $\varphi \wedge \neg\varphi \vdash \gamma$, for arbitrary γ. When $\neg$ is replaced by the co-Heyting $-$, $\varphi \wedge -\varphi$ is initial no longer, corresponding rather to the boundary of φ, which in general is non-null.

This is also of interest for minimal logic, which is weaker than IL_0 in having only $\varphi \wedge \neg\varphi \vdash \neg\gamma$.

If rather than $\bot$, a relativized false statement dependent on φ is taken, the suggested distinction is further refined in precategories, by inverting the order of quantifiers in the customary Π_3-definition of identities (and the other basic categorical notions too), thus passing to a Π_2-definition, which makes them relative to the given maps. In place of identity maps a precategory has only endomaps $u_f : A \to A$ and $v_f : B \to B$ (taken as idempotent) depending on the particular $f : A \to B$, so that $fu_f = f$ and $v_f f = f$ but in general $u_f \neq u_g$ and $v_f \neq v_g$, for $g : A \to B$ with $g \neq f$, see (Peruzzi, 1994).

If **P** is supposed to be a poset category (i.e., having only one arrow $\leq$ for $\vdash$ between any two objects as elements of the poset) the direct way to add modalities is by taking a subcategory **M** of **P**, with the inclusion map $i : \mathbf{M} \hookrightarrow \mathbf{P}$ having a left adjoint p (for possibilization) and a right adjoint n (for necessitation), so that the composites $pi = \diamond$ and $ni = \Box$ are defined. By adjunction, $\Box\varphi \vdash \varphi$, $\varphi \vdash \diamond\varphi$ and iterated modalities reduce to just $\Box$ and $\diamond$, which after all is in agreement with common sense.

Modal logicians who dislike such reduction will find more palatable the presheaf approach which increases the subtlety of analysis to a degree unknown in standard modal logic, as, rather than one or more *specified* accessibility relations between two indexes (thought of as indexing "possible worlds"), there is now a family of maps to be considered as a whole, so that soundness and completeness must take the variability of transitions into account. The standard case in which accessibility is $\leq$ (as for IL_0) corresponds to a poset-category of states (say, of increasing information) ignoring different transition processes between states.

The above constructions by adjoints make it apparent that LEM is a special addition that narrows the range of categories required for adequate analysis of the structure of reasoning in the presence of incomplete information. Previous constructive rejection of LEM gets more specification as the data space is no longer given in terms of point-set topology: boundaries matter, whereas LEM collapses the process of identifying points into a mere assumption that they exist, see (Vickers, 1988).[11]

Taking definability by adjunction as the principle, the categorical properties needed to reach full intuitionistic first-order logic IL_1 are trickier to list than those for higher-order logics, which are just those defining a topos (cartesian closedness plus representability of subobjects). The properties of IL_1 are in fact obtained by subtraction, as witnessed by the very name, as the corresponding level of structure is identified as that of a *pretopos*, after (Artin et al., 1972), as a category with finite limits, initial object (to match $\bot$), binary coproducts and (effective) coequalizers of equivalence relations (though there are more elegant definitions of a pretopos).

[11] Already in standard topology what generally matters is whether or not connectedness, rather than total disconnectedness, holds.

In a sense, the pre-eminence assigned to first-order in applications of logic in the analysis of language and the methodology of science must be rethought.

However mathematically substantive, the mere analysis of connectives by means of categories would not have made much of a difference in logic. The key was the definition of quantifiers as adjoints along a map f, whereas such f had traditionally remained implicit. The simplest example is given in the case of poset-categories, such as the $\subseteq$-lattice of parts of any given set. Given sets X, Y and a map $f : X \to Y$, for any $A \subseteq X$ and $B \subseteq Y$,

$$\frac{A \subseteq f^{-1}(B)}{\exists_f(A) \subseteq B} \qquad \text{and} \qquad \frac{f^{-1}(B) \subseteq A}{B \subseteq \forall_f(A)}$$

When X and Y are taken as types in extension, substitution along f replaces a Y-term by an X-term. By the very consideration of the map along which substitution occurs, first-order logic becomes many-sorted by default, and even in the case $X = Y$, explicit recording of the given f is still crucial for specifying images and their Galois dual. In more general categories, the place of inclusion is taken by an arbitrary morphism, with subobjects replacing subsets. The two equivalences above are natural, thus $\exists_f \dashv f^{-1} \dashv \forall_f$.

For any category $\mathbf{C}$ and X any object of $\mathbf{C}$, one can form $\mathbf{C}/X$ as the so-called "slice" category, with objects all $\mathbf{C}$-maps to X and morphisms from $h : A \to X$ to $k : A' \to X$ those $\mathbf{C}$-maps $g : A \to A'$ such that $h = kg$. When $\mathbf{C}$ is a topos, $\mathbf{C}/X$ is too and f^{-1} induces the functor $f^* : \mathbf{C}/Y \longrightarrow \mathbf{C}/X$, while the two functors $\exists_f, \forall_f : Sub(X) \longrightarrow Sub(Y)$ induce a pair of functors $\Sigma_f, \Pi_f : \mathbf{C}/X \longrightarrow \mathbf{C}/Y$ such that $\Sigma_f \dashv f^* \dashv \Pi_f$. It is essentially as a result of this analysis of quantifiers as adjoints to substitution that the equational approach to logic started by Boole and interrupted by Frege gained a new lease on life, one which later matched the development of λ-calculi.

Efforts by logicians to control the apparently innocuous substitution procedure, under way since Frege's *Grundgesetze* and Whitehead-Russell's *Principia*, reached a first precise formulation in Hilbert-Bernays' *Grundlagen der Mathematik*. Then an autonomous theory was obtained through the investigations of Schönfinkel, Curry and Church. However, it was only through Lawvere's understanding of the logical aspects of the change-of-base technique of algebraic geometers that the priority of substitution to quantification acquired mathematical substance.

To give a categorical treatment of recursion, one must add the notion of a "natural numbers object" (Lawvere), i.e., an object N, with $0 : 1 \to N$ and $s : N \to N$ such that for any element $x : 1 \to A$ and any endomap $f : A \to A$, there is exactly one map $h : N \to A$ which makes the resulting diagram commute, namely $x = f0$ and $fh = hf$. The existence of such an object N in a cartesian closed category $\mathbf{C}$ is determined by the forgetful functor $U : \mathbf{C}^{\circlearrowleft} \longrightarrow \mathbf{C}$ having a left adjoint. In a topos the existence of N implies the usual axioms for arithmetic. Starting from N, the number systems can be axiomatized, with increased "resolving" power of the formal setting, as shown by the difference between Cauchy and Dedekind reals and the distinction of various concepts of finiteness. The smooth topos for synthetic differential geometry is a prime instance of a model for a theory of the continuum which has no models even in Boolean extensions.

As for computability, typed λ-calculus, as a prototype of a functional programming language, is by now a central topic in the convergence between logic and theoretical computer science. This convergence finds its medium precisely in cartesian closed categories, since such a category **C** provides the ambient surroundings needed to relate product types and power types in a purely functional language, as the adjunction $\times A \dashv (-)^A$ matches "currying": the two basic β and η equations being given, respectively, by $ev_{A,B} \cdot \hat{f} \times id_A = f$ and $(ev_{A,B} \cdot (g \times id_A))^\wedge = g$, for $f : C \times A \to B$ fixed, and g any map $C \to B^A$.

The Curry-Howard principle, "objects as types, morphisms as terms," expresses this link in compact form: terms in simply typed λ-calculus $\sim$ proofs in IL_0-natural deduction, so that "p is provable" is read as if it said p is an inhabited type. In particular, it was by means of the categorical notion of **C**-monoid, corresponding to an object X for which $X^X \cong X$, that models of *untyped* λ-calculus were obtained, whereas any such X in the universe of classical sets could only be a singleton. But category theory provides the tools for representing higher-order versions of the λ-calculus too. At its root, λ-calculus is a formalism both for representing functions and for making "function" the ground notion of mathematics, but this very ambition calls for a proper rendering of functions. As concerns logical structure, there are two aspects of concern to take into account.

First, it is usually forgotten that the idea of "functions first" was behind Church's attempt at formulating an intensional logic that develops the Fregean distinction of sense and reference differently from Carnap's method of intension and extension. The contributions by Carnap and Church were not intended to confine intensional aspects to face-value syntax—otherwise, the only candidate for synonymy, as an equivalence relation finer than extensional identity, would ultimately produce singleton quotients in the definition of *meaning* by abstraction. In the 1960, possible worlds semantics appeared as the best (workable) approximation. Such a framework to enrich Tarski-style formal semantics, by now well-known, was of a modal nature and proved efficient in the semantics for intuitionistic logic. But, by taking functions as set-theoretically defined, cartesian products are presupposed and thus forced to be given in extension, rather than by their universal property (in terms of functions), and due to this presupposition the framework could not avoid collapsing each process with its outcomes. A categorical treatment can be developed to define a finer equivalence of propositions, different from $[[\varphi]] = [[\psi]]$ on the collection of indexed domains, for $[[-]] : Ob(\mathbf{P}) \longrightarrow \mathbf{H}$, with **H** a Heyting algebra, for this does not take into account the possibly different state transitions. By taking these into account, as the notion of sheaf permits, a concept of local intension (as a germ) can so be defined, see (Peruzzi, 1991b), which avoids the extensional collapse.

Second, equivalence of proofs is trivial if the category **P** is treated as just a poset: that the structure of actual deductive steps matters as much as the form of propositions is a natural requirement for a theory purportedly aimed at grasping form and structure by means of the patterns of their variation; and as a bonus for this refinement of logical analysis, one can also see that categorical proof theory for classical logic is not trivial at all. Thus the above adjunction diagram for **P**, though with its explicit notation for proofs in IL_0, described as a cartesian closed category with finite colimits, does not fill the gap by itself.[12] If one ends up with the same set of semantically valid sentences as for Kripke semantics, then the added structure is redundant and a poset will be sufficient. Vice versa: if there are logics which pay attention to constraints on parallel state-transitions (as the motivations for quantum, relevant and paraconsistent logics suggest), a poset will not be sufficient. Therefore, as far as mathematical understanding of processes (computational steps, transition-states, ...) counts for logical validity, a theory designed to capture universality cannot forget differences among them in the case of proof structure either. If reference is a function of the procedures involved in the construction of sense, then more than the existence (or not) of any such procedure is needed, hence the demand for an adequate formal treatment already for **P**.

Unfortunately, for any propositional theory T, the syntactic category C_T is a preorder. In fact, it coincides with the Lindenbaum algebra of T. Thereby, any $\varphi \rightarrow \varphi$ is the identity. Once again the only proof endomap on any given φ is id_φ, whereas, as non-contractible closed paths in homotopy teach, the non-null loops can contain precious information, which can be taken into account in a kind of "logical genus" for theories as homotopy defines the topological genus of a space.

On passing to first order theories, the link between syntax and semantics becomes stricter, as the two are functorially related. What in classical model theory had to be proved, by taking into account the form of axioms for a given theory, is directly granted by the preservation-properties of models as functors. The obstacle to the prenex normal form theorem in a constructive setting does not so much concern $\exists$, in view of the Frobenius law, i.e., $\varphi \wedge \exists y \psi \vdash_{\mathbf{x}} \exists y(\varphi \wedge \psi)$ (provided y is not in $\mathbf{x}$) and commutativity of $\exists$ with $\vee$. Rather the obstacle concerns $\forall$: though $\forall$ trivially commutes with $\wedge$, the sequent $\forall y(\varphi \vee \psi) \vdash_{\mathbf{x}} \varphi \vee \forall y \psi$ is classical. This obstacle is bypassed by considering a hierarchy of signatures and theories of increasing expressive power, with corresponding structure-preserving functors between the models at each layer.

In doing this, categorical logic endorses the requirement that inferential schemes cannot be superimposed, by an "external observer" so to speak, on a given universe of discourse and even less can it be developed in isolation, as if it were the expression of the (possibly constructive) way thought works independently of any subject matter. Nor, for this very reason, does logic undergo a fission, in order

[12] What ultimately matters in representing logic by posets is whether there is a map $1 \rightarrow A$ or not (for completeness, whether $\top \rightarrow \varphi$ or $\varphi \rightarrow \bot$) and the only map $A \rightarrow A$ that matters is identity. Therefore, the map $p_2 < f, g >: A \rightarrow A$, obtained by pairing $f, g : A \rightarrow A$ to have $< f, g >: A \rightarrow A \wedge A$ and then by composing with $p_2 : A \wedge A \rightarrow A$, is trivial, as is the composition of $tw_1 : A \wedge B \rightarrow B \wedge A$ and $tw_2 : B \wedge A \rightarrow A \wedge B$. Whereas in a general category with 1, from the existence of $x : 1 \rightarrow A$ it doesn't follow that $A \cong 1$, the existence of only $id_A : A \rightarrow A$ collapses A onto 1, "the" absolutely true proposition (given the composition $!_A \cdot x \cdot !_A$, necessarily $!_A \cdot x = id_A$, and if $id_A = x \cdot !_A$, then $A \cong 1$). So, if there is more than one true proposition, there is at least one $f : A \rightarrow A$ such that $f \neq id_A$, but there is no trace of any such f when a deductive system collapses to a poset.

to match an unprincipled range of contexts of use, each with its own axioms and rules. Rather, logical analysis proceeds in agreement with mathematical architecture, starting from categories with products and images of morphisms, to deal with n-ary predicates and existential formulae, up to categories with all finite limits—to interpret terms and predicates of n arguments, equations between terms and the substitution procedure (by pullbacks). Accordingly, increasingly powerful theories are internalized and their various forms are unified under adjunction principles expressing universality conditions. Variation of logic is approached as intrinsically principled, avoiding any clash of "-isms" precisely because variation is one of the main themes of categorical thought just as characteristic as the search for invariants.

From this perspective, the boundary line separating full propositional from full predicate logic turns out to be secondary with respect to "mixed" layers of structure. To simplify the picture, here only four layers of increasing expressive power will be mentioned. Categorical logic is sequent-oriented and each sequent comes with its explicit environment of declared variables, so that in writing $\varphi \vdash_{\mathbf{x}} \psi$, the list $\mathbf{x}$ contains the free variables in both φ and ψ, in order to specify the domain of definition of formulae. Semantically, this constraint is motivated by the need to define the "extension" of predicates with respect to one well-defined product as their environment. Four main layers respectively correspond to theories in sequent form which are *cartesian*, expressed by sequents relating only formulae built from atomic formulae $R(t_1, \ldots, t_n)$, equations, $\wedge$ and $\top$, plus $\exists x\varphi$, provided its uniqueness is provable; *regular*, in a language built up from regular formulae, i.e., atomic formulae $R(t_1, \ldots, t_n)$, equations, $\wedge$, $\top$, and $\exists x(\varphi)$; *coherent*, that is, regular with also $\perp$ and $\vee$; and *geometric*, that is, coherent with $\bigvee_{i \in I} \varphi_i$, provided the whole set of free variables is finite.

Since sequents obtained by composition or substitution now come with explicit names of proof-maps, further control on proof-trees is achieved. Soundness and completeness are provided by the fact that, if T is a cartesian (regular, coherent, geometric) theory over a signature Σ and $\mathbf{M}$ is any model of T in a cartesian (regular, coherent, geometric, resp.) category $\mathbf{C}$, then if σ is a cartesian (regular, coherent, geometric. resp.) sequent over Σ, $\mathbf{M} \models \sigma$ iff $T \vdash \sigma$. Accordingly, cartesian logic, regular logic, coherent logic and geometric logic are identified.

In particular, coherent logic is geometric logic restricted to finitary disjunctions. As mentioned above, geometric logic was identified as a crucial logical layer in view of the preservation properties of geometric morphisms between Grothendieck toposes. Why? The motivation for the name of such functors lies in that, given topological spaces X, Y with their respective lattice of open sets $O(X)$, $O(Y)$, any continuous map $f : X \to Y$, f induces a pair of adjoints between sheaves $f_* : \mathbf{Sh(X)} \longrightarrow \mathbf{Sh(Y)}$, $f^* : \mathbf{Sh(Y)} \longrightarrow \mathbf{Sh(X)}$, such that if $p : E \to Y$ is *étale* (i.e., a local homeomorphism), f^* gives the pullback of p along f and f_* acts by composition. Namely, if $F : O(X)^{op} \longrightarrow \mathbf{Set}$ is a sheaf on X and U belongs to $O(Y)$, $(f^*F)(U) = F(f^{-1}U)$, while if U' belongs to $O(X)$, $(f_*F)(U') = F(fU')$. One crucial property of f^* is that it inherits from f^{-1} the preservation of $\bigcap$ and $\bigcup$ and thus is left exact. Hence the definition of a *geometric morphism* in general as a map $f : \mathbf{E} \longrightarrow \mathbf{E}'$ of toposes such that $f^* \dashv f_*$ and f^* is left exact. Since any left adjoint preserves colimits, the condition of left-exactness implies distributivity. Apart from (coherent) sites for which any coverage is finitely-generated, distributivity concerns infinitary disjunctions, therefore, in view of the fact that the other categorical properties match the needed properties of $\wedge$, $\top$, $\exists$, $\bigvee$, $\bot$, geometric logic is thus determined by invariance for geometric morphisms.

As noted above, we get a categorical semantics for IL_1 through the notion of pretopos. A pretopos is Heyting when, given any pullback square $gh = fk$, it satisfies (i) the Beck-Chevalley condition $g^{-1}\exists_f = \exists_h k^{-1}$ and (ii), for any two objects A and B and any morphism $f : A \to B$, $f^* : Sub(B) \longrightarrow Sub(A)$ has both a left and right adjoint, $\exists_f \dashv f^* \dashv \forall_f$. Conditions (i) and (ii) do not necessarily come together. In addition to the Beck-Chevalley condition, a key role is played by the Frobenius reciprocity law, which is a property, commonly taken as a trivial theorem in IL_1, the utility of which as an axiom (since its proof in IL_1 depends on $\Rightarrow$) is shown in categorical logic: Frobenius reciprocity is sufficient to yield regular logic from cartesian logic, and, in presence of distributivity, to yield coherent logic.

The last topic of significance for grasping the actual content of categorical logic is *internalization*. Of its many aspects, especially for models of type theories, only the general strategy will be described in relation to the layers of logical structure identified above. In order to specify the internal language of a category **C**, we define a system of terms and types which is canonical for the translation of theories interpretable in **C** and for model-variation $\mathbf{C} \longrightarrow \mathbf{C}'$.

Given **C**, its canonical signature $\Sigma_{\mathbf{C}}$ is given by the following assignment: for any object A a type $\overline{A}$; for any morphism $f : A_1 \times A_2 \times A_n \to B$ a function symbol $\overline{f}$ with arguments in types $\overline{A_1}, \overline{A_2}, \ldots, \overline{A_n}$ and values in type $\overline{B}$; for any subobject R of $A_1 \times A_2 \times \cdots \times A_n$ a relation symbol $\overline{R}$ defined over corresponding types. To this one can add product and exponential types in correspondence with the kind of category and respective terms for the maps required to define products and exponentials. The definition of signature is turned into a set of type-axioms stating which sorts (as basic types), which type-constructors and which term-constructors there are. Note that $\Sigma_{\mathbf{C}}$ has more types than the objects in **C**. For instance, the type $\overline{A \times B}$ is, at face value, different from $\overline{A} \times \overline{B}$ and likewise $P(\overline{A}) \neq \overline{P(A)}$ but we will identify them.

Higher-order logic in its many-sorted, intuitionistic and free version is revealed, via the notion of topos, as the underlying logic of a "local set theory." Basic types correspond to 1, Ω and possibly further ground types, on which to define product– and power-types; corresponding terms are defined as usual, but note the special term-assignments: $* : \overline{1}$; if $\varphi : \overline{\Omega}$ and $x : \overline{A}$, then $\{x : \varphi\} : \overline{P(A)}$; if $t, s : \overline{A}$ then $t = s : \overline{\Omega}$; if $t : \overline{A}$ and $s : \overline{P(A)}$ then $t \in s : \overline{\Omega}$. In particular, it turns out that $true$ is definable as $* = *$. The **C**-sets are the $\sim$-equivalence classes of terms t of power-type, where $[t]/_\sim = \{t' : t' = t\}$ is internally provable in $L(C)$ and functions from one such set X to another Y are the f for which it is provable that $f \in Y^X$. Together, they form a category which is a topos.

In a topos **E**, $Sub(A)$ is a Heyting algebra, for any object A. Because of the presence of partially defined singular terms, the support of which is a proper subobject of 1, the cut rule applies to $\Gamma \vdash \varphi$ and $\varphi, \Gamma \vdash \psi$ provided the variables free in φ are free in Γ or ψ. Thus, for the *modus ponens* ($\Gamma = \emptyset$), the variables free in φ must be free in ψ and thus $\exists x\varphi$ cannot be inferred from $\varphi(x)$ and $\varphi(x) \vdash \exists x\varphi$.

The main difference with respect to simple type theories of the past is given by the existence of the truth-value object Ω as a basic object, to which a basic type-of-propositions corresponds. In particular, propositional quantifiers are present in the internal logic of a topos and actually the definitional resources at hand in a topos reduce the logical machinery to sequents between equations about Ω-terms, by which connectives and propositional quantifiers can be equationally defined, as shown in (Bell, 1988, Ch. 3).

Then, by adding the required syntax, a (typed) language $L(\mathbf{C})$ on $\Sigma_{\mathbf{C}}$ is fully defined as are, by the presence of typed $\in$, also the set-theoretic operations (which can now be applied to "sets" of the same type). Once a model structure for $L(\mathbf{C})$ is defined, $\mathbf{C}$ is recovered up to equivalence. Both passages, from categories to theories and vice versa, turn out to be fruitful. The first way, for example from categories with finite limits, allows identifying special kinds of logic. The second way, for example from theories in a fragment of IL_1 or versions of the λ-calculus, allows identifying special kinds of categories. By composition of the two ways, $\mathbf{C}_{Th(C)} \equiv \mathbf{C}$ and $Th(\mathbf{C}_T) \approx T$ (up to translation). This is also called the "Equivalence Theorem."[13]

The usefulness of $L(\mathbf{C})$ lies in that, by using intuitionistic reasoning (possibly restricted to one of the above fragments), we get a category of $\mathbf{C}$-*sets* the objects of which can be treated as sets, the morphisms of which act as functions and the subobjects as subsets, with equations between terms (also of power types in toposes) relativized to their common type, and finally $\in_A$ is defined as a relation between terms of type $\overline{A}$ and sets of type $\overline{PA}$. Thus, for instance, $f = 1_A$ iff $T \vdash_x \overline{f}(x) = x$. Conversely, given a theory T, we can associate with it a *syntactic* category $\mathbf{C}_T$ as already described. Within such a category $\mathbf{C}_T$, there is the generic model of T canonically defined by the same kind of assignment of objects to sorts, maps to function symbols and subobjects to relation symbols as in the case of the internal language. It can be proved that, in bijective correspondence with the four forms of T described above, there is an equivalence between functors $Cart(\mathbf{C}_T, D)$, ..., $Geom(\mathbf{C}_T, D)$ and the respective categories of T-models in D, for *any* D, thus even beyond the standard case $D = \mathbf{Set}$. Mimicking in $\mathbf{C}_T$ the usual term-model out of T-syntax produces the generic model G_T of T, i.e., a model such that

(∗) $\varphi \vdash_{\mathbf{x}} \psi$ holds in G_T iff it is provable in T

and all possible T-models in any other category are images of G_T under functors of the same kind as the theory (cartesian, regular, coherent, geometric). G_T lives in a category designed to match the form of T and which, in general, is different from Set. (Models in **Set** suffice if T is cartesian.)

[13] This notion of theory-equivalence requires some care, as Johnstone explains in (Johnstone, 2002, vol. 2).

The two-way procedure mentioned above is thus at hand: the first goes from categories **C**, through the internal language $L(C)$, to theories $Th_{L(C)}$, the second one goes backwards from C_T to T. If T is cartesian and **D** a category in which T is interpreted, the category $T - Mod_D$ of T-models in **D** "coincides" with the functor category $Cart(C_T, D)$, ..., and finally if T is geometric, $T - Mod_D$ "coincides" with $Geom(C_T, D)$.

Thus a model of T in **D** is (up to iso) just a functor $\mathbf{C}_T \longrightarrow \mathbf{D}$ of the appropriate kind (adequate to T-structure). This systematic match extends to full first order theories, in which case any given category $T - Mod$ is taken with elementary morphisms. Moreover the category of functors $\mathbf{C}_T \longrightarrow \mathbf{Set}$ which preserve the form of T is jointly conservative, though this does not extend to theories in full first order logic, see (Freyd and Scedrov, 1990).

Category theory also offers a previously undetected formulation of theories, namely as a *sketches*, see (Johnstone, 2002, vol. 2, Ch. D2). Although sketches have only proved to be useful for algebraic theories, the approach by means of sketches identifies a dimension in the analysis of theories which is intermediate between the logical notion of a theory T and C_T. From the model-theoretic point of view there is an equivalence between sketches and theories of special kind, the main link being expressed in a theorem proved in (Makkai and Paré, 1989).

Now, let us consider an arbitrary (consistent) theory T and its models in toposes. $\mathbf{C}_T$ is also the underlying category of a site $(\mathbf{C}_T, J)$ with suitable coverage J, called the "syntactic site." There is an equivalence between functors from $\mathbf{C}_T$ to **E** and geometric morphisms from **E** to $\mathbf{Sh}(\mathbf{C}_T, J)$ in S-toposes, and the latter can then be identified with the *classifying* topos **Set** [**T**]. The result is canonical, since for any (cartesian, ...) theory T, the topos $\mathbf{Sh}(\mathbf{C}_T, J)$—with J suitable for the different forms of T—contains a generic T-model G_T. Hence $(*)$ holds and any model of T in **E** is obtained, up to equivalence, as $f^*(G_T)$ for $f : \mathbf{E} \longrightarrow \mathbf{Sh}(C_T, J)$.

For geometric theories, there is a result of particular importance: models in Boolean toposes suffice, that is, if T is geometric and σ is a geometric sequent $\varphi \vdash \psi$, if the sequent holds in any T-model in Boolean toposes, then it is provable in T. The crucial step is that for any Boolean topos **B** there is a surjective geometric morphism $f : \mathbf{B} \longrightarrow \mathbf{Set}\ [T]$, with *surjective* meaning that f^* is faithful (i.e., it is injective on maps). For, by hypothesis $f^*(G_T) \models \sigma$ and thus $G_T \models \sigma$, hence $T \vdash \sigma$. Therefore, any classical proof of σ can be turned into an intuitionistic one. There is no classifying topos for full intuitionistic (possibly infinitary) first order theories, but there is a canonical model: for any such T there is Grothendieck topos **E** and a T-model N_T such that $T \vdash \sigma$ iff $N_T \models \sigma$.[14]

[14] Among many other related topics are categorical versions of definability theorems (Makkai and Reyes, 1977), and a refined version of completeness, called "conceptual completeness" (Johnstone, 2002, D3.5).

5 Further Connections with Type Theories and Concluding Remarks

The above hierarchy of logics, with their categorical counterparts, is more than updated formalistic tinkering in a constructive setting. As Mac Lane once claimed that topos theory lets Brouwer the topologist finally meet Brouwer the intuitionist, so one could add that categorical logic constrains axiomatic bricolage by representability conditions within (pre)sheaf-theoretic constructions, see (Troelstra and van Dalen, 1988), and it is just as a particular byproduct of this constraint that constructive reasoning re-emerges from variation and cohesion patterns of structure.

One advantage of the topos-theoretic medium is the existence of a certain special topos. The *free* topos **F**, generated by the free type theory, with only ground types for 1 and Ω, is initial, in the sense that it has a unique logical morphism into any any linguistic topos, as generated by a type theory. **F** can be considered as an ideal world for philosophical convergence, as it satisfies intuitionistic constraints such as "$\exists x A(x)$ is assertible only if there is a closed term t for which $A(t/x)$" and "$\varphi \vee \psi$ is assertible only if such is either φ or ψ," respectively corresponding to the external property that 1 is projective (i.e., given any epimorphism $f : A \twoheadrightarrow B$, for every $x : 1 \rightarrow B$ there is a $y : 1 \rightarrow A$ such that $x = fy$) and 1 is indecomposable (i.e., is not the union of any two proper subobjects). In view of such properties of the free topos (possibly with the addition of a natural numbers object N), there would be an ideal universe of discourse on which the three traditional approaches to foundations agree, as argued in (Lambek and Scott, 1986).

Another pro is that the same medium provides a unified framework for comparing the range of forms taken by the concept of choice and the relationships between choice and constructivity. On one side the standard form of choice says, in categorical language, that epimorphisms split: any $f : A \twoheadrightarrow B$ has a section $s : B \rightarrow A$, for which $fs = 1_B$; which in a topos is equivalent to saying every object is projective. This splitting implies Booleanness. On the other side, choice is implicit in the constructivist picture when the assumption "For all x there is a y such that $\varphi(x, y)$" is taken constructively. Taken constructively, the assumption itself says there is a construction of a suitable y for each x, and that very construction determines a function f where $f(x) = y$ is the suitable y for x.

A similar phenomenon occurs with extensionality, in the form of wellpointedness, which together with such form of choice is enough to characterize **Set**, but which in the internal language holds trivially, for if $f \neq g : A \rightrightarrows B$, there is a B-element x that distinguishes them, namely the generic $x = 1_A$. Here the subtle analysis of external (sheaf-theoretic) and internal properties of a topos is efficient in clarifying the real power of assumptions about extensionality, bivalence and choice. For instance, it can be proved that a topos is (externally) wellpointed iff (internally) $\forall x \varphi$ follows from $\varphi(t/x)$, for all t closed and x the only free variable in φ. More generally, categorical methods refine checking the relative strength of different principles or of different formulations of one principle, since many of them prove no longer to be equivalent as in the classical universe and weak versions of them (as in the case of the axiom of choice) turn out to be sufficient for most purposes.

The use of categories also helps to radically rethink diagonal arguments (and fix-point theorems). The Gödel and Tarski Theorems look, in fact, different from the received view taking them as prescriptions against paradoxes. Categorical rewriting makes them special instances of a positive, structural, statement. Thus the question as to which consequences of incompleteness and the indefinability of truth can be drawn from the analysis Lawvere started in (Lawvere, 1968), in view of the larger range of models provided by toposes different from **Set**, calls for further attention. It is one of the subjects on which categorical logic is expected yet to achieve more systematic results to improve the received view of self-reference (and also to avoid ad hoc systems with hierarchical truth-predicates that, though ingenious, lack mathematical relevance). This line is far from uncontroversial but makes it clear that categories are not so flexible as to legitimate a sort of metamathematical "anything goes," as if the limitations of the classical universe were replaced by a no commitment attitude. Though what has been described in Section 4 already provides evidence for the contrary, let us make it more explicit: if the plurality of notions of constructivity and computability benefits from categorical logic, constraints are added, as shown in the axiomatization of logical systems since introduction and elimination come together as far as definition by adjoints is the underlying principle, from which the needed equations follow (as a counterweight to just "up to iso" conditions).

A byproduct of interest for realizability is the step from notions of local character in spatial sheaves ($\exists$-equivalence relations on overlappings) to partial equivalence relations. As indicated above, the same technique is efficient also in intensional logic, for it leads to defining *local intensions* by abstraction as germs, which in the case of sheaves over a site of indexes (possible worlds, information stages) can be glued together into a unique notion, thus avoiding the difficulties of both the global concept of intension associated, after Carnap and Montague, with relational semantics (but functorially undefined in terms of accessibility) and the index-relative character of a similarity relation (family-resemblances), see (Peruzzi, 1991b).

The inconsistent belief that the intended universe of discourse X is an unqualified whole and at the same time a very specific one led many to take as a defect of set-theoretic semantics that, given two predicates φ and ψ over the domain X, if any X-element satisfying φ also satisfies ψ and vice versa, then extensionality implies the equivalence of φ and ψ.[15] Now, if X is really "the whole," it also contains as a sub-domain the language in use to which φ and ψ belong, thus the equivalence is innocuous; if X is one model among others, the equivalence does not follow unless

[15] In most applications of set-theoretic semantics, one argues as if $\{a\} \neq \{b\}$ for $a \neq b$, but in **Set** the two singletons are isomorphic and that's all. Hence either urelements are given or some content foreign to explicit semantics is used, to confirm the claim that meaning is in the eye of the beholder: in other words, one supposes to know more than what is allowed by the theory. On the other hand, the very notion of singleton is far from trivial in categories corresponding to constructive reasoning, see (Fourman et al., 1979).

we make explicit the underlying theory which is supposed to specify the meaning of predicates. At first sight, since category theory is designed to identify properties up to $\cong$ between objects, the resulting semantics should be even *more* extensional. But this is not the case, because the interpretation of a predicate (or a term) is a map, and in cases of interest it is between objects with structure rather than mere sets, thus the information to be taken into account is expressible within X as a category with further structure. The interpretations of φ and ψ in X may be pointless spaces which still have many variable and partial elements making the equivalence between them a much stricter relation. Similarly, if models are functors, at first sight a theory should have fewer models than those allowed by set-theoretic semantics. But, since models can now be taken in categories different from **Set**, there are more models and in fact some theories have models in them only.

By taking a theory as a category, term models become more than a completeness device, as they identify initial algebras. If only in this respect, categorical semantics is revealed as essential. In fact, the interaction of categorical logic and higher-order programming has led to the formation of a vast area of research, within which the resources employed in modelling constructive reasoning by dependent and predicative type theories are further refined. The resulting framework is centered on fibrations, which provide additional flexibility and unification, acting as a red thread among logics in computer science. Though topos theory and the associated higher-order intuitionistic logic do not need such a complex framework, they find their place within it as a particular case, one with an actual mathematical content still to be reached by such an even more general framework.

Polymorphism uses type variables in addition to function variables and uses predicative/impredicative definition of functions, the evaluation of which is independent of specific data types. In dependent type theory, types depend on terms and change with term-values. Predicates (as indexed propositions) and types are indexed by contexts, which declare the types of free variables, where a context is a sequence of variable declarations. Categorically, there is a constraint or parametric polymorphism, as objects lie within one category or are functorially related, and impredicativity inheres by default in universal constructions, even though these are localized to a given category. The intertranslation of topos-theoretic, type-theoretic and set-theoretic statements becomes subtler in Martin-Löf constructive type theory, which also admits a presentation as a functional programming language. It is no accident that models in locally cartesian closed categories, introduced by Robert Seely to deal with dependent types (Seely, 1984), have been elaborated as categories $\mathbf{C}$ of contexts with a functor $\mathbf{C}^{op} \longrightarrow Fam(\mathbf{Set})$ where the families (or attributes) (Jacobs, 1999, Ch. 1 and 11), come with adequate closure conditions for dependent sums and products.

A logical counterpart of the notion of genericity in passing from $Mod_{Set}(T)$ to $Mod_{E}(T)$ is not optional, but intrinsic to untyped λ-calculus ($\mathbf{C}$-monoids) and polymorphic λ-calculus (Girard, Reynolds), as made apparent in the Moggi-Hyland model. Here the crucial process is multiple indexing, as some types depend on term-variables the types of which depend on ... (recursively). So the hierarchy of terms and types is defined by simultaneous induction. Index types have a structure of their own and can be assumed to form a category I. The overall structure is captured precisely by means of the concept of *fibration*. Whereas an indexed category $I \longrightarrow X$ is focused on the single fibres, the converse picture $X \longrightarrow I$, with X generalizing the disjoint union $\bigcup_{i \in I} X_i$, assigns the global structure a central role. A functor $F : \mathbf{E} \longrightarrow \mathbf{B}$ is a fibration if for any $\mathbf{E}$-object E and $\mathbf{B}$-morphism u to its F-value $F(E)$, the $\mathbf{E}$-morphism f such that $u = Ff$ satisfies a unique factorization condition (f is "cartesian"), see (Jacobs, 1999, p. 27) for details. As Benabou argues, fibrations can be exploited to deal with foundational problems, since they turn the dependence on set-theoretical assumptions and limitations of size into functorial dependence on other categories. Together with Lawvere's description of the category of categories, the fibrational approach is a viable answer to doubts about the real autonomy of category theory and shows how to control hierarchical type-dependent structure, well beyond classical logic as encoding principles of thought for constancy and discreteness.[16]

The general idea is that logic is determined by a fibration above a type theory, which in turn is given as a fibration. A suitable base-structure has more than one logic over it, corresponding to different fibrations. One can investigate the structure of fibrations over fibrations, corresponding to double indexing, and the composition of fibrations (by substitution, in logical terms). Both involve category theory in extensive and essential way. Doctrines and hyperdoctrines, IL_1 as well as topos logic turn out to be determined by certain fibrations with structure.

That conditions such as those on $\exists$ and $\vee$ must work as a criterion was tied to a constructive theory of meaning of a subjective nature; that the free topos satisfies both argues that no such commitment is necessary, for their validity depends on the initiality of $\mathbf{F}$, and initiality does not count as a "good" property in virtue of subjective constraints. Is it then for its formal role within the given framework? But then one could proceed likewise with any adequacy criterion and ultimately the very notion of structure would be *implicitly defined*, which means that it should be freed from any previous understanding. To be consistent, such liberality should receive the same (metatheoretic) treatment, which brings us back to Buridan's paradox. True, the free topos has nice properties. However, a system built out of syntax is the result of objectifying language as a mathematical structure among others. Were the above equivalence of categories to mean that anything needed is "essentially" of linguistic nature, this would be either trivially true or fallacious. It is trivially true insofar as mathematics manipulates symbols for spaces, flows and tile patterns, etc. It is fallacious insofar as the virtues of what is syntactic are identified through what is not (or is not supposed to be) and precisely for this reason the equivalence is far from trivial. Though the idea of language as all-pervading may be widespread, the comfort it gives is either trivial or fallacious; and as the same line applies to any

[16] In this regard, the equiconsistency of the elementary topos axioms with Z_0, and the internalization of a topos into a local set theory, can be both misleading. What is achieved by means of categories is a theory of variable and cohesive sets. In particular classical sets appear as the limiting case of vanishing variation and cohesion. Languages *qua* special mathematical structures have no ontological priority.

representation theorem, some doubts about the comfort are in order. That everything is formal in mathematics does not entail that everything is (or can safely be taken as) linguistic. Were it so, the difference between syntax and semantics would vanish and the relevance of the properties defining the free topos would reduce to a matter of convention.

A different option is endorsed by constructivism. The hard constructivist claims there is one ideal ground-logic and it has such and such axioms and rules. The soft constructivist investigates the pros and cons of different candidates and is ready to let the selection depend on formal reasons, or to dispense with the need of a choice. The strong constructivist's claim is simply the mirror image of the pretension of (Platonic, by default) realism associated with the claim that logic is Boolean and bivalent, and the mirror image extends to indispensability arguments in the light of sheaf theory. The logic of variation also embodies the case in which no variation is present. In particular, also the logicist version of realism has now a non-classical mirror image, concerning the ontology of the thinking subject. The soft constructivist renounces uniqueness of logic and envisages an open plurality of universes of discourse, each one with its internal patterns of reasoning.

To the formalist's eye, the controversy attests to the cunning of reason: the cluster of content-laden motivations addressed by the contenders only serves to enlarge the web of formal notions and formal systems; the tools offered by category theory can thus be exploited to refine and extend the spirit of formalism, now freed from classical metatheory. Motivations forgotten, structure shines through. By avoiding any commitment to identify a minimal substantive ground, sufficiently rich in mathematical content, what remains is a new version of logical relativism; but what it may contribute to the understanding of the constraints on the range of a priori possible type-theoretic conventions is not clear. *Unless further qualified*, even the adequacy criterion consisting in the "reflection" of patterns of reasoning in the object language and in the metalanguage, is insufficient to determine one system (since both intuitionistic and classical metatheory, as well as set-theoretic and categorical metatheory, are supported). Were it even sufficient, with proper qualification, the formalist would count it as just one criterion among others: there are many categories and many corresponding logics; toposes provide an excellent frame for intuitionistic theories, but there are also Boolean toposes.

Now, as a slice topos $\mathbf{E}/A$ is still a topos, toposes are locally cartesian closed categories, but in general the latter are not cartesian closed and precisely those which are not serve as models (actually, classifying categories) for the constructive reasoning expressed by dependent type theories, such as Martin-Löf's. When the left adjoint to the "hom"-functor which internalizes the function-space construction is a tensor product (not necessarily cartesian, because just associative), we have monoidal closed categories, and the latter (with further conditions) provide a semantics for linear logic. The world of categories seems to be sufficiently generous to accommodate these and other logics as well. Presumably, further similar correspondences will be added in the future. But this is no argument supporting the idea that "anything goes," i.e., that categorical thought is compatible with any kind of structure. Conventionalism ends at the resources which create the space for conventions.

On the other hand, even assuming that the ground logic manifests the constructive structure of an Ideal Subject and also assuming that this can be uniquely characterized as a universal in categorical terms, there is no proof that the latter is an *autonomously* identifiable source of mathematical reasoning. Bell proposed looking at topos theory as signifying the passage from absolute to local mathematics: from the Newtonian world of constant (classical) sets to the invariants from one topos to another. And one might extend this proposal beyond toposes, but already in their case the obstacle is that Lorentz transformations form a group, whereas geometric morphisms are not generally invertible. At the very least, even were the variety of all possible logics to form one hierarchy of increasingly general invariants (in analogy to more general groups), no corresponding relativity theory for logics exists as yet.

Thus it seems we are back to saving the phenomena by resorting to a pluralistic formalism, reminiscent of Carnap's Principle of Tolerance, now sufficiently flexible to accommodate constructive constraints too. This is the "no solution" solution. However comfortable, it is not the only, or the last, option: for it not only leaves the range of kinds-of-structure accessible to human thought unexplained, but makes it a pseudoproblem. Is it really so? After the age of incompleteness, with its presumably paradigmatic moral that the limitations inherent to a kind of self-encoding formal systems have the effect of a cosmic sentence, categorical analysis allows rethinking the matter at its mathematical roots. Time is ripe to revive Hilbert's dictum: *wir müssen wissen, wir werden wissen*. It is obvious that philosophy has already made its entrance here; it's not obvious that it was already there in disguise. Though there is no unique philosophy adapted to categorical logic, there are (mutually opposite) philosophical objections to it. These include the "essentialism" charge leveled by (Girard, 2004, p. 152), Martin-Löf's criticism of definitions of entities "up to isomorphism," rejection of the idea that types come first, terms after, as well as both circularity and "abstract nonsense" arguments by supporters of $\in$-based set theory as a foundation. Such objections can be met only if the escape into a sort of neo-formalism is avoided. The above emphasis on geometric logic as well as the logical import of adjoints and fibrations serves this aim.

As for the very definition of a category, the objects are what they are and not isomorphism-classes, unless the category is skeletal (which still presupposes the data for a quotient). To work up-to-iso from top to bottom, one should not only skip identities but also to reformulate composition accordingly, and it is difficult to see how to have $g \cdot f$ before the equation $dom(g) = cod(f)$. (Resort to partiality just shifts the problem.) Thus the objection that when we assert φ, we mean φ at face value, and not φ up-to-iso, is misleading insofar as it suggests that synonymy classes are singletons. Ernst Cassirer's reconstruction of the process leading modern science and mathematics from *Substanzbegriff* to *Funktionbegriff* is not cast aside by characterizing logical operations by means of universal properties. Vice versa, the priority of structure over object-substance does not commit us to a formalist version of logicism in a type-theoretical setting. Up-to-iso sounds like an admission of original sin: even though the objects of a category **C** are supposed to be given, what matters is their structure and their structure is defined (as the Yoneda Lemma witnesses) by the **C**-morphisms to and from any other object in **C**. But if there is any place where canonical constructions, providing unique existence and naturality, are really fundamental it is category theory. At present, categorical logic is the most general framework for a variational study of principles of reasoning and also the view of logical problems according to which if there is a solution, it is a universal solution. It is unique existence up-to-iso expressed by Π_3-sentences and it is a virtue that can be perfected, rather than ignored.

Doubtless, categorical logic can be treated as just a formal device for securing preservation of some proof-structure and model-theoretic structure. As such, it already repays the effort to learn it, but its very source in geometry provides better understanding and draws attention to the space-laden nature of the roots of thought. Syntactic constructions are no exception; our understanding of algebraic manipulations of symbols is, as the verbs for such actions reminds us, only possible because those roots are never uprooted. When the language of categories makes the mathematical content of proofs, computations, and the syntax-semantics interface explicit, it points to a concrete background of cohesive objects and patterns of actions by which to investigate the most sophisticated type systems. Semantics as commonly understood preserves a few fossils of content, while much such content is carried by logical syntax. So the meaning often ascribed to the internal language and to the Equivalence Theorem can be misleading. More than shifting the demarcation line between syntax and semantics, the point is to have invariants which are independent of given theory-presentations. One great value of these invariants is their use in overcoming the contrast between the axiomatic and the phenomenological method, as opposed to supporting a neo-formalist view. It enables us to take the givenness of objects and maps at face value, as manifested in concrete patterns of action.

On the one hand logicism seems vindicated, provided its original demands are relaxed and finally the tension with constructive reasoning is overcome within a purely "structural" horizon. On the other hand logicism must be enriched, for it is the inherent structure of the types-and-terms system which grounds the constructive sense of logic. So, if one intends to avoid dependence of logic on a prior formal ontology, typing has to be taken as a basic logical notion, while the nature of the ontology remains indeterminate, in agreement with formalism. The plurality of such systems of types-and-terms does the rest. But then it is problematic to grasp the real effect of the dictum "a logic is always a logic over a type theory," meaning that $\vdash$ is relative to which system of types allows for which variable declarations. The idea agrees with an underlying theme of the categorical approach to logic since its beginning, but it may also be taken as support for reviving logical conventionalism *à la* Carnap: "in logic there are no morals" now becomes "choose the type system you need." In this sense, if internalization relative to a multiple hierarchy of types only refines the lesson already drawn from the Skolem paradox applied to the semantics of classical set theory, the gain is small.

As for the internal/external distinction, given **C** as a mathematical universe of discourse, the logic-of-**C** is that which results from the common structure of objects, is preserved by morphisms and coincides with the structure of valid inferences as representable within **C**. Its constructive nature is measured by the kinds of variation and cohesion present. (Arguments for constructivity find their ultimate motivation in the fact that what becomes—be it the very thinking subject through its states—is endowed with dynamic structure.) This very idea needs qualification, in view of the ambiguity between the internal logic $L_i(\mathbf{C})$ and the external logic $L_e(\mathbf{C})$. Though there are categories the structure of which is insufficient to represent internally the logic used to describe the properties of their objects and maps, the internal language of a topos is sufficient to express local properties. Yet there are also global properties, externally detectable, that are not represented internally and they cannot be taken as logically irrelevant. To reduce the gap between internal and external, in the presence of representability, one can consider the strongest logic L for which $L_i(\mathbf{C}) \equiv L_e(\mathbf{C})$. Were $\mathbf{C}$ taken as the whole of possible mathematical universes-of-discourses, it would be improper to ask for such logic, since there is no external logic. When we talk about logic as a whole, either it is really abstract nonsense or it is something internally representable. Thus, the issue is not whether all logical properties can be collectively represented into one definite universe of discourse, but rather whether there is a finite range of patterns that, suitably fine-tuned (by adjunctions), can generate all possible variations. Categorical logic supports a positive solution to this issue.

Bibliography

Artin, M., Grothendieck, A., and Verdier, J. L. (1972). *Séminaire de Géometrie Algébrique, IV: Théorie des Topos et Cohomologie Étale des Schemas (1964).* Lecture Notes in Mathematics volumes 269, 270, 305. Springer, Berlin.

Asperti, A. and Longo, G. (1991). *Categories, Types and Structures.* MIT Press, Cambridge, MA.

Bell, J. L. (1986). From absolute to local mathematics. *Synthese*, 69:409–426.

Bell, J. L. (1988). *Toposes and Local Set Theories: An Introduction.* Clarendon Press, Oxford.

Bell, J. L. (1993). Hilbert's ε-operator and classical logic. *Journal of Philosophical Logic*, 22:1–18.

Bell, J. L. (1996). Logical reflections on the Kochen-Specker theorem. In Clifton, R., editor, *Perspectives on Quantum Reality*, pages 227–235. Kluwer, Amsterdam.

Bell, J. L. (1998). *A Primer of Infinitesimal Analysis.* Cambridge University Press, Cambdrige.

Bell, J. L. (2005). The development of categorical logic. In Gabbay, D. and Guenthner, F., editors, *Handbook of Philosophical Logic*, volume 12, pages 279–361. Springer, New York, NY.

Bell, J. L. (2006). Cover schemes, frame-valued sets and their potential uses in space-time physics. In Reimer, A., editor, *Spacetime Physics Research Trends. Horizons in World Physics*, volume 248. Nova Science, New York, NY.

Benabou, J. (1985). Fibred categories and the foundations of naive category theory. *Journal of Symbolic Logic*, 50:10–37.

Boileau, A. and Joyal, A. (1981). La logique de topos. *Journal of Symbolic Logic*, 46:6–16.

Borceux, F. (1994). *Handbook of Categorical Algebra.* Cambridge University Press, Cambdrige.

Cartier, P. (2001). A mad day's work: From Grothendieck to Connes and Kontsevich. The evolution of concepts of space and symmetry. *Bulletin of the American Mathematical Society (New Series)*, 38(4):389–408.

Dummett, M. (1975). The philosophical basis of intuitionistic logic. In Rose, H. and Sheperdson, J., editors, *Logic Colloquium '73*, pages 5–40. North-Holland, Amsterdam. Reprinted in *Truth and Other Enigmas*, Duckworth, 1978, pp. 215–247.

Eilenberg, S. and Mac Lane, S. (1945). General theory of natural equivalences. *Transactions of the American Mathematical Society*, 58:231–294. Reprinted in S. Eilenberg and S. Mac Lane, *Collected Works*. Academic Press, New York, NY, 1986.

Fourman, M., Mulvey, C., and Scott, D., editors (1979). *Applications of Sheaves. Proceedings of the London Mathematical Society Durham Symposium 1977*, volume 753 of *Lecture Notes in Mathematics*. Springer, Berlin.

Fourman, M. and Scott, D. (1979). Sheaves and logic. In Fourman, M., Mulvey, C., and Scott, D., editors, *Applications of Sheaves*, pages 302–401. Springer, Berlin.

Fourman, M. (1977). The logic of topoi. In Barwise, J., editor, *Handbook of Mathematical Logic*, pages 1053–1090. North-Holland, Amsterdam.

Freyd, P. and Scedrov, A. (1990). *Categories, Allegories*. North-Holland, Amsterdam.

Freyd, P. (1972). Aspects of topoi. *Bulletin of the Australian Mathematical Society*, 7:1–76, 467–480.

Girard, J.-Y. (2004). *Logique à Rome*. Università di Roma Tre, Roma.

Goldblatt, R. (1979). *Topoi: The Categorial Analysis of Logic*. North-Holland, Amsterdam. Revised edition, 1984.

Grothendieck, A. and Dieudonné, J. (1971). *Éléments de géometrie algébrique I*, volume 166 of *Die Grundlehren der mathematischen Wissenschaften in Einzeldarstellungen*. Springer, Berlin.

Grothendieck, A. (1970). *Catégories fibrés et descente (1960/61)*, volume 224 of *Lecture Notes in Mathematics*. Springer, Berlin.

Hatcher, W. (1968). *Foundations of Mathematics*. WB Saunders, Philadelphia, PA.

Hyland, M. (1982). The effective topos. In Troelstra, A. and van Dalen, D., editors, *The L.E.J. Brouwer Centenary Symposium*, pages 162–217. North-Holland, Amsterdam.

Jacobs, B. (1999). *Categorical Logic and Type Theory*. Elsevier, Amsterdam.

Johnstone, P. (1977). *Topos Theory*, volume 10 of *London Mathematical Society Monographs*. Academic Press, New York, NY.

Johnstone, P. (1982). *Stone Spaces*. Cambridge University Press, Cambridge.

Johnstone, P. (2002). *Sketches of an Elephant. A Topos Theory Compendium*. Clarendon Press, Oxford.

Kan, D. (1958). Adjoint functors. *Transactions of the American Mathematical Society*, 87: 294–329.

Kock, A. and Reyes, G. (1977). Doctrines in categorical logic. In Barwise, J., editor, *Handbook of Mathematical Logic*, pages 284–313. North-Holland, Amsterdam.

Lambek, J. and Scott, P. (1986). *Introduction to Higher-Order Categorical Logic*. Cambridge University Press, Cambridge.

Lambek, J. (1968). Deductive systems and categories I. *Journal of Mathematical Systems Theory*, 2:278–318.

Lawvere, F. W., Maurer, C., and eds., G. W. (1975). *Model Theory and Topoi*, volume 445 of *Lecture Notes in Mathematics*. Springer, Berlin.

Lawvere, F. W. and Rosebrugh, R. (2003). *Sets for Mathematics*. Cambridge University Press, Cambridge.

Lawvere, F. W. (1963). *Functorial semantics of algebraic theories*. PhD thesis, Columbia University, New York, NY.

Lawvere, F. W. (1964). An elementary theory of the category of sets. *Proceedings of the National Academy of Sciences USA*, 52:1506–1511.

Lawvere, F. W. (1966). The category of categories as a foundation for mathematics. In Eilenberg, S., Harrison, D., Mac Lane, S., and Röhrl, H., editors, *Proceedings of the Conference on Categorical Algebra, La Jolla*, pages 1–21. Springer, Berlin.

Lawvere, F. W. (1968). Diagonal arguments and cartesian closed categories. In *Category Theory, Homology Theory and their Applications, II*, volume 92 of *Lecture Notes in Mathematics*, pages 134–145. Reprinted in *Theory and Applications of Category Theory*, No.15 (2006), pp. 1–13. Springer, Berlin.

Lawvere, F. W. (1969). Adjointness in foundations. *Dialectica*, 23:281–295.

Lawvere, F. W. (1971). Quantifiers and sheaves. In *Actes du Congrès International des Mathématiciens*, volume 1, pages 329–334. Gauthier-Villars, Paris.

Lawvere, F. W., editor (1972). *Toposes, Algebraic Geometry and Logic*, volume 274 of *Lecture Notes in Mathematics*. Springer, Berlin.

Lawvere, F. W. (1973). *Teoria delle Categorie sopra un Topos di Base*. Universitá di Perugia, Perugia.

Lawvere, F. W. (1975). Continuously variable sets: algebraic geometry = geometric logic. In Rose, H. and Sheperdson, J., editors, *Logic Colloquium '73*, pages 135– 153. North Holland, Amsterdam.

MacIntyre, A. (1973). Model-completeness for sheaves of structures. *Fundamenta Mathematicae*, 81:73–89.

Macnamara, J. and Reyes, G., editors (1994). *The Logical Foundations of Cognition*. Oxford University Pres, New York, NY.

Mac Lane, S. and Moerdijk, I. (1992). *Sheaves in Geometry and Logic: a First Introduction to Topos Theory*. Springer, New York, NY.

Mac Lane, S. (1935). A logical analysis of mathematical structures. *The Monist*, 45:118–130.

Mac Lane, S. (1986). *Mathematics: Form and Function*. Springer, Berlin.

Mac Lane, S. (1997). Despite physicists, proof is essential to mathematics. *Synthese*, 111:147–154.

Makkai, M. and Paré, R. (1989). *Accessible Categories: the Foundations of Categorical Model Theory*, volume 104 of *Contemporary Mathematics*. American Mathematical Society, Providence, RI.

Makkai, M. and Reyes, G. (1977). *First Order Categorical Logic*. Springer, Berlin.

Makkai, M. (1987). Stone duality for first order logic. *Advances in Mathematics*, 65:97–170.

Mangione, C. (1976). Logica e teoria delle categorie. In Geymonat, L., editor, *Storia del Pensiero Filosofico e Scientifico*, volume 7, pages 519–653. Garzanti, Milano.

Mayberry, J. (1994). What is required of a foundation for mathematics? *Philosophia Mathematica*, 2:16–35.

McLarty, C. (1992). *Elementary Categories, Elementary Toposes*. Clarendon Press, Oxford.

McLarty, C. (2005). Saunders Mac Lane (1909–2005). His mathematical life and philosophical works. *Philosophia Mathematica*, 13:237–251.

Mitchell, W. (1972). Boolean topoi and the theory of sets. *Journal of Pure and Applied Algebra*, 2:261–274.

Moore, E. (1908). On a form of general analysis with application to linear differential and integral equations. In *Atti del IV Congresso Internazionale dei Matematici*, pages 98–114. Accademia dei Lincei, Roma.

Osius, G. (1974). The internal and external aspects of logic and set theory in elementary topoi. *Cahiers de Topologie et Géométrie Différentielle*, 15:157–180.

Peruzzi, A. (1991a). Categories and logic. In Usberti, G., editor, *Problemi Fondazionali nella Teoria del Significato*, pages 137–211. Olschki, Firenze.

Peruzzi, A. (1991b). Meaning and truth: the ILEG project. In Wolf, E., Nencioni, G., and Tscherdanzeva, H., editors, *Semantics and Translation*, pages 53–59. Moscow Academy of Sciences, Moscow.

Peruzzi, A. (1994). On the logical meaning of precategories. *http://www.philos.unifi.it/persone/peruzzi.htm*.

Peruzzi, A. (2000a). Anterior future. *Rendiconti del Circolo Matematico di Palermo*, 64:227–248.

Peruzzi, A. (2000b). The geometric roots of semantics. In Albertazzi, L., editor, *Meaning and Cognition*, pages 169–201. John Benjamins, Amsterdam.

Pitts, A. (2001). Categorical logic. In Abramsky, S., Gabbay, D. M., and Maibaum, T. S. E., editors, *Handbook of Logic in Computer Science*, volume 5, Logic and algebraic methods, pages 39–123. Oxford University Press, Oxford.

Rasiowa, H. and Sikorski, R. (1963). *The Mathematics of Metamathematics*. Polish Scientific Publishers, Warsaw.

Reyes, G. (1974). From sheaves to logic. In Daigneault, A., editor, *Studies in Algebraic Logic*, volume 9 of *MAA Studies in Mathematics*, pages 143–204. The Mathematical Association of America, Providence, RI.

Seely, R. (1984). Locally cartesian closed categories and type theory. *Mathematical Proceedings of the Cambridge Philosophical Society*, 95:33–48.

Seely, R. (1987). Modelling computations: A 2–categorical framework. In *Proceedings of the Second Symposium on Logic in Computer Science*, pages 61–71. IEEE Computer Science Press, Trier.

Troelstra, A. S. and van Dalen, D. (1988). *Constructivism in Mathematics: An Introduction*, volume 1–2. North-Holland, Amsterdam.

Vickers, S. (1988). *Topology via Logic*. Cambridge University Press, Cambridge.

Yanofsky, N. (2003). A universal approach to self-referential paradoxes, incompleteness and fixed points. *arXiv:math.LO/0305282 v1*. May 19.

Zangwill, J. (1977). Local set theory and topoi. MSc thesis, Bristol University, Bristol.

Part IV
Philosophy of Science

Chapter 16
Gunk, Topology and Measure

Frank Arntzenius

1 Introduction

It is standardly assumed that space and time consist of extensionless points. It is also a fairly standard assumption that all matter in the universe has point-sized parts. We are not often explicitly reminded of these very basic assumptions. But they are there. For instance, one standardly assumes that one can represent the states of material objects, and of fields, by functions from points in space and time to the relevant point values. Electric fields, mass densities, gravitational potentials, etc.... are standardly represented as functions from points in space and time to *point* values. This practice would seem to make no sense if time and space did not have points as parts.

There is an alternative that has not been much explored. The alternative is that space and time and matter are "pointless," or "gunky." The idea here is not that space and time and matter have smallest finite-sized bits, that space and time and matter are "chunky." Rather the idea is that every part of space and time and matter has a non-zero, finite, size, and yet every such part can always be subdivided into further, smaller, parts. That is to say, the idea is that every part of space and time and matter has a non-zero size, and yet there is no smallest size.

Let me emphasize how radical this idea is. It is very natural to think that any thing decomposes into some ultimate collection of fundamental parts. And it is very natural to think that the features of any object are determined by the way that object is constructed from its ultimate parts, and by the elementary features of these ultimate parts. Indeed, much of the history of science can be seen as an attempt to break down complex objects and processes into ultimate parts, and to find the laws that govern these ultimate parts. But if there are no smallest regions, and if there are no smallest parts of objects, then a spatial or temporal decomposition of a region, and of an object, cannot bottom out at an ultimate level. The idea that the features of large regions and large objects are determined by the features of minimal-sized regions and minimal-sized objects cannot work if space and time, and the objects in it, are gunky, i.e. pointless. Space, time, and objects would simply not have ultimate parts.

F. Arntzenius (✉)
Sir Peter Strawson Fellow in Philosophy, University College, Oxford, UK
e-mail: frank.arntzenius@philosophy.ox.ac.uk

D. DeVidi et al. (eds.), *Logic, Mathematics, Philosophy: Vintage Enthusiasms*, The Western Ontario Series in Philosophy of Science 75,
DOI 10.1007/978-94-007-0214-1_16,

There would just be an infinite descending chain of ever-smaller parts. A somewhat dizzying prospect.

Well, let's not get ahead of ourselves. Not only would it require a fairly radical revision of our atomistic intuitions, it would also require a fairly radical and extensive re-working of standard mathematical methods for doing physics. If we cannot use real numbers to coordinatize locations in space and time, what can we use? If we cannot use ordinary functions to describe the states of things, what can we use?

All things in good order. We will get started on the business of re-writing physics a bit later on. First we will consider arguments for undertaking this seemingly mad enterprise. To preview: we will find no utterly compelling arguments against the existence of points. But we will find non-compelling reasons to explore the mathematics of gunky space and time.

2 The Possibility of Motion and Determinism

Zeno argued that if time consists of instants of zero duration, then during each such instant an object cannot move. But if time consists entirely of a series of such instants then objects can never move. In view of this problem Aristotle proposed that there are no instants, no 0-sized intervals of time, indeed no smallest sized, atomic, intervals of time. Rather, time consists of smaller and smaller intervals. To put it another way: the world is a true movie, not a sequence of snapshots. To put it even more suggestively: becoming is not reducible to being.

One may not be impressed by Zeno's argument. One may for instance respond, as did some commentators in the Middle Ages, that to be in motion is just to be at different locations at different times, so that it simply is not true that just because one occupies only one location at one time one never moves.

Indeed, this is a perfectly coherent way to respond to Zeno's problem. However one can then formulate a new worry, which is closely related to Zeno's worry. For if motion is just a matter of being at different locations at different times, then the intrinsic state of an object at an instant does not include its velocity. How then does an object at an instant "know" in which direction to continue and at which speed? Less anthropomorphically: if the instantaneous states of objects do not include their velocities, then how could the instantaneous state of the world determine its subsequent states? That is, how could determinism hold? The world may in fact develop in a deterministic fashion, and it may not, but surely whether it does, or does not, should depend on the character of the laws of evolution of the world, rather than that the atomicity of the structure of time alone should imply that the world cannot be deterministic.[1]

---

[1] Well, it could still be deterministic if the equations of motion were first-order, as they are in quantum mechanics. Still, one might like to think that even if the equations of motion are second-order, as they are in classical mechanics, the world could be deterministic.

One might attempt to respond to this argument by claiming that even if time consists of 0-sized instants, nonetheless the intrinsic state of an object at a time does include a velocity. Such an "intrinsic velocity" would not be defined (as in ordinary calculus) in terms of (limits of) the position development of an object. Rather, it would be a primitive intrinsic feature of an object at a time, which *causes* the object to subsequently move in the direction in which the intrinsic velocity is pointing.

This does not strike me as a plausible response. For according to this response the intrinsic velocity at an instant and the direction in which the trajectory in space continues are not definitionally related, but are merely causally related. So it should be logically possible to have an object whose spatial trajectory continues in a direction that differs from the direction in which its intrinsic velocity points. Now, one might claim that such a bizarre non-alignment of direction of trajectory and intrinsic velocity is ruled out by the laws of nature. However, the mere conceptual possibility of such a misalignment seems puzzling, to say the least. Furthermore, if the laws of nature are to connect the directions in which primitive velocities point and the directions of trajectories in space, as they must do, then there is going to have to exist some further primitive relation, "parallelhood," which obtains, or fails to obtain, between intrinsic velocities and spatial directions. Other things being equal it seems undesirable to add "intrinsic velocities" and "parallelhood" to one's stock of primitive quantities and relations, when one has no real need for them. In short, this response on behalf of points does not seem plausible to me. (For more detail on this line of argumentation, see (Arntzenius, 2000).)

However, a more plausible response to Zeno can be made on behalf of points. For one could simply claim that determinism should not be understood as the idea that the state at an instant determines states at all other times. Rather it should be understood as the idea that *any finite history* of states determines states at all other times.

Indeed, it seems to me that Zeno's arrow provides no compelling argument against point-sized instants. Let's turn to another argument.

3 Cutting Things in Half

If space consists of points then one cannot cut a region exactly in two halves. For if one of the two regions includes the point on the cutting line, i.e. if it is closed at the cut, then the other does not include the points on the cutting line, i.e. it is open at the cut. Imagine, for instance, that we have x and y coordinates which are parallel to the sides of a rectangle. Suppose that the horizontal, x-coordinate, of the rectangle runs from 0 to 2, and suppose that we cut the rectangle at $x = 1$. The question then arises: do the points that have x-coordinate=1 belong to the left hand side after we have made our cut, or to the right hand side? If they belong to the left hand side, then the left half is closed at the cut, and the right half is open. If vice versa, then the left side is open. So the two parts would not be identical. So one cannot cut a region exactly in half if regions are composed out of points. One might reasonably conjecture that such a difference between open and closed regions is an artifact of our mathematical representation of regions, that does not correspond to a difference in reality. I, for one, find it hard to believe that there really are distinctions between open and closed regions in nature. But I agree that this is hardly a knock-down argument.

4 Paradoxes of Size

If there exist points in space, and space is continuous, then it can be shown that there must be regions that have no well-defined size. For instance, there will be a part of any wall in any room such that it has no well-defined size. If you wanted to paint such a part of a wall in your house blue, there would be no possible answer to the question: "How much paint will I need to paint that part of my wall blue?" The problem is not that you would not know how much paint you would need, or that you would need 0 quarts of paint. Rather, the problem is that there just exists no quantity r of paint such that you would need exactly that quantity to paint that region. Let me be a bit more precise.

One can prove that in a continuous, pointy space there must exist regions that have no well-defined measure, if one assumes the axiom of choice and one assumes that the measure is countably additive. One can also prove that in a continuous pointy space of three or more dimensions there must exist regions that have no well-defined measure, if one assumes the axiom of choice and one assumes that the measure is finitely additive and invariant under (distance preserving) translations and rotations.[2]

There is even more weirdness about points and sizes: Banach and Tarski have shown that the existence of points implies cost-free guaranteed increases in size. That is to say, they showed that in a continuous pointy 3-dimensional space one can take a sphere, break it into a finite number of pieces (five pieces in fact), move those pieces around rigidly (i.e. while preserving distances between the parts of the pieces), and re-arrange those pieces to form a sphere of twice the size! That is to say, by breaking an object into five parts, and merely re-arranging these parts spatially, without any stretching or changing of shapes, one can make an object larger or smaller, as one desires.

There is in fact a close relationship between this result and the fact that there are regions which have no well-defined size. Some of the parts into which we must break the sphere must have no well-defined sized. It is not hard to see that this must be so, for rigid motions preserve size, and the size of an object that consists exactly of five non-overlapping parts is just the sum of the sizes of those parts. So Banach and Tarski's result depends essentially on the existence of size-less regions.

How might one respond on behalf of points? Well, in the first place, one might simply deny the axiom of choice. This is an issue that could take us deep into

[2] See, e.g., (Skyrms, 1983; Wagon, 1985).

philosophy of mathematics and mathematical physics, to which I have nothing new to contribute. I merely wish to point out that denying the axiom of choice implicitly commits one to (being part of) a large project, namely that of re-writing that part of mathematics and mathematical physics that one wants to retain, in such a way that it makes no use of the axiom of choice. My only comment on this project is that I am interested in a different project, namely that of doing physics without points. My project has several independent motivations, only one of which concerns the measure theoretic paradoxes.

A second possible response to the measure theoretic paradoxes is: who cares? Surely we will not as a practical matter get our hands on measureless parts of objects. Surely size-altering de-compositions and re-compositions are not practically achievable. So why worry?

Indeed, since one needs the axiom of choice to prove the existence of measureless regions, one cannot have explicit constructions of measureless, or of measure-altering, de-compositions and re-compositions. Nonetheless the mere existence of regions and/or parts which have no measure, and the mere possibility of size-altering de-compositions and re-compositions, remains rather bizarre, and prima facie implausible.

A third possible response to the measure theoretic paradoxes (on behalf of points) starts by making a distinction between sets of points in space-time, which are mathematical entities, and physical regions. One could, e.g., suggest that all physical regions are Borel regions.[3] If that is so, then all physical regions are (Lebesque) measurable, and no size-altering de-compositions and re-compositions are possible.

Indeed, one could say this. But note that this means that regions fail to satisfy the standard axioms of mereology. For one is denying that the fusion of any arbitrary collection of regions is a region. (Some collections of points are such that their fusion is a non-Borel region.) It seems hard to motivate this failure independently.

Nonetheless, yet again, we have found no devastating argument against points. We have simply found one more reason to try to see how far we can go without points. Let us turn to another argument against points.

5 Quantum Mechanics and Points

In non-relativistic quantum mechanics one can represent the state of a single particle by a wave-function. The probability that a particle will be found in a particular region upon measurement is given by the integral of the square of this wave-function in this region. If one has two functions whose values differ on a set of points of measure 0, then integrating them over any region will always yield identical results. Thus, as far as probabilities of results of measurements are concerned, functions that

[3] Borel regions of the real line: start with the collection of all open intervals, then close this collection up under countable union and intersection, and complementation.

differ on a set of points of measure 0 are equivalent. This provides motivation for the claim that functions that differ on a set of points of measure 0 correspond to the same wave-function, i.e. the same quantum state.

A slightly more formal motivation for this derives from the fact that in a Hilbert space there is a unique null vector, a unique vector whose inner product with itself is 0. Thus, if one wishes to represent vectors in a separable Hilbert space (with a countable infinity of dimensions) by (complex) functions on space (or configuration space), and one wishes to represent the inner product of vectors by integration of the corresponding functions, then one has to represent vectors not by functions, but by equivalence classes of functions whose values differ on up to (Lebesque) measure 0 points. Indeed, although it is not often brought to the fore, it is a standard assumption in quantum mechanics that wave-functions correspond to equivalence classes of (square integrable) functions that differ up to Lebesque measure 0.

This ignoring of measure 0 differences between regions in space suggests that quantum mechanics should be set in a gunky space, not in a pointy space. (We will flesh out this claim in more detail when we examine the measure theoretic approach to gunk.) But, as always, there are responses possible on behalf of the point lover.

In the first place one might respond that the above is a false claim: quantum mechanics standardly uses wave-functions that are eigenfunctions of position, so-called "delta functions," which differ from each other only on measure 0 sets of points. This line of response takes us into a tricky area. So-called "delta functions" are not functions at all. Indeed position operators, on the standard separable Hilbert space approach to quantum mechanics, simply cannot have eigenstates. Nonetheless, it is true that there are (non-standard) ways of rigorizing the notion of an eigenstate of position, and thereby sanctioning states that in a clear sense are confined to a single point, while departing from the standard formalism of separable Hilbert spaces. (See, e.g., (Bohm, 1978; Halverson, 2001).) Not only does one have to depart from the standard formalism of separable Hilbert spaces in order to do so, but position eigenstates also have the feature that observables such as momentum and energy have no well-defined expectation values in such position eigenstates. In (Arntzenius, 2004) I have discussed whether it is worth paying this price for the acquisition of position eigenstates, and argued for a cautious "no." Let me here merely say that it is far from clear that it is worth paying this price, and leave it at that.

There is of course another possible response that can be made on behalf of the point lover. One could simply accept that quantum mechanics happens not to make use of measure 0 differences, and argue that this is all good and well, but this does not mean that such differences do not exist. Not every theory needs to make use of all the features that nature has on offer.

Indeed, I agree that both of the above two responses (on behalf of points) are perfectly coherent and possible. Nonetheless it seems to me that nature is piling up the hints that there just might be no points out there in space and time. Let's look at one more problem with points.

6 Contact Between Objects

In the nineteenth century some people started worrying about the possibility of contact between solid objects if space consists of points. Here is a sketch of such worries. Let us suppose that solid objects cannot interpenetrate, i.e. that solid objects can not occupy overlapping regions. Now consider two solid objects which always occupy closed regions, i.e. regions which include their own boundary. Such objects can never be in contact, for closed regions either overlap or are a finite distance apart. In order to avoid interpenetration such objects must decrease their velocities when they are still a finite distance apart, so some kind of action at a distance would have to occur. It seems strange and objectionable that the mere existence of solid objects should imply action at a distance. Alternatively suppose that solid objects occupy open regions. Then there must always be at least one point separating them. So they still cannot be in genuine contact, and they still must change their velocities without ever being in genuine contact.

The impossibility of genuine contact seems to provide an objection to the existence of points. However, there are a number of decent responses that one can give on behalf of points.

In the first place, one could respond that one would not want such "genuine contact" anyhow, since collisions would lead to sharp, undifferentiable, kinks in the trajectories of objects. One could plausibly argue that a more realistic physics has objects interacting through fields. Then there will never be "genuine contact," so there is no "problem of contact." One could amplify this line of thought by claiming that it is even more realistic to suppose that quantum mechanics, with an ontology of wave-functions (or perhaps wave-functions plus point particles), is correct, and that given such an ontology there is no problem of contact.

Secondly, one could argue that even if one wants to countenance solid objects which interact by contact, one could just have a slightly different account of what it is to "be in contact" and what it is to "interpenetrate." One could, e.g., just say that two objects are "in contact" if and only if the boundaries of the regions that they occupy overlap. (A point p lies on the boundary of region R if any open set containing p intersects both R and the complement of R.) And one can say that objects do not "interpenetrate" unless they overlap on more than just their boundaries. Physics can then proceed as usual. Of course, this would mean that objects occupying open regions (in a 3-dimensional space) that are separated by a two-dimensional surface are in contact, and that bodies occupying closed regions which overlap on a two-dimensional surface do not interpenetrate. But so what? It does not lead to any trouble in formulating physics, or any trouble with experiment. It only leads to trouble with philosophers who think that it is a priori that "genuine contact" is possible, where "genuine contact" means having not even a single point in between, and who think it is a priori that "interpenetration" is not possible for solid objects, where "interpenetration" means not overlapping even on a single point. I don't know whether to respond to such philosophers that in a Newtonian collision world there are, in their sense of "solid," no solid objects, or whether to respond that in their sense of "genuine contact" there is no genuine contact, and in their sense of "interpenetration" there is interpenetration. But one can do Newtonian collision physics when one defines contact as having overlapping boundaries, and interpenetration as overlapping on more than a boundary.

Both of the above responses on behalf of points seem adequate. Nonetheless note that neither of the responses requires a physics that makes essential use of points, or of measure 0 differences. So one is still left with the suspicion that points, and measure 0 differences, are artifacts of the mathematics, and do not exist in reality.

7 Now What?

It appears that every problem associated with the existence of points can be overcome; there appears to be no single devastating argument that space and time (or matter) have to be gunky. Nonetheless it remains of interest to examine the possibility of doing physics in gunky space and time in more detail.

There have been a number of approaches to the mathematics of gunky spaces. These approaches divide into three categories: the measure theoretic approach (Skyrms, 1993; Sikorski, 1964), the topological approach (Roeper, 1997), and the metric approach (Gerla, 1990). In this paper I will not look in any detail at the metric approach. The reason I will not is that in the metric approach one assumes from the start as fundamental notions the notion of the "diameter" of a region and the notion of the "distance" between regions. This approach is prima facie ill-suited for the purposes of modern physics since in general relativity the notion of distance is local and path-dependent (rather, it is not a non-local path-independent) relation between regions. It would seem preferable to first be able to build a pointless differentiable manifold, and then to be able to put a metric tensor field on such a differentiable manifold. The measure-theoretic and topological approach to gunk are prima facie more amenable to this idea, since they do not start by presupposing the existence of non-local metric structure. Let's look at these two approaches in more detail and let's start with the topological approach.

8 The Topological Approach to Pointless Spaces

My strategy for constructing a pointless topological space will be as follows. I will start with an ordinary pointy topological space. I will then put on blurry spectacles which wash out differences in regions which, intuitively speaking, are differences in the (pointy) mathematical representation of space that do not correspond to differences in actual physical space. This will yield a pointless topology. Once I have a pointless topology, I, of course, no longer have ordinary (point to point) functions. But there are still maps from pointless regions to pointless regions. We will see that a rather natural set of such maps corresponds one-one to pointy functions that map regular closed region to regular closed regions. Unfortunately this does not include functions which are constant on a finite region, so that we do not appear to have enough materials to do physics with. Furthermore, one would like to be able to put a measure on a pointless topological space. We will find that there is also a problem in putting a measure on a pointless topological space. I will therefore advocate switching to the measure theoretic approach. But first some of the details of the topological approach.

Let us start with an ordinary pointy topological space which is a "locally compact T_2 space." A topological space is a "T_2 space" if and only if for any distinct points x and x' there are disjoint open subsets O and O' containing x and x' respectively. This is a very mild separability condition. A topological space is "locally compact" if and only if for every point x there exists a "compact" closed set C such that x lies in the "interior" of C. A set S is "compact" if and only if for every collection of open sets $\{O_a\}$ such that S is a subset of the union of these open sets, $S \subseteq \bigcup\{O_a\}$, there is a finite subcollection of these open sets such that S is a subset of the union of that subcollection: $S \subseteq O_{a_1} \cup O_{a_2} \cup \cdots \cup O_{a_n}$. The demand that a space be locally compact is very mild and roughly speaking amounts to the demand that each point is contained in an open set whose closure is not "very large." (The closure of a set S is the union of S with its boundary.)

Now let us put on our blurry spectacles, and ignore differences between sets that "differ only on their boundaries." We will say that two sets A and B "differ only on their boundaries" if and only if the closure of the interior of A is equal to the closure of the interior of B, i.e. if $\mathbf{Cl\,Int}(A) = \mathbf{Cl\,Int}(B)$. (The interior of a set consists of the points of that set that do not lie on its boundary.) Here are a couple of examples of sets that, by this definition, differ only on their boundaries. Any set and its interior differ only on their boundaries. ($\mathbf{Cl\,Int\,Int}(A) = \mathbf{Cl\,Int}(A)$.) Any set consisting of finitely many points and any other set consisting only of finitely many points differ only on their boundaries, since the closure of the interior of each of them is the empty set.

Now, let us divide up all pointy regions (all sets of points) into equivalence classes R of regions that differ only on their boundaries. The motivation for doing this is that our "blurry glasses" cannot distinguish regions that are in the same equivalence class, so we can regard these equivalence classes as corresponding to pointless regions. (From here on the symbols "R" and "R_i" will be always taken to denote pointless regions rather than pointy regions.)

Now let us give these equivalence classes R mereological structure. (This mereology will be standard except that it will include a "null region," i.e., it will be a complete Boolean algebra.) In order to do this, let me first note that every equivalence class of pointy regions will include exactly one "regular closed" pointy region, where pointy region S is said to be "regular closed" if and only if $\mathbf{Cl\,Int}(S) = S$. For, take pointy region S in some equivalence class. Now consider $\mathbf{Cl\,Int}(S)$. It will be in the same equivalence class as S, since $\mathbf{Cl\,Int}(S) = \mathbf{Cl\,Int\,Cl\,Int}(S)$. For the same reason $\mathbf{Cl\,Int}(S)$ is regular closed. It is also the *only* regular closed pointy region in that equivalence class. For suppose S' is regular closed and in the same equivalence class as S. Then $\mathbf{Cl\,Int}(S) = \mathbf{Cl\,Int}(S') = S'$, so S' is the same as $\mathbf{Cl\,Int}(S)$. So there is a one-one correspondence between pointless regions R, and regular closed pointy regions PR. So we can define a mereological structure on the equivalence classes R by defining a mereological (Boolean) structure ($\leqslant, \neg, \&, \vee$) on the corresponding regular closed pointy regions PR. This we can define in the following way:

1. The empty set is the null region.
2. $PR_i \leqslant PR_j$ if and only if $PR_i \subseteq PR_j$. ("$\leqslant$" stands for the "part of" relation.)
3. $\neg PR = \mathbf{Cl}(\mathbf{Co}(PR))$. ($\mathbf{Co}(PR)$ is the set theoretic complement of PR.)
4. $PR_i \vee PR_j = PR_i \cup PR_j$.
5. $PR_i \,\&\, PR_j = \mathbf{Cl}(\mathbf{Int}\, PR_i \cap \mathbf{Int}\, PR_j)$.
6. If S is a set of regular closed pointy regions then $\bigvee S = \mathbf{Cl} \bigcup \{PR | PR \in S\}$.
7. If S is a set of regular closed pointy regions then $\bigwedge S = \mathbf{Cl}\,\mathbf{Int} \bigcap \{PR | PR \in S\}$.

Next let us give the pointless regions topological structure. The topological structure we will give pointless regions cannot be given in the same way that we gave pointy spaces topological structure, namely in terms of a distinction between open and closed regions. For that is exactly the kind of distinction that we do not believe exists if reality is pointless. Instead we will give the topological structure of pointless regions in terms of the primitive notions of "part of," "connectedness" and "limitedness." And again, we will use the one-one correspondence with regular closed pointy regions to determine the topological structure of the pointless regions. In particular, we stipulate that

1. Two pointless regions are "connected" if and only if the closed regular pointy regions that they correspond to have non-empty intersection.
2. A pointless region is "limited" if and only if the closed regular pointy region that it corresponds to is compact.

Now we can make use of a result that Peter Roeper proved in (Roeper, 1997). He has shown that any collection of pointless regions that is constructed in the above way (i.e. by taking equivalence classes of pointy regions in a locally compact T_2 space which differ only on their boundaries) will satisfy the following axioms of pointless topology:

A_1 If pointless region A is connected to pointless region B, then B is connected to A.

A_2 Every pointless region that is not the pointless "null region" is connected to itself. (The pointless "null region" corresponds to the equivalence class of regions which differ only on their boundaries from the null set.)

A_3 The null region is not connected to any pointless region.

A_4 If A is connected to B and B is a part of C then A is connected to C.

A_5 If A is connected to the "fusion" of B and C, then A is connected to B or A is connected to C. (The "fusion" of B and C is the smallest pointless region that has B and C as parts.)

A_6 The null region is limited.

A_7 If A is limited and B is a part of A then B is limited.

A_8 If A and B are limited then the fusion of A and B is limited.

A_9 If A is connected to B then there is a pointless limited region C such that C is a part of B, and A is connected to C.

A_{10} If A is limited, B is not the pointless null region, and A is not connected to the "complement" of B, then there is a pointless region C which is non-null and limited, such that A is not connected to the "complement" of C, and C is not connected to the "complement" of B. (The "complement" of a pointless region A is the pointless region $-A$ such that A and $-A$ have no parts in common, and every non-null pointless region has some part in common either with A or with $-A$.)

From a philosophical point of view it might seem that it would have made more sense to start with the axioms of pointless topology, and then to explain that any collection of pointless regions which satisfies these axioms will correspond one-one to equivalence classes of pointy regions in the unique corresponding locally compact T_2 space. After all, I certainly do not want to say that pointless regions *just are* equivalence classes of pointy regions, for that would mean that pointless regions just are mathematical constructions out of entities (pointy regions) which I believe not to exist. And that would not make much sense. No, the pointless view that I am here exploring is that pointy regions really do not exist, let alone that equivalence classes of them exist. The things that really exist are pointless regions, the primitive predicates and relations that are needed are the "part of" relation, the "limitedness" predicate and the "connected to" relation, and the axioms that characterize the true topology of space are A_1 through A_{10}. However, not only is it much easier to introduce the machinery of pointless topologies via a construction out of pointy topologies, it is also very important to see that pointless regions behave exactly the way that our blurry spectacle motivation wants them to behave. That is why I constructed pointless topologies in the way that I did. OK, on to the next tasks: placing material objects and fields in such a pointless topological space, and giving this space more structure than topological structure.

9 Objects in a Pointless Topology

If space is pointless then one cannot specify the locations of material objects by indicating for each point in space whether that object occupies it or not. So how should we conceive of the locational properties of objects in a pointless space? Well, here's a suggestion. We specify the locational state of a material object by specifying for every pointless region whether the object is entirely contained in that region or not.

This suggestion is problematic. The problem is that, despite the fact that space is pointless, one could nonetheless have point particles if one followed this suggestion. How? Well, imagine that a material object is such that it is entirely contained in each of a collection of smaller and smaller pointless regions. Now, if for any pointless region within which the object is contained there is an arbitrarily small pointless subregion within which the object is contained, the object could not have any finite size. So it must have size zero.[4] This is surprising. For it means that one can have pointless space containing point particles! However, allowing such a thing seems to defeat most of the reasons we started on this whole business of gunk. We wanted neither points in space nor point particles.

Moreover, allowing such point particles also leads to a formal feature that seems objectionable, namely a violation of "countable additivity." Here's what that means in this context and why it fails. Consider the following plausible looking principle: if an object is wholly outside each of a countable collection of regions R_i, then it is also wholly outside the fusion of these regions. Now consider our example. If a particle is entirely contained in each of a collection of converging regions, then it is wholly outside the complements of these regions. Now consider the fusion of the complements of these regions. Intuitively speaking the only thing that this fusion does not contain is the point that the converging collection is converging to. But remember that we are in a pointless space, and there exists no such point. So one should expect that the fusion of this countable collection is the whole space, for there is no pointless region that it misses out on, as it were. And indeed, this is correct: the fusion of these complements is the whole space. But a material object cannot lie entirely outside the whole of space. So we have a countable collection of regions such that the object lies wholly outside each one of the regions in the collection, but is wholly contained in their fusion. This is a failure of countable additivity, and seems bizarre and objectionable. So it seems that one should not allow a specification of the locational properties of a material object by specifying for each region whether it is entirely contained in it or not.

The obvious alternative is the following. One specifies the locational properties of a material object by specifying which region the object exactly fills. It will then, of course, be entirely contained in any region that includes this region, etc. But it could not be entirely contained in a converging collection of regions, for there is a minimal region, such that it is not contained in anything smaller than that region. No problem.

10 Fields in a Pointless Topology

How about the states of a field such as the electric field in a pointless topology? Here's a very natural suggestion. We specify the state of a field by specifying for each pointless region in space the exact range of values that the field obtains in that region. This brings up a further issue. Should we think of the possible ranges of values of the field as pointless ranges or as pointy ranges? Should we think that fields can have exact point values, or that the value spaces of fields are as gunky as the physical space that they inhabit? I don't know. In what follows I will make the

[4] This doesn't quite mean that it has to be a point particle, since it could still be a line, or an infinitely thin surface. But one can define a notion of "a converging set of regions" in such a way that the particle does indeed have to be a point particle if it is entirely within each of the regions in the "converging set of regions."

weaker assumption, i.e. I will assume that the value space of a field is a pointless space, and see how far we get.

Following Peter Roeper (Roeper, 1997), let us call a map h from a pointless physical space S to a pointless field value space VS a "bounded continuous mereological" map, if it satisfies the following constraints:

1. $h(R)$ is the null region in VS if and only if R is the null region in S.
2. If R_1 is part of R_2 then $h(R_1)$ is part of $h(R_2)$.
3. If V is non-null and part of $h(R_1)$ then there exists a non-null R_2 such that R_2 is part of R_1 and $h(R_2)$ is part of V.
4. If R_1 is connected to R_2 then $h(R_1)$ is connected to $h(R_2)$.
5. If R is limited, then $h(R)$ is limited.

One can prove (see (Roeper, 1997)) that there is a one-one correspondence between bounded continuous mereological maps (between two pointless spaces) and continuous pointy functions (between the two corresponding locally compact T_2 spaces) which map regular closed sets of points to regular closed sets of points. That is to say, if we specify the state of a pointless field in a pointless space by means of a bounded continuous mereological map h, then this is equivalent to specifying the corresponding pointy field in the corresponding pointy space by means of a function f from points in space to pointy field values, where f must satisfy the constraint that it maps regular closed sets of points in space to regular closed sets of field values.

Now, suppose that it were the case that every pointy field function f that one ever is likely to need when doing standard pointy physics has the feature that it sends regular closed sets to regular closed sets. Then one could suggest that even though physical space and field value spaces in fact are pointless, one can still continue the standard practice of using ordinary pointy functions f when doing one's calculations in physics, since the possible gunky field states in gunky space correspond one-one to such pointy functions in pointy space.

Unfortunately, though, this is not true. For consider a pointy function f that has a fixed constant value v over some pointy region PR. It will map every subset of PR, and hence every regular closed subset of PR, to the singleton set $\{v\}$. And a singleton is not a regular closed set. So a function that is constant over some (finite) region PR does not preserve the property of being regular closed. But clearly physics needs to make use of such functions. So we have a problem.

And there is more trouble. It seems clear that we will need to put a measure on pointless regions. For how else are we going to able to talk of the sizes of regions, and how else are we going to be able to do the pointless analogue of the integration of functions, something that we surely have to be able to do? Unfortunately when one tries to put a measure on a pointless topological space one will run into difficulties that appear to be insurmountable.

Let me start on the project of putting a measure on a pointless topological space by considering a very simple case. We know that there is a one-one correspondence between pointless topological spaces and pointy locally compact T_2 spaces. Let us now consider the pointless topological space that corresponds to the pointy

1-dimensional continuum, i.e., the real number line. We know that there is a one-one correspondence between the pointless regions R in the pointless 1-dimensional continuum and the regular closed sets of real numbers. Given this fact, the obvious way to try to put a measure on the pointless regions in the pointless continuum is to identify the measure of any pointless region R with the Lebesque measure of the corresponding closed regular set of real numbers PR. The problem now is that this will turn out to yield a measure on the pointless regions which violates countable additivity. We can see this by looking at a "Cantor-set," or rather, the complement of a Cantor set.

Start with the set [0, 1]. Call this set S_0. It is a regular closed set with Lebesque measure 1. Now consider the middle quarter of this set, i.e., the set $[3/8, 5/8]$. Call this set S_1. S_1 is a regular closed set with Lebesque measure 1/4. Now consider the set S_2 which consists of two parts which fill the middle of the gaps left by S_1 and which has Lebesque measure 1/8. That is to say $S_2 = [7/32, 9/32] \cup [23/32, 25/32]$. Keep on doing this. That is, set S_n has parts which are slotted halfway between all the parts of all the previous sets, and S_n has half the Lebesque measure of set S_{n-1}. Since each set S_n is a regular closed set, each such set corresponds to a pointless region R_n in the pointless 1-dimensional continuum. Let us now ask what the fusion $\bigvee\{R_n\}$ of all these pointless regions is. Well, by our previous account of the mereology of pointless regions, this is going to be the unique pointless region that corresponds to the regular closed pointy region $\mathbf{Cl} \Cup \{Sn\}$. The union of all pointy regions S_n is dense on the set [0, 1]. So its closure is just [0, 1]. So the pointless region $\bigvee\{R_n\}$ corresponds to the equivalence class of regions that differs by measure 0 from the pointy region [0, 1].

Now we can see why we are in trouble if we assign measures to pointless regions by assigning them the Lebesque measure of the unique regular closed regions that they correspond to. For $\bigvee\{R_n\}$ will be assigned measure 1 by this method, while the measures of the R_n will sum to 1/2. That is to say, this measure will not be countably additive. This is a terrible problem, for we need a countably additive measure.

Now one might suggest that the problem here is that I simply suggested the wrong rule for assigning measures to pointless regions. However, not only is there no other obvious candidate for such a measure, one can in fact prove that there can be no such measure. That is to say, one can prove that there cannot exist a countably additive measure that is defined on every element of an algebra if that algebra is isomorphic to the algebra of closed regular regions of a continuum.[5] So our attempt to do physics in this kind of pointless topological space is in big trouble. Combined with the implausibility of our account of the possible states of fields in this kind of

[5] This is so because the algebra of closed regular regions of the real line is not "weakly distributive," and one cannot have a "semi-finite" countably additive measure that is defined on every element of an algebra that is not weakly distributive. A measure is said to be "semi-finite" if every element that has infinite measure has a part that has finite measure. For the definition of weak distributivity and a proof of the fact that one cannot put a semi-finite measure on an algebra that is not weakly distributive, see Chapters 32 and 33 of Fremlin (2002).

topological space, this provides us with a good reason to try our luck instead with the measure theoretic approach to pointless spaces.

11 The Measure-Theoretic Approach to Pointless Spaces

Let's concentrate on a simple case: the real number line. As before, we are going to create a pointless space by putting on blurry glasses. On the measure theoretic approach what we are going to blur out is differences of Lebesque measure 0. In order to do that, we first have to restrict ourselves to Lebesque measurable sets. So let's start by restricting ourselves to the Borel sets. One gets the collection of all Borel sets on the real line by starting with the collection of all open intervals (open sets of the form (a, b) for any real numbers a and b), and closing this collection up under complementation, countable union and countable intersection. Now let us put on our blurry glasses and define pointless regions R of the pointless real line to be equivalence classes of Borel sets of the pointy real line that differ up to Lebesque measure 0. Note that forming such equivalence classes preserves complementation, countable union and countable intersection. Indeed one can show that the algebra of such equivalence classes is a complete Boolean algebra, i.e., a standard mereology (with a null region) which is closed under arbitrary fusion. (See Sikorski, 1964, pp. 73–75.)

As before we would like to be able to recover standard physics, and we would therefore like to be able to recover a large collection of pointy functions from pointy space to a pointy field value space from some suitable collection of mappings between the corresponding pointless spaces. Luckily there already exists a nice and well-known account of how to do this. In particular, one can prove the following (see Sikorski, 1964, § 32). There exists a one-one correspondence between equivalence classes of pointy "Borel-measurable" functions *from* real line A *to* real line B that differ on up to Lebesque measure 0 sets of points, and "σ-homomorphisms" *from* pointy regions on the pointy real line B *to* pointless regions on the pointless real line A. A function is said to be "Borel measurable" if it sends Borel sets to Borel sets. A mapping h between Boolean algebras that are closed under countable union and intersection is said to be a "σ-homomorphism" if:

1. $h(\neg R) = \neg h(R)$
2. $h(\bigvee R_i) = \bigvee h(R_i)$, for any countable collection $\{R_i\}$
3. $h(\bigwedge R_i) = \bigwedge h(R_i)$, for any countable collection $\{R_i\}$.

That is to say, if we make the very simple and natural assumption that the state of a pointless scalar field in a pointless continuum (the above generalizes to n-dimensional continua) can be given by a σ-homomorphism from pointy value ranges to pointless regions in space, then we can recover all Borel measurable pointy functions (including highly discontinuous ones) up to differences of Lebesque measure 0. This is a great result. Not only can one recover all the functions that one could reasonably be expected to ever need in physics, one can also only recover these functions up to the kind of differences that one would expect not to correspond to real differences in nature.

What about topology though? We have just put a measure on an atomless mereology of pointless regions, but that tells us nothing about which pointless region is connected to which pointless region. For, loosely speaking, cutting out a segment of the real line, and pasting it in somewhere else along the real line does not alter the mereology of the real line, nor the measure-theoretic structure of that mereology. So we need to add a topology separately. How could we do that? Well, what we could try to do is to start with the pointy real line, and then use its pointy topology to define a topology on the pointy real line which is invariant under differences of up to Lebesque measure 0. Let's try that.

Let's say that pointy Borel sets A and B are "connected" if and only if there exists a point p such that any open set containing p has an overlap of non-zero measure both with A and with B. And let us say that pointy Borel set A is "limited" if and only if for some compact Borel set B we have that $A \cap \text{Complement}(B)$ has measure 0.

Clearly these definitions are invariant under differences in regions A and B up to measure 0. So we can use it to define a topology on the pointless regions of the pointless real line. The resulting structure will satisfy Roeper's axioms A_1 through A_9, but it will violate axiom A_{10}. Let me remind you what this axiom says: If R_a is limited, R_b is not the pointless null region, and R_a is not connected to the complement of R_b, then there is a pointless region R_c which is non-null and limited, such that R_a is not connected to the complement of R_c, and R_c is not connected to the complement of R_b.

To see that this axiom fails consider a Cantor-type pointy set, for instance the pointy set $B = (0, 1) \cap \text{Complement} \bigcup(S_n)$ where the S_n are the gap-filling sets that I defined in the previous section. Set B is a measure 1/2 Borel set, so we can consider the corresponding non-null pointless region R_b to which it corresponds. Now let R_a be the null region. R_a is limited and not connected to the complement of R_b since the null region is not connected to anything. So there should be a non-null and limited R_c such that R_c is not connected to the complement of R_b. Now the complement of R_b is the union of three pointless regions: $\{-\infty, 0\}$, $\bigcup R_n$, and $\{1, \infty\}$. Now, any pointy non-null open set has an overlap of non-zero Lebesque measure with any pointy set in the equivalence corresponding to this pointless region, so this pointless region is connected to every non-null pointless region. So there cannot be such an R_c.

The problem is the following. The basic idea of axiom A_{10} is that there is a topological notion of pointless region R_1 being "strictly inside" pointless region R_2. The idea is that R_1 is strictly inside R_2 if R_1 is disconnected from the complement of R_2. And then the idea of "pointlessness," or the idea of "non-atomicity," suggests that if R_1 is strictly inside R_3 then there ought to be an R_2 such that R_1 is strictly inside R_2 and R_2 is strictly inside R_3. In particular for any non-null R there should be a non-null region R' which is strictly inside R. This axiom fails given the way that I have defined connectedness on the measure theoretic approach, since there are Cantor type non-null regions such that there are no regions that are strictly inside such a Cantor type region, since the complement of such a Cantor type region is connected to every non-null region.

Now one might think that the failure of this axiom shows that we do not really have a *pointless* space. However, the fact that our space is pointless is still unambiguously represented in two different ways:

1. The algebra of regions is non-atomic.
2. Other than the null region, every region has non-zero measure, and for every non-zero measure, no matter how small, there are regions that have that measure.

So I am not terribly worried about the failure of axiom A_{10}. However, it is interesting to note that the fact that there exists a pointless region R_b corresponding to a pointy Cantor set shows that one should not think that each pointless region can be decomposed into a collection of extended "solid islands." The pointless region R_b, for instance, is not so decomposable. Ah well, so be it.

There is a question that I have not yet answered. Namely: to what extent does a measure algebra with a topology satisfying axioms A_1–A_9 uniquely correspond to a pointy topological space plus measure? Part of the answer is well known: every atomless separable measure algebra is isomorphic, and hence corresponds uniquely to the mereology of the continuum with the Lebesque measure on its Borel algebra. (See Royden, 1968, Ch. 15.) But I do not yet know to what extent the pointless topological structure uniquely determines the corresponding pointy topological structure. So there is interesting work left.

And, of course, this is only a beginning. We also need to add differential structure and then metric structure in order to be able to do modern physics. But that is work for the future. For now let me simply conclude that the measure theoretic approach to gunky, or pointless, spaces is the most promising.

Bibliography

Arntzenius, F. (2000). Are there really instantaneous velocities? *The Monist*, 83:187–208.

Arntzenius, F. (2004). Is quantum mechanics pointless? *Philosophy of Science*, 70:1447–1457.

Böhm, A. (1978). *The Rigged Hilbert Space and Quantum Mechanics*. Springer, Berlin.

Fremlin, T. (2002). Measure theory, vol. 3. Self-published, available at www.essex.ac.uk/maths/staff/fremlin/mtsales.htm.

Gerla, G. (1990). Pointless metric spaces. *Journal of Symbolic Logic*, 55:207–219.

Halverson, H. (2001). On the nature of continuous physical quantities in classical and quantum mechanics. *Journal of Philosophical Logic*, 30:27–50.

Roeper, P. (1997). Region-based topology. *Journal of Philosophical Logic*, 26:251–309.

Royden, H. (1968). *Real Analysis*. MacMillan, London.

Sikorski, R. (1964). *Boolean Algebras*. Springer, Berlin.

Skyrms, B. (1983). Zeno's paradox of measure. In Cohen, R. S. and Laudan, L., editors, *Physics, Philosophy and Psychoanalysis*, Boston Studies in Philosophy of Science, pages 223–254. Reidel, Dordrecht.

Skyrms, B. (1993). Logical atoms and combinatorial possibility. *Journal of Philosophy*, 90 219–232.

Wagon, S. (1985). *The Banach-Tarski Paradox*. Cambridge University Press, Cambridge.

Chapter 17
Pitholes in Space-Time: Structure and Ontology of Physical Geometry

Robert DiSalle

1 Introduction: The Philosophy of Space-Time and the Foundations of Mathematics

The philosophy of space and time did not begin with Newton and Leibniz, but there are perfectly good reasons why contemporary discussions see their origin in the controversy between those two. On the one hand, the issues explicitly raised between them—especially, and most obviously, the epistemological and methodological questions surrounding Newton's theory of absolute space and motion—have never lost their relevance to the continuing evolution of physics. On the other hand, in different but equally unprecedented ways, they saw the question of the nature of space and time as part of a larger set of deeply interconnected questions, not only in the foundations of physics, but also in metaphysics, epistemology, and the foundations of mathematics.

To approach these questions as a seamless whole is a doubtful prospect. In the case of Newton and Leibniz, it involved theological and even psychological controversies, among others, with which one would hardly wish to burden a twenty-first century discussion of the philosophy of natural science. Kant, perhaps, had a similarly broad perspective on the epistemology and metaphysics of mathematics and physics; his transcendental account of space and time enabled him to exclude some of the philosophical questions (particularly concerning God and the soul) that had preoccupied Newton and Leibniz, and to integrate the scientific and philosophical aspects of space and time on the basis of sensible intuition. It is arguable, however—if not obvious—that the very simplicity of Kant's view is one of its fatal limitations. A view so neatly bound up with Euclidean geometry and Newtonian physics, and with the intuitive constructive procedures that form the epistemological basis of both, could hardly survive the dramatic evolution of mathematics and physics in the nineteenth and twentieth centuries. Given this situation, the dominant line of philosophical response to non-Euclidean geometry and post-Newtonian physics seems in

R. DiSalle (✉)
Professor of Philosophy, University of Western Ontario, London, ON, Canada
e-mail: rdisalle@uwo.ca

D. DeVidi et al. (eds.), *Logic, Mathematics, Philosophy: Vintage Enthusiasms*,
The Western Ontario Series in Philosophy of Science 75,
DOI 10.1007/978-94-007-0214-1_17, © Springer Science+Business Media B.V. 2011

retrospect to have been a natural one: the mathematical theory of space and time came to be seen as the theory of a certain kind of formal structure, independent of its empirical origins or, indeed, any intuitive content; the connection of such uninterpreted structures with experience came to be seen as determined by conventional choice.

This view dominated the philosophy of space and time in the twentieth century, to the extent that its proponents, the logical empiricists and their sympathizers, were the dominant voices in the philosophy of science generally. But it was not the only significant view. Hermann Weyl, in particular, suggested that the space-time geometry of general relativity, as represented in the theory of Riemannian manifolds, retained a profound link with intuitive conceptions of space and motion, a link which could be illuminated by deeper analysis of the nature of the continuum. In this respect, Weyl can be seen as revisiting some of the epistemological and metaphysical aspects of the philosophy of space and time that were of such concern to Newton and Leibniz, and as integrating concerns in the foundations of physics and mathematics in a similar way.

Even after the logical empiricist view has receded from prominence, an approach such as Weyl's remains outside the main currents of the philosophy of physics. Yet the study of the continuum, in light of continuing work in the foundations of mathematics as well as physics, may yet offer some insight into the metaphysics of space and time. It is, moreover, a project that has been taken up, with characteristic insight and enthusiasm, by John Bell. It has been the third most important subject of conversation between us over many years—after atonal music and *film noir*—and it is a privilege to be able to offer some critical reflections on its early history, in John's honour.

My aim is to revisit some Leibnizian themes regarding the ontology of space and time and their connection with his study of the continuum. Considered as arguments against Newton's theory of absolute space, Leibniz's celebrated arguments against the reality of space are not as cogent, or even as apposite, as they have generally been taken to be. Considered as arguments from the foundations of mathematics, however, bearing on the general idea of space as a substance, they have profound and surprising implications for ongoing discussions about the ontology of space-time in modern physics.

2 Origins of an Ontological Problem

The philosophical controversy between Newton and Leibniz may be said to have begun with Descartes, since it was against Descartes' account of space and motion—of space as essentially identical with body, and of motion as displacement relative to contiguous bodies—that Newton was chiefly reacting. But the classic philosophical question about space and time, the one that came to dominate subsequent discussion, was framed most explicitly by Leibniz: Are space and time real things, or merely ideal, abstracted from the spatial and temporal relations among real things and events? The distinction appears to be a fairly clear one. We ought to be able to distinguish between the real existence of, say, two individual brothers, and the merely ideal existence of "the pair" of them, or of the abstract relation "brotherhood." In the philosophy of modern physics, at least since the time of Mach, it has seemed obvious that this question corresponds to an empirically motivated question of reduction: space and time, especially as they appear in Newton's theory of absolute space and time, are essentially hypothetical theoretical entities whose evidential basis is the set of spatial and temporal relations that we directly observe and measure; the question is whether these entities are simply reducible to the empirical relations. No one who views the issues from a modern perspective, however, can see the question in such simple terms.

For one thing, it was easier to believe before the nineteenth century that spatial and temporal relations are simply given to us as raw material, as unproblematic facts from which to abstract a spatial or a temporal structure. Through the work of Helmholtz and Poincaré, however, it became clear that a conception of the spatial relations among things is already a fairly theoretical conception. It is made possible only by simple, but nonetheless peculiar, assumptions about the possibilities of certain kinds of motion. Then, too, it was easier in the nineteenth century to think of spatial relations as, at least, epistemologically unproblematic, than it became in the aftermath of special relativity. With the advent of special relativity, those relations, along with temporal intervals, were shown to depend on empirically dubious assumptions about simultaneity. In short, to the extent that the "relationalist" view of space and time was supposed to be an epistemological thesis, arguing from the supposition that geometrical relations are better and more immediately known than the geometrical structure of space itself, it has to be seen as a naive precursor to, if not a historical casualty of, the emergence of modern mathematics and physics.

For another thing, to translate the classic Leibnizian question into twentieth or twenty-first century terms is not so easy as it is sometimes taken to be. Setting aside the epistemological difficulties for the moment, it is at least straightforward to ask whether "absolute space" can be reduced to spatial relations. But if there is a meaningful question whether "absolute space-time" (however one understands this) can be reduced to spatio-temporal relations, it is not necessarily analogous to the older question. The older question made sense because the structure in question involved assumptions that went beyond what is contained in the (supposedly fundamental) underlying relations: to the structure of momentary relative positions, the structure added the notions of same position at different times, or, in the weaker case of a Newtonian space-time, sameness of velocity over time. In other words, the structure known as "absolute space" is essentially a space-time structure over and above the set of merely spatial relations at any given temporal instant. The structure that defines the spatio-temporal relations in a relativistic space-time, by contrast, is the space-time itself. In the Newtonian case, a class of geometrical relations could be defined by a structure that did not, by itself, impose any kind of dynamical constraint on possible motions (though one could argue that even those geometrical relations impose upon the motions of physical things in ways that hardly seem compatible with the notion that the structure is only an abstraction from some actual relations: consider, for example, that there is a unique straight path connecting any two points). Therefore the structure in virtue of which relative motions are well defined stood independent of the structure on which privileged states of motion could be defined. But in relativistic space-time, the structure in virtue of which objects have spatio-temporal relations is inseparable from the dynamical structure in virtue of which they have states of motion. This is a specific form of a more general point about space-time theory as it has developed from Newtonian physics through general relativity: that space-time structure is an essential part of dynamical theory, as understood classically; and that if it were to be eliminated, it would not be through reduction to an epistemologically more basic set of relations, but through replacement by another sort of theory altogether, such as a quantum theory, whose fundamental structure is of a radically different sort.

3 The Ontological Problem in Its Present State

The last point is not particularly new, and was pointed out very clearly in (Weyl, 1927); see also (DiSalle, 2006). It suggests that the most-discussed arguments of Leibniz against Newton, from Leibniz's correspondence with Clarke, are completely misdirected—at least, if they are judged as arguments against the theory of space and time presented in Newton's Scholium on space, time, and motion. That theory defines absolute space as a structure essentially connected with time, in such a way that motion is distinguished from rest, i.e. that one can determine *the same place* in absolute space at different times, and therefore the absolute translation of a body from one place to another. But the familiar "indiscernibility" arguments urged by Leibniz in his correspondence with Clarke (Leibniz, 1716, p. 364) concern only spatial symmetries: the universe would not be different if the locations of all its contents were shifted by some given distance, or reflected East to West, or rotated about a given axis. Indeed, Newton's theory, in principle, strictly requires such symmetries, since they are the symmetries of the Euclidean space with which the theory begins. The crucial question about absolute space is whether, in fact, it makes a difference whether bodies remain at the same place through time, or what their velocities are. And this is a question whose answer must come from the theory of motion, that is, from an inherently spatio-temporal theory, not from general considerations on the symmetry properties of space in itself. It is Newton's own theory of motion—not the uniformity of space—that, by virtue of its implicit relativity principle, makes motion and rest "indiscernible" and absolute space a questionable structure.

For our present purpose, it is useful to rehearse the evident defects of the classic Leibnizian arguments in order to get a clearer view of the genuine insight behind them. Newton's arguments in the Scholium and Leibniz's arguments to Clarke are essentially at cross-purposes: Leibniz is concerned with an ontological matter which, to Newton, was entirely separable from the structural question that the Scholium tries to answer, i.e., what sort of spatio-temporal structure is presupposed by the dynamical theory. No doubt part of the reason for confusion is the weakness in Newton's answer, which consists in the assertion that "absolute space" is necessary for a theory that, in fact, only requires the weaker structure that we know as "Newtonian space-time" (cf. (Stein, 1967)). Thus an equivalence class of spaces (the inertial frames) ought to replace the single privileged "immobile" space. But the more interesting reason is Leibniz's preoccupation with the notion of substance, and his concern with whether space can possibly satisfy this notion—and, if it can't satisfy the notion of substance, whether there is any sense in the claim that it has a real existence.

One could dismiss these concerns, with a certain degree of plausibility, by tracing them to the peculiarities of Leibniz's own account of substance. On that account, a genuine substance is a complete individual with a complete individual concept: the specification of an individual is the specification of everything that is truly predicated of that individual. The true concept of an individual, then, is the set of all predicates that distinguish it from every other actual or possible thing, and therefore specifies, in effect, its relations to the entire universe, past, present, and future. On this account it is indeed difficult to see how the indistinguishable parts of space could possibly qualify as substances, but the account of substance is itself not so easy to accept. It places the "principle of the identity of indiscernibles," which we are inclined to treat as a reasonable epistemological condition on any scientific metaphysics—no real distinction where there is no empirically discernible difference—in a fairly peculiar light, as denying the possibility of real things that differ only numerically. The modern physicist might, after all, reject the identity of indiscernibles in the form that Leibniz intended it; it is one thing to eliminate theoretical distinctions that make no discernible difference, but it is quite another thing to insist that no two things differ only numerically: arguably a central feature of the atomic theory is to reduce the qualitative differences among types of matter to numerical differences of atomic structure, or arrangements of generally indiscernible fundamental particles.

It is less straightforward, however, to dismiss Leibniz's scruples about whether space deserves to be called a substance, and the continuum of space to be thought of as a collection of individual points. For these arise from the fundamental nature of the continuum, and from Leibniz's appreciation of the distinction that this implies between space and any of the things that we are inclined to think of as real. For a real thing is a whole composed of its actual parts, whereas the parts of space are divisions that we create in thought:

> But space, like time, is not something substantial, but ideal, and consists in possibilities, or in an order of coexistents that is in some way possible. And thus there are no divisions in it except those made by the mind, and the part is posterior to the whole. (Leibniz, 1705, p. 278)

It follows that the points of space, in particular, must be thought of as mere extremities, or limits, of lines in space, and emphatically not as the elementary parts of space. Understanding them in the latter sense leads inevitably to paradoxes like those of Zeno:

> Regarding *indivisibles*, when by that is meant the simple extremities of time or of a line, we are unable to conceive new extremities in them, nor parts either actual or potential. Thus points are neither large nor small, and no leap is needed to pass them. The continuum, however, though it has such indivisibles everywhere, is not composed of them, as seems to be supposed by the objections of the skeptics [i.e. the paradoxes], in which, in my opinion, there is nothing insurmountable. (Leibniz, 1692, p. 416)

From reflections of this sort we gather that, in opposing "absolute space," Leibniz was not directly concerned with the theory of space and time, and of the dynamical connection of space with time, that is proposed in Newton's Scholium; indeed, through his own dynamical assumptions he may be thought of as presupposing such a theory himself (cf. (DiSalle, 2002)). His concern is that "absolute space" is assumed to be a real thing, in spite of having the properties of something ideal—a totality of parts, in spite of having the properties of a genuine continuum.

> One deceives oneself in wishing to imagine an absolute space that is an infinite whole composed of parts; there is no such thing, it is a notion which implies a contradiction, and those infinite wholes, with their opposed infinitesimals, are only in place in the calculations of geometers, just like imaginary roots in algebra. (Leibniz, 1704, p. 145)

Beyond the empirical indiscernibility of spatially distinct states of affairs, Leibniz is pointing to the mathematical and conceptual incoherence of any view of space as a genuine substance. Of course this is not a novel observation about Leibniz's views of the continuum. It is worth noting, however, that the problem of the composition of the continuum has a bearing on the physics of space and time that transcends its role in Leibniz's controversies with Newton. For this problem is the point on which Leibniz's view engages, not the empirical arguments of the Scholium, but that very conception of the reality of space that characterizes modern substantivalism: the differentiable manifold as the ontological basis of space-time.

It is on this last point, moreover, that the "substantivalist" of modern times goes beyond anything that Newton himself was willing to defend. While the Scholium carefully avoids such ontological matters, focusing only on the structural claims that Newton (not entirely correctly) regarded as underpinning the laws of motion, the essay *De Gravitatione*, where Newton directly addresses such matters, casts doubt on their relevance. His understanding of the nature of space focuses on those of its features that experience and geometrical reasoning suggest to us, whether or not we understand what kind of thing it is. What we learn from geometrical reasoning is that the points of space are not the parts of space; when Newton speaks of the parts of space, he means "places," which are, appropriately, not very carefully delineated. In fact the delineation of places is simply by the construction of figures, whose boundaries—planes, lines, and, especially points—are mere extremities or "limits" (Newton, 2005, p. 22). It would appear, then, that Newton's view has more in common with Leibniz's than is generally supposed (cf. (DiSalle, 1994)). It is true that, while Newton does not, any more than Leibniz, think of space as a substance composed of its points, he does not therefore conclude that space is something ideal. He concludes, instead, that space has "its own manner of existing that fits neither substances nor accidents." He even suggests that the usual notion of substance is "unintelligible." In this way the unpublished *De Gravitatione* reinforces the view that appears to be implicit in the Scholium, that the theory of absolute space—considered simply as a theory that space-time has a certain "absolute" structure—is independent of a substantivalist or any other interpretation of its ontological foundation. It can be defended, therefore, much as Newton defended his theory of gravity against scruples regarding action at a distance: as a theoretical description of a notable feature of nature, a description that can be empirically applied and evaluated even in the absence of any satisfactory view of its ultimate metaphysical basis. Its modern counterpart is, perhaps, "structural space-time realism" (cf. (Dorato, 2000)) rather than substantivalism. What the latter has especially in common with Newton's view is the emphasis on the epistemology of physical geometry (cf. (DiSalle, 1995)): questions about the nature of space-time are, ultimately, questions about the empirical meaning of claims that the world has a spatio-temporal structure of one kind or another, and about the empirical investigation of the truth of such claims.

One might plausibly argue, however, that such as position is not a considered "middle course" between substantivalism and relationalism, as is sometimes suggested. Especially (but not only) in the case of Newton himself, it could be seen as the dismissive move of one criticizing a philosophical tradition from a perspective that fails to take the traditional questions seriously; perhaps one should say that Newton merely threw up his hands at a difficulty that, to someone more seriously interested in ontological questions, could not be honestly ignored. The basis for the continuing ontological discussion, then, is not a debate between modern-day Leibnizians and modern-day Newtonians; rather, it is more like a debate between the modern day Leibnizian and a sort of philosopher—the substantivalist—who shares some of Leibniz's ontological concerns in a way, and to a degree, that Newton did not. On both sides, it is assumed that a realist view of space requires a coherent account of its ontological foundation. The substantivalist therefore faces a challenge that Newton's theory does not, namely, to defend the individuality of space-time points against Einstein's "hole argument," the modern counterpart of Leibniz's indiscernibility argument (cf. (Earman and Norton, 1987; Earman, 1989; Rynasiewicz, 1992)).

This is one reason for the emergence of "structural space-time realism" as an alternative. Another alternative is a "sophisticated" version of substantivalism that simply accepts the equivalence of indiscernible spatio-temporal situations. It is quite understandable that a modern sympathizer with relationalism would interpret this move as essentially a concession of defeat, a "pale imitation of relationalism" (Belot and Earman, 2001); for similar reasons, though with even less justice, "structural space-time realism" has been referred to as "relationalism in disguise" (cf. (Dorato, 2006)). While it hardly seems worthwhile or relevant to dispute about the names given to such positions, I think that there are good historical and conceptual reasons to resist going so far. It makes sense to say that there are varieties of realism about space-time that a relationalist *ought* to be able to accept. But to put it this way is to acknowledge that accepting such a position requires a significant concession from relationalism as well, from the perspective of the historic motivations for a relationalist view. In fact, the "relationalist" who can see her own position reflected in "structural space-time realism" must have conceded a great deal already. For there are three classical philosophical elements of relationalism that this acceptance of structure implicitly denies:

Abstraction There is no reasonable sense in which a relativistic space-time structure can be thought of as a mere abstraction from a collection of actual or possible spatio-temporal relations, as originally demanded by Leibniz. Space-time is the structure in virtue of which certain relations can be considered possible or actual.

Reduction there is no reasonable sense in which the structure is reducible to actual or possible relations, for the same reasons which told against the claim of abstraction.

Relativity There is no reasonable sense in which relationalism, taken in this sense, underwrites those arguments for the relativity of motion that were, after all, the paramount scientific theme of relationalism in its classical sense. Indeed, on this account of relationalism, one can perfectly well have a relationalistic theory of Newton's "absolute space" and "absolute motion," in which neither is reducible to more basic relations, but both are fundamentally incorporated into a relational structure.

Rather than say that one position is an imitation of the other, one ought to say that the correct position contains essential reasonable elements of each classical position. But to call this just a reasonable middle ground between the classical positions does not do justice to this view itself, nor to the nature of the difficulties with the classical positions. For the reasonable elements are those which the classical positions distorted in order to satisfy their respective peculiar ontological demands. In each case there are ontological arguments being made that only distract from the problem of understanding how space-time structures function in physics.

4 The Structure of Space-Time and the Ontology of the Space-Time Manifold

From considerations like the foregoing, it appears that a realistic view of space-time can be, and perhaps should be, detached from a substantivalist ontology. It is worth considering, therefore, why the latter in some form or other remains central to defending realism against relationalism in its classic form. There is at least one good philosophical reason, and one good mathematical reason. On the philosophical side, there is a characteristic modern account of the ontology associated with a physical theory that goes beyond the general methodological point that we should take the world to be more or less as the best physical theory says it is (a view compatible with "structural space-time realism" in any case); on this account, a theory implies—or at least entitles us to—specific ontological commitment to the entities that the theory talks about. In short, if the theory is true, then the terms that occur in the theory must have genuine referents (cf. (Quine, 1953)). From this philosophical point of view, the mathematical fact that general relativity treats space as a differentiable manifold (assuming that general relativity is a successful theory) establishes the differentiable manifold as a legitimate entity, and the fact that the theory quantifies over space-time points establishes the latter as part of the theory's ontology. That is to say, in a mathematical framework in which any number of space-time theories—Newtonian or relativistic—can be expressed in the form of tensor fields, then it would seem to be sensible to regard the generic manifold as representing space-time itself, i.e., as the object whose structure is the common subject-matter of those theories just because it is the domain on which the tensor fields are defined.

From this point of view, it seems obvious that if the general theory of relativity is our best theory of space-time, and as such the most reliable guide to the metaphysics of space-time, then the center of our metaphysical attention ought to be the differentiable manifold, which has at least a prima facie claim to be the ontological basis of the theory. But if we look back over the evolution of space-time theory, and reconstruct earlier theories (Newtonian and Minkowskian space-time theories) in the formalism of differentiable manifolds, then it seems easy and natural to identify space-time with the manifold, and to interpret the question, "What is the structure of space-time?" as the question, "What geometrical objects does physics attribute to the differentiable manifold?" The claim of "structural space-time realism," that assertions about the real structure of physical geometry can be interpreted realistically on independent grounds, does not by itself determine the significance of the manifold on which our mathematical account of such structures depends.

It seems reasonable, then, given the history and continuing prominence of Leibnizian indiscernibility arguments such as the "hole argument," to pursue an understanding of the differentiable manifold, and the individuality of its points, that answers the relationists' epistemological challenge. One can consider whether the points of the space-time manifold can find a principle of individuation in the structures imposed on the manifold, that is, can "inherit" an individual identity from the structures that are built upon them. This effort has an obvious and sound motivation, reminiscent of Newton's attempt to argue that the parts of space get their identity from the order of situation of which each forms a part. This is something more subtle than "manifold plus metric substantivalism": it seeks to show that geometrical structures determine individuating properties of space-time points in a way that defeats the arguments from indiscernibility. "Metric essentialism" is one plausible form that such an effort might take, suggesting that points of the space-time manifold have essential metrical properties (cf. (Maudlin, 1990, 1993; Butterfield, 1989)). Alternatively, (Stachel, 2002, 2005) proposes that the structures defined on the space-time manifold individuate the points of the manifold which, in themselves, have no individuating features:

> The points of space-time have quiddity as such, but only gain haecceity (to the extent that they do) from the properties they inherit from the metrical or other physical relations imposed upon them. In particular, the points can obtain haecceity from the inertio-gravitational field associated with the metric tensor: For example, the four non-vanishing invariants of the Riemann tensor in an empty space-time can be used to individuate these points in the generic case. (Stachel, 2005, p. 7)

Such proposals share a sound motivation with "structural space-time realism," namely, to defend a realistic view of space-time against the more doubtful claims of relationalism.

Yet one may doubt whether this building-up of an individuating structure—as one might say, heaping Pelion upon Ossa in order to ascend from a featureless point in the manifold to a determinate element in a relational structure—actually can perform the task required of it. The various arguments from indiscernibility suggest that it is not points, but equivalence classes of points, that are specified by such means, that is, identifying points that belong to equivalence classes under the relation of indiscernibility. In that case the sort of specification that the structure makes possible hardly seems to bestow an individual identity on the points. I think, however, that the indiscernibility arguments are not the most important neo-Leibnizian arguments. Even if they could be overcome by some method of individuation, the deeper ontological question would remain: can such a specification of points ever give us an ontological basis for the space-time continuum?

Here is another way to raise the issue. Philosophical debates over indiscernibility, whatever confusions have surrounded them, have one fairly clear question at their heart: how large is the set of equivalent structures, or, how wide is the group of transformations that respects whatever structure is supposed to be invariant? If by an individuating structure, one means something like the way in which parts of space-time can differ in the curvature assigned to them by the local variation of the curvature tensor, then one can say at least of a variably-curved space-time that its parts have individuating features. If one, going further, took the existence of such features as necessary for thinking of something as substantially existing, then perhaps one could say that a curved space-time is immune to a certain line of relationalist argument that tells against spaces with non-trivial symmetries. One might therefore suggest that a non-symmetric space-time has a claim to be a real substance, while the properties of symmetric spaces require us to treat them as merely ideal. Yet there is both a Newtonian and a Leibnizian argument to resist this way of thinking. On the Newtonian side—that is, on the side of physical geometry as an object of scientific knowledge, independent, as Newton suggested, of how well we understand its ultimate ontological significance—whether space has any non-trivial symmetries ought to be considered an empirical question. If we identify gravity and inertia, and therefore re-interpret the Newtonian gravitational field as the variation of space-time curvature with mass-energy distribution, surely this is because of the striking empirical fact about gravitation identified in Einstein's equivalence principle. Similarly, if we expect of a quantum theory of gravity that it maintain the identity of gravity and inertia, instead of treating gravity as a field (like other quantum fields) defined against a flat space-time background, our motivation ought to be this same empirical fact. By the same token, an experimental violation of the equivalence principle would provide at least a prima facie motivation to treat gravity on a footing with the other quantum fields. One might be tempted to think of the latter as a philosophically retrograde step, but it would make sense if the empirical basis of Einstein's philosophical argument for space-time curvature—the equivalence principle—turned out to be merely approximate.

To see the Leibnizian reason, we must look beyond the familiar epistemological arguments, to one arising from Leibniz's arguments about the nature of the continuum. The question is not merely the one that seems so evidently relevant to the philosophy of physics, whether the empirical indistinguishability of the points of space-time is compatible with the idea that they form a substance in the sense of the contemporary debate. It is, rather, whether the individuation of the parts of the continuum makes mathematical sense, or provides a sensible ontological interpretation of the space-time manifold. The original significance of Riemann's notion—that which made it the mathematical basis for the theory of variably-curved spaces—is not so much connected with the notion of a collection of points, such as one begins with in the treatment of a differentiable manifold from a set-theoretic point of view. Rather, the significance of the manifold is connected with the Riemann's notion of a "domain of variability," or of a collection of domains of variability comprising the separate dimensions of the space (cf. (Riemann, 1867)). In other words, the Riemannian idea is that of a multiplicity (aggregate, variety—or manifoldness, to use Clifford's literal translation of *mannigfaltigkeit*) of independent but conjoined continua. Arguably, the set of locations that are thus constituted—the ordered n-tuples of real numbers that make up the points of the manifold—are not conceptually or ontologically prior to these continuous domains of variability. If the manifold is characteristically thought of as a particular kind of topological space, with topological features defined on an underlying set of objects, this is not because the latter is taken as a well-understood metaphysical foundation for physical geometry. Rather, it is because it is on the basis of set theory, and the set-theoretic reconstruction of the differential calculus, that the usual notions of differential geometry are given a rigorous formulation. As Russell noted (Russell, 1903), it is this rigorization that first placed the study of the continuum on a logically coherent conceptual foundation, and set aside those puzzles about the continuum with which Leibniz had to contend. But this fact does not by itself require us to interpret the point-set literally as an ontological foundation for space or space-time.

From these Leibnizian reflections, we are led to ask the question whether the space-time continuum has an ontological basis in the collection of its points. It is an obvious and useful way of speaking to say, of any Riemannian manifold, that it has a certain metrical structure at any point. Then, it would seem, one ought to be able to speak of the points as individuated by their "essential metrical properties," as suggested above. But what are these essential properties that distinguish the points of space-time? If they are metrical properties, they correspond to the metric "at a point," which, intuitively, yields the inner product of "infinitesimal" vectors at any point. We know, however, that *at a point* there are, strictly, no nonzero vectors. Hence the familiar mathematical way of speaking, in which the metric is said to operate upon vectors in the tangent space to a given point. In the case of a sufficiently uniform space, such as Minkowski space-time, since we can treat the entire space-time as a vector space, no ambiguity need arise: the metric determines the interval between any two space-time points, and thereby the length of a vector lying in the space. But in a non-uniform space-time of the sort that is typical of general relativity, the fact that the metric operates in the tangent spaces has a deeper significance. Intuitively, the metric of such a space exhibits its "local" structure and its variation over finite regions. In mathematical strictness, however, all points are exactly the same. So it is not in the space-time manifold, but in and among its tangent spaces, that the interesting work of classical differential geometry is done. It is striking to find, as John Bell has often noted, that a classic text on general relativity notes this very distinction, and (rightly!) warns the reader not to misinterpret the rigorous description of the space-time metric as a literal account of vectors *in* the manifold (cf. (Bell, 2005, p. 317; Misner et al., 1973)).

This is one compelling motivation for the exploration of synthetic differential geometry, whose conceptual structure and philosophical significance John Bell has done so much to illuminate (Bell, 1998, 2005). By analogy to the role that non-standard analysis plays in relation to the differential calculus, synthetic differential geometry treats the curved space-time manifold, not through the tangent spaces to its points, but "intrinsically" through the behavior of its smallest "flat" parts. The metric thus operates on actual infinitesimal vectors lying *in* the space-time, which thereby exhibits the structure of a true continuum. Of course, the usual rigorous presentation may well remain the standard one, and perhaps even the most useful one, in spite of the greater intuitive plausibility of synthetic differential geometry. But the mere possibility of such an alternative formalism ought to give us pause, when we return to the question of space-time ontology. It suggests that the space-time point is an ideal limit—much as Leibniz suggested—rather than a genuine part of space-time. If we take this representation too literally, we are forced to ask, where is the variable structure of this space, all of whose parts are exactly the same? Indeed, such a situation raises something like Zeno's "arrow" paradox for substantivalists: the curvature of space-time is the variation of its metric from point to point, yet the metric is precisely identical *at* every point. This paradox is avoided by the synthetic representation, in which curvature is directly exhibited by variations in infinitesimal structures. Perhaps more simply, we can avoid the paradox by carefully distinguishing the mathematical representation from the structure that it seeks to represent. In either case, the important lesson is that metrical structure is an essential property of space-time; it is not an essential property of the points of which we may take space-time to be composed. In Leibniz's words, "Several people who have philosophized in mathematics about the point and unity have become confused, for the failure to distinguish between resolution into notions and division into parts" (Leibniz, 1715, p. 583; cf. Bell, 2005, pp. 86–90).

To accept this lesson is not to endorse the Leibnizian view as a whole. If Newton was at fault for failing to provide a coherent ontology for absolute space, Leibniz must be faulted for confusing the problem of the ontology of space with the problem of physical geometry—allowing questions about the composition of the continuum to obscure the Newtonian investigation of the geometrical presuppositions of dynamics. In the case of Leibniz, the fact that one could think clearly about what one might call the essential properties of space, but not about the individuality of its parts—that, indeed, the lack of individuation of its parts is one of those essential properties of space—implies that one ought to regard it as a merely ideal entity. Newton, apparently aware of and convinced by these same considerations, is also convinced by considerations of a quasi-transcendental character that space cannot be merely "nothing," but must be some aspect of the real world. Therefore in his case this peculiar combination of features requires us to acknowledge that space has a peculiar kind of existence. Here again is an instructive analogy to the case of universal gravitation: the mode of action of gravity, as proposed by Newton, has features that on Leibniz's view cannot belong to any genuine physical interaction; for Newton, however, these features suggest that gravity may be a natural "power" of a hitherto-unexpected kind, and that our conception of what is physically intelligible may have to be revisited. A similar revisiting of metaphysical categories is occasioned by the need to understand spatio-temporal structure as a real feature of the world, and as a genuine object of empirical study.

A further lesson is to be cautious in applying philosophical criteria for "what there is." When we ask whether the terms that occur in the theory "really refer," we should be careful to ask, which terms really do occur in the theory? Which occur, instead, only its mathematical representation? More precisely, which terms refer to theoretical objects of which the theory gives us, at least in principle, some empirical grasp, as genuine features of the natural world? Which terms, instead, are merely descriptive devices that make possible the representation of features of the world, without themselves being taken literally as features of the world? The fact that "our best physical theory" employs or quantifies over some set of objects or structure does not necessarily provide, in any straightforward sense, grounds for taking the relevant objects or structure to be something real. Philosophically, one might demand, further, that in addition to invoking or "referring to" that structure, the theory provide some empirical access to it—provide some empirical means of individuating its parts, or an empirical description of its structural features. The potential difficulty of providing such an empirical connection—that it doesn't *necessarily* emerge from the fact that the theory "refers to" or quantifies over such things—can be seen from some familiar problems in the interpretation of modern physics, for example in discussions of the status of the state vector in quantum mechanics. In that case it seems fairly obvious that the mathematical apparatus of a physical theory can make use of objects, perhaps even find them indispensable, without thereby settling the question of their place in the theory's ontology. In the case of space-time, it seems less obvious, because of the longstanding familiarity of the notion of a space-time point, and because of the rigorous account of the differentiable manifold as a point-set. But the implication should be equally clear. The subject-matter of space-time geometry (in general relativity) is the continuous variation of the metric over space-time; the points at which its "local" values are given are indispensable elements of the standard framework in which this subject-matter is described.

5 Conclusion: Structure from a Pithole

I would like to close with a note on the title of this paper, and its bearing on the philosophical theme. John Bell and I once spent a weekend ambling around northeastern Pennsylvania, discussing the individuation of space-time points in an atmosphere free of distractions. But we could not help being distracted by a roadside sign indicating "the phenomenon that was Pithole": a nineteenth century ghost-town, relic of the heroic days of Pennsylvania's oil industry. Our minds were prepared with Hollywood images of ghost-towns: deserted store-fronts, saloon doors swinging open and shut in the breeze, and tumbleweeds doing that for which tumbleweeds are justly renowned. We were unprepared, therefore, to find that the trace of Pithole was in fact nothing at all—not a town abandoned to ghosts, as it were, but the ghost of a town whose features had completely vanished. It was amusing to note the historical markers scattered about the indifferent grass, suggesting that such and such a structure might have been here or there; we began to suspect that the phrase "the phenomenon that was Pithole" had a metaphysical resonance that eluded the local historians who had coined it.

This incident seems to have at least a metaphorical resonance for the metaphysics of space-time. In the long struggle toward a scientific grasp of space-time structure, the concept of the differentiable manifold obviously deserves its pre-eminent place—even if, as seems likely, the successor to general relativity turns out to require a fundamental structure of a radically different sort. Its conceptual significance lies, however, in its illuminating treatment of the relations between the infinitesimal and the large-scale structure, or the local and the global structure, of a space that is continuously variable. The collection of space-time points, along with the tangent spaces on which we construct a rigorous formulation of the varying metrical structure, is the mathematical basis for an extremely fruitful representation—not the ontological basis for our scientific understanding of space-time. In itself, the point is essentially a "pithole" in space-time, a place from which all structure has vanished, an entity whose "essential properties" are generic ones that offer no clue to the physical geometry that the formalism builds upon them. It is space-time itself that has "essential properties": one of these is the metrical structure whose symmetries, or inhomogeneities, can be objects of physical investigation; and one of these is the continuity in virtue of which we can pursue the analysis of its structure down to the level of the infinitely small—ideally, to the level of its structure "at a point." Even if his criticisms of Newtonian space and time were somewhat misdirected, then, Leibniz's strictures against taking this idealization too seriously are instructive. If there is to be a good reason to regard "the space-time continuum" as a collection of individuals, it ought to come from physics, e.g., from a quantum theory of gravity or a theory that incorporates a discrete space, rather than from a peculiarly literal interpretation of some aspect of the mathematical formalism. This is an especially compelling point in light of the efforts, of John Bell and others, to show that an alternative, "pointless" characterization of the continuum is a live possibility, and, perhaps, even a tool for the eventual construction of quantized space-time (cf. (Bell, 2005, pp. 317–318)).

It would be an odd, if not an ironic, outcome if what we took to be the ontological essence of space-time, in general relativity, turned out to be that very feature of space-time that the successor theory dispenses with. It does not make much sense to me, I admit, to try to interpret general relativity in anticipation of the theory that must replace it; it seems to me that this must be done after the fact, when we know what that theory is, and can then consider in what relation to it general relativity stands—as some form of limiting case, for example. One could argue, as is done persuasively in (Belot and Earman, 2001), and as (among others) (Smolin, 2005) exemplifies, that particular interpretations of general relativity serve as heuristic principles and motivations in various quantum gravity research programs, but that is a different matter altogether. What does make sense, at the moment, is to ask in what ways the empirical success of general relativity—in whatever ways and to whatever extent it has been a success—may constrain its possible replacements. One could say, in all strictness, that the only rigid constraint is the demand to reproduce the successful predictions of general relativity, or even to improve upon them. Alternatively, one could argue that certain physical or philosophical insights associated with general relativity ought to be preserved in any sound successor to it, such as general covariance or background independence. But from the perspective of physical geometry, we can suggest that general relativity describes an interdependence between the structure of space-time and the distribution of matter that—barring violations of the equivalence principle, or similarly momentous empirical surprises—a future theory might incorporate, even with a radically different account of the ontology of space-time than anything suggested by the concept of a differentiable manifold.

Bibliography

Bell, J. L. (1998). *A Primer of Infinitesimal Analysis*. Cambridge University Press, Cambridge.

Bell, J. L. (2005). *The Continuous and the Infinitesimal in Mathematics and Philosophy*. Polimetrica, Milan.

Belot, G. and Earman, J. (2001). Pre-Socratic quantum gravity. In Callender, C. and Huggett, N., editors, *Physics Meets Philosophy at the Planck Scale*, pages 213–255. Cambridge University Press, Cambridge.

Butterfield, J. (1989). The hole truth. *British Journal for the Philosophy of Science*, 40:1–28.

DiSalle, R. (1994). On dynamics, indiscernibility, and spacetime ontology. *British Journal for the Philosophy of Science*, 45:265–287.

DiSalle, R. (1995). Spacetime theory as physical geometry. *Erkenntnis*, 42:317–337.

DiSalle, R. (2002). Newton's philosophical analysis of space and time. In Cohen, I. and Smith, G., editors, *The Cambridge Companion to Newton*. Cambridge University Press, Cambridge.

DiSalle, R. (2006). *Understanding Space-Time: The Philosophical Development of Physics from Newton to Einstein*. Cambridge University Press, Cambridge.

Dorato, M. (2000). Substantivalism, relationalism, and structural spacetime realism. *Foundations of Physics*, 30:1605–1628.

Dorato, M. (2006). Is structural spacetime realism relationalism in disguise? The supererogatory nature of the substantivalism/relationalism debate. In *Proceedings of the Second Montreal Conference on the Ontology of Spacetime, Concordia University, June 2006*. http://philsci-archive.pitt.edu/archive/00003393/01/struct_spacfin.doc

Earman, J. (1989). *World Enough and Spacetime: Absolute and Relational Theories of Motion.* MIT Press, Cambridge, MA.

Earman, J. and Norton, J. (1987). What price spacetime substantivalism? The hole story. *British Journal for the Philosophy of Science*, 38:515–525.

Leibniz, G. (1692). Letter to S. Foucher. In Gerhardt, C., editor, *Die philosophischen Schriften von Gottfried Wilhelm Leibniz (1960)*, volume I, pages 415–416. Georg Olms, Hildeshiem.

Leibniz, G. (1704). Nouveaux essais sur l'entendement. In Gerhardt, C., editor, *Die philosophischen Schriften von Gottfried Wilhelm Leibniz (1960)*, volume V, pages 39–509. Georg Olms, Hildeshiem.

Leibniz, G. (1705). Letter to B. de Volder. In Gerhardt, C., editor, *Die philosophischen Schriften von Gottfried Wilhelm Leibniz (1960)*, pages 278–279. Georg Olms, Hildeshiem.

Leibniz, G. (1715). Letter to L. Bourguet. In Gerhardt, C., editor, *Die philosophischen Schriften von Gottfried Wilhelm Leibniz (1960)*, volume III, pages 578–583. Georg Olms, Hildeshiem.

Leibniz, G. (1716). Correspondence with S. Clarke. In Gerhardt, C., editor, *Die philosophischen Schriften von Gottfried Wilhelm Leibniz (1960)*, volume VII, pages 345–440. Georg Olms, Hildeshiem.

Maudlin, T. (1990). Substances and space-time: What Aristotle would have said to einstein. *Studies in History and Philosophy of Science*, 21:531–561.

Maudlin, T. (1993). Buckets of water and waves of space: Why space-time is probably a substance. *Philosophy of Science*, 60, 183–203.

Misner, C., Thorne, K., and Wheeler, J. (1973). *Gravitation.* W.H. Freeman, New York.

Newton, I. (2005). On the gravity and equilibrium of fluids. In Janiak, A., editor, *Newton's Philosophical Writings*. Cambridge University Press, Cambridge.

Quine, W. (1953). On what there is. In *From a Logical Point of View*. Harper, New York.

Riemann, B. (1867). Ueber die Hypothesen, welche der Geometrie zu Grunde liegen. In Weber, H., editor, *The Collected Works of Bernhard Riemann*, pages 272–287. B.G. Teubner, Leipzig. Published 1902. Reprint, New York: Dover Publications, 1956.

Russell, B. (1903). *The Principles of Mathematics.* Cambridge University Press, Cambridge.

Rynasiewicz, R. (1992). Rings, holes, and substantivalism: On the program of Leibniz algebras. *Philosophy of Science*, 59:572–589.

Smolin, L. (2005). The case for background independence. Technical Report 0507235, arXiv: High Energy Physics—Theory.

Stachel, J. (2002). The relations between things versus the things between relations: The deeper meaning of the hole argument. In Malament, D., editor, *Reading Natural Philosophy: Essays in the History and Philosophy of Science and Mathematics to Honor Howard Stein on His 70th Birthday*, pages 231–266. Open Court Press, Chicago IL.

Stachel, J. (2005). Structure, individuality, and quantum gravity. Technical Report 0507078v2, arXiv: General Relativity and Quantum Cosmology.

Stein, H. (1967). Newtonian space-time. *Texas Quarterly*, 10:174–200.

Weyl, H. (1927). Philosophie der Mathematik und der Naturwissenschaften. R. Oldenburg Verlag, Munich and Berlin.

Chapter 18
A Silly Answer to a Psillos Question

Elaine Landry

In this paper I offer an answer to a question raised in (Psillos, 2006): How can one speak of structures without objects? Specifically, I use category theory to show that, *mathematically speaking*, structures do not need objects. Next, I argue that, *scientifically speaking*, this category-theoretic answer is silly because it does not speak to the scientific structuralist's appeal to the appropriate kind of morphism to make precise the concept of shared structure. Against French et al.'s approach,[1] I note that to account for the scientific structuralist's uses of shared structure we do not need to formally frame either the structure of a scientific theory or the concept of shared structure. Here I restate my (Landry, 2007) claim that the concept of shared structure can be made precise by appealing to a kind of morphism, *but*, in science, it is methodological contexts (and not any category or set-theoretic framework) that determine the appropriate kind. Returning to my aim, I reconsider French's example of the role of group theory in quantum mechanics to show that French already has an answer to Psillos' question but this answer is not found in either his set-theoretic formal framework or his ontic structural realism. The answer to Psillos is found both by recognizing that it is the context that determines what the appropriate kind of morphism is and, as Psillos himself suggests,[2] by adopting a *methodological* approach to scientific structuralism.

E. Landry (✉)
Associate Professor of Philosophy, University of California, Davis, CA, USA
e-mail: emlandry@ucdavis.edu

With sincere sentiment and gratitude for my mentor and my friend, I dedicate this paper to John; a man who really is in a category of his very own! As well, I wish to thank Katherine Brading, Anjan Chakravartty, Steven French, Antigone Nounou, Stathis Psillos and Dean Rickles for valuable discussions and comments. Research for this paper was supported by a Social Sciences and Humanities Research Council of Canada grant.

[1] Here I have in mind the approaches of (da Costa and French, 1990; da Costa et al., 1997; French, 1999b, a, 2000).

[2] See, for example, (Psillos, 2006, p. 18), where he attempts to show that "there is an important methodological insight in Worrall's suggestion ...."

D. DeVidi et al. (eds.), *Logic, Mathematics, Philosophy: Vintage Enthusiasms*,
The Western Ontario Series in Philosophy of Science 75,
DOI 10.1007/978-94-007-0214-1_18, © Springer Science+Business Media B.V. 2011

1 A Psillos Question

In his most recent paper aimed at challenging the various versions of scientific structuralism,[3] Psillos claims that "(a) structures *need* objects and (b) scientific structuralism should focus on *in re* structures" (Psillos, 2006, p. 3). Borrowing from the philosophy of mathematics literature, Psillos begins by taking a "system" to be a collection of objects with certain properties and relations, so that a "structure" is the abstract form of this system (p. 4). Accordingly, this yields the *basic structuralist postulate* that

> [a]mong the many structures that can characterize a system some is privileged. This is *the* structure of this system as specified by the relationships among the objects of the system. It's this postulate that renders talk about *the* structure of system meaningful .... This issue then is what exactly makes a structure privileged. (Psillos, 2006, pp. 3–4)

To speak to this issue of privileging, Psillos next considers the distinction (borrowed from Shapiro, 1997) between *ante rem* and *in re* structuralism, whereby

> *[a]nte rem* structuralism has it that structures are abstract, freestanding, entities, they exist independently of systems, if any, that exemplify them ... [and] *in re* structuralism takes systems as being ontologically prior to structures: it denies that structures are freestanding entities. Structures are abstractions out of particular systems .... (Psillos, 2006, p. 5)

In re claims about structure are thus to be understood in one of two ways: they are about *a* privileged system or about *all* systems that are so privileged because they share structure (e.g., are isomorphic). So far so good. However, Psillos next claims that, in contrast to the *ante rem* position,

> [a]ccording to *in re* structuralism, there are no extra objects [such as Shapiro's "places"] which 'fill' the structure. It's obvious then that the objects that 'fill' the *in re* structures have more properties than those determined by their interrelationships in the structure. They are given, and acquire their identity, independently of the abstract structure they might be taken to exemplify. (Psillos, 2006, p. 5)

I take it Psillos means to point out that, for example, when considering the various systems that have the natural number structure, say the models of von Neumann ordinals and the Zermelo numerals, the object 2 will have some properties in one system that it does not have in the other, for example the property $2 \in 4$. But as Benacerraf has shown, this does not mean, as it does for the Fregean, that such objects have or acquire their identity *qua* natural numbers independently of their position in a structure. Indeed for the structuralist[4] this means that one ought not to countenance such objects as things that have independent identity conditions[5];

[3] Specifically, the versions he considers are: structural empiricism, epistemic structuralism and ontic structuralism.

[4] I am not claiming that Benacerraf advocated a structuralist position; see (Benacerraf, 1991, pp. 284–294), where he refers to his position as that of a "formist" and distinguishes it both from the position of the formalist and the Fregean.

[5] Indeed the conclusion for (Benacerraf, 1991) is that if by "object" we mean thing or individual that has independent identity conditions, then we ought not to countenance numbers as objects at all.

rather, objects are nothing but "positions in a structure," whatever we then mean by the term "structure." So it is simply a mistake to claim that "[m]athematical structuralism, then does not view structures without objects. It's not revisionary of the underlying ontology of objects with properties and relations" (Psillos, 2006, p. 6). It certainly is revisionary of an ontology of objects *qua* individuals,[6] and really, that is the whole point. For example, the point is to show that all there is to 2, or all we need to know of 2, is that it is a position in an abstract system (or structure) that has the appropriate kind of structure, i.e., that satisfies the Peano axioms. Let me explain.

Both the *in re* and *ante rem* structuralist take objects to be nothing but positions in an abstract system (or structure), as having no other identity conditions, and so no other relevant properties, than those specified by what is taken to characterize the structure of that abstract system (or structure). Really, that's the *raison d'être* of mathematical structuralism; that it can forego traditional debates about the independent existence, and/or absolute criterion of identity, of mathematical objects in favor of discussions about objects *qua* positions in an abstract system (structure) whereby objects are fully characterized by their shared structure (are characterized up to isomorphism). For example, for the *ante rem* structuralist the natural number 2 is nothing but a position in a structure that satisfies the Peano axioms, where "structure" means a freestanding ontological entity and "position" means a "bare position" or "place" in this structure. For the *in re* structuralist, the number 2 is likewise nothing but a position in an abstract system that satisfies the Peano axioms, where "abstract system" means *a* privileged system or *all* (possible) systems that are privileged because they have the same structure (e.g., are isomorphic). In any case, the only "objects" that the *ante rem* structuralist is committed to are structures, and the only "objects" that the *in re* structuralist is committed to are (possible) systems. Of course, both of these commitments are problematic; that is, *ante rem* structuralism commits us to the actual (metaphysical) existence of structures and *in re* structuralism commits us to the possible (modal) existence of systems, but *neither commits us to the existence of objects*. So it is simply a mistake to draw, from the claim that mathematical structuralism is not revisionary of ontology, the "immediate moral" that "structures need objects" and that "This holds for both *ante rem* and *in re* structuralism" (Psillos, 2006, p. 6).

Psillos next changes his line of attack; he uses the above "mathematical moral" to argue against any version of scientific structuralism that does not countenance independently existing objects. He does this by showing that when it comes to considering physical systems:

1. *Ante rem* structuralism is ill-motivated. This because "finding the structure of a natural system is an a posteriori (empirical) enterprise. Its structure is *in re* (Psillos, 2006, p. 6). And,

[6] I use the term "individual" to mean, minimally, an object that has or acquires its identity independently of its position in a structure.

2. *In re* structuralism must concede that there is more to the world than structure. This because

> [a physical system] is a natural structure in the sense that is captures the natural (causal-nomological) relations among the objects of the systems. It is *the* structure that delimits a certain domain as possessing causal unity. Hence, it is grounded on the causal relations among the elements of the domain. (Psillos, 2006, p. 6)

Psillos further claims that, for both the *ante rem* and *in re* interpretations of scientific structuralism, "[o]bjects are needed in either case" (p. 7). This because "[p]laces in structures and formal relations do not cause anything at all. It's the "fillers" of places [individuals] and concrete (*in re*) relations that do." When considering physical systems, Psillos thus concludes that since

> the discovery of the *in re* structure is an empirical matter, it may not be isomorphic to any of a set of antecedently given *ante rem* structures. Second, even if the discovered *in re* structure turns out to be isomorphic to an *ante rem* one, the order of ontic priority has been reversed: the *ante rem* structure is parasitic on the *in re*; it's an abstraction from it. (Psillos, 2006, p. 7)

I will challenge both of these claims (that the discovery of the *in re* structure is an empirical matter and that an abstract system of any kind need be a "bottom-up" abstraction from a concrete *in re* structure). For now, however, I will offer a "top-down" version of an *in re* interpretation of mathematical structuralism that is not committed to *any* independently existing "objects,"[7] and that does not begin with concrete systems and work "bottom-up" to the concept of an abstract system (or structure). I will then reconsider French's example from quantum mechanics to show that, as a matter of methodological fact, science works both "top-down" and "bottom-up." Thus, while French is wrong to presume that science works only "top-down" and so can cut the world by structure, Psillos is wrong to presume that it only works "bottom-up" and so is bound to a world of individuals. Nonetheless, French does have an answer to offer Psillos, but it is one that arises neither from his formal framework nor from his ontic structural realism; it arises from structuralist considerations as made within a methodological context.

2 A Category-Theoretic Answer to the Mathematical Structuralist

Before providing my category-theoretic answer to Psillos' question, it is necessary to provide a clear distinction between the interpretations and varieties of mathematical structuralism.[8] There are two levels at which one finds structuralist ideas: the abstract and the concrete. At the *concrete level* there are two

[7] As I will explain, by "object" I intend to include Psillos' objects *qua* individuals, Shapiro's structures as actual objects and Hellman's systems as possible objects.

[8] See (Parsons, 1990; Dummett, 1991; Hale, 1996) for excellent overviews of the interpretations and varieties of mathematical structuralism, and for thoughtful analyses of the problems of each.

interpretations: the formalist and the model-theoretic. The formalist interpretation[9] forgoes both semantic and ontological questions in favor of providing a methodological account of how mathematics is done as opposed to considering what mathematics is about. Thus, the formalist has nothing to say about the semantic or ontological status of either objects or structures. The model-theoretic interpretation[10] concerns itself primarily with semantic issues; that is, the existence of an object *qua* a position in a model is a consequence of the truth of a statement in which it occurs.[11] This is because structures are taken as models and objects as positions in models; again, either in some privileged model or in all models that have the same structure (that are isomorphic). As noted by Dummett, model-theoretic structuralism at the concrete level is not really mathematical structuralism, because

> [t]here is an unfortunate ambiguity in the standard use of the word 'structure,' which is often applied to an algebraic or relational system—a set with certain operations or relations defined on it, perhaps with some designated elements; that is to say, a model considered independently of any theory which it satisfies. This terminology hinders a more abstract use of the word 'structure'; if, instead we use 'system' for the foregoing purpose, we may speak of two systems as having an identical structure, in this more abstract sense, just in case they are isomorphic. The dictum that mathematics is the study of structure is ambiguous between these two senses of 'structure.' If it is meant in the less abstract sense, the dictum is hardly disputable, since any model of a mathematical theory will be a structure in this sense. It is probably usually intended in accordance with the more abstract sense of 'structure'; in this case, it expresses a philosophical doctrine that may be labeled 'structuralism.' (Dummett, 1991, p. 295)

Moving then to the *abstract level* we find two interpretations and several varieties of what is typically understood as mathematical structuralism. As already noted, the two interpretations are the *ante rem* and the *in re*. The *ante rem* interpretation comes in two varieties: the set-theoretic and the structure-theoretic. The *set-theoretic* interpretation takes a structure to be an abstractly considered set-structured system, i.e., an abstract system *qua* a set/collection/class with certain elements and first-order functions/relations and, perhaps, second-order predicates, typically symbolized as $S = \langle e, f, P \rangle$. The abstract structure of any system is expressed in these terms so that an object, *qua* a position in a structure, is a type of set-structured object,

---

[9] I mention here the formalist position because, while it is not typically characterized as a mathematical structuralist position, it does bear upon the reading in (Psillos, 1995) of the structure/object debate as similar to, if not the same as, the form/content or structure/nature debate. As we will see, at least for the mathematical structuralist, the meaning of a term or expression or the existence of an object is fixed fully by its position in a structure, so really there is no form/content debate. See (Landry, 1999) for a discussion of why mathematical structuralism bypasses this debate.

[10] Again, while it is not typically taken as a mathematical structuralist position, I mention the model-theoretic interpretation because of its association with the semantic view of scientific theories, with Putnam's internal realism, and also with Putnam's Paradox. See, however, Dummett's point, as noted above, which has been missed by many philosophers of science, that the move from the model-theoretic interpretation to full-blown structuralism involves a conflation of concrete and abstract levels of analysis.

[11] Note also the connection here to the slogan from (Quine, 1948), that to be is to be a value in the range of a bound variable.

i.e., is an abstractly considered element, function or predicate. The problem with any such version is that the point of (Benacerraf, 1991) applies as much to set theory as it does to arithmetic: Which set theory acts as the background theory and so captures the structure of (all) such set-structured objects/systems/structures?

The *structure-theoretic* variety of *ante rem* structuralism, as best exemplified by the account in (Shapiro, 1997), seeks to forego the question of what provides the background theory by taking structures themselves at face value, that is, by taking structures as "freestanding abstract objects" so that objects are "bare positions" or "places" in such structures. Though we might begin with concrete systems, from these we abstract away the non-essential (non-structural) properties so that we are left with objects as bare positions in a structure. The problem with the structure-theoretic version is that, instead of reifying objects *qua* independently existing things (as with standard mathematical Platonism), it reifies structures *qua* independently existing things. That is, it yields an ontology of structures *qua* individuals for which, just as if we had an ontology of objects *qua* individuals, we need identity conditions and face the problem of epistemic access. Finally, there is the problem, shared by both the set- and structure-theoretic versions of *ante rem* structuralism. This arises from the question: What is meant by "abstractly consider"? Typically, the *ante rem* structuralist of either variety replies with the "ladder analogy": concrete systems act as rungs on ladder, which we use to climb "bottom-up," via the process of abstraction, towards all systems that have the same structure. But once we reach the top, i.e., once we arrive at the structure, we kick away the ladder. The problem is that we are now owed an account of the process of abstraction, and one that does not come up against Plato's "regress problem."[12]

The *in re* option likewise comes in two varieties: the eliminativist and the schematic. Each eschews talk of *both* objects and structures *qua* individuals, i.e., as independently actually existing things. The *eliminative* variety is best characterized by Hellman's[13] eliminative modal-structuralism, wherein talk of both objects and structures is eliminated in favor of talking about all possible systems that have a structure. The terms of a theory are not genuine singular terms (as they are with *ante rem* structuralism), but neither are they purely schematic or variable; rather, they are terms of "modalized assertions" that range over all possible systems that share structure (i.e., are isomorphic). This, however, seems to force us to choose from among three problems: (i) we must assume the concepts of possibility/necessity as primitive, or (ii) we must adopt a possible world semantics, or (iii) we must assume that there are enough possible objects to make up such systems.

The *schematic* variety of *in re* structuralism takes the terms of theory as purely schematic or variable and is expressed in two ways: informally and formally. *Informally*, as characterized by (Parsons, 1990), it holds that the most fundamental concept of structure for this purpose is meta-linguistic; the structure of an

[12] See (Psillos, 2006, p. 6) for an account of the regress problem as applied to the *ante rem* structuralist.

[13] See (Hellman, 1996, 2001, 2003).

abstract system *qua* a "domain" is given by a predicate and the relations and functions between domains by further predicates and functors. The problem is that predicates, functions and functors are themselves taken either as set-theoretic or as quasi-concrete objects, and so cannot be accounted for in purely structuralist terms. Thus to account for these objects we seem to run the risk of requiring set theory or logic as a background theory or of requiring physicalism/mentalism to underpin their nature/construction. Accepting the latter as untenable, the former leaves us asking, *pace* Benacerraf, which set theory, which logic?

I have argued elsewhere (see Landry and Marquis, 2005) that one can use category theory to *formally* frame a schematic *in re* interpretation of mathematical structuralism. What the informal version demonstrates is that if the concept of an abstract system (a "domain") is to be taken as both meta-linguistic and formally precise we, likewise, need a linguistic framework that *begins with* the schematic concept of an abstract system. Unlike the *ante rem* or informal schematic *in re* varieties, this version is "top-down"; it does not seek to *build up* abstract systems from either (quasi-) concrete objects or concrete systems, it begins with the concept of an abstract system characterized in terms of the cat-structure of "objects" and "arrows."[14] Yet, unlike the *ante rem* approaches or the modal *in re* approaches, it eschews taking structures or systems as actually or possibly existing "objects"; abstract systems, as cat-structured systems, are schemata and so require no conditions of possibility or actuality.

It is in this sense that a category provides the schemata for talking about the shared structure of objects and abstract systems in terms of the arrows (kinds of morphisms) between them. Because a category is taken as a schema for what we say, as opposed to being taken as an object-holder for what exists, we need only take category theory as a linguistic framework; we need not take category theory as a background theory, or foundation,[15] for mathematics. That is, because no single category or type of category is taken as privileged, Benacerraf's point does not apply. A category, then, is not constitutive of what an abstract system (or a structure) is; rather it provides a schema for what we say about the shared structure of both objects and abstract systems. Moreover, because we work top-down and so *begin with* the concept of an abstract system, the "objects" and "arrows" that define a category are themselves taken to be schematic or variable. We need not work bottom-up from anything that makes-up "objects" and "arrows," and so we need not take them or their constituents as quasi-concrete objects.

So how, in mathematics, can we speak of structure without objects? First, as structuralists we construe objects as nothing more than positions in abstract systems

---

[14] In a similar vein, see (Bell, 2006) for a "top down" account of abstract sets, i.e., for an account that begins with the notion of an abstract set characterized in terms of a type of category-theoretic structure, e.g., in terms of topos-theoretic structure. From there, he goes on to account for the notion of a variable set characterized in terms of different types of category-theoretic structure, e.g., in terms of the category of bundles and sheaves.

[15] See (Bell, 1981; Feferman, 1977; Mayberry, 1994; Landry, 2006) for criticisms of taking category theory as a foundation.

(structures) that have a structure. For example, 2 *qua* natural number is nothing more than a position in any or all systems that satisfy the Peano axioms. Second, we conceive of abstract systems themselves as schemata for what we say about the shared structure of objects, as opposed to conceiving of them as freestanding structures or possibly existing systems. Finally, we use category theory as a linguistic framework, as opposed to using set theory, structure theory or modal logic as a background theory, to express what can be said of the shared structure of both objects and abstract systems. For example, we use the category of sets, *Set*, to talk about the object "set" in terms of the morphisms between them, and this regardless of what a set is. And, likewise, to talk about the shared structure of, for example, the abstract systems of groups and sets, we use the forgetful functor between the category of groups, **Grp**, and the category of sets, **Set**. Thus, for a structuralist, objects are nothing but positions in abstract systems, and for a formal *in re* schematic structuralist, abstract systems are not independently existing "objects" (actual or possible); rather their structure, too, is formally schematized by category theory. What we have then is a version of *mathematical* structuralism that provides an answer to Psillos' question: How in *mathematics* can we speak of structures without objects? But can the same category-theoretic framework be used by the *scientific* structuralist? In the next section, I will answer: No.

3 Why This Answer Is Silly for a Scientific Structuralist

Recent semantic approaches to scientific structuralism, aiming to make precise the concept of shared structure between physical systems *qua* models, formally frame a model as a type of set-structure. This set-theoretic framework is then used to provide a semantic account of (a) the structure of a scientific theory, (b) the applicability[16] of a mathematical theory to a physical theory, and (c) the structural realist's appeal to the structural continuity between predictively successful successive physical theories. In (Landry, 2007), I challenged the idea that, to be so used, the concept of a model[17] and so the concept of shared structure between models must be formally framed in set-theoretic terms.

For those who thought that I intended to argue, or at least make way for the claim, that it ought to be framed in category-theoretic terms, let me now be clear: what I was then, and am now, arguing is that for the scientific structuralist to use the concept of shared structure between physical systems *qua* models (both theoretical

---

[16] In accounting for applicability, this use also attempts to frame the concept of "surplus structure" as in (Redhead, 1975, 1980, 1995).

[17] Note here that the level of analysis is not, in the first instance, intended to be at the concrete level so that "model" is not intended to be understood in the Tarskian model-theoretic sense; rather the analyses that I consider (for example, French and da Costa's partial structures account) are intended to be at the abstract level. In particular, they are accounts that, being Bourbakian in spirit, are in line with the set-theoretic *ante rem* version of mathematical structuralism. See also the next note.

models and phenomenological/data models), *no formal framework is needed.* What is needed, however, is the concept of an appropriate kind of morphism, where both the determination of the appropriate kind and the meaning of the term "morphism" are fixed by some context. Let me explain.

In (Landry, 2007), I first challenged the Bourbaki-inspired assumption that scientific models *qua* structures are types of abstractly considered set-structured systems and next considered the extent to which this problematic assumption underpins both Suppes' and French et al.'s[18] semantic view of the structure of a scientific theory. I pointed out that, mathematically speaking, there is no reason for our continuing to assume that types of abstract systems (structures) and/or the morphisms between them are "made up" of set-theoretic elements and functions. What I did not make explicit is that, in the same vein, I was arguing that there is no reason to assume that types of abstract systems and/or the morphisms between them are "made up" of category-theoretic objects and arrows. Thus, to account for the fact that two physical systems *qua* models share structure, one does not have to specify what models are *qua* types of set- or cat-structures. It is enough to say that, in the context under consideration, there is a morphism between the two systems, *qua* mathematical or physical models,[19] that makes precise the claim that they share the appropriate kind of structure. I then used this investigation to show that when it comes to *using* the concept of shared structure—to account for the structure of scientific theories, the applicability of a mathematical theory to a physical theory, and the structural continuity between predictively successful successive theories—there is no need to agree with French that "without a formal framework for explicating this concept of 'structure-similarity' it remains vague, just as Giere's concept of similarity between models does ..." (French, 2000, p. 114). The meaning of the concept of shared structure need not remain vague; it can be made precise by appealing to a kind of morphism, *but* it is the context (and not any set– or category-theoretic type) that determines the appropriate kind for its use.[20] To demonstrate this I considered the

[18] Seen in the light of the set-theoretic version of *ante rem* structuralism, Bourbaki structures are *types of* abstractly considered set-structured systems and, similarly, for French and da Costa models *qua* partial structures are Bourbaki structures. Finally, like Shapiro's structure-theoretic *ante rem* account of mathematical objects as "bare positions," (French, 2006), uses his ontic structural realism to conceive of physical objects as purely structural "nodes."

[19] By speaking only of mathematical and physical models I do not mean to exclude iconic models. An iconic model, taken in terms of (Hesse, 1963; Achinstein, 1968; Suppe, 1977; Redhead, 1980), is a concrete physical system that functions as an icon for another system in such a way that the properties that hold of the first can be said to hold, perhaps by analogy, of the second. For example, the solar system is often taken as an iconic model for the orbital theory of the atom. See also (Suppes, 1962, pp. 290–291), where he characterizes the concept of an iconic model as a "physical model" or, equally, as the "physicists' concept" of a model.

[20] I thus take the concept of a morphism at face value, i.e., as a map between two kinds of structured systems, *qua* mathematical or physical models, where, as explained in note 5, there are at least three options other than taking models themselves as types of set-structures. For example, one could use category theory to provide a formal framework for the concept of an abstract structured system and too for the concept of a morphism as a map between such cat-structured systems, *but* again this is quite beside my point here. My point is that, *regardless* of formal frameworks, it will

example given in (French, 1999a) from the development of quantum theory to show that, as witnessed by both Weyl and Wigner's programmes, it was the foundational and representational contexts of considering the "relevant symmetries" that determined that the appropriate kind of morphism was the one that preserved the shared Lie-group structure of both the theoretical and data models.[21]

Why, then, can't category theory be offered as a formal framework for a "top-down" *in re* interpretation of scientific structuralism and so be used to answer Psillos' question: "How in science can we speak of structures without objects?" Because although, mathematically speaking, we can offer up a category-theoretic answer, scientifically speaking, the answer is given at the wrong level. Recall that debates about *in re* and *ante rem* structuralism and, likewise, debates concerning those varieties of mathematical structuralism that depend on a set-theoretic background theory or category-theoretic linguistic framework are at the abstract level. That is, they are about formally framing the concept of shared structure between models themselves formally framed as *abstract mathematical* systems or "structures." In contrast, for Psillos, debates of the versions of scientific structuralism concern the shared structure between models as *concrete physical systems*[22] and, as we have seen, this is structuralism at the concrete (model-theoretic) level. Moreover, as noted by Dum-

---

be a specific context that determines what kind of morphism is appropriate. For example, in the context of speaking about the shared structure of systems structured by space-time theories the appropriate kind of morphism is a diffeomorphism, and this regardless of what a diffeomorphism *is*, i.e., regardless of whether it is a function between set-theoretic elements or an arrow between category-theoretic objects. Note also that if we narrow the context we must likewise narrow the kind of morphism. For example, while for generally relativistic theories the morphism between the dynamically possible models will be any diffeomorphism, for special-relativistic theories the morphism between models will be a restricted kind of diffeomorphism called a Poincaré transformation and for Newtonian Mechanics it will be another restricted kind diffeomorphism called a Galilean transformation, the groups of Poincaré and Galilean transformations being subgroups of the diffeomorphism group. Thanks to Dean Rickles for pressing me to spell out the details here.

21 The attention to use/context is thus intended to be amenable to the (Cartwright et al., 1995) view of the importance of "phenomenological" models, but, contra (Suárez, 2003, 2006), it does not go as far as rejecting "isomorphism" accounts of shared structure. Rather it seeks to present a "morphism" account of shared structure wherein the morphism that preserves the appropriate kind of structure is determined by the use/context and not by a presumption that any one type of morphism is, or is not, *the* type that makes precise the concept of shared structure and, in so doing, fixes its meaning/use for any context.

22 There are several ways to go here in explaining what is meant by the term "model" in the concrete, physical, sense. First, one might take model itself at face value, viz., to be understood, in the Tarskian sense, as an interpretation that makes a set of sentences true *without* adding that a model is a set-theoretic entity. Second, one might offer, along the lines of van Fraassen, a state-space account of what is meant by a mathematical model, so that a physical model is a model (in the Tarskian sense) of a mathematical model *qua* a state-space. Finally, one might forego giving an account of models *per se* and instead offer up an account of what is meant by the term "system" so that one may then consider a concrete physical system as (or as represented by) a model (again, in the Tarskian sense) of a mathematical model *qua* a kind of abstract mathematical system. In any case, I am not suggesting that these are the only ways or that one of these ways is preferable to the other; I am only pointing out that none of these requires us to take physical models as types of set- or cat-structures or as types of set- or cat-structured systems.

mett, since any model, in the model-theoretical sense of the term, of a mathematical theory will be a structure in this concrete sense, this is hardly structuralism in the mathematical sense of the term. Indeed, whether we interpret theories semantically or syntactically, this shows why when, in a scientific setting, we attempt to move from the concrete to the abstract level of analysis,[23] we encounter problems such as Putnam's paradox and the Newman problem.[24]

Put in other words, using any version of mathematical structuralism to formally frame the concept of shared structure will not assist with questions about the structure of scientific theories (as represented by theoretical models), the structure of physical systems (as represented by mathematical/physical models), or the structure of the phenomena (as represented by data models), unless we presume that the abstract mathematical structure of a theory matches the concrete physical structure of the world. So too our category-theoretic framework cannot be used to answer Psillos's scientific question unless we presume that "the world," or what we know of it, is cut into category-theoretic kinds. Moreover, to borrow from Newton, this presumption, or any like it, is far removed from the scientific task of both reasoning to and from the phenomena.

Without this presumption, the fact that we can, in mathematics, speak of structures without objects, does not answer the question of how this can be so in science. Simply, the reason is as follows: in mathematics, the axioms alone determine whether a system has a structure of the appropriate kind, and so determine whether any position is an object in the structuralist sense of the term. For example, the Peano axioms fully determine whether a system has a structure of the natural numbers, and so determine whether the 2 of the Zermelo numerals is an object called a natural number, i.e., is a position in any or all systems that have the same structure. In science, by contrast, the determination of the appropriate kind of structure is made only methodologically, that is, is made by considering the structure of a system in what are taken to be significant contexts. For example, as we will see, the group-theoretic structure of quantum-mechanical systems (both theoretical and phenomenological) was determined by considering the "relevant symmetries" between such systems in both foundational and representational contexts.

---

[23] Note also French's claim that Psillos' comparison between mathematical and scientific structuralism is misleading because "the central claim of OSR [Ontological Structural Realism] is that it is the structure [and not the *in re* concrete system composed of objects] that is both (ultimately) ontically prior and also concrete" (French, 2006, p. 8). But as we shall see, French's version of scientific structuralism, in so far as it is formally framed by set theory, needs to presume that "the world" is set-structured to make the connection between the abstract mathematical structure of a theory and the concrete physical structure of the world.

[24] See (Psillos, 2006, pp. 3–4) for a precise account of what gives rise to these problems, viz., that "if we start with the claim that a certain domain D has an arbitrary structure W, and if we posit another domain D' with the same cardinality as D, it follows as a matter of logic that *there is* a structure W' imposed on D' which is isomorphic to W. This claim has been the motivating thought behind Newman's critique of Russell's structuralism and of Putnam's model-theoretic argument against metaphysical realism ...."

As my nod to Newton suggests, and as the quantum mechanical history shows, the methodology of science is such that this determination of the appropriate kind of structure is made *both* by working down from the structure of the theory to the structure of the phenomena *and* working up from the structure of the phenomena to the structure of the theory. Unfortunately, just as French's attempts to formally frame the structure of scientific theories presume that a scientific structuralist story can be told by working *only* "top-down" from the abstract set-structure of scientific theories, so Psillos' criticisms presume that one must work *only* "bottom-up" from the concrete (set-structure) of the phenomena.

The French story goes something like this: structures are formally framed as types of set-structures, models are structures, physical theories and physical systems are (or are represented by) models, so the shared structure between models *qua* types of set-structures is formally framed by a type of set-theoretic morphism (homeomorphisms, partial isomorphisms, partial homomorphisms, etc).[25] Thus, the scientific structuralists' uses of shared structure are gleaned by working "top-down" from the abstract set-structure of a theory: the structure of a scientific theory, as characterized by its models, is given by its set-structure; applicability is explained in terms of a type of set-theoretic morphism between mathematical and physical models; and continuity of structure is, likewise, explained in terms of a type of set-theoretic morphism between the models of predictively successful successive theories.

Psillos' story differs not by its set-theoretic presumption but rather because it works "bottom-up" from the concrete set-structure of the world, i.e., it presumes both that "finding the structure of a natural system is an a posteriori (empirical) enterprise" and the world is "made up" of objects *qua* individuals *qua* elements of a domain (see Psillos, 2006, p. 6).[26] Psillos works up from concrete physical systems to *in re* structures so that

> there are objects that 'fill' the structures, these objects have [non-structural] properties over and above those that are determined by their interrelationships within the structure ... it's in virtue of these properties that they have causal unity .... (Psillos, 2006, p. 9)

In any case, what is not answered by French or Psillos is: Why do we presume either that scientific theories of physical systems are set-structured or that the phenomena are cut into objects *qua* elements and properties *qua* functions between them? And too, why do we suppose that the methodology of science must work *only* "top-down" or "bottom-up"?

---

[25] Specifically, formal accounts of shared structure have been made in terms of: homeomorphisms between models *qua* types of lattice structures (da Costa et al., 1997); partial isomorphisms between models *qua* partial structures in a function-space (French, 1999a); and, partial homomorphisms between models *qua* partial structures (French, 2000). See also (Redhead, 1975) for the development of the "function space" approach and (Redhead, 1980) for the use of this approach for analyzing the role of models in physics.

[26] It is the latter presumption that has Psillos claiming that relations require relata, etc.

While the first presumption may be made by appeal to some set-theoretic foundationalism (as was done by Suppes and as is adopted by French et al.[27]), it is the second presumption that leaves both the scientific structuralist and the structural realist open to the Psillos' question: How can one speak of structures without objects? That is, if the formal framework for scientific structuralism is taken to be set theory, Psillos' question translates (perhaps without remainder) to the question: How can one have relations (functions) without relata (elements)? Now this question might not be a problem for the *epistemic structural realist*, who, while allowing for talk of objects/elements/individuals, claims that all we know is their structure, but it certainly is a problem for the *ontic structural realist*, like French, who claims that *all there is* is structure.[28]

4 An Answer for the Methodological Scientific Structuralist

I now reconsider French's example of the role of group theory in quantum mechanics to show that French already has an answer to Psillos' question, *but* this answer is not found in his set-theoretic formal framework or his ontic structural realism. Regardless of any presumption of the structure of a theory or the structure of the world, what the history of science shows us, and what the scientific structuralist relies upon, is that the phenomena are structured for and by mathematical theories, so that mathematical models and physical models share structure, and, furthermore, so that the models of predictively successful successive theories share structure. But what this history (and philosophy) also shows us is that it is *contexts*, as set within the methodology of reasoning to and from the phenomena, and not any formal framework or any causal-nominological world that tells us what the appropriate kind of structure is. Let me explain.

We first consider Psillos's analyses of the various positions of the scientific structuralist. He considers three: the ontic structuralist, the empirical structuralist and the

---

[27] See Suppes' claim that "... there is no theoretical way of drawing a sharp distinction between a piece of pure mathematics and a piece of theoretical science. The set-theoretical definitions of the theory of mechanics, the theory of thermodynamics, the theory of learning, to give three rather disparate examples, are on all fours with the definitions of the purely mathematical theories of groups, rings, fields, etc. From the philosophical standpoint there is no sharp distinction between pure and applied mathematics, in spite of much talk to the contrary" (Suppes, 1967, pp. 29–30). See (French, 1999a, p. 201) where this quote from Suppes is used to explain the role of models in science and see also (French, 2000, p. 104), where this same quote is used to claim that "[W]ithin such a [Suppesian set-theoretic] framework the applicability of mathematics to science comes to be understood in terms of the establishment of a [type of set-theoretic] relationship between one kind of structure and another." As noted, in my 2007 paper I challenge the idea that accounts of shared structure need be given in such set-theoretic terms.

[28] Certainly, since formally speaking arrows do not require objects, one could give a category-theoretic account of physical objects in terms of arrows only, that is, one could construe physical objects themselves as arrows. This would certainly avoid reference to objects *qua* elements and perhaps could even be used to formally frame French's 2006 notion of an object *qua* node. But, again unless we presume that "the world" is cat-structured, what would we gain?

epistemic structuralist. Against the position of the ontic structuralist, who holds that structure is all there is, he says:

> [I]f we take structures to be causal, we should not look to their structural properties for an underpinning of this causal unity and activity. This is not surprising. Places in structures and formal relations do not cause anything at all. It's the 'fillers' of the places and concrete (*in re*) relations that do.

This with the "moral" that

> OS [ontic structuralists] will take structures to be either *ante rem* or *in re*. Objects are needed in either case. If OS reifies structures, their causal unity, role and efficacy are cast to the wind. If OS gives ontic priority to *in re* structures, there is more to the world than structure. (Psillos, 2006, p. 7)

Let's pause here to consider an example by analogy. Let the notion of mathematical necessity[29] be taken as analogous to the notion of physical causality. To say that 2 is necessarily greater than 1, do we need 1 and 2 as independently existing objects? Not for the structuralist; indeed, the structure of the natural numbers as determined by the Peano axioms and the successor relation are all that are needed to guarantee that, whatever (or even whether) 1 or 2 are, as positions in any system that has a natural number structure, 2 is greater than 1. So *it is* objects *qua* positions and their structural properties and not objects *qua* individuals and their non-structural properties that underpin necessary relations. Likewise, why can't the scientific structuralist argue that two objects, *qua* positions in a system, stand in some causal relation in virtue of their structural properties and/or some (perhaps higher-order) relations that result from their shared structure?

Against the position of the structural empiricist, like van Fraassen, who allows that the structure of the phenomena *qua* appearances can be known, but denies that science should or need aim at knowing more, Psillos notes that "structural empiricism buys into a substantive metaphysical assumption: that its at least possible that the structure of the appearances is causally connected to a deeper unobservable structure" (Psillos, 2006, p. 8). Indeed, such "excess structure" is possible, but the

---

[29] Ladyman is developing an alternative modal form of ontological structural realism which aims to account for the notion of physical necessity in purely structural terms. Adopting this modal stance, one may say that the structure of a physical system specifies the necessary properties associated with what it is for a particular physical object to be an object of the appropriate kind. To explain what might be meant here, again consider our example from mathematics: it may be said that while $2 \in 4$ is a possible property of the natural numbers, it is not a structural, i.e., a necessary, property because $2 \in 4$ is not true for all systems that have a natural-number structure. Modal structural realism is, therefore, at once both more modest and more ambitious than other varieties of structural realism. Unlike the standard ontic version, it does not aim to capture *all* the properties of physical objects, but it does aim to capture their *necessary* properties. The necessary properties transfer, via the shared structure of those systems that have the appropriate kind of structure, to the objects *qua* positions, thus representing the necessary relations between objects. See too his claim that "the abstract mathematical structures it [the theoretical part of a theory] employs ... must have some grip on reality. It is clear that the "grip on reality" in question must go beyond a correct description of the actual phenomena to the representation of modal relations between them" (Ladyman, 1998, p. 418).

question is whether it is needed to tell the structuralist story of either the structure of a scientific theory, the applicability of mathematics to a scientific theory, or the continuity of structure over theory change. Psillos has not demonstrated that it is; other than simply stating that is it needed to tell a "causal" story.

Against the epistemic structuralist, who claims that all we know is the structure of the phenomena but leaves open that we might know too the structure of their (perhaps unobservable) causes, Psillos notes that this marks an improvement over structural empiricism "only if it is accepted that the world has already built into it a natural structure .... This is a non-structural principle" (Psillos, 2006, p. 9). Yet, as we have seen, Psillos himself holds that the world has a "natural structure," only for Psillos it is a causal-nomological structure. In any case, according to Psillos, both the structural empiricist and epistemic structuralist positions fail because

> given that we talk about *in re* structures, there are objects that 'fill' the structures, these objects have properties over and above those that are determined by their interrelationships within the structure ... in any case, these *in re* structures are individuated by their non-structural properties since it's in virtue of these (non-structural) properties that they have causal unity and are distinguished from other *in re* structures. (Psillos, 2006, p. 7)

But what if this causal-nomological structure can be fully captured by structural properties? Again, by analogy, let's suppose an *in re* interpretation of the natural numbers. There are things (von Neumann ordinals or Zermelo numerals) that fill the structure, but *as objects* (as natural numbers) these things are only positions in any, or all, systems that have the appropriate structure. They might have non-structural properties (such as $2 \in 4$) but, as objects *qua* positions, they are not individuated by, nor do they bear the necessary relations they do (such as $4 > 2$) in virtue of any non-structural properties. It remains to be argued, then, that not only is it possible but it is necessary that physical objects, and so concrete *in re* structures, are individuated by their non-structural properties and too that it is precisely these properties that underpin causality. Psillos has argued that it *might be* the case that non-structural properties and, therefore, the objects *qua* individuals that support them, are needed to get at causes, but he has not demonstrated that it *must be* the case. Thus, it might be the case, as (French, 2006) suggests it is,[30] that to tell a causal-structuralist story about what cuts the world into its "natural structure" we do not require objects *qua* individuals.

---

[30] French, for example, uses his ontic structural realism to respond, claiming that "there should be no physical properties that cannot be captured in structural terms, since any such property worth its salt, as it were, would feature in the relevant causal-nomological relations and would thus be incorporated into the structural description" (French, 2006, p. 8). Moreover, in reply to Chakravartty, he has argued that, to best explain why it is that whatever conception of causality we presume (productive or Humean), we find properties in packages, we do not need objects *qua* individuals. Rather, something like the "bundle view" of objects can be adopted as a metaphysical frame for understanding his "nodes" view of objects. He has further explained how the "nodes" view might mark an improvement over the "bundle" view: "the structuralist could simply avail herself of a structural analog to compresence which similarly 'ties together' aspects of different structures. Indeed, it would be some such principle that metaphysically, as it were, gives rise to the 'nodes' in the world structure that we identify as electron, quarks etc." (p. 17).

Indeed, French does have an answer to Psillos' question of how, in science, we can speak of structure without objects (*qua* individuals), but it is an answer that arises neither from his set-theoretic framework nor from his ontic structural realism. (French, 2000) discusses two "contexts of applicability" of group theory in development of quantum mechanics: one relating to foundations and the other to the representation of physical phenomena. To exemplify both, he considers the role of group theory for both the "Weyl programme" (which was concerned with the group-theoretic elucidation of the foundations of quantum mechanics) and the "Wigner programme" (which was concerned with the utilization of group theory in the application of quantum mechanics itself to physical phenomena). Of the Wigner programme, French notes:

> Wigner himself emphasized the dual role played by group theory in physics; the establishment of laws—that is, fundamental symmetry principles—which the laws of nature have to obey; and the development of "approximate" applications which allowed physicists to obtain results that were difficult or impossible to obtain by other means. (French, 2000, p. 107)

French then details the history of Wigner's programme as motivated by the search for the mathematics that would represent the needed symmetries—permutation and rotation—and which would further take account of spin. He next recalls a point he made about Weyl's programme, viz., that "behind these 'surface' relationships there may lie deeper, mathematical ones" (p. 109). One such deeper relation is the reciprocity between the permutation and linear groups that Weyl refers to as "the guiding principle" of his work and also as the "bridge" within group theory. It is in considering this "bridge" that French concludes:

> Thus with regard to the construction of the 'bridge' between the theoretical and the mathematical structures, represented by T and M', on the quantum mechanical side we have the reduction of the state space into irreducible subspace and on the group theoretical side we have the reduction of representation. It is here we have the (partial) isomorphism between (partial) structures, (weakly) embedding T into M' .... Interestingly, then, the construction of this bridge ... crucially depends on a further one within group theory itself—the bridge that Weyl identified between the representations of the symmetry and unitary groups as expressed in the reciprocity laws. (French, 1999a, pp. 198–99; see also French, 2000, pp. 109–110.)

Where, however, are French's partial structures/partial isomorphisms, and so his set-theoretic framework, doing any real work? It seems to me that there are two contexts that determine the appropriate kind of morphism and in each it is group-theoretic morphisms that do the work. The first context, of reasoning from the phenomena, is exemplified by Weyl's programme; it uses the "relevant symmetries" to "work up" from the concrete structure of the phenomena, via quantum mechanical principles and/or experimental results expressed as group-theoretic symmetries, to *present*[31] the abstract structure of the theory. The second context, of reasoning

[31] See (Brading and Landry, 2007) for a more detailed discussion of the distinction between presenting and representing, and for a consideration of what this distinction implies for both the structural realist and structural empiricist positions.

to the phenomena, is exemplified by Wigner's programme; it uses the "relevant symmetries" to "work down" from the abstract mathematical theory, via the group-theoretic formalism and corresponding "internal bridges," to *represent* the concrete structure of the phenomena. In either case, what does the *real work* is the group-theoretic morphisms that underwrite, in both the foundational and representational contexts, the "relevant symmetries," and so serve to tell us what the appropriate kind of structure is for both presenting the structure of quantum theory and for representing the structure of quantum phenomena.

In choosing to consider group theory to be "the appropriate language" for quantum mechanics, French believes that he is left to face a Psillos-style question:[32] How can we talk of a group if we have done away with the elements that are grouped? French's reply is as follows:

> [w]e begin with a conceptualization of the phenomena ... informed by a broadly classical metaphysics ... in terms of which the entities involved are categorized as individuals. That categorization is projected into the quantum domain, where it breaks down and the fracture with the classical understanding is driven by the introduction of group theory; the entities are classified via the permutation group which imposes perhaps the most basic division into 'natural kinds,' namely bosons and fermions. It is over this bridge that group theory is related to quantum mechanics as indicated above. (French, 1999a, p. 204)

Later, making his ontic structural realist conclusions more explicit, French notes that

> [t]he introduction of group theory into quantum mechanics provides a useful example, and one that has important ... implications. Metaphysical: the very basis of the applicability of group theory lies in the non-classical indistinguishablity of quantum particles so that their permutation can be treated as a symmetry of the system. Furthermore, this emphasis on symmetry and invariance subsequently led to a metaphysical characterization of elementary particles as, ontologically, nothing more than sets of invariances. Epistemologically: this latter characterization can be adduced as a further aspect of an ontological version of structural realism which claims not simply that all we can *know* about the world is its structure but all that there *is* about the world is this structure .... (French, 2000, p. 103)

But where is either French's set-theoretic framework or his ontic structural realism doing any scientific structuralist work? In French's story, clearly what "drives" and "imposes" our quantum-mechanical "natural kinds" is the shared Lie-group structure. So why do we need the additional assumptions that (i) such group-theoretic systems *are* set-theoretic types, and (ii) that quantum particles *are* nothing but sets of invariances? What do we gain from these additional assumptions and, perhaps more importantly, what do we lose?

What we lose is precisely an answer to Psillos' question; even if French could use the "ladder analogy" to kick away concrete objects/individuals/elements,[33] if groups

[32] See (Psillos, 1995, 2001) for exact details of his criticisms of any structural realist attempt to separate nature from structure, content from form, and relata from relations.

[33] See (French, 1999a), where he argues that objects *qua* individuals are only heuristic devices; they are to be used only to introduce structures and so can be "kicked away" once they have served this purpose. See also (Psillos, 2001), where he criticizes just this claim.

really are sets, and if sets require elements, then to represent quantum phenomena as group-theoretic sets of invariances we need relations among elements, so we need objects *qua* individuals. But if we drop both the set-theoretic framework and the ontic structural realism, we can answer Psillos as follows: as the example of the role of group theory in quantum mechanics shows, the determination of the appropriate kind of structure both for presenting the structure of a scientific theory and for representing the structure of the phenomena is fixed by those contexts in which we reason both to and from the structure of the phenomena. Thus, contra Psillos, the phenomena might be represented "purely structurally," i.e., might be represented by working "top-down" from the abstract structure of a mathematical theory, and so might not require objects/individuals/elements to drive/impose/cut the world into its "natural" (causal-nomological) kinds. But, contra French, this "naturality" is taken as a fact of our scientific method as employed in some context; it is neither a consequence of a formal set-theoretic reading of the structure of a scientific theory nor an ontic structural realist reading of the structure of "the world."[34]

In sum, *against* Psillos, mathematical structuralism does not require objects or structures as actually or possibly existing objects *qua* things or individuals that have independent identity conditions. And, even if scientific structuralism focuses on *in re* structures, we can work "top-down" from the structure of the theory to represent the structure of the phenomena, and so we need not begin with objects *qua* individuals. But, *against* the structural realist, this "cutting" of the world using structuralist scissors is a methodological fact about how we *present* the structure of a theory or *represent* the structure of the phenomena; it is not an epistemological fact about what we know or an ontological fact about what there is. Thus, the answer to Psillos is found both by recognizing that contexts determine what the appropriate kind of morphism is for representing the structure of the phenomena and, as Psillos himself suggests, by adopting a *methodological* approach[35] to scientific structuralism.

[34] As (French, 2006) suggests, we can use kinds of properties to cut "the world" into kinds of objects. For example, we can use group-theoretic properties to cut the phenomena into quantum-mechanical kinds of objects. That is, "[t]he kinds of properties that feature in [the reconceptualization of objects] will be those group-theoretic invariants described in terms of the relevant symmetry principle (see for example Castellani, 1993). Thus proceeding down the 'natural' kind structure, a particle will be understood as a fermion, say, in terms of the relevant (anti-symmetric) representation of the permutation group ... and as an electron in terms of the properties of mass and spin associated with the relevant irreducible representation of the Poincaré group and so on. It is in such terms that structuralism secures objectivity and these symmetry principles have been viewed as 'higher rules and principles' imposed on the laws. What this gives us is a multi-layered, or multi-aspected, kind of structures involving 'webs or relations'—as represented by the relevant laws, etc.—tied together, as it were, by higher order symmetry principles representing the invariant in terms of which the 'nodes' in this structure can be described" (French, 2006, pp. 5–6). What French does not mention, however, is that while Castellani's structuralist account is used to account for quantum-mechanical objects it is not used to cut "nature" at its joints and so it brings with it no ontic structural realist aim of reconceptualizing objects as mere "nodes."

[35] See (Brading and Landry, 2007) for just such an approach, which we call "minimal scientific structuralism."

Bibliography

Achinstein, P. (1968). *Concepts of Science: A Philosophical Analysis*. Johns Hopkins University Press, Baltimore MD.

Bell, J. L. (1981). Category theory and the foundations of mathematics. *British Journal for the Philosophy of Science*, 32:349–358.

Bell, J. L. (2006). Abstract and variable sets in category theory. In *What is Category Theory*, pages 9–16. Polimetrica Monza.

Benacerraf, P. (1991). What numbers could not be. In Benacerraf, P. and Putnam, H., editors, *Philosophy of Mathematics*, pages 272–294. Cambridge University Press, New York, NY, 2nd edition. Originally published 1965.

Brading, K. and Landry, E. (2007). Scientific structuralism: Presentation and representation. *Philosophy of Science*, 73:571–581.

Cartwright, N., Shomar, T., and Suárez, M. (1995). The toolbox of science. In Herfel, W., Krajewski, W., Niiniluoto, I., and Wójcicki, R., editors, *Theories and Models of Scientific Progress*, pages 137–149. Rodopoi, Amsterdam.

Castellani, E. (1993). Quantum mechanics, objects and objectivity. In Garola, C. and Rossi, A., editors, *The Foundations of Quantum Mechanics — Historical Analysis and Open Questions*, pages 105–114. Kluwer, Dordrecht.

da Costa, N. C. A. and French, S. (1990). The model-theoretic approach in the philosophy of science. *Philosophy of Science*, 57:248–265.

da Costa, N. C. A., Bueno, O., and French, S. (1997). Suppes predicates for space-time. *Synthese*, 112:271–279.

Dummett, M. (1991). *Frege and Other Philosophers*. Oxford University Press, Oxford.

Feferman, S. (1977). Categorical foundations and foundations of category theory. In Butts, R. and Hintikka, J., editors, *Foundations of Mathematics and Computability Theory*. Reidel, Dordrecht.

French, S. (1999a). *Models and Mathematics in Physics: The Role of Group Theory*. Cambridge University Press, Cambridge.

French, S. (1999b). Theories, models and structures: Thirty years on. *Philosophy of Science, Supplement: Proceedings of PSA 1998*.

French, S. (2000). The reasonable effectiveness of mathematics: Partial structures and the application of group theory to physics. *Synthese*, 125:103–120.

French, S. (2006). Structure as a weapon of the realist. *Proceedings of the Aristotelian Society*, 106(2):1–19.

Hale, B. (1996). Structuralism's unpaid epistemological debts. *Philosophia Mathematica*, 4: 124–143.

Hellman, G. (1996). Structuralism without structures. *Philosophia Mathematica*, 4:100–123.

Hellman, G. (2001). Three varieties of mathematical structuralism. *Philosophia Mathematica*, 9:184–211.

Hellman, G. (2003). Does category theory provide a framework for mathematical structuralism? *Philosophia Mathematica*, 11:129–157.

Hesse, M. (1963). *Models and Analogies in Science*. Oxford University Press, Oxford.

Ladyman, J. (1998). What is structural realism? *Studies in the History and Philosophy of Science*, 29:409–424.

Landry, E. (1999). Category theory: The language of mathematics. *Philosophy of Science*, 66: S14–S27.

Landry, E. (2006). *Category Theory as a Framework for an* In Re *Interpretation of Mathematical Structuralism*. Kluwer, Dordrecht.

Landry, E. (2007). Shared structure need not be shared set-structure. *Synthese*, 158:1–17.

Landry, E. and Marquis, J. P. (2005). Categories in context: Historical, foundational and philosophical. *Philosophia Mathematica*, 13:1–43.

Mayberry, J. (1994). What is required of a foundation for mathematics? *Philosophia Mathematica*, 2:16–35.

Parsons, C. (1990). The structuralist view of mathematical objects. *Synthese*, 84:303–346.

Psillos, S. (1995). Is structural realism the best of both worlds? *Dialectica*, 49:15–46.

Psillos, S. (2001). Is structural realism possible? *Philosophy of Science, Supplement: Proceedings of PSA 2000*, pages 13–24.

Psillos, S. (2006). The structure, the whole structure and nothing but the structure? *Philosophy of Science*, 73:560–570.

Quine, W. (1948). On what there is. *Review of Metaphysics*, 2:21–38.

Redhead, M. (1975). Symmetry in inter-theory relations. *Synthese*, 32:77–112.

Redhead, M. (1980). Models in physics. *British Journal for the Philosophy of Science*, 31: 145–163.

Redhead, M. (1995). *From Physics to Metaphysics*. Cambridge University Press, Cambridge.

Shapiro, S. (1997). *Philosophy of Mathematics: Structure and Ontology*. Oxford University Press, Oxford.

Suárez, M. (2003). Scientific representation: Against similarity and isomorphism. *International Studies in the Philosophy of Science*, 17:225–244.

Suárez, M. (2006). An inferential conception of scientific representation. *Philosophy of Science, Supplement: Proceedings of PSA 2004*.

Suppe, F. (1977). Alternatives to the received view. In Suppe, F., editor, *The Structure of Scientific Theories*, pages 119–232. University of Illinois Press, Chicago IL.

Suppes, P. (1962). Models of data. In Nagel, E., Suppes, P., and Tarski, A., editors, *Logic, Methodology and Philosophy of Science*, pages 252–261. Stanford University Press, Stanford.

Suppes, P. (1967). Set theoretical structures in science. Technical report, Stanford. Mimeograph.

Chapter 19
The Vacuum in Antiquity and in Modern Physics

Michael Redhead

In antiquity there were two diametrically opposed views about the vacuum, or the void as it was often called.

The first view, associated with the ancient atomists, was happy to embrace the conceptual possibility of the void, and indeed to assert that it actually "existed," allowing room, so to speak, for identifying change in the physical world with the changing configuration of the (unchanging) atoms within the void. There was some confusion here as to whether the void existed only *between* the atoms or actually penetrated *inside* the atoms. This problem was not properly cleared up until the time of John Philoponus (sixth century AD), who was the first person to draw a clear distinction between space and a body occupying the space.

The second view, essentially the majority view, was that the void was a paradoxical, even contradictory, concept. The basic thought was that if the void was identified with "nothing" then it could not exist, and conversely if it did exist it could not be nothing! If this Parmenidean style of argument did not convince, then Aristotle could argue that if the void existed then motion through it would be impossible since a material medium would be necessary to sustain motion. But equally, if one did not accept that argument, then motion would be infinitely fast, since there would be no resistance to the motion, and actual infinities were, of course, rejected by Aristotle. (Aristotle himself seems to have achieved the remarkable intellectual feat of maintaining both arguments simultaneously!)

With the scientific revolution in the seventeenth century AD, the arguments continued. Descartes, for example, believed that extension was a necessary concomitant of matter (*res extensa*), so if you tried to remove all the matter from inside a flask, you could not possibly succeed, because you would end up with the flask containing an empty volume, which Descartes regarded as a *reductio ad absurdum*. On the other hand, Pierre Gassendi revived the views of the ancient atomists, and people like Robert Boyle and, most famously, Isaac Newton entertained the corpuscular

M. Redhead (✉)
Centennial Professor, London School of Economics, London, UK; Emeritus Professor of History and Philosophy of Science, Cambridge University, Cambridge, UK; 119 Rivermead Court, Ranelagh Gardens, London SW6 3SD, UK
e-mail: mlr1000@cam.ac.uk

D. DeVidi et al. (eds.), *Logic, Mathematics, Philosophy: Vintage Enthusiasms*, The Western Ontario Series in Philosophy of Science 75,
DOI 10.1007/978-94-007-0214-1_19, © Springer Science+Business Media B.V. 2011

hypothesis concerning the ultimate nature of matter. For Newton, in particular, the void in the guise of absolute space was given a theological interpretation as the sensorium of God. But many people objected to Newton's account of gravitation, and other influences such as electric and magnetic effects, as acting at a distance across a vacuum. In the famous queries appended to his *Opticks*, Newton speculated about a subtle medium, the aether, as ultimately responsible for transmitting these influences. But the aether itself might be corpuscular (which would just reset the problem of how action could be transmitted) so, the alternative view was of "effluvial" theories modelled on Newton's own conception of a corpuscular theory of light.

It was the rise of the wave theory of light at the beginning of the nineteenth century that led back to the view that truly empty space did not exist. If you could empty a flask of aether it would become opaque to light and no such effect had ever been observed, even with the best available vacuum pumps! Somehow the aether was so subtle it could just flow back unimpeded through the walls of the flask.

But then came the famous negative result of the Michelson-Morley experiment (1887) designed to measure the velocity of the earth through the aether, and with Einstein's special relativity theory (1905) the rejection of the whole concept of a material aether. So were we back to the void of antiquity? There were two reasons why this was not so.

1. In Einstein's general relativity theory (1915) space (or more accurately space-time) became a dynamical object in its own right, acting (gravitationally) on matter and being reacted on by matter. So the void was causally efficacious, a totally different view from the unchanging, featureless backdrop against which events unfold contemplated with Newtonian absolute space.
2. But the new quantum theory, introduced by Max Planck in 1900, was to lead to an even more dramatic revision of the physicist's conception of the vacuum. To understand this there are two basic ideas that we need to get across:

   (a) *Wave-particle duality*: Material particles like electrons are to be thought of as possessing a wave-like aspect as well as a particle-like aspect.

   Perhaps the best way of thinking of this is to think of an electron as a "discrete" excitation of a continuous matter field spread throughout space. The idea of a *discrete* excitation is where the quantum mechanics comes in; the excitations are "quantized" as one says, so you can't have a half electron or a quarter electron, but only a whole electron as a possible excitation of the field. It is the same with all the other particles like protons and neutrons (now thought to be made up of still more fundamental particles, the quarks). And also the fields of force between the material particles, like the electromagnetic field, which, reciprocally, have also a particle aspect, so the interactions can be given effectively the old-fashioned effluvial interpretation, i.e., as mediated by streams of particles (in the case electromagnetism these are called photons). But remember these "particles" have also a wave aspect, allowing for interference and diffraction effects, so there is also a continuum (i.e., plenum) way of talking about the interactions!

If you find all this confusing, don't worry, famous physicists have wrestled with these novel ideas for the past 70 years or so, and although the empirical predictions are wonderfully vindicated, there is still a deal of argument about what the equations mean!

(b) *The uncertainty principle*: Consider a pendulum in a grandfather clock. If you set the bob in motion it acquires kinetic energy, but as it swings up to the endpoint of its arc the kinetic energy has all been converted into potential energy, and then as it swings down the potential energy gets converted again into energy of motion, i.e., kinetic energy. But in the absence of friction the total energy remains constant—the famous law of the conservation of energy.

Suppose you now try to bring the pendulum to rest, with the bob hanging vertically, so classically it would have zero energy.

According to quantum theory *this is not possible to do*. If you try to reduce the potential energy by moving the bob to the vertical position, you inevitably introduce "fluctuations" in the speed of the bob so the kinetic energy goes up. On the other hand, if you try to bring the bob to rest so that the kinetic energy is reduced, "fluctuations" will occur in the position of the bob, so increasing its potential energy. This is an example of the celebrated Heisenberg uncertainty principle, which says roughly that doing one sort of thing to a physical system prevents you at the same time doing other sorts of thing.

In the case of the pendulum there is effectively a trade-off between reducing the two sorts of energy, so that the minimum total energy of the bob is not zero, as we would expect classically, but is given by the famous formula $^1\!/\!_2\, h\, f$, where f is the frequency of the pendulum, the number of complete oscillations it makes in 1 s, and h is known as Planck's constant. Now h is very small indeed—e.g. with a pendulum making one oscillation per second, so $f = 1$, the minimum, often called the zero-point energy, comes out at approximately 10^{-34} J. For comparison the typical energy of the pendulum in a grandfather clock is about $1/10$ J, so it is not surprising that for macroscopic objects like a pendulum in a grandfather clock, the zero-point energy can safely be ignored!

But when we are dealing with atomic particles like electrons, the zero-point fluctuations become very important. From the field point of view we can say that the oscillations of the field can never be brought entirely to rest. The vacuum in quantum theory is defined as the state of lowest energy for the field. Classically this would be the state where the field was not oscillating at all. But quantum-mechanically there is no such state. We are always going to be left with the zero-point energy.

Let us look at the situation from the particle point of view. The vacuum fluctuations in the field and the associated zero-point energy can now be described in terms of the creation and annihilation of so-called "virtual" particles. These are literally created out of nothing—well it would be better to say "out of the vacuum of the quantum field." But according to Einstein a particle of mass m carries energy

mc^2, where c is the velocity of light. So where has the energy come from to create the virtual particle? The answer lies again in another application of the uncertainty principle. In the quantum theory energy need not be conserved in creating a particle of mass m, so long as it is "paid back" over a time not longer than h/mc. To conform to this principle virtual electrons, for example, have to annihilate, i.e., "disappear" (back into the vacuum) in a time which comes out numerically at about 10^{-21} s. But if the electron is moving very fast, near to the speed of light, then it can cover a distance of about 10^{-11} cm during its lifetime, and this is detectable since it lies between the scale of atomic dimensions (10^{-8} cm) and nuclear dimensions (10^{-12} cm). So virtual particles can produce all kinds of important effects in atomic and nuclear physics. Theoretical predictions of these effects are in amazingly close agreement with experimental data. Indeed the predictions have been verified up to ten significant figures in the most favourable case (the anomalous magnetic moment of the electron), which stands as one of the most outstanding triumphs of the theory of quantum physics.

If we did the same calculations for virtual billiard balls, the lifetimes come out at around 10^{-48} s, far beyond the scope of any possible detection, so the possibility of virtual billiards can definitely be ignored in everyday life!

So, back to the problem of trying to empty a flask. We can, in principle take out all the "real" particles, but we are always going to be left with the seething activity of the virtual particles which we cannot get rid of, unless, *per impossibile*, we pumped out the fields themselves. But, remembering that the gravitational field, and possibly other fields as well, are part of the geometry of space, this would mean pumping space out of space, so to speak, and that sounds like a conundrum for the ancients. Which is where we came in ....

Acknowledgment I am grateful to Richard Sorabji for helpful advice, and Gresham College where the paper was originally presented.

Chapter 20
Falsifiability, Empirical Content and the Duhem-Quine Problem

Elie Zahar

1 Historical Background

Having been struck by the steady growth of mathematical knowledge, Duhem wondered whether a similar cumulative pattern had also been achieved in the development of physical science. In this respect, his conclusions were largely negative: as long as explanatory science relies on metaphysics and as long as the latter is and must remain unstable, progress in physics cannot possibly be viewed as cumulative. But provided physical theories be restricted to their purely representative parts, then thanks to the Correspondence Principle, the mathematical structure of scientific systems can be seen to evolve continuously: the old equations, though strictly incompatible with the new ones, constitute limiting cases of the latter (Duhem, 1954, Part 1, Ch. 3).

Let us examine more closely Duhem's negative conclusions regarding the envisaged parallel between mathematics and physics. Duhem took mathematics to consist of *synthetic* theories whose certainty flows from two facts. First, all mathematical axioms, e.g. those of arithmetic and of Euclidean geometry, turn out to be very simple and therefore afford a direct and infallible insight into their intended domains. Secondly, the rules of inference used by mathematicians are all deductive, hence infallibly transmit truth from the premises to conclusions. It follows that mathematical theorems are established once and for all. Old mathematical truths are never revised, they are simply added to (Duhem, 1954, Part 1, Ch. 1, § 3).

This way of viewing mathematics was questioned first towards the end of the nineteenth, and then during most of the twentieth century. Most logical empiricists took mathematical propositions to be logically true and hence vacuous. Since mathematical truths were considered empty, there could no longer be any question of a genuine progress in mathematical *knowledge*, but at best of an increasing *psychological* awareness of the tautologous character of certain statements. This view now lies in ruins; for the most fundamental mathematical system, namely set theory,

---

E. Zahar (✉)
Reader Emeritus in Logic and Scientific Method, London School of Economics, 21 City Road, Cambridge, CB1 1DP, UK

D. DeVidi et al. (eds.), *Logic, Mathematics, Philosophy: Vintage Enthusiasms*, The Western Ontario Series in Philosophy of Science 75,
DOI 10.1007/978-94-007-0214-1_20,

makes essential use of axioms like those of choice and of the power set which are known to be synthetic. (We shall see that this logical fact strongly supports Quine's approach to the Duhem-Quine dilemma. See below, Section 6). As a result, Duhem's assessment of the status of mathematics seems nowadays less wrong-headed than it did 50 years ago: the certainty of mathematics appears to be due, on the one hand, to the deductive nature of its inference rules, and on the other, to the alleged perspicuity of axioms which articulate the meanings of some basic concepts like those of "class" and of "belonging to a class" ($\in$). Mathematics is therefore synthetic, so the difference in certainty between its propositions and those of physics is one of degree rather than kind; but where does this difference really lie? The Duhemian answer has implicitly been given above: physical hypotheses are not only synthetic but also complex and far from self-evidently true; so they provide us with no direct insight into their intended domain. Moreover, their rules of inference are not all deductive; for unlike mathematics, empirical theories must in some sense be based on observation. The question is: *in exactly what sense*?

According to Duhem: despite their empirical character, the methods of induction and of crucial experimentation were—wrongly—supposed to parallel the mathematical methods of direct proof and of "reductio ad absurdum"(Duhem, 1954, Part 2, Ch. 6 §§ 3,6).

2 Induction and Direct Proof

Let us now address a highly controversial question, namely that of the allegedly heuristic role of induction in the construction of scientific laws. In a direct mathematical proof, we begin by positing a set of axioms from which we then derive, step by step, a sequence of theorems. As for the inductive process, it supposedly starts from a set of indubitable factual statements from which a general hypothesis is then inferred. Duhem showed this method to be invalid on at least two counts. First, unlike their common-sense counterpart, the empirical results on which induction rests are "symbolic" and theory-laden, hence fallible. For example, in the absence of an electrical theory, a statement like "the current is on" would be meaningless and hence devoid of truth-value (Duhem, 1954, Part 2, Ch. 6). Thus we have to face the so-called vertical transcendence of all factual scientific propositions, as distinct from common-sense statements like "there is a white horse in the street"; for according to Duhem, the latter can be both realistically interpreted and infallibly known to be true. Note that Duhem's fallibility thesis concerning all *objective* empirical statements has a lot to commend it, even though his view about the incorrigibility of common-sense propositions be highly dubious. There is secondly the well-known Humean or horizontal transcendence of any universal law with respect to any of its instances.

As is well-known, Descartes made a clear distinction between *autopsychological* sentences like "It seems to me that I hear a noise" and any *transcendent* proposition which refers to some objective "external" reality (Descartes, 1986, Meditation 2). For example, corresponding to the autopsychological report just mentioned, we have the objective statement that there is—in physical reality—a sound-wave.

Descartes acknowledged that only autopsychological propositions are indubitable. Going back to Duhem, let p and p' be defined as follows: $p \equiv$ (the current is on), $p' \equiv$ (I see—what I take to be—the pointer of the galvanometer move). Thus p', but not p, is or might be incorrigible. Now note that in inducing laws from basic statements, we need to start from singular propositions having the form, not of p', but of p. For example: I need to know that iron, and copper, and aluminum, etc., conduct electricity, not that *it seems to me* that they do so. Duhem was therefore right about physical induction being neither certain nor theory-independent; or rather about its being fallible precisely because it is theory-dependent; for even the singular basic statements on which it relies in order to arrive at generalisations are dubitable.

We have already mentioned that experimental laws like "All metals conduct electricity" are fallible for a second and more prosaic reason: since they proceed from a finite to a potentially infinite sample, they run the risk of being refuted by the next recalcitrant observation. So induction seems to lead from theory-dependent to other fallible statements. A very different conclusion, namely a hypothetico-deductivist one, can however be drawn from this state-of-affairs. We can give up the unacceptable idea of directly inducing laws from facts and look upon an autopsychological proposition like p' as the only allowable type of basic statement. There is nothing to prevent us from deducing such level-0 reports from high-level theories *taken in conjunction both with boundary conditions and with psycho-physical assumptions*. We shall see that this approach increases the complexity of the Duhem-Quine problem; but in return, indubitable descriptions of sense-experience would constitute what Quine himself regarded, at least initially, as the fixed periphery of our system of knowledge (Quine, 1980). Anyway, having provisionally dealt with induction as a heuristic tool, let us now turn to what Duhem calls the indirect method of proof.

3 Indirect Method and Crucial Experiments

Duhem maintained that:

> Those who assimilate experimental contradiction to reduction to absurdity imagine that in physics we may use a line of argument similar to the one Euclid employed so frequently in geometry. Do you wish to obtain from a group of phenomena a theoretically certain and indisputable explanation? Enumerate all the hypotheses that can be made to account for this group of phenomena; then, by experimental contradiction, eliminate all except one; the latter will no longer be a hypothesis, but will become a certainty ...; but the physicist is never sure he has exhausted all the imaginable assumptions. (Duhem, 1954, Part 2,Ch. 6)

I shall use Popper's demarcation criterion to support Duhem's intuitive argument. Let us recall that a proposition is to be considered scientific if it is universal, non-verifiable and empirically refutable. In the quotation above, Duhem maintains—without any real justification—that no disjunction $H_1 \vee H_2 \vee \cdots \vee H_n$ of scientific theories can be known to be true; he simply invokes the obvious psychological fact that the physicist can never be certain of having exhausted the set of all possible hypotheses. Using Popper's criterion, it can however easily be shown that any disjunction like $H_1 \vee H_2 \vee \cdots \vee H_n$ is scientific and hence unverifiable—in the objective sense.

Without loss of generality, let us restrict ourselves to the case where $n = 2$. Being scientific and hence universal statements, H_1 and H_2 can be written in the form:

(i) $H_1 \equiv (\forall x)B_1(x)$ and $H_2 \equiv (\forall y)B_2(y)$.

Since the two variables x and y can always be chosen to be distinct, we have:

(ii) $\vdash [(H_1 \vee H_2) \leftrightarrow (\forall x)(\forall y)(B_1(x) \vee B_2(y))]$;

i.e., $H_1 \vee H_2$ is universal.

If we suppose that H_1 is consistent, then so is the weaker proposition $H_1 \vee H_2$. In order to prove that $H_1 \vee H_2$ is empirically refutable, let a_1 and a_2 be potential falsifiers of H_1 and H_2 respectively. That is:

$\vdash a_1 \rightarrow \neg H_1, a_2 \rightarrow \neg H_2$. Therefore,

(iii) $[(a_1 \; \& \; a_2) \rightarrow \neg(H_1 \vee H_2)]$.

It can generally be assumed that a_1 and a_2 are logically compatible sentences: given the universality of both H_1 and H_2 with respect to the time parameter, a_1 and a_2 can be taken to describe events occurring at different moments; so that $a_1 \; \& \; a_2$ is a consistent and empirically decidable statement. By (iii), $(H_1 \vee H_2)$ is therefore experimentally falsifiable. $(H_1 \vee H_2)$ is finally unverifiable; for any empirical result verifying $(H_1 \vee H_2)$ would have to verify either H_1 or H_2 or both; which, given the scientific character of both H_1 and H_2, is impossible. Thus the scientist is never in a position to know that a disjunction of empirical theories must be true; and this independently of his powers of imagination.

4 Indirect Method and the Duhem-Quine Problem

Let us turn to the problem posed by the refutability of scientific theories. It is obvious that the truth of any scientific theory or of that of any disjunction $H_1 \vee \cdots \vee H_n$ of physical hypotheses must always remain uncertain; but could we not at least be certain about an isolated theory H_i being definitively falsified by experience? Duhem rightly drew our attention to the fact that a falsifying experiment undermines, not an isolated theory but a whole system including the theory in question. Quine reinforced this ungainsayable Duhemian thesis by claiming—for reasons which will be examined below—that a falsifying experiment challenges the whole of science (Quine, 1980, § 5).

Note that even in the absence of Quine's globalist thesis, serious problems confront the falsificationist viewpoint. Lakatos showed that certain boundary conditions, e.g., the assumption that only the gravitational field acts within a given region, do not have the character of singular statements but that of universal and hence unverifiable hypotheses. This objection however constitutes no damaging criticism of Popper's falsificationist criterion; for we can consider all unverifiable boundary conditions as part of the system undergoing the test. This would complicate the Duhem-Quine problem, without however essentially changing its nature.

Lakatos also defended the view that a theoretical system S normally involves a *ceteris paribus* clause, i.e., the caveat: other things being equal (Lakatos, 1970). This situation can be rendered more precisely as follows. We have seen that the hypothetico-deductive model of explanation can be described by means of the scheme: $\vdash (S \rightarrow E)$, where: $E \equiv (B \rightarrow P)$, S is a theory, B the description of boundary conditions and P a prediction. According to Lakatos, S consists of some core hypothesis H taken together with an indefinite set Δ of propositions of the form: on some specific occasion, only gravitational forces acted on the test body, there were no random variations in the weather conditions, etc. Thus, we have: $H, \Delta \vdash E$. Were E to be falsified, then we should allegedly not know whether to blame H or any member of the possibly infinite set Δ. This objection to falsificationism need not however be taken too seriously. As long as we carry out our deductions in a first-order language, then by the compactness theorem, E will logically follow from some finite subset of $\{H\} \cup \Delta$. Hence $H \,\&\, A \vdash E$ will hold, where A is a conjunction of finitely many elements of Δ. To the extent that $\neg E$ can be observationally verified, $(H \,\&\, A)$ can be said to have been refuted. This might admittedly leave us with an aggravated Duhem-Quine problem; for we cannot a priori decide whether H or any conjunct in A is false. All the same, the single finite proposition $S \equiv (H \,\&\, A)$ will have been experimentally refuted.

The Duhem-Quine problem can furthermore be partially solved as follows. Let $H \,\&\, A$ be an empirically falsified conjunction. If successive variants $A_1, A_2, \ldots, A_n$ of A lead to the refutations of $H \,\&\, A_1$, $H \,\&\, A_2, \ldots$ and $H \,\&\, A_n$, then according to both Duhem and Popper, it can reasonably be conjectured that the fault lies with H. Popper did not however admit that such reasonableness rests—as it clearly does—on the following intuitive probabilistic argument: if, despite all the negative outcomes just mentioned, we decide to adhere to H, then each of $A, A_1, \ldots, A_n$ must be considered false; which yields the unique assignment $(t, f, \ldots, f)$ of truth-values to $(H, A, A_1, \ldots, A_n)$. But should we be prepared to give up H, then each of $A, A_1, \ldots, A_n$ could, for all we know, be either t or f; which yields 2^{n+1} assignments compatible with all the experimental results. Then, provided $H, A, A_1, \ldots, A_n$ be mutually independent, the chances are that H is false. $A, A_1, \ldots, A_n$ may of course share a core G such that $H \,\&\, G$ is testable and G actually false; which might account for the successive refutations of $H \,\&\, A, \ldots, H \,\&\, A_n$. This is why a caveat of mutual independence has to be entered. Be it as it may, the above—admittedly crude—piece of probabilistic reasoning provides a rationale for our feeling that, barring miracles, H must be the culprit.

Despite this—tentative—solution which is aimed at reducing the scope of the Duhem-Quine problem, many questions concerning the conjunction $H \,\&\, A$ remain unanswered. In Duhem's time the validity of Logic, which was usually identified with the Aristotelian syllogism, was not called into question; and its schemes almost never explicitly figured among the premises yielding empirical predictions. Logic has since been both enriched and formalised; so that A might well be a logical principle, or more generally an analytic proposition. In such a case, we should normally feel entitled to regard A as immune to refutation. The Duhem-Quine problem would then automatically reduce to the question of the falsifiability of H, where the latter might also turn out to be a conjunction. Quine however rejects the analytic-synthetic distinction. It would thus follow that A cannot be hived off even though A might normally be regarded as analytic. Thus H & A would still have to face the verdict of experience *en bloc*. Quine claims that even mathematics and logic can be revised in the face of recalcitrant empirical evidence; with the result that the Duhem-Quine problem can no longer be localised. This is why we ought to examine the possibility of demarcating between, on the one hand, all the analytic statements including the principles of logic and mathematics and, on the other, the domain of synthetic propositions including all scientific hypotheses.

5 Kripke's Challenge

Empiricist methodology faces two threats coming from opposite directions. First: Quine regards no component of our scientific knowledge as being immune to revision. Hence observational anomalies could conceivably call into question any of our principles, whether these be empirical, mathematical or even logical. This Quinean view which, if correct, would render the Duhem-Quine problem unmanageable, will be examined in the next section. There is secondly the threat posed by Kripke's thesis that there exist statements which are both necessary and a posteriori, i.e., whose necessary truth could be empirically established. "Hesperus is Phosphorous," "Cicero is Tully" and "Heat is a motion of particles" are supposed to be examples of such propositions. Thus, in respect of any testable conjunction of premises H_1 & ... & H_n, one can no longer infer from the a posteriori character of some H_i that H_i can in principle be rejected in the light of evidence undermining H_1 & ... & H_n; for despite being a posteriori, H_i might, unbeknownst to us, prove metaphysically necessary (Kripke, 1972). Contrary to Quine's proposal, this Kripkean thesis dramatically, but also mysteriously and unpredictably, restricts the scope of the Duhem-Quine problem. It moreover flies in the face of normal scientific practice. Showing that Kripke's conclusions are at best irrelevant to science, more particularly to the Duhem-Quine problem, is therefore an important task.

Let us start by granting that the truth of some analytic, and hence necessary statements may come to be known a posteriori. Without the use of a pocket calculator, I might have been unaware that $2^9 = 512$; but this in no way implies that the validity of this equation is contingent. Needless to say, Kripke's thesis goes well beyond this truism. So let us consider identities between singular terms like "Hesperus = Phosphorous." In this particular case, Kripke concedes that "Hesperus" and "Phosphorous" could be regarded as definite descriptions. We would thus be dealing with an identity of the form: $[\iota x H(x) = \iota y P(y)]$, where: $H(x) \equiv$ (x is the Evening Star); $P(y) \equiv$ (y is the Morning Star); so that $H(x)$ and $P(y)$ have different Fregean senses. After carrying out a Russellian analysis, the above identity proves equivalent to the sentence $[(\exists x)(\forall z)(H(z) \leftrightarrow (z = x))$ & $(\forall y)(P(y) \leftrightarrow H(y))]$, which is clearly contingent.

We are however told that "Hesperus" and "Phosphorous" could also be regarded as proper names; so that "Hesperus=Phosphorous" assumes the form $[a = b]$, where "a" and "b" are now distinct primitive individual names. Kripke maintains that under such a construal, "$[a = b]$" must, if true, be necessarily so. But this *prima facie* startling claim turns out to be a trivial consequence of a stipulation, namely that every proper name "a" is to be regarded as a rigid designator, i.e., that if "a" has a referent, then the latter will be the same in all (metaphysically?) possible worlds. More precisely: let "a" and "b" be two distinct primitive symbols, e.g., let "a"="Cicero" and "b"="Tully." J.S. Mill reasonably maintained that a name might possess a referent but certainly not any (Fregean) sense. The above analysis in terms of definite descriptions such as $[\iota x H(x)]$ and $[\iota y P(y)]$ cannot therefore be applied to the identity "$[a = b]$." The latter will hold on one and only one condition, namely that "a" and "b" denote the same object. Assume this condition to be satisfied in our world. Kripke's astonishing conclusion is that "$[a = b]$" is then necessary in the sense of holding in all possible worlds. Such a bold claim is nonetheless trivial since it follows from an arbitrary decision.

> Possible worlds are *stipulated*, not *discovered* by powerful telescopes. There is no reason why we cannot *stipulate* that in talking about what would have happened to Nixon in a certain counterfactual situation, we are talking about what would have happened to *him*. (Kripke, 1972, p. 44)

Thus Kripke demands, by *fiat*, that "a" and "b" be treated as rigid designators; i.e., the class W of possible worlds is in effect defined in such a way that our world lies in W and each of "a" and "b" denotes the same entity in all members of W; from which it trivially follows that if "$[a = b]$" holds in our world, then it will be true throughout W. Furthermore, this definition of *metaphysical* possibility seems doctored to yield this rather uninteresting result.

Scientists have shown little interest in metaphysical, as opposed to logical or mathematical, necessity. As admitted by Kripke himself, "$[a = b]$" is not an analytic or tautological assertion. There obviously exist possible interpretations whose domains contain more than one element and where "a" and "b" denote different individuals, thus making $[a = b]$ into a false sentence. And though nothing stops a scientist from postulating $[a = b]$ where "a" and "b" are simple names, there is hardly any reason for him to do so; for he might just as well systematically substitute "a" for "b," thus greatly simplifying the presentation of his system.

There remains the question concerning the sense in which "$[a = b]$," as opposed to the identity "$[a = a]$," might have some informative content, no matter how minimal. From what has been said, it would seem to follow that "$a = b$" asserts no more than "$a = a$," i.e., than the statement that a is identical with itself; which is highly counter-intuitive since we have the unerring conviction that "$a = a$" is totally devoid of content. This is however only because, when reading a text, we do not remain within the bounds of the object-language but are also meta-linguistically aware of the latter's syntax. Thus the sequence of *symbols* "$a = b$" tells us that the same entity is (contingently) denoted by the two distinct names "a" and "b"; while "$a = a$" does nothing but remind us of a convention adopted once and for all, namely that independently of context, the same primitive sign always refers to the same object. This is one reason why Frege initially held "$a = b$" to be not only about the individuals denoted by "a" and "b," but also about the symbols "a" and "b" themselves. No matter how mistaken this approach might be, it still reveals an important psychological fact: our reading of a formula is directed not only at its referents, but also at the sequence of signs constituting the formula. As a proposition of the object-language, the identity "$a = b$" does not tell us that "a" $\neq$ "b" since it does not talk at all about "a" or "b"; but "$a = b$" exhibits the scriptural difference between "a" and "b," thus showing us that "$a = b$" could prove false under certain interpretations and must therefore be synthetic. Once again, a Kripkean analysis turns out to be superfluous. The air of necessity surrounding "$a = b$" flows from Kripke's misleading intuition that every equality "$a = b$" is either necessary or false.

Let us now turn to more serious examples which allegedly demonstrate the a posteriori necessity of some theoretical identities (or identifications):

> ... One might very well discover essence empirically (p. 110).... Philosophers have, as I've said, been very interested in statements expressing theoretical identifications; among them, that light is a stream of photons, that water is H_2O, that lightning is an electrical discharge, that gold is the element with the atomic number 79 (p. 116) ... So if this conclusion is right, it tends to show that such statements representing scientific discoveries about what this stuff is are not contingent truths but necessary truths in the strictest possible sense (p. 125). (Kripke, 1972)

Let us explain why, despite its seemingly scholastic and hence innocuous character, Kripke's thesis about the necessity of certain a posteriori propositions threatens the empirical testability of scientific theories. This can most clearly be seen by drawing a parallel between Kripke's position and Kant's views concerning the status of such synthetic a priori principles as the law of the conservation of substance, which turns out to be that of the conservation of matter. Kant admittedly regarded his principles as a priori propositions while the Kripkean identities are rightly held to be a posteriori and hence synthetic. Still, both "Matter is conserved" and "Light is a stream of photons" are respectively declared by Kant and by Kripke to be necessarily true (Kant, 1957, § III, Thm 2). Under no circumstances can such propositions be legitimately regarded as having been empirically refuted. Let us once again note that this conclusion runs counter to the intuition of working scientists. In the post-Einsteinian era, it is well-known that the conservation of matter—that is: should the latter be distinguishable from energy—can be rejected without rendering science impossible. As for Kripke's views, they have the following paradoxical consequence: given that light *is* a particulate stream of photons, then by proposing that light is a wave process, Huygens, Fresnel and Maxwell were—albeit unbeknownst to themselves—going against a "necessary truth[s] in the highest possible sense"; i.e., they were violating something akin to a logical principle; which also entails that Kripke forecloses the very possibility of accounting for optical processes in continuous field-theoretic terms.

Kripke might of course have meant that *only if true* would a theoretical identity be necessary. So it looks as though his theses can always be affirmed with total impunity; and this for two reasons. First: since the identities in question are universal synthetic propositions, they cannot be strictly verified, so none of them can definitively be claimed to be true. Secondly, even if they could somehow be ascertained as true, then they would by definition hold good in all possible worlds, for the latter are defined in such a way that if the identity "Water is H_2O" proves true in one of them then, thanks to rigid designation, it will automatically obtain in all of them. But Kripke is not out of the woods yet, for his account of essences is too geared to subject-predicate logic, too *monadic*, to be entirely satisfactory. He neglects relational properties. For example: it clearly follows from Kripke's theses that the elementary particles constituting H and O—i.e. the electrons, protons, etc.—must both belong to every possible world and bear to one another the same relations which hold in ours. In other words: to the extent that the electron possesses an essence, the latter must include the relations it bears to that of the proton and of other constituents of matter, as well as of all ambient fields; for what would the *essence* of the electron be in the absence of positively charged particles which attract it? Every possible world ought therefore to be vastly enriched; it must end up being governed by the same fundamental laws as ours, while differing from the latter only through its boundary conditions, where the form of these conditions might in turn be circumscribed by the laws. Thus Kripke's seemingly bold theory reduces to the classical picture adhered to by most working scientists, that is, essentially *one* world governed by a fixed system of basic laws but in which different sets of boundary conditions can be envisaged, the latter being chosen so as to be compatible with the laws. Moreover, the entailed generalisations which are sensitive to changes of boundary conditions can be considered "contingent," and the remaining ones can be dubbed "necessary." As for Kripke's identity statements, they ought to be treated as nominal definitions, i.e., as mere abbreviatory devices. For example: most of the stuff which had previously been labeled "water" on the basis of certain phenomenal properties was subsequently found to be composed of two atoms of hydrogen (H_2) and of one atom of oxygen (O). On the basis of such a composition and of atomic theory, of physiology and of certain boundary conditions, the previously ascertained phenomenal qualities of water can be accounted for, at least in principle. This might have led scientists to adopt a new nominal definition according to which "water" from then on stood for: substance consisting of H_2 and of O; i.e., "water" was treated as an abbreviatory device. This admittedly turns "Water is H_2O" into a necessary statement, albeit into a merely analytic one, while the connection between H_2O and the phenomenal properties of water, though predictable, remains synthetic. All this goes to show that the only notion of necessity needed in the sciences is the strictly logical one.

6 Quine's Holism and the Analytic-Synthetic Distinction

Rather than start by giving a definition of the terms "analytic" and "synthetic," let us consider an example given by Quine himself. We shall then analyse certain features of this concrete case in order to arrive at a general characterisation of analyticity. Intuitively, the example must of course be unambiguously either analytic or synthetic. Consider the proposition "No bachelor is married" which, according to Quine himself, is held to be quintessentially analytic, i.e., true by definition, or true exclusively by virtue of the meanings of its terms. For if one were to look up the meaning of "bachelor" in a dictionary, then one might well find: bachelor ≡ (male and unmarried). Through the replacement of "bachelor" by its verbal or dictionary definiens, the above statement is transformed into: "There exist no unmarried males who are married," which is logically true and hence true independently of all empirical states-of-affairs. So Quine rightly claimed that *if* there were any necessary propositions, then these would have to be analytic and hence escape all empirical control. We shall now examine his thesis that no such propositions exist.

Let us go back to his concrete example about bachelors being unmarried, while provisionally accepting the following stipulation: a true proposition is analytic if its truth depends exclusively on the meanings of its (non-logical) constituent terms. Thus, by means of the verbal definition of "bachelor":

α "No bachelor is married"

is reducible to:

β "No male unmarried person is married,"

i.e., to $\neg(\exists x)[P(x)\ \&\ \neg M(x)\ \&\ M(x)]$, where: $P(x) \equiv$ (*x* is a male person); $M(x) \equiv$ (*x* is married). This last formula is logically true, i.e., true independently of the meanings of its descriptive components P and M. Since logical principles are not only particular cases but also the most important instances of analytic truths, we appear to have reached a paradoxical conclusion, namely that $\neg(\exists x)[P(x)\ \&\ \neg M(x)\ \&\ M(x)]$ holds both independently of, and by virtue of the meanings of its terms. This "paradox" flows from the ambiguity of the notions of "meaning" and of "definition." The dictionary or verbal meaning of a symbol is the sequence of primitive signs of which the symbol constitutes an abbreviation. Note that only non-primitive symbols, i.e., those not belonging to the basic vocabulary of a language, possess verbal definitions. In going from [α] to [β], we replaced "bachelor" by its dictionary meaning, i.e., by "male and unmarried," thus obtaining the logically true sentence [β]. An expression however possesses another "meaning" or Fregean *Bedeutung*, namely a referent. This is the—normally extra-linguistic—entity denoted by the expression. It will henceforth be called its *ostensive meaning* since it is often, though not always, fixed by ostension. For example: tables, colours and even elementary particles can be pointed to with the finger, but though performing an objective function, a logical connective cannot be singled out by ostension; it can, at best, be partially determined by a number of postulates circumscribing its usage. A Platonist might maintain that even a logical constant refers to an entity inhabiting some World of Forms, which is far from being an absurd claim. None of my theses however rests on the assumption that abstract terms possess real referents, which is why I chose the agnostic expression "ostensive meaning" to subsume the referents of all names of physical entities as well as the objective meanings of logical constants and of mathematical terms. In what follows, all that matters is our ability clearly to distinguish between the ostensive meanings of words on the one hand, and the dictionary or verbal definitions of nonprimitive expressions on the other.

Quine seems initially to have carried out his critique at the verbal level. We have seen that $[\alpha]$ turns out to be analytic provided "bachelor" stands for "male and unmarried person." Quine however asks: how could we possibly know, with any degree of certainty, that we have here the real definition of "bachelor"? We might of course consult a dictionary, but all dictionaries suffer from one basic defect: since a natural language contains only finitely many words, dictionaries are necessarily circular. No dictionary of a natural language can therefore be relied upon to give us the primitive meaning of a word like "bachelor." We could of course go round asking people whether by "bachelor" they really mean "unmarried male"; but the answers might well vary from one person to the next, so that the status of "All bachelors are unmarried" will always remains uncertain.

Though interesting as questions of applied linguistics, these Quinean considerations do not bear on any logical or epistemological problems. In its—admittedly idealised—form, theoretical science starts by laying down a primitive vocabulary all of whose terms need not even be ostensively defined. Some molecular expressions may subsequently prove useful and are therefore fixed by means of verbal definitions, i.e., of abbreviations. Thus all propositions can—in principle—be reformulated in terms of the basic vocabulary; so they will turn out, whether effectively or not, to be either analytic or inconsistent or synthetic. Hypotheses are then formulated and basic statements derived. Since the latter are intended to be empirically decidable, their terms must be observational and hence defined directly by ostension. Note that throughout these operations, a unique logic and the *fixed ostensive meanings of its connectives* are presupposed. (Remember that these meanings are embodied in certain principles as well as in the rules of inference.) Thus $[\beta]$ is a logical truth provided the universal quantifier, the conjunction and the negation keep their classical (ostensive) meanings. It should be remembered that $(\neg\neg p \rightarrow p)$ and $(\forall x)[\neg\neg M(x) \leftrightarrow M(x)]$ cease to be analytic in intuitionistic logic. In other words: once properly expressed in the primitive vocabulary, an analytic statement will be true independently of the ostensive meanings, *not of its logical constants* but of its descriptive terms.

So far, nothing has been said in defence of globalism, i.e., of the thesis that experience challenges the *whole* fabric of our knowledge. At this point however,

Quine's critique takes on a very radical form. Quine maintains that *any* hypothesis, whether analytic, mathematical, logical or contingent, can be thrown into doubt by an experimental refutation. For example: as a result of empirical findings, we might decide to reject a *prima facie* analytic premise like "No bachelor is married." As explained above, this is tantamount to giving up the verbal definition according to which "bachelor" is shorthand for "male unmarried person." Thus we might enrich our basic vocabulary by adjoining to it the word "bachelor" as a primitive predicate, which would enable us to deny [α] without fear of contradiction. Such an ad hoc move would however be irrelevant to the problem of holism, for it comes down to weakening our original theory while leaving its underlying logic intact: treating "bachelor" as a primitive symbol is in effect equivalent to giving up a trivial premise, namely $(\forall x)[(x \text{ is a bachelor}) \leftrightarrow ((x \text{ is male}) \ \& \ (x \text{ is not married}))]$. Let us note that science hardly ever makes use of such premises; moreover, although weakening a hypothesis might ward off an empirical refutation, such a move would leave Quine's globalist thesis unsupported. In short: altering the verbal definition of words will not lead, beyond Duhem, to the Duhem-*Quine* problem. But what about modifying the *ostensive* meanings of certain terms?

Note that changing the referents of descriptive *non-observational* terms does not stave off empirical falsification. By definition, a relation of the form $S \vdash p$ in no way depends on the ostensive meaning of the descriptive symbols occurring in S or in p. By changing the referents of certain terms, we may nevertheless change the meaning of p in such a way that p is no longer undermined by experience. But if p is empirical, then it contains exclusively observational predicates, so that only by altering the latter could we conceivably save S from refutation. Observational notions ought however to possess meanings fixed in advance of theory. Should this condition be met, then we shall again face the refutibility of S alone, i.e., of the conjunction of a well-circumscribed number of assumptions. The issue of holism is consequently left untouched. This seems to be one reason for Quine's apparent readiness to give up—if need be—a logical truth like $(\neg\neg p \rightarrow p)$ or $[(p \ \& \ (q \vee r)) \leftrightarrow ((p \ \& \ q) \vee (p \ \& \ r))]$. But such a move would come down to altering not the verbal but the ostensive sense of certain connectives. We would therefore have modified our logic and, from a Platonist viewpoint, the referents of our logical constants. Thus any parallel between the rejection of a banal proposition such as [α] and that of a logical principle vanishes. We might similarly be led to negate certain mathematical postulates like the axiom of choice or the continuum hypothesis, which would again be tantamount, not to some verbal redefinition, but to the assertion or denial of a substantive existence claim. *Since logic and—to a lesser extent—mathematics are common to all scientific hypotheses, questioning these two disciplines might put our whole system of knowledge in jeopardy.* This is in effect the only acceptable version of Quine's holistic thesis, which has however been vindicated at the very high cost of modifying either logic, or mathematics, or both. Given his attachment to classical first-order logic, Quine himself ought to have found such a cost prohibitive; while most working scientists would be loath to resort to such drastic measures.

7 Empirical Content and the Duhem-Quine Problem

In this section it will be shown that a number of epistemological confusions can be traced back to a misunderstanding, or rather to a faulty diagnosis, of problems of the Duhem-Quine type. In previous works, I claim to have vindicated the view that scientific systems can be effectively refuted by level-0 reports. Against this naive falsificationist attitude, Lakatos had argued that a theory H cannot be regarded as having been undermined by a factual proposition $(p \ \& \ \neg q)$ until and unless a rival hypothesis H^* has been proposed, where H^* both yields $(p \rightarrow \neg q)$ and is incompatible with H.

Paul Feyerabend also mounted an attack on falsificationism; his starting point was however different from Lakatos's. He correctly maintained that if methodologies are to be useful, then they ought to have prescriptive import; i.e., they should be normative in the sense of guiding the scientist in his construction and in his choice of hypotheses. Of course, an alternative definition can be adopted according to which the only task of a methodology is to provide a system for the post hoc appraisal of available theories. But this sort of academic exercise gives no guideline for any effective action and is therefore contemptuously dismissed by Feyerabend. Should prescriptive import be granted, then Feyerabend claims that all *extant* methodologies contain directives which, if systematically followed, would have inhibited the progress of science *at some points* of its historical development. For example, Feyerabend attacks Newtonian empiricism, where the latter is interpreted as implying that hypotheses can legitimately be criticized by way not of any theoretical considerations, but only of empirical results. Thus Newtonianism entails that in order to be acceptable, every law should be induced from the facts and held to be true—or approximately true—until such phenomena are discovered as either make the law more precise or else constitute exceptions to it. Feyerabend rightly maintains that this inductivist methodology is reactionary in that it discourages the formulation of new theories as long as the reigning paradigm encounters no empirical difficulties.

In order to combat the conservatism inherent in empiricist philosophy, Feyerabend advanced the view that the empirical content of a theory generally depends on the existence and nature of its rivals. Should this thesis prove tenable, then it would hit not only inductivism but also the very notion of falsifiability; for the latter relies on the existence, for each theory H, of a well-circumscribed set Σ of potential falsifiers such that, should any member of Σ be verified, then H must be held to have been refuted. Popper called Σ "the empirical content" of H. But if we are to believe Feyerabend, Σ will in principle depend on all the rivals of H, whether these be actual or potential, so that the appellation "falsifier of H" ceases to have any clear meaning. Hence it no longer makes sense to accept—even provisionally—a theory H on the grounds that unlike its rivals, H is unrefuted; for H might be falsified by some fact which, though well-known, belongs to Σ only by virtue of some rival hypothesis H^* which is yet to be proposed.

In (Feyerabend, 1975), Feyerabend put forward two examples which seem *prima facie* to establish his thesis: those of Brownian motion and of Mercury's perihelion. With some historical justification, he maintained that only after it had been explained in 1905 by the Einstein-Smoluchowski theory was Brownian motion regarded as having refuted classical thermodynamics. And yet Brownian motion had been observed long before 1905, i.e., at a time when many physicists looked upon thermodynamics as the very paradigm of a successful hypothesis. Feyerabend concluded that Brownian motion formed part of the (falsifying) empirical content of classical physics only *after* the emergence of the Einstein-Smoluchowski theory. Similarly: the anomalous precession Mercury's perihelion had been noticed and described long before Einstein put forward the General Theory of Relativity (GTR), but only after GTR appeared on the scene did the motion of Mercury constitute a refutation of Newtonian physics.

It seems to me that Feyerabend put his finger on a major difficulty resulting from the Duhem-Quine problem—a difficulty which he mis-identifies as pertaining to the empirical content of scientific theories. Let us examine the general test-structure which allegedly applies to the case of Brownian motion. Consider a system T which implies a statement P about some microprocess, where the latter is correctly described by some proposition P' which is incompatible with, yet observationally indistinguishable from P; and let M' be an observation statement which, though easily verifiable and in fact already verified, remains ignored until a new theory T' is proposed, where T' is such that:

$$\vdash [T' \rightarrow (P' \rightarrow M')] \text{ and } \vdash [T' \rightarrow P']; \text{ whence } \vdash [T' \rightarrow M'].$$

According to Feyerabend, the appearance of T' changes the methodological situation in a dramatic way: M' confirms T' *and hence* refutes T. Thus the presence of T' has increased the empirical content of T by adding M' to the set of falsifiers of T. As an example, M' can now be interpreted as a sentence describing the motion of a Brownian particle, P' as a proposition about molecular processes, T' as the Einstein-Smoluchowski theory and T as Classical Thermodynamics.

Despite its superficial plausibility, this example can be shown to be irretrievably flawed. To begin with, Feyerabend commits a simple logical error; for there are only two possibilities:

1. Either M' fails logically to conflict with T, i.e., Not[$\vdash (T' \vdash M')$]; in which case M' refutes T neither before nor after T' was discovered; or
2. For some observation statement M, M is both incompatible with M' and entailed by T, in which case M' refutes T *both before and after* T' was proposed.

Of course, as a matter of psychological fact, people might have realized the relevance of M' only after the "refuting" theory T' was proposed; but this fact is hardly relevant either from the logical or from a methodological viewpoint.

These considerations suffice to refute Lakatos's and Feyerabend's general theses. But though largely correct, our logical analysis has so far overlooked an important Duhemian point, which also escaped Feyerabed's attention, namely that by themselves, i.e., without the adjunction of auxiliary assumptions, such high-level theories as T and T' entail no observation statements. Thus M' cannot possibly follow from T' alone. And the fact that Feyerabend mentions Mercury's perihelion and the Brownian motion in the same breath suggests that he might well have had the following situation in mind.

Let T be an old theory, T' a new rival of T, A a conjunction of (appropriate) auxiliary assumptions, p a sentence describing some boundary conditions; and finally let m and m' be two predictions. Suppose that:

3. $\vdash [p \rightarrow (m \ \& \ m')]$; $\vdash [(T \ \& \ A) \rightarrow (p \rightarrow m)]$ and $\vdash [(T' \ \& \ A) \rightarrow (p \rightarrow m')]$.

In other words: given the same initial conditions described by p, m and m' express incompatible predictions, while $(p \ \& \ m)$ and $(p \ \& \ m')$ are potential falsifiers of $(T \ \& \ A)$ and of $(T' \ \& \ A)$ respectively. Note that in view of the incompatibility of m and m', $(p \ \& \ m')$ is also a potential falsifier of $(T \ \& \ A)$. That is :

4. $\vdash [(T \ \& \ A) \rightarrow \neg(p \ \& \ m')]$.

Now assume that $(p \ \& \ m')$ is verified, so that $(T' \ \& \ A)$ is confirmed and $(T \ \& \ A)$ refuted. With regard to $(T \ \& \ A)$, we therefore confront a typical Duhem-Quine situation: we know that $(T \ \& \ A)$ is false but not which of the two conjuncts to blame. There may in fact exist another set of auxiliary hypotheses, A_0 say, such that:

5. $\neg[(T \ \& \ A_0) \rightarrow (p \rightarrow m')]$.

It might now look as though $(T \ \& \ A_0)$ is confirmed by $(p \ \& \ m')$ just as strongly as is $(T' \ \& \ A)$. This is however to forget that A_0 might be a complex assumption whose sole function is to save T from refutation. Compared with A, A_0 may not only be unnatural but also unsupported by independent evidence. Still, it cannot legitimately be maintained that T alone has been experimentally falsified; for the evidence $(p \ \& \ m')$ conflicts with the whole of $(T \ \& \ A_0)$. Now suppose that some revolutionary hypothesis T' is put forward which, in conjunction with the "natural" auxiliary assumption A, yields $(p \rightarrow m')$; i.e. that:

6. $\neg[(T' \ \& \ A) \rightarrow (p \rightarrow m')]$.

Only then would Feyerabend—as well as Lakatos—accept the claim that T has been refuted. In other words: because *both* of the presence of T' and of the truth of $(p \ \& \ m')$, we are entitled to hold A to be true and hence, in view of (4), T to be false. Thus the crucial step consists in deciding, in the light of (6), that A is true. In other words: given the availability of the new theory T' which, together with A, is corroborated by $(p \ \& \ m')$, we can now conclude that A holds good. Neither Lakatos nor Feyerabend seems however to realize that we are here confronting a classical Duhem-Quine problem. As an example, consider the concrete case of Mercury's perihelion, where: $T =$ (Newtonian Gravitational Theory); $T' \equiv$ (General Relativity (GTR)) and $A \equiv$ (The Sun is a point mass). Let A_0 be some complicated assumption about the uneven distribution of mass density within the sun, p a statement about the mass, initial position and velocity of Mercury, and finally, m' a description of the precession of Mercury's perihelion. Then the relations (3)–(6) will be satisfied—naturally within the limits of observational error (For more details, see (Zahar, 1989, Ch. 8).)

It is probably a historical fact that the precession of Mercury's perihelion was regarded as having refuted Newton only after GTR was proposed. But without violating any logical rules, we can account for this state-of-affairs far more satisfactorily than does Feyerabend or Lakatos. Although $(p \ \& \ m')$ is verified, (6) and (4) do not imply that A is true and hence T false; for $(p \ \& \ m')$ belongs to the empirical content, not of T, but of the conjunction $(T \ \& \ A)$. In other words: $(p \ \& \ m')$ is not some new falsifier of T created by the emergence of T'. We can nonetheless explain why, other things being equal, $(T \ \& \ A_0)$ derives little support from the verification of $(p \ \& \ m')$: $(T \ \& \ A_0)$ is ad hoc with respect to $(p \ \& \ m')$ since A_0 was "cooked up" in order to accommodate Mercury's perihelion, i.e., to fit $(p \ \& \ m')$. But suppose—counterfactually—that A_0 subsequently receives some measure of independent factual support; i.e., assume that independently of Mercury's perihelion, we have some reason for holding the sun's density not to be evenly distributed throughout its volume; the situation will have changed dramatically, for we should then consider $(T \ \& \ A_0)$ to have become a serious rival of $(T' \ \& \ A)$.

Thus Feyerabend's thesis, though psychologically illuminating, depends on the faulty analysis of a historical accident.

8 Is There a Bayesian Solution of the Duhem-Quine Problem?

The Bayesians rightly insist on the fact that probabilistic arguments are at the heart of scientific reasoning—a view strongly defended by Leibniz, who realized that Logic alone could not solve all epistemological problems. The starting point of Bayesianism is Bayes's theorem, i.e., a straightforward consequence of Kolmogorov's axioms which, in its strong form, asserts that:

$$\begin{aligned} P(h|e) =_{\text{def}} P(h \ \& \ e)/P(e) &= P(e|h) \cdot P(h)/P(e) \\ &= P(e|h) \cdot P(h) \Big/ \left[\sum_j P(e \ \& \ h_j) \right] \\ &= P(e|h) \cdot P(h) \Big/ \left[\sum_j P(e|h_j).P(h_j) \right]; \end{aligned}$$

where $P(\bigvee_j h_j) = 1$ and $P(h_j \ \& \ h_u) = 0$ for $j \neq u$.

I have written down this cascade of trivial equations in order to underline the fact that Bayes's theorem is an uncontroversial, not to say a trivial consequence of Kolmogorov's axioms. Most philosophers—and this includes Popper—accept these axioms and hence the relations written down above. So the difference between the

Bayesians and their opponents cannot possibly consist in the former's unique acceptance of Bayes's Theorem. From reading various texts, it appeared clear that this difference lies in the subjective interpretation given by the Bayesians to the notion of probability—and this even though, if we are to believe Colin Howson, Bayesianism is compatible with objectivism. But if this were really the case, then Bayesianism would lose all claim to originality, for the difference between the Bayesians and the genuine objectivists would at best be one of degree, not of kind.

Let me give an example which illustrates this point. In (Bovens and Hartmann, 2003), the authors cite an example in which they claim that $P(R_1, R_2) = 0.10$ and $P(R_1, \neg R_2) = 0.01 = P(\neg R_1, R_2)$, where $R_1 \equiv$ (the culprit spoke with a French accent), and $R_2 \equiv$ (the culprit drove off in a Renault car). These figures seem plausible, so no attempt is made to justify them. The context nevertheless indicates not only that these probabilities reflect objective states-of-affairs but also that they could have been obtained by means of a poll. Clearly, the total number of French citizens, of Renault owners and even of all the people inhabiting a given region could easily be determined. So we have here to do with objective frequencies lying at the basis of allegedly subjective assessments of probabilities. However, the authors also make use of such relations as $P(\text{Rep} \cdot R_j | R_j) = p_j$ and $P(\text{Rep} \cdot R_j | \neg R_j) = q_j$, where $\text{Rep} \cdot R_j$ is the proposition that there exists a report according to which R_j is true (Bovens and Hartmann, 2003, pp. 12–14). But then it is difficult to see not only how p_j and q_j could be objectively estimated but also how they could reasonably be postulated in all circumstances. Consider the case where the truth value of R_j is unknowable. R_j might, e.g., denote some high-level quantum theory, the reporters being Einstein and Heisenberg. Even if both reporters knew the extent to which R_j was empirically adequate, $P(\text{Rep} \cdot R_j | R_j)$ will, in all probability, receive widely divergent values; for, unlike Heisenberg, Einstein is not likely to accept reporting R_j as true. Hence a minimal psychologistic element must be allowed for by all consistent Bayesians; this is probably why Howson and Urbach wrote:

> The other notion of probability is epistemic. This type probability is, to use Laplace's famous words, "relative in part to our ignorance, in part [to] our knowledge. It expresses numerically degrees of uncertainty in the light of data. (Howson and Urbach, 1993, p. 22)

It seems to me that we have here a confusion between two theses:

1. The use of probabilities arises from our uncertainty as to the actual truth-value of some propositions. In other words, had we been in a position to know the truth value of certain statements, then we should never have resorted to the probability calculus;
2. Probabilities express or refer to the degree of our certainty regarding these propositions.

This type of reasoning is on a par with the following obvious *non sequitur*: Logic owes its origin to our need to learn something about the connections between various state-of-affairs; *ergo*, Logic expresses our desire to acquire knowledge. There is related *non sequitur*—of a more sophisticated kind—we often use logical rules in order to infer some truths from other truths; hence Logic is a description of the psychological rules of thought followed by the (healthy) human mind. This is of course the well-known psychologistic fallacy.

Be that as it may, the thesis that probability "expresses numerically degrees of uncertainty in the light of the data" is admittedly consistent. It is therefore time to examine how the Bayesians use this approach in order to solve the Duhem-Quine problem. Howson and Urbach consider an example cited by Lakatos, namely that of Prout's theory, which will henceforth be denoted by H. H asserts that the atomic weight of every element is an integral multiple of that of hydrogen. As for the auxiliary hypothesis A implicated in the testing of H, it consists of laws about some measuring techniques, about the reliability of certain instruments, about the purity of the chemicals employed, etc. Let e be the verified report that the measured weight of chlorine is 35.83. Thus $e \vdash \neg(H \ \& \ A)$, which means that $(H \ \& \ A)$ is empirically refuted. Yet the scientists of the day took e to have undermined A rather than H, a fact which Howson and Urbach set out to explain. They write:

> It seems that chemists of the early nineteenth century, such as Prout and Thomson, were fairly certain about the truth of t [$=H$], but less so of a [$=A$], though more sure that a is true than that it is false. ... For these reasons, we conjecture that $P(a)$ was of the order of 0.6 and that $P(t)$ was around 0.9, and these are the figures we shall work with. (Howson and Urbach, 1993, p. 138)

Following Jon Dorling, the two authors go on to assume that H is independent of A; they then lay down, without any justification, the following relations: $P(e|\neg H \ \& \ A) = 0.01 = P(e|\neg H \ \& \ \neg A)$ and $P(e|H \ \& \ \neg A) = 0.02$. From all these assumptions they finally deduce $P(A|e) = 0.073 \lll 0.878 = P(H|e)$.

These operations are however nothing but a series of blatantly ad hoc adjustments of parameters carried out in order to obtain a foreknown result, namely that in the Prout case, the Duhem-Quine problem was unexpectedly resolved in favour of the high-level theory H rather than in that of the auxiliary hypothesis A. It can of course hardly be doubted that many historical events can be "modelled" through assignments of probabilities to various sentences. Such modelling however hardly constitutes an explanation. It even seems to me that Howson, Urbach and Dorling confuse the *explanans* with the *explanandum*. That many chemists attributed to $P(H)$ a value higher than that of $P(A)$ is not an *explanans* but the very fact which, if the Duhem problem is to be solved, ought to be explained. And the same goes for all the other ad hoc assumptions made by the authors. Yet there are at least two satisfactory methods which might explain why most, though by no means all, the scientists of the day were prepared to accept H rather than A. The first consists in showing that, when conjoined with auxiliary assumptions different from A, H had previously been confirmed on numerous occasions, which takes us back to the conclusions reached in Section 4 above. The second method consists in demonstrating that H had introduced unity into a scientific system by establishing unexpected connections between some of its hitherto disparate elements. The scientists would consequently have refused to regard such unity as being due to the existence of a malicious demon bent on deceiving them. All this underlines the relevance not only of objective, but also of metaphysical considerations to the solution of the Duhem-Quine problem.

Let me end this section by showing that Bayesianism is fundamentally incapable of solving an epistemological problem arising from the ad hoc feature of certain hypotheses, namely of those obtained by ad hoc maneuvers. None of the statements H, A, e, etc. entering as arguments into the probability functions considered by the Bayesians bears any trace of the methods by which these propositions have been constructed. In other words, a cooked up law is treated on a par with an ad hoc one. As an illustration, consider an urn Δ containing an infinite number of pairwise incompatible hypotheses, each one of which involves a large number of parameters (i.e., the conjunction of any two hypotheses is inconsistent, and this independently of the values assigned to the parameters). Hence at most one member of Δ can be true. Now let E be any limited set of verified factual statements. Assume that every hypothesis belonging to Δ is known to yield E. No matter how sophisticated it might become, Bayesianism has no machinery for discriminating between the various elements of Δ; and yet if any hypothesis belonging to Δ were to be chosen at random, its chances of being true would clearly be vanishingly small. The Bayesians may of course decide to allot higher probabilities to the non-ad hoc members of Δ, but they are incapable of defending such a decision on Bayesian grounds—all of which is nothing but the generalised version of the argument fielded against the Bayesian treatment of the Proutian case, as described above.

9 Conclusion

This paper dealt first with Duhem's ungainsayable but modest claim that an experimental refutation undermines not one theory, but a whole cluster of hypotheses no member of which is singled out as the culprit. Secondly, Kripke's thesis that there exist necessary a posteriori truths was discussed; it was shown that in all cases relevant to the empirical sciences, there was an excluded middle between metaphysical (i.e., untestable yet fallible) hypotheses and scientific (i.e., refutable and hence also fallible) propositions. Thirdly, Quine's holistic thesis that "the unit of empirical significance is the whole of science" was examined. It was found that throwing doubt on supposedly analytic statements like "No bachelor is married" comes down to abandoning merely verbal definitions which do not seriously bear on any aspect of knowledge. It is only when one moves towards the centre of the "field" of science, i.e., towards mathematics and logic, that holism begins to acquire some clout. It is certainly the case that the mathematical and, even more so, the logical axioms are common to almost all physical theories. Altering logic—or even mathematics—will therefore have repercussions in *all* branches of knowledge. Thus holism comes into its own *only* when, as a result of empirical difficulties and in order to increase the simplicity of a system, logic is tampered with; for by its very nature, logic permeates the *whole* fabric of our knowledge. What is however puzzling is that Quine consistently maintained that only one logic, namely classical complete first-order logic, is legitimate. But then one fails to see in what sense Duhem-Quine goes beyond Duhem.

Bibliography

Bovens, L. and Hartmann, S. (2003). *Bayesian Epistemology.* Oxford University Press, Oxford.

Descartes, R. (1986). *Meditations on First Philosophy.* Cambridge University Press, Cambridge. transl. J. Cottingham.

Duhem, P. (1954). *The Aim and Structure of Physical Theory.* Princeton University Press, Princeton.

Feyerabend, P. (1975). *Against Method.* Humanities Press, London.

Howson, C. and Urbach, P. (1993). *Scientific Reasoning: The Bayesian Approach.* Open Court, La Salle, IL.

Kant, I. (1957). *Metaphysische Anfangsgründe der Naturwissenschaft.* Insel Verlag, Wiesbaden.

Kripke, S. (1972). *Naming and Necessity.* Basil Blackwell, Oxford.

Lakatos, I. (1970). *Falsification and the Methodology of Scientific Research Programmes*, pages 91–196. Cambridge University Press, Cambridge.

Quine, W. V. O. (1980). Two dogmas of empiricism. In *From a Logical Point of View.* Harvard University Press, Cambridge, MA, second (revised) edition.

Zahar, E. (1989). *Einstein's Revolution: A Study in Heuristic.* Open Court, La Salle, IL.

Part V
Decision Theory and Epistemology

Chapter 21
Can Knowledge Be Justified True Belief?

Ken Binmore

1 Fallibilism or Skepticism?

The view that knowledge can usefully be interpreted as justified true belief has fallen into disfavor in recent times. David Lewis observes that the use of such a definition seems to require an apparently impossible choice between the rock of fallibilism and the whirlpool of skepticism, but that we can—just barely—escape both perils by steering with care (Lewis, 1996). This paper offers a more radical defense of the same conclusion.

A standard objection to the traditional definition has been voiced in (Gettier, 1963). In a bowdlerized version of his story, Boris and Vladimir have both proposed marriage to the beautiful Olga. She blushes with pleasure when Vladimir pays her compliments, but seems not to remember Boris at all. Boris is not surprised, because he is poor and Vladimir is rich. He thinks his knowledge of the world justifies his believing that Olga will marry money. When Olga surprises Boris by accepting his proposal, his belief turns out to be true because, unknown to anyone, a long-lost uncle has died and left Boris a fortune. But do we want to say that Boris therefore *knew* that Olga would marry a rich man?

The traditional definition can be defended against such attacks by challenging the standard of justification that is employed. In our Russian story, Boris should perhaps have been more realistic about his own inexperience in matters of the heart, and sought advice from an agony aunt. He would then have learned that beautiful maidens sometimes pretend indifference with a view to fanning the flames of a favored suitor's ardor.

This paper avoids such disputes over what counts as adequate justification by treating the justification process as algorithmic. The general problems that arise in treating knowledge algorithmically are surveyed in (Binmore and Shin, 1992). This paper confines its attention to investigating the formal implications of maintaining

K. Binmore, CBE, FBA (✉)
Economics Department, University College London, Gower Street, London WC1E 6BT, UK
e-mail: k.binmore@ucl.ac.uk

D. DeVidi et al. (eds.), *Logic, Mathematics, Philosophy: Vintage Enthusiasms*,
The Western Ontario Series in Philosophy of Science 75,
DOI 10.1007/978-94-007-0214-1_21,

that knowledge must simultaneously be justified and true. The findings depend on the manner in which the underlying decision problem is framed.

In formulating what is nowadays called Bayesian decision theory, Leonard Savage distinguished between large and small worlds (Savage, 1951). His *Foundations of Statistics* says at one point that it would be "ridiculous" and at another that it would be "preposterous" to apply his theory in a large world, but no formal criteria are offered to distinguish between small-world decision problems and large-world decision problems (Binmore, 1992a).

This dimly recognized distinction between large and small worlds in decision theory echoes a distinction in proof theory made precise by Gödel. A mathematical system large (or complex) enough to include arithmetic cannot be simultaneously consistent and complete. This paper adapts the halting argument for Turing machines in defending the claim that decision theory needs to recognize a similar distinction.

In a context sufficiently far removed from the small worlds to which Bayesian decision theory properly applies, the knowledge assumptions that the theory implicitly takes for granted can no longer be sustained. But we can still steer between the rock of fallibilism and the whirlpool of skepticism by explicitly building into our framing of the underlying decision problem the possibility that any such framing may fail to capture something significant.

2 Small World Assumptions

Bayesian decision theory takes for granted that a decision-maker's knowledge at any time partitions the universe of discourse into a collection of disjoint possibility sets. The partitioning property of these possibility sets is then inherited by the information sets that von Neumann introduced into game theory (Binmore, 1992c, p. 454). An assumption of "perfect recall" is then usually made to ensure that a player's knowledge partition is always a refinement of his previous partitions. This section reviews the linkage between such knowledge partitions and the idea of a knowledge operator.

We identify an event E with a subset of a given universe of discourse denoted by Ω. The event in which Boris knows that E has occurred is denoted by $\mathcal{K}E$, where $\mathcal{K}$ is his knowledge operator. The event in which Boris thinks it possible that E has occurred is denoted by $\mathcal{P}E$, where $\mathcal{P}$ is his possibility operator.

If we make the identification $\mathcal{P} = {\sim}\mathcal{K}{\sim}$, then we establish a duality between $\mathcal{K}$ and $\mathcal{P}$. Either of the following lists will then serve as a rendering of the requirements of the modal logic S5:

(K0)	$\mathcal{K}\Omega = \Omega$	(P0)	$\mathcal{P}\emptyset = \emptyset$
(K1)	$\mathcal{K}(E \cap F) = \mathcal{K}E \cap \mathcal{K}F$	(P1)	$\mathcal{P}(E \cup F) = \mathcal{P}E \cup \mathcal{P}F$
(K2)	$\mathcal{K}E \subseteq E$	(P2)	$\mathcal{P}E \supseteq E$
(K3)	$\mathcal{K}E \subseteq \mathcal{K}^2 E$	(P3)	$\mathcal{P}E \supseteq \mathcal{P}^2 E$
(K4)	$\mathcal{P}E \subseteq \mathcal{K}\mathcal{P}E$	(P4)	$\mathcal{K}E \supseteq \mathcal{P}\mathcal{K}E$

The "infallibility" axiom can be taken to be either (K2) or (P2). The seemingly innocent (K0) and (P0) will be called "completeness" axioms.

These ideas are linked with knowledge partitions by defining the possibility set $P(\omega)$ to be the set of states that Boris thinks is possible when the true state of the world is ω. In Bayesian decision theory, the minimal requirements for such possibility sets are usually taken to be:

(Q1) The collection $\{P(\omega) : \omega \in \Omega\}$ partitions Ω
(Q2) $\omega \in P(w)$

The second of these assumptions is the "infallibility" requirement.

To establish an equivalence between the two approaches, it is only necessary to define $\mathcal{P}$ and P in terms of each other using the formula:

$$\omega \in \mathcal{P}E \quad \leftrightarrow \quad P(\omega) \cap E \neq \emptyset\,. \tag{21.1}$$

With this definition, (P0)–(P3) can be deduced from (Q1)–(Q2) and vice-versa. However, the role of the completeness axiom (P0) is peripheral. If we dispense with (P0) and redefine both (P1)–(P3) and (21.1) so that they apply only to non-empty events, then (new P1)–(new P3) are equivalent to (Q1)–(Q2).

It is significant that (P0) can be eliminated, because the result of the next section can be regarded as saying that (P0) and (P2) cannot both hold in a large enough world when the possibility operator is algorithmic.

3 Justification

The process of justification is abstracted away in the previous section. It is understood to be somehow built into the knowledge or possibility operators. We now unpack this black box by postulating that justification is actually carried out by a "Leibniz engine" J that makes judgements on what events are possible.

The assertion that justification is algorithmic is interpreted to mean that J is a Turing machine that sometimes answers NO when asked questions that begin:

Is it possible that ... ?

Issues of timing are obviously relevant here. How long does one wait for an answer before acting? Such timing problems are idealized away by assuming that Boris is able to wait any finite number of periods for an answer.

As in the Turing halting problem, we suppose that $[\,N\,]$ is some question about the Turing machine N. We then take $\{M\}$ to be the question:

Is it possible that M answers NO to $[\,M\,]$?

Let T be the Turing machine that outputs $[\,x\,]$ on receiving the input $\{x\}$, and employ the Turing machine $I = JT$ that first runs an input through T, and then runs the output of T through the justification machine J. Then the Turing machine I responds to $[\,M\,]$ as J responds to $\{M\}$.

An event E is now defined to be the set of states in which I responds to $[\,I\,]$ with NO. We then have the following equivalences:

$$\begin{aligned}\omega \in E &\leftrightarrow I \text{ responds to } [\,I\,] \text{ with NO}\\ &\leftrightarrow J \text{ reports it to be impossible that } I \text{ responds to } [\,I\,] \text{ with NO}\\ &\leftrightarrow \omega \notin {\sim}\mathcal{P}E\end{aligned}$$

It follows from (P2) that

$${\sim}\mathcal{P}E = E \subseteq \mathcal{P}E\,.$$

This identity only holds when $\mathcal{P}E = \Omega$. Since $E = {\sim}\mathcal{P}E$, it follows that $E = \emptyset$, and so $\mathcal{P}\,\emptyset = \Omega$. That is to say, we are led to the following apparent contradiction:

Proposition *If the states in Ω are sufficiently widely construed and knowledge is algorithmic, then infallibility implies that the decision-maker always thinks it possible that nothing will happen.*

If one seeks to maintain (P0) or (K0) in a world large enough for our use of the Turing argument to make sense, this proposition puts paid to Lewis's attempt to steer between the rock of fallibilism and the whirlpool of skepticism. But why should we hang on to these hard-to-interpret completeness axioms in a large world?

4 What Is an Event?

I think the apparent contradiction built into the preceding proposition signals a failure of the model to capture the extent to which familiar assumptions from small-world decision theories need to be modified when moving to a large world.

For example, we think of an event E as having occurred if the true state ω of the world has whatever property defines E. But how do we determine whether ω has this property?

If we are committed to an algorithmic approach, we need an algorithmic procedure for the defining property P of each event E. This procedure can then be used to interrogate ω with a view to getting a YES or NO answer to the question:

Does ω have property P ?

We can then say that E has occurred when we get the answer YES, and that ${\sim}E$ has occurred when we get the answer NO.

But in a sufficiently large world, there will necessarily be properties for which the relevant algorithm sometimes will not halt. Our methodology will then classify

ω as belonging neither to E nor to $\sim E$. Our inadequate formalism then forces us to place ω in $\emptyset$—although we can no longer interpret this as the set with no elements.

A more satisfactory analysis would perhaps appeal to some appropriate version of constructivist or intuitionistic logic,[1] but I hope the preceding remarks will at least make it plausible that we can sustain the conclusion of the proposition of the previous section in a large world. To say that $\mathcal{P}\emptyset = \Omega$ can be interpreted to mean that Boris necessarily thinks it possible that the true state of the world will remain unclassified according to any of the properties recognized by his algorithmic classification system.

5 All-Encompassing Worlds

Robert Aumann, in (Aumann, 1987), has pioneered an approach to the foundations of game theory in which the states of the world in the model are to be thought of as encompassing absolutely everything that could conceivably happen—including Boris's states of mind and behavior. I have used Gödelian arguments elsewhere to criticize Aumann's use of Bayesian decision theory in such a large-world context (Binmore, 1987, 1992b). But what if Aumann's visionary approach is employed, using a decision theory that is suited to large-world applications?

On this subject, I shall only point out that the argument of Section 3 survives allowing the justification machine J to depend on the true state of the world. Boris's justification algorithm is then one of the many properties of whatever the true state ω turns out to be. When Boris interrogates ω, he will then sometimes be asking a question about the workings of his own cognitive processes.

The non-halting argument of Section 3 is based on precisely this self-referential possibility. The argument can therefore be seen as another telling of the tale that Boris cannot operate an algorithmic model that is always successful in predicting the workings of his own mind. Still less can Boris operate an algorithmic model that is always successful in predicting what the workings of his mind would be if the true state were not ω, but some other state ζ. It obviously makes no sense to postulate (P2) or (P3) in such a large world, but these assumptions are needed if knowledge is to be modeled in terms of the possibility partitions of Bayesian decision theory.

6 Conclusion

There is no problem in requiring knowledge to be both justified and true in a small world. This paper argues that the same may be possible in large worlds, but only at the expense of abandoning much of the structure that Bayesian decision theory takes for granted.

---

[1] Note that possibility in Section 3 is taken to be the failure to get a NO when the justification machine is asked whether something is possible. But this is not the same as getting a *yes* to the same question.

Bibliography

Aumann, R. (1987). Correlated equilibrium as an expression of Bayesian rationality. *Econometrica*, 55:1–18.

Binmore, K. (1987). Modeling rational players I. *Economics and Philosophy*, 3:9–55.

Binmore, K. (1992a). Debayesing game theory. In Skyrms, B., editor, *Studies in Logic and the Foundations of Game Theory: Proceedings of the Ninth International Congress of Logic, Methodology and the Philosophy of Science*. Kluwer, Dordrecht.

Binmore, K. (1992b). Foundations of game theory. In Laffont, J. J., editor, *Advances in Economic Theory: Cambridge, 1992. Sixth World Congress of the Econometric Society*. Cambridge University Press, Cambridge.

Binmore, K. (1992c). *Fun and Games*. D. C. Heath, Lexington, MA.

Binmore, K. and Shin, H. (1992). Algorithmic knowledge and game theory. In Bicchieri, C. and Chiara, M., editors, *Knowledge, Belief and Strategic Interaction*. Cambridge University Press, Cambridge.

Gettier, E. (1963). Is justified true belief knowledge? *Analysis*, 23:121–123.

Lewis, D. (1996). Elusive knowledge. *Australian Journal of Philosophy*, 74:549–567.

Savage, L. (1951). *The Foundations of Statistics*. Wiley, New York, NY.

Chapter 22
The Stochastic Concept of Economic Equilibrium: A Radical Alternative

Moshé Machover

1 Introduction

My aim in this article is to present the gist of some ideas first proposed in (Farjoun and Machover, 1983): that the main economic quantities—such as the unit price of any type of commodity, and the rate of profit—at any time should be modelled not as determinate numerical magnitudes but as random variables; and that at equilibrium these quantities have characteristic *distributions* rather than determinate numerical values. I add some methodological remarks about mathematical models and about what economics can borrow from physics.

I am not an economist but a mathematician, whose knowledge of academic mainstream economics is quite patchy. I made my first acquaintance with it in the early 1960s, when I was assigned the task of teaching some courses of mathematics to students of economics. In order to find examples of applications and problems that would be close to the interests of these students, I decided to have a look at some books on mathematical economics.

My attention was attracted by a recent book (Schwartz, 1961), because it was by a well-known mathematician, Jacob T. Schwartz, co-author (with Nelson Dunford) of an important monograph on linear operators. I found it very interesting and instructive. In particular, I was fascinated by Part A of the book, entitled "The Leontief Model and the Technological Basis of Production." I realized immediately that this was just the right framework for formalizing Marxian economics (with which I was familiar). Of course, this was no accident, as Leontief's input–output matrix formalism was a direct descendant of Marx's "schemes of reproduction."[1]

M. Machover (✉)
Emeritus Professor of Philosophy, King's College; Voting Powers and Procedures, Centre for Philosophy of Natural and Social Sciences, London School of Economics, London, UK
e-mail: moshe.machover@kcl.ac.uk

An earlier version of this paper was presented at the Seminar on Dissent in Science, CPNSS, LSE; 16 March 2004.

[1] I found out later that before emigrating from the USSR in 1931, Wassily Leontief had worked in GOSPLAN, the Soviet Economic Planning Board, which used Marxian economic theory. Leontief's work, for which he was awarded the Nobel Prize in 1973, was one of the channels through which Marxian theory exercised its—largely unrecognized—influence on mainstream economics.

D. DeVidi et al. (eds.), *Logic, Mathematics, Philosophy: Vintage Enthusiasms*, The Western Ontario Series in Philosophy of Science 75,
DOI 10.1007/978-94-007-0214-1_22, © Springer Science+Business Media B.V. 2011

I played around with this formalism in order to analyse the so-called *transformation problem* of Marxian economics (of which more anon), unaware that I was duplicating other people's work. I didn't get very far, and let the matter rest.

In the late 1970s, my interest was rekindled by my friend Emmanuel Farjoun. He got involved in the controversy between Sraffians and Marxists that flared up following some publications by the former (see, in particular, Steedman's book (Steedman, 1977)). The Sraffians showed that the transformation problem—as commonly understood—is not solvable. Hence they concluded that Marx's labour theory of value is wrong and worthless. The Marxists sprang to the defence of this theory. (Some Marxist responses are collected in (Langston et al., 1984).)

At the heart of this controversy was a notion of equilibrium which was shared by both sides.

1.1 The Equilibrium Price–Profit Equation

Let me outline the simplest form of the linear equilibrium model of prices and profit in the Leontief formalism.

We consider a (closed) capitalist economy, in which n types of commodity (excluding labour power), say $C_1, \ldots, C_n$, are produced. In this simple model it is assumed that each type of commodity is produced by a unique technical process, and there are no by-products.[2] More importantly, it is assumed that each type of commodity has a determinate equilibrium unit price[3]; and that all types of production yield the same equilibrium rate of profit.

If p^i is the price of one unit of C_i and ρ is the rate of profit, then

$$p^i = \sum_{j=1}^{n} \left(a^i_j + \rho k^i_j\right) p^j, \quad i = 1, \ldots, n. \tag{22.1}$$

Here a^i_j is the amount of C_j consumed (i.e., used up) as input in the production of one unit of C_i; and k^i_j is the amount of C_j employed (i.e., used but not necessarily used up) in the production of one unit of C_i multiplied by the number of units of time during which it is so employed.[4,5]

---

[2] These two assumptions can be, and indeed have been, challenged as unrealistic. But this issue is irrelevant to the present discussion.

[3] These prices are determined only up to an arbitrary factor of proportionality; thus, it makes no difference if all unit prices are multiplied by the same positive number. Alternatively, we can put, arbitrarily, $p^1 = 1$, and then all unit prices are completely determined.

[4] Thus, if time is measured in years and x units of C_j are employed during a year in producing a total of y units of C_i, then $k^i_j = x/y$.

[5] Note that labour does not occur in (22.1); this is because it has been eliminated from the accounting by incorporating into the a^i_j the wage goods consumed by the workers who supply the labour power used in producing C_i. This elimination is possible due to the fact that the price of labour

The system (22.1) of n equations can be written in matrix form:

$$\mathbf{p} = (\mathbf{A} + \rho\mathbf{K})\mathbf{p}, \tag{22.2}$$

where $\mathbf{p}$ is a column n-vector and $\mathbf{A}$ and $\mathbf{K}$ are $n \times n$ matrices with non-negative elements. The unknowns here are $\mathbf{p}$ and ρ. Since $\mathbf{p}$ is determined only up to proportionality,[6] the number of unknowns is n, the same as the number of equations. To make sense, a solution $\mathbf{p}$ and ρ should be positive.

Under very reasonable conditions,[7] there is indeed such a solution, and it is unique.

1.2 The Uniformity Assumption

Like all mathematical models, the one just described makes various assumptions that simplify reality. Here I would like to draw attention to a very fundamental conceptual assumption: at equilibrium, the unit prices of all types of commodity produced by the economy have (up to proportionality) determinate numerical values; and the rates of profit accrued by capital in all productive units are equal.

Clearly, this assumption does not purport to describe a real-life state of affairs in a capitalist economy. Everyone knows that if you shop around you will find that the same type of commodity is sold at the same time by different sellers at a variety of unit prices; and rates of profit vary greatly both within industries and between them. So the equilibrium that the input–output model describes is an ideal one. However, it is implied that the real economy is driven by market forces, the forces of competition, towards an ideal equilibrium of this sort, and are only prevented from actually reaching it by various *disequilibrating* forces, that act as "noise."

(Schwartz, 1961, p. 9) provides the following justification for the assumption regarding the rate of profit:

> We have here taken an *essential* step in assuming the rate of profit, ρ, to be the same for all types of production. This corresponds to the *ordinary assumption*, in the theory of prices, of "free competition"; it can be justified *in the usual way* by arguing that a situation in which the production of different commodities yields different rates of profit cannot be stable, since investments would be made only in the industry yielding the highest rate of profit to the exclusion of other commodities yielding lower rates of profit. Long-term equilibrium, of which our simple theory is alone descriptive, would be reached only when all such rates of profit became equal (My emphases).

---

power does not contain any direct profit, but is just the total price of the wage goods consumed by the workers.

[6] See footnote 3.

[7] The economic meaning of these conditions is that the economy is capable of producing a physical surplus; and that it cannot be partitioned into two or more closed sub-economics. For details, see (Schwartz, 1961, Lecture 2). The mathematical tool used here is the Frobenius theory of the eigenvalues and eigenvectors of non-negative matrices.

Note that Schwartz refers to the uniformity assumption as "ordinary" and to his justification of it as "the usual" one. They are indeed both common and time-honoured: they go back at least to Adam Smith, and have been accepted and repeated (with some variations) by many authors of various schools, including Marx.[8]

Note also that Schwartz does not bother to justify the assumption that at equilibrium the unit prices are determinate, although a similar justification (in terms of competition) might surely be offered. Apparently he (like many others) regards this as more or less obvious.

1.3 The Transformation Problem

Marx had not one but two sets of ideal prices or price-like quantities. He starts off, as a first approximation, with the notion of *exchange value* (to which I shall refer here briefly as "value"). The value of a given commodity is, roughly speaking, the total amount of labour (measured, say, in labour hours) necessary to produce it, including not only the labour used directly in its production but also the labour used in producing its material inputs, as well as the inputs of these inputs, etc.[9]

But values cannot serve as equilibrium prices in an (ideal!) state of affairs in which the rate of profit is uniform. This is because if commodities were to exchange at their values, then a commodity whose production process has higher "organic composition" (roughly speaking, smaller labour intensity) would yield a lower rate of profit than a commodity produced by a process with lower organic composition (greater labour intensity).

Since Marx—following the classical economists, especially Adam Smith and David Ricardo—subscribed to the uniformity assumption, he introduced what he called "prices of production," which are ideal equilibrium prices corresponding quite closely to the p^i of Section 1.1.[10]

Marx's price–profit system of equations, connecting the ideal equilibrium rate of profit and prices of production was very similar to the Leontief equation (22.2), of

---

[8] See (Farjoun and Machover, 1983, pp. 14–16) for quotes from Smith and Marx, as well as from a later article by Schwartz.

[9] In the formalism of Section 1.1, let v^i be the value of one unit of C_i, and let l^i be the amount of labour used *directly* in producing it. Further, let c^i_j be defined like the a^i_j of (22.1), except that they *do not* incorporate the wage goods consumed by the workers who supply the labour power used directly in producing C_i (cf. footnote 5). Then the v^i are the unique solution of the system of n linear equations:

$$v^i = l^i + \sum_{j=1}^{n} c^i_j v^j, \quad i = 1, \ldots, n.$$

[10] Both values and prices of production must be distinguished from what Marx called "market prices": these are real-life prices, whose relationship to the prices of production is as described in Section 1.2.

which it was in fact a direct ancestor.[11] However, Marx postulated an additional constraint: the equilibrium ideal rate of profit must, according to him, equal the *average* rate of profit that would obtain if all commodities were priced at their values. Thus, if S is the total value of the surplus produced during a unit of time,[12] and K is the total value of the goods employed (but not necessarily used up) in production during this period, then the ideal rate of profit ought to be

$$\rho_{\mathrm{M}} := \frac{S}{K}. \tag{22.3}$$

Through this link, values are "transformed" into prices of production.

But Marx was unable to solve [his rudimentary form of] (22.2) with the added constraint $\rho = \rho_{\mathrm{M}}$. In fact, we now know that this so-called "transformation problem" is in general not solvable—at least not in the sense in which it has commonly been understood. As pointed out by the Sraffians, the unique ρ solving (22.2) need not in general equal ρ_{M} as defined by (22.3): there are reasonable counter-examples, cases of the model outlined in Section 1.1, whose solution fails to satisfy $\rho = \rho_{\mathrm{M}}$.

Emmanuel Farjoun, who got involved in the controversy around the transformation problem (and who contributed to Langston et al. (1984)), eventually came up with a radical idea: the point was not whether the Sraffians were right or wrong about the Leontief model, or whether the model could be tweaked in some way so as to satisfy Marx's postulate; rather, it was whether that model—or indeed anything like it—was a reasonable way of theorizing equilibrium. An analogy with statistical mechanics suggested very strongly that it was not: the uniformity assumption (which both sides in the transformation controversy subscribed to) was all wrong. We elaborated this idea together in Farjoun and Machover (1983).

2 Stochastic Equilibrium

The main aim of Farjoun and Machover (1983) was to amend and reconstruct the Marxian labour theory of value, preserving what we regard as its invaluable core, while ditching the concept of prices of production (which we regard as unnecessary and mistaken) and avoiding altogether the transformation problem as commonly understood (which we regard as a non-problem).

But the book's basic methodological message—which is my present topic here—is much more general. It concerns the notion of economic equilibrium used in several economic theories of various schools. In this connection, Marx's theory and the Leontief model discussed in the Introduction serve as a mere illustration.

---

[11] Cf. footnote 1.

[12] This is obtained from the value of the whole product by deducting from it the value of the non-labour inputs consumed in its production and the value of the goods consumed by the workers engaged in this production.

The methodological message is, briefly, that the concept of equilibrium in which unit prices and "the" rate of profit are determinate quantities is fundamentally erroneous. This is suggested by analogy with statistical mechanics, a theory underlying thermodynamics, founded by Boltzmann and Maxwell in the 1870s. This analogy is explained in some detail in Farjoun and Machover (1983), and here I shall only highlight a few points.

2.1 Equilibrium in Statistical Mechanics

In a volume of gas enclosed in a closed container, the molecules are in constant motion (the total kinetic energy of this motion is what constitutes the heat energy stored in the gas). In this motion, the molecules collide with one another and as a result the more energetic molecules tend to slow down (by imparting some energy to slower molecules with which they collide); conversely, the less energetic molecules tend to speed up.

However, this does not mean that at equilibrium all molecules reach an equal level of kinetic energy.[13] The point is not merely that such a state of uniformity does not actually occur; but that if it ever did it could not last for a split second. If it were miraculously brought about through the mechanism of incessant collision, then *this very same mechanism* would instantly disturb it. Note also the use of the word "tend" in the description of this mechanism: what is meant by it is that a more energetic molecule is *more likely* to slow down than to speed up, not that it will *always* do so.

Another important feature of this physical system is the distinction between macro-state and micro-state. A macro-state can be described by some global data, such as the total energy of the gas, its volume, as well as some statistical data that will be mentioned below. A micro-state is described by an enormous number of data: the position of each molecule (given by its three space coordinates) and its momentum (given by three independent numerical components). The number of these data is called the *number of degrees of freedom* of the system. And the system in question has a great many. Thus, to each macro-state there corresponds a very large set of micro-states.

The notion of equilibrium of such a system refers to the macro level. It does not imply that all the molecules are motionless—they never can be, for this would happen only at 0°K, which is unattainable—but that the macro-state remains unchanged unless perturbed by external forces.

Among the data characterizing a macro-state are the statistical distributions of the individual molecules' energies, speeds and positions. In particular, an equilibrium

[13] This false assumption—analogous to the uniformity assumption discussed in Section 1.2—was actually made by J.P. Joule before the advent of statistical mechanics. This is discussed by us in Farjoun and Machover (1985).

macro-state is characterized by specific distributions, which are supposed to stay unchanged in the absence of external perturbation.[14]

Even such a macro-equilibrium does not exist in actual reality, because no system can in practice be perfectly isolated from external interactions. Nevertheless, fairly close approximations to it—good enough for practical and even for many theoretical purposes—do actually exist: this is what insulation is all about.

2.2 Economic Equilibrium

The analogy between the kind of system just described and an economy need hardly be spelled out.

The fundamental error of the uniformity assumption is that it conceptualizes the market forces driving the economy towards an ideal equilibrium as endogenous, while the disequilibrating forces perturbing the system are conceptualized as exogenous. But the latter surely include also market forces, the forces of competition. This is an untenable logical inconsistency.

The trouble with the kind of ideal equilibrium assumed in Section 1.1 is not that it is purely ideal but that it is prevented from occurring by the very forces that are supposed to drive the economy towards it.

The distinction between macro-state and micro-state is surely valid in economics no less than in statistical mechanics. And if a model of a capitalist economy is to have any verisimilitude, it must possess a large number of degrees of freedom: a micro-description must include detailed data on the simultaneous states of a great multitude of agents and the transactions between them.

The concept of economic equilibrium surely makes sense only as a macro concept, which is compatible—indeed presupposes—great mobility at the micro level. Basic economic quantities such as the rate of profit of an enterprise and unit prices must be conceptualized as random variables, which at equilibrium have specific distributions rather than determinate values.

Such a representation is needed even if in the end one is only interested in relations between the equilibrium mean values of these quantities. These mean values cannot, generally speaking, be taken in advance as reasonable approximations for the random variables themselves. The reason for this is that a given functional relation that holds between random variables does not, in general, hold between their means.[15]

Finally, simple observation suggests that a real economy of a country is most of the time not all that far from what common sense would regard as macro-equilibrium. Macro quantities, such as annual GNP, the level of employment, or

---

[14] Note that this is quite another matter from the micro-state staying unchanged. Similarly, if the age distribution of a population is unchanged, it does not follow that each member of it remains of constant age.

[15] For example, if X, Y and Z are random variables such that $\mathsf{XY} = \mathsf{Z}$, it is not in general true that $\mathrm{E}\,\mathsf{X} \cdot \mathrm{E}\,\mathsf{Y} = \mathrm{E}\,\mathsf{Z}$, where "E" denote the mean operator.

the statistical distribution of incomes, do not change very rapidly except at some critical moments, when instability becomes evident. It is therefore unreasonable to use a theoretical concept of equilibrium that is not even approached, let alone attained, by a real economy.

3 Methodological Comments

Let me end with a couple of methodological comments.

First, a mathematical model of a complex real system need not—indeed cannot—be realistic in every way. It is not a duplicate of reality but a simplified simulacrum of it. But the vital question is what simplifications are acceptable.

A model can work very well in simulating real behaviour even if it makes quite drastic simplifications of some aspects of reality. But it must fail if it simplifies away essential aspects of the reality under investigation. Of course, what counts as an essential aspect depends on the specific phenomena being investigated and on the purpose of the investigation.

The thesis that I have argued in this paper is that a non-stochastic concept of equilibrium is inappropriate for modelling the behaviour of prices and profits in a capitalist economy.

On the other hand, Ian Wright's paper (Wright, 2005) illustrates how a model that is in many ways very simplistic can nevertheless display in a surprisingly realistic way various phenomena of a capitalist economy. Wright's model does not impose a deterministic concept of micro-equilibrium; rather, the macro-equilibrium that emerges from it is stochastic.[16]

Second, it is quite fruitful for economic theory to look for concepts it can usefully borrow from natural science, particularly physics. Of course, by no means all physical concepts can be borrowed, or have useful analogues in economics.

Some specifically physical concepts—such as mass, force field, three-dimensional physical space, four-dimensional space-time, mechanical micro-equilibrium, and perhaps even energy (in the sense in which it features in the law of preservation of energy)—need not have useful applications or analogues in economic theory. On the other hand, more general concepts that originated in physics and engineering—such as feedback, degrees of freedom, steady state, macro-equilibrium, and perhaps entropy—seem to be quite usefully applicable.

[16] I am grateful to Ian Wright for the following comment on a draft of the present paper: "You may want to note that Econophysics is a relatively new field that has extensively used statistical equilibrium concepts with some success (particularly in providing very simple, abstract and satisfying explanations of the detailed income distribution). Your book (Farjoun and Machover, 1983) presaged this development by some years."

Wright's work is presented and amplified in a forthcoming book (Cottrell et al., 2006). For other recent work making use of ideas proposed by Farjoun and Machover (1983), see the Probabilistic Political Economy website: `https://www.jiscmail.ac.uk/cgi-bin/webadmin?A0=ppe`.

Bibliography

Cottrell, A., Cockshott, P., Michaelson, G., and Wright, I. (2006). Information, money and value. Draft manuscript.

Farjoun, F. and Machover, M. (1983). *Laws of Chaos: A Probabilistic Approach to Political Economy*. Verso/NLB, London.

Farjoun, F. and Machover, M. (1985). Probability, economics and the labour theory of value. *New Left Review*, 152:95–108.

Langston, R., Mandel, E., and Freeman, A. (1984). *Ricardo, Marx, Sraffa : The Langston memorial Volume*. Verso/NLB, London.

Schwartz, J. (1961). *Lectures on the Mathematical Method in Analytical Economics*. Gordon and Breach, London.

Steedman, I. (1977). *Marx after Sraffa*. NLB, London.

Wright, I. (2005). The social architecture of capitalism. *Physica A*, 346:589–622.

Chapter 23
Collective Choice as Information Theory: Towards a Theory of Gravitas

George Wilmers

The present paper introduces a new approach to the theory of voting in the context of binary collective choice, which seeks to define a dynamic optimal voting rule by using insights derived from the mathematical theory of information. In order to define such a voting rule, a method of defining a real-valued measure of the weight of independent opinion of an arbitrary set of voters is suggested, which is value free to the extent that it depends only on probabilistic information extracted from previous patterns of voting, but does not require for its definition any direct information concerning either the correctness or incorrectness of previous voting decisions, or the content of those decisions. The approach to the definition of such a measure, which I call *gravitas*, is axiomatic. The voting rule is then defined by comparing the gravitas of the set of those voters who vote for a given motion with the gravitas of the set of those who vote against that motion.

1 What Can We Learn from a Council of Elders?

As a motivating thought experiment[1] let us consider a precapitalist tribal society governed by a hereditary chief who takes all decisions *de jure*, but who is advised by a council of elders M which he chairs. Custom has determined that, after due deliberation, but prior to the chief making a final decision on any resolution before M, each elder must pronounce a declaration of opinion for or against the resolution: abstentions are not permitted. Let us imagine that M is considering a particular resolution. If the chief is wise then he will listen carefully to the advice he is given by the elders on the resolution; but how should he evaluate it? He may perhaps reason that, since his position is hereditary, he is unlikely to be wiser than the average

G. Wilmers (✉)
Lecturer in Mathematics and Member of the Mathematical Logic Research Group,
University of Manchester, Manchester, UK
e-mail: George.Wilmers@manchester.ac.uk

[1] The present section arose from my reflections on a conversation in 1968 between John Bell and my late father, recorded in John's autobiography *Perpetual Motion*.

D. DeVidi et al. (eds.), *Logic, Mathematics, Philosophy: Vintage Enthusiasms*,
The Western Ontario Series in Philosophy of Science 75,
DOI 10.1007/978-94-007-0214-1_23, © Springer Science+Business Media B.V. 2011

elder, even though he happens to be possessed of a certain mathematical knowledge and ability; hence his best policy may be to efface entirely all his own subjective judgments about the matters under deliberation both now and previously, and also to efface all his personal opinions about the value of the previous judgments of the elders. However if the chief is to eliminate from consideration all such personal judgments, then he must find some objective way to compare the weight of opinion of the set of those elders who are in favour of the particular resolution against the weight of opinion of those who are against the resolution. How are these two weights of opinion to be measured?

Our chief could of course simply count up those in favour and those against the resolution, and compare the resulting cardinalities, as the leaders of the great western democracies would surely enjoin him to do,[2] but his mathematical learning makes him extremely reluctant to throw away the extensive objective information which is contained in the pattern of advice given to him by the elders concerning previous resolutions. Also he has noticed that in the past the members of certain groups of elders have generally voted together in a rather predictable manner, so that he does not feel it appropriate to count their votes as if they represented quite separate opinions. He reflects that he would prefer to have at his disposal a measure of weight of *independent* opinion of each of the two sets of elders representing opposing viewpoints on the merits of approving the resolution. So, in order to ensure that his approach is truly objective, the chief decides to erase from his memory all details about the actual content of previous resolutions and of any advice he has been given previously, and to treat in a formal mathematical manner the information contained in the resulting abstract matrix of the elders' declarations for or against all previous proposals. The chief's mathematical problem is now how to extract from this remaining information two weights of independent opinion for comparison. This problem is in essence the subject matter of the present paper.

2 The Notion of Gravitas: A Preliminary Discussion

Our formal starting point is a fixed assembly M of n voters in a binary choice context, where M is endowed with a probability distribution σ on the set of possible of divisions $D(M)$ of M. Formally a *division* α of M is just a map from M to $\{0, 1\}$ which represents the *event* that the members of M vote on the latest motion before the assembly in such a manner that for all $a \in M$, a votes yes if $\alpha(a) = 1$ and no if $\alpha(a) = 0$. Since we are identifying divisions with events, our notation will allow logical disjunctions of divisions also to be treated as events, so that, e.g., for $\alpha, \beta \in D(M)$, $\alpha \vee \beta$ is the same event as $\beta \vee \alpha$.

We may think of σ as derived by some statistical rules from the evidence of previous voting records. We shall not concern ourselves here with exactly what

[2] ... provided, of course, that the results of such a calculation were likely to be consistent with their own assessments of the correct decision.

statistical procedures are used to derive such an a posteriori probability distribution on $D(M)$, but will instead take it as given. Thus we start with a mathematical idealization of the problem in the previous section. In general σ will be dependent on time since it will change as further information of the voting records of the members of M is accumulated. In the discussion below however we shall mostly treat σ as if it were fixed at a particular moment in time, and will take it as given at that moment in time, even though the concepts defined below should properly be thought of as defined relative to $\sigma(t)$ and variable time t.

Our fundamental question can now be phrased as follows. Suppose a new motion is presented to M and a given subset A of voters of M vote one way on the motion while the complement of A in M, A^c, vote the other way. Does there exist some natural *measure* which we can define in order to compare the "weight of independent opinion" of A with that of A^c?

In Section 3 we shall approach this question from an axiomatic standpoint, and we shall call this idea of the weight of independent opinion the *gravitas* of A, denoted by $G^\sigma(A)$. We formulate strong natural axioms for gravitas for arbitrary σ, which generalise the special classical situation in which σ is taken a priori to be the uniform distribution on $D(M)$, i.e. where the voters are a priori considered to vote independently with each voter voting yes with probability $\frac{1}{2}$. In particular the quantity $G^\sigma(A) - G^\sigma(A^c)$, or *gravitas margin*, generalises the classical notion of margin. We show that there exists a measure which satisfies the axioms, which we call polarity-free entropy (PFE). Although it does not seem easy to find an alternative measure to PFE which satisfies the given axioms for gravitas (other than a trivial translation by a constant), it is as yet unclear if there exist *intuitively convincing* additional axioms which would make PFE the unique solution for G^σ.

Given a notion of gravitas G^σ, we may define a voting rule R_{G^σ} by setting, for any division α such that the set of those who vote yes in α is A,

$$R_{G^\sigma}(\alpha) = \begin{cases} 1 & \text{if } G^\sigma(A) > G^\sigma(A^c) \\ 0 & \text{otherwise} \end{cases}$$

R_{G^σ}, which we call the *gravitas majority rule*, may be regarded as a conceptual generalization of the simple majority rule for the case in which the information contained in σ is available. It may be described as a realization of the intuitive concept of *rule by weight of independent opinion*. Furthermore, by analogy with the classical margin $|A| - |A^c|$, we will call the quantity $G^\sigma(A) - G^\sigma(A^c)$ the *gravitas margin*; the intuitive idea of the gravitas margin is to provide an indicator of reliability of a judgment arrived at by applying the rule R_{G^σ}. In the remainder of this section we shall consider briefly the philosophical background to these ideas.

The axiomatic approach which we are adopting differs considerably from that of more traditional semantic constructions which are used to interpret the meaning of the act of voting. In particular we make no a priori assumption that in the act of voting the individual voters are expressing personal opinions or personal preferences which are in any sense independent of the opinions or preferences of other voters.

Our intuitive philosophical focus is rather on analysing the properties of the *sets* of temporarily like-minded voters A and A^c which form when the assembly M is considering a particular motion, and on treating these subsets as being the important collective actors in an information theoretic analysis of voting.

There exists a large corpus of scholarly work on the mathematics of democratic choice, most of which can trace its philosophical origins either to the (quite separate) work of the eighteenth century luminaries (Condorcet, 1785; Rousseau, 1762), or to the twentieth century game theoretic considerations of social choice theorists arising from the celebrated impossibility theorem of (Arrow, 1963). In the case of unicameral binary choice the former tradition, which we may loosely call the *epistemic* tradition,[3] has been concerned primarily with the problem of examining the mathematical conditions under which a majority decision rule can be theoretically justified in the context where an objectively correct answer is assumed to exist,[4] while the latter tradition is concerned with the reconciliation of individual subjective preference orderings and seeks typically to examine under what conditions decision rules can avoid certain types of paradox or inconsistency. However, to our knowledge, there has been no work done on the axiomatic or mathematical foundations of a theory which would attempt to generalise the classical ideas of either Condorcet or Rousseau to the situation in which *extra objective information* is available in the form of the probability distribution σ.

The notion of gravitas which we present here and its associated decision rule R_{G^σ} could naturally be seen as belonging to the epistemic tradition. However the author believes that the notion of gravitas is relevant not just to a "Condorcet jury" type of context where an objectively correct answer is assumed to exist, but to a much more general context in which we require only that a correct answer to a motion put before M is accepted as existing with a *normative but probabilistic sense given to the meaning of the word correct, as being defined relative to certain limited but precisely defined information.* In the present case the limited information is taken to consist of σ together with the actual division of the voters on the given motion. Thus correctness in such a sense is an information-theoretic and relative notion: on the basis of certain symmetry and information theoretic principles, if we strictly limit the information available as above, then a particular outcome is deemed probabilistically correct *in that context.* Such a notion of probabilistic correctness relative to a precise informational framework may be viewed as an attempt to transcend the traditional distinction between epistemic and proceduralist interpretations of voting, and to provide a common epistemic analysis of voting theory.[5] This is however an

---

[3] See (Cohen, 1986; Coleman and Ferejohn, 1986) for philosophical discussions of the concept of an epistemic justification of democracy, and also for a critical discussion of proceduralist approaches.

[4] See e.g. (Grofman et al., 1983; Ladha, 1993; Boland, 1989; List and Goodin, 2001; List, 2004) for details.

[5] It may be noted that the philosophical idea of a separate notion of probabilistic correctness relative to limited information makes sense even in the case when we suppose that there exists an "objectively true" answer. For example, in a jury trial, the criterion for conviction is typically that

idea of for epistemic interpretation of voting which is very different from the usual notion of what might constitute an epistemic explication of voting.[6] We will not pursue the analysis of this idea further here, since it is peripheral to the development of the main ideas of this paper.

The general theory of voting is associated with probability theory in various ways, notably in the classical theory of voting power, and in Condorcet style justifications of majority decision rules. However we may reasonably ask the question why there has been so little theoretical work done at a foundational level on optimal collective decision rules in a context where additional objective information concerning prior individual voting records of members of an assembly is available, and in particular why that most powerful tool of mathematical reasoning under uncertainty, information theory,[7] has been so strikingly absent from deliberations. There are two related reasons for this situation, both of which have their origins in the tradition of centuries. The first of these reasons is that the foundational principle of "one person one vote" (OPOV), however hierarchically modified, underlies in some form or other all modern institutional collective forms of decision making; thus since the academic field of study of collective decision making is dominated by a consideration of *existing* types of institution, rather than a study of what might be possible, the consideration of fundamentally more complex decision rules invoking the use of additional information is normally ruled out a priori.[8] The other, related, reason is that, despite its rather weak theoretical justification, OPOV and its natural corollary of majority rule are ideologically so closely associated with the contemporary political concept of democracy, that any suggestion that some other conflicting principle might be more profound, more equitable, and might produce better collective judgments, is likely to meet with incredulity at best. In addition there are two further technical reasons why the point of view advocated in the present paper might not have appeared worthy of consideration until relatively recently. On the one hand the appropriate mathematical ideas from information theory have only been current

guilt is proved "beyond reasonable doubt." If therefore we make the reasonable assumption that all judgments in such trials are *de facto* made on a probabilistic basis, then, given that the information which can be made available to a jury is of necessity limited, a jury (or indeed an individual jury member) may in fact make a decision which is probabilistically *correct on the basis of the evidence available*, but which is nonetheless *incorrect in an absolute sense*. Our restriction of the admissible information available to the decision rule to σ together with the actual division of the voters, may in this case be interpreted as a uniform (or fair) method of reifying the information contained in the accumulated subjective judgments of jury members on the evidence available to them. (Of course this presupposes that an estimate for σ is actually available, which would not be the case for a one-time only jury).

[6] See (Cohen, 1986) for an account of the latter.

[7] In particular Shannon's notion of entropy (Shannon and Weaver, 1964); see e.g. (Paris, 1994) for a modern, detailed axiomatic presentation of the use of entropy in probabilistic reasoning.

[8] We may note here that types of information other than that encoded in σ might in principle also be recorded and used in the calculations of a decision rule; for example, normalised information about the strength of conviction which individual voters attach to individual judgments could be recorded and used in some way. A closely related point is made in Dummett's discussion of Arrow's theorem in (Dummett, 1984).

in the last half century, while on the other hand the necessary technology of instant communication, and the computational power necessary to process the raw voting data in order to estimate numerical values for a notion of gravitas have only become available within the last 20 years.[9]

In the general context of the idealised Condorcet jury, where there exists a clearly defined objectively correct answer associated with each motion put before the assembly, it will in many cases be possible to carry out experiments to determine empirically whether or not a rule R_{G^σ} associated with a particular definition of gravitas G compares favorably in the decisions it generates when comparison is made with the simple majority rule, or indeed with a rule $R_{G'^\sigma}$ where G' is some other notion of gravitas. For example, where a disease can be infallibly diagnosed by some laboratory test, a panel of medical experts could be asked to evaluate for a lengthy sequence of patients whether or not, on the basis of clinical evidence, each of them had the disease. The votes of the experts would in each case be aggregated separately using R_{G^σ} and the simple majority rule, and a comparison of the results could then be made with the objectively correct answers which would be supplied by applying the laboratory test. Extensive experiments of this kind over a variety of different scientific domains could be used to provide strong evidence for or against the efficacy of a rule R_{G^σ} for a particular definition of G. If it turns out that a particular definition of G can be uniquely characterised by a convincingly natural set of axioms, then empirical evidence of the above kind could provide powerful independent evidence in favour of adopting R_{G^σ} as a decision rule in a more general context.

We should remark that, although in the Condorcet jury context there has been considerable research carried out concerning the performance of various voting rules which employ extra information concerning the *competence* of individual voters, such voting rules have an entirely different nature to the approach adopted in the present paper, since they depend on information concerning the correctness of previous judgments, whereas our framework of analysis makes no assumption that such information is available.

3 Axioms for Gravitas

Before we state our axioms we need to establish some simple notation. The probability distribution σ on $D(M)$ extends naturally to a probability function on the set of disjunctions of elements of $D(M)$ and we shall identify σ with this extension, so that for example if $\alpha, \beta \in D(M)$ with $\alpha \neq \beta$, then $\sigma(\alpha \vee \beta) = \sigma(\alpha) + \sigma(\beta)$. Also for every $A \subseteq M$, σ induces a probability distribution σ_A on $D(A)$, the set of divisions of A. In fact for any $\alpha \in D(A)$ $\sigma_A(\alpha) = \sigma(\alpha)$.

---

[9] I do not intend by this statement to minimise the difficulty of the computational problems involved, which I have not addressed here, and which would certainly be substantial in the case of a large electorate.

We now introduce our axioms, and explain briefly the motivation behind them. It is understood that the axioms should hold for all possible M and σ. We also assume that the gravitas function G^σ takes real number values in the interval $(0, \infty)$.

Continuity Axiom For any $A \subseteq M$, $G^\sigma(A)$ is continuous as a function of σ.

This axiom simply expresses mathematically the intuitive idea that the gravitas function should be a smooth function with respect to changes in σ: although gravitas might change quite rapidly as σ changes, there should be no sudden jumps in its values.

Locality Axiom For every $A \subseteq M$, $G^\sigma(A)$ is a function of σ_A alone.

This axiom expresses the intuitive idea that the gravitas of the set of voters A should depend only on the behaviour of the voters in A, and should in particular be independent of how the remaining voters of M vote. While this property is very natural, there does exist however an alternative natural point of view, and we shall return to this in our considerations later.

Voter Renaming Axiom Let π be a permutation of the voters of M which, given σ, induces the probability distribution σ^π on M defined by $\sigma^\pi(\alpha^\pi) = \sigma(\alpha)$ for each $\alpha \in D(M)$, (where α^π denotes the obvious permutation of the division α). For any $A \subseteq M$ let A^π denotes the image of A under π.

Then $G^{\sigma^\pi}(A^\pi) = G^\sigma(A)$.

This axiom is just a version of the familiar idea of anonymity; the gravitas of A should not depend on the names which the elements of A happen to possess but only on their properties as determined by σ.

Monotonicity Axiom For any $A \subseteq M$ and $b \in M$, $G^\sigma(A) \leq G^\sigma(A \cup \{b\})$.

This axiom expresses the idea that adding a new member to a set of voters A cannot decrease the gravitas of A, given that the voting behaviour of the other members of A remains unchanged. Note that this natural assumption immediately implies that the voting rule R_{G^σ} is monotone.

Clone Axiom For any $A \subseteq M$, if $a, b \in A$ are distinct voters such that the probability (calculated using σ) that a votes the same way as b is 1,

Then $G^\sigma(A) = G^\sigma(A - \{b\})$.

This axiom just expresses the idea that if two voters in A behave identically, then one of them is redundant in calculating the gravitas of A since the two voters vote systematically as if they were of one opinion. The axiom reflects the intuitive idea that in calculating gravitas we are seeking to count not voters, but independent points of view.

For any $A \subseteq M$ and $\alpha \in D(A)$, let $\overline{\alpha}$ denote the dual division to α in which each member of A votes the opposite way to the way they voted in α. We can now state our next axiom.

Polarity Free Axiom For any $A \subseteq M$, $G^\sigma(A)$ depends only on the values $\sigma(\alpha \vee \overline{\alpha})$ where $\alpha \in D(A)$.

This axiom needs some explanation. The idea here is that the *actual* direction (for or against motions) in which voters vote is immaterial in calculating a measure of their independence: all that matters is their voting patterns *relative to each other*. So if σ were altered because a proportion of motions were arbitrarily replaced by their negations, this should not affect the value of $G^\sigma(A)$, assuming that the voters would reverse their votes in line with their beliefs. Obviously this axiom represents a strengthening of the Locality Axiom which could have been included in it. However because of its different and less obvious status, we have separated it from the Locality Axiom.

Let us denote by σ_A^* the probability distribution which is obtained from σ by considering just the set of events of the form $\alpha \vee \overline{\alpha}$ where $\alpha \in D(A)$. Thus the Polarity Free Axiom asserts that $G^\sigma(A)$ depends only on the information in σ_A^*. In the case when A is M we will write σ^* to denote σ_M^*.

We shall call an event of the form $\alpha \vee \overline{\alpha}$ a *polarity free division of* A. More generally any disjunction of polarity free divisions of M may be referred to as a *polarity free* event. Trivially, polarity free events are closed under Boolean operations.

Clearly σ^* contains less information than σ. However the information which it contains has an interesting epistemological status as we now explain.

For the purpose of the present discussion let us now assume the existence of some objective notion of correctness for all motions presented to M. Then by complete analogy with σ there exists another probability distribution τ which encapsulates information about the *correctness* of the previous voting of voters in M. To make this precise we define for each division $\alpha \in D(M)$ an analogous event $\widehat{\alpha}$ by replacing "voting yes" by "voting correctly" and "voting no" by "voting incorrectly" in the definition of α. We call $\widehat{\alpha}$ a *truth-division of* M and we let $D^T(M)$ be the set of all truth-divisions of M. Also, given some $\widehat{\alpha} \in D^T(M)$, we may define its dual truth-division $\overline{\widehat{\alpha}}$ to be $\widehat{\overline{\alpha}}$. Again by analogy we call an event of the form $\widehat{\alpha} \vee \overline{\widehat{\alpha}}$ a polarity free truth-division of M. Then analogously to σ there exists a probability distribution τ on $D^T(M)$ which could be defined in the same way from records as to whether voters voted correctly or incorrectly, *if such records existed*, as σ is defined from records of actual yes or no votes cast. Of course, unlike σ, the distribution τ will not in general be accessible to us, since except under rather special circumstances we will not have access to the data from which τ would be constructed.

$\tau(\widehat{\alpha})$ tells us the probability of the event $\widehat{\alpha}$ occurring, based solely on the record of correctness of the previous votes of members of M. However we may now notice that for any M and any $\alpha \in D(M)$ the events $\alpha \vee \overline{\alpha}$ and $\widehat{\alpha} \vee \overline{\widehat{\alpha}}$ are extensionally identical, and hence, provided that enough previous votes are taken into account, it will be the case that σ and τ nearly coincide for such events, i.e., that

$$\sigma(\alpha \vee \overline{\alpha}) \simeq \tau(\widehat{\alpha} \vee \overline{\widehat{\alpha}})$$

for any polarity free division $\alpha \vee \overline{\alpha}$ of M, where the approximation tends to equality as the number of previous votes taken into account increases. So *if we have no access to the records of correctness* from which τ would be constructed, we should regard the function σ restricted to these polarity free events as the best approximation available, say ρ^*, to the values which τ would give to the polarity free truth divisions; i.e. we define ρ^* by

$$\rho^*(\widehat{\alpha} \vee \overline{\widehat{\alpha}}) = \sigma^*(\alpha \vee \overline{\alpha})$$

for any polarity free division $\alpha \vee \overline{\alpha}$ of M.

Two linked foundational questions now arise. The first question may be stated as follows: if we are given *only* the information contained in σ^*, how should ρ^* be extended to a probability distribution ρ defined on the whole of $D^T(M)$ in such a manner that, from a purely information-theoretic standpoint, ρ is the best estimate of τ we can make in the absence of any other information? The second, related, question is: if instead we are given all the information contained in σ, how should we extend σ to a natural joint distribution on the Boolean algebra of events generated by $D(M) \cup D^T(M)$? These questions will not be considered further here but will be pursued in a later paper.[10]

Our last two axioms generalise properties of the classical notion of margin. The absolute values of $G^\sigma(A)$ are intuitively less important than a comparison of the values of $G^\sigma(A)$ and $G^\sigma(A^c)$. For any measure of gravitas G we let $Mar_{G^\sigma}(A)$ denote the G^σ*–margin* of A in M, i.e., we define

$$Mar_{G^\sigma}(A) = G^\sigma(A) - G^\sigma(A^c).$$

Now the classical margin of A (over A^c) is of course just $|A| - |A^c|$. So if $Mar_{G^\sigma}(A)$ is to generalise the classical margin we should expect that the two notions would coincide for the paradigm case of the uniform distribution on $D(M)$. Accordingly we may now state the following

Classical Margin Axiom Let *unif* denote the uniform distribution on $D(M)$. Then for any $A \subseteq M$

$$Mar_{G^{unif}}(A) = |A| - |A^c|$$

In the case of the simple majority rule, the classical margin $|A| - |A^c|$ provides, in a Condorcetian analysis, an indicator of the probability that the majority decision is correct; although this analysis is dependent on absurdly idealised assumptions concerning voters' independence, nevertheless under these special conditions the classical margin possesses certain attractive invariance properties.[11] So it is natural that if we are seeking to generalise the concept of margin to $Mar_{G^\sigma}(A)$ for arbitrary σ, then we should seek to ensure that this generalisation possesses a strong conceptual stability. Our final axiom below should be interpreted with this in mind.

For the purposes of defining our final axiom we will assume that $G^\sigma(A)$ satisfies the Locality, Polarity Free, and Voter Renaming axioms. Recall that we have

---

[10] We confine ourselves here to noting that the first of these questions can reasonably be considered an analogue in collective choice theory of the problem in uncertain reasoning (by a single agent) of choosing a canonical probability distribution from a set of possible distributions constrained by certain data. The method of choice for solving the latter problem is the use of the maximum entropy principle (see, e.g., (Paris, 1994)), but without additional insight maximum entropy appears powerless to help in solving the former problem. The author believes however that the notion of gravitas can be used to provide the appropriate missing idea necessary to partially solve this problem.

[11] The importance of invariance properties of the notion of margin in a classical Condorcetian analysis of voting has been emphasized by List (2004).

insisted by the Locality and Polarity Free axioms that $G^{\sigma}(A)$ depend only on σ_A^*; in particular $G^{\sigma}(A)$ therefore depends only on the probability of polarity free events, i.e., of events which refer only to how the voters vote relative to each other, not to which way they actually vote. However, if we now take as given some such G and consider instead as a possible alternative notion of gravitas the expected value of G on A with σ_A^* conditioned upon the polarity free divisions corresponding to every possible way in which the members of A^c could divide, then we obtain a rather natural, but *not* locally defined quantity, which we will define below and will denote by $\mathcal{E}_{G^{\sigma}}(A)$. We call the function $\mathcal{E}_{G^{\sigma}}$ of subsets A of M the *polarity free expectation (over M)* of G^{σ}.

There is a slight notational difficulty in formally defining $\mathcal{E}_{G^{\sigma}}$. This difficulty arises because if we consider a polarity free division $\alpha \vee \overline{\alpha}$ where $\alpha \in D(A)$ and condition this event on the polarity free event $\beta \vee \overline{\beta}$ where $\beta \in D(A^c)$, then we need to consider the conditional probabilities of two possible alternative polarity free divisions of M, namely $\alpha\beta \vee \overline{\alpha\beta}$ and $\overline{\alpha}\beta \vee \alpha\overline{\beta}$. The respective conditional probabilities of these events are given by

$$\frac{\sigma(\alpha\beta \vee \overline{\alpha\beta})}{\sigma(\beta \vee \overline{\beta})} \quad \text{and} \quad \frac{\sigma(\overline{\alpha}\beta \vee \alpha\overline{\beta})}{\sigma(\beta \vee \overline{\beta})} \quad .$$

Note that if $|A| = m$ say, then for each event $\beta \vee \overline{\beta}$ as above there are 2^m disjoint atomic polarity free divisions of M as above and 2^m corresponding conditionalised probability values summing up to 1. If we denote by $\sigma^*_{A|\beta\vee\overline{\beta}}$ the probability distribution just described, then this distribution may be thought of as a distribution on the 2^m polarity free divisions of the set $A \cup \{d\}$, where $d \notin A$ is treated as a placeholder element whose value in a division indicates the relative polarity of $\beta \vee \overline{\beta}$ to $\alpha \vee \overline{\alpha}$ corresponding to the two forms above. To be precise: given a fixed $\beta \in D(A^c)$, let $\gamma \in D(A \cup \{d\})$ be such that $\gamma(d) = 1$. Let $\alpha \in D(A)$ be defined by $\forall a \in A\ \alpha(a) = \gamma(a)$. Then the polarity free event $\gamma \vee \overline{\gamma}$ is identified with $\alpha\beta \vee \overline{\alpha\beta}$.

With this interpretation we may define for non-empty A and A^c

$$\mathcal{E}_{G^{\sigma}}(A) = \frac{1}{2} \sum_{\beta \in D(A^c)} \sigma(\beta \vee \overline{\beta})\ G^{\sigma^*_{A|\beta\vee\overline{\beta}}}(A \cup \{d\}).$$

The factor of $\frac{1}{2}$ is present since otherwise because of the notation each event $\beta \vee \overline{\beta}$ would be counted twice. For the special cases when A or A^c is empty, it is natural to define $\mathcal{E}_{G^{\sigma}}(A) = G^{\sigma}(A)$.

Now, given some notion of gravitas G^{σ}, one can reasonably argue that, despite the nonlocality of its definition, $\mathcal{E}_{G^{\sigma}}(A)$ has an almost equally good claim to be considered as a measure of gravitas as $G^{\sigma}(A)$, since intuitively it just represents the expected value of $G(A)$ conditionalised on all possible appropriate events corresponding to the behaviour of the rest of the assembly, A^c. At first sight it would be nice therefore if G^{σ} and its polarity free expectation $\mathcal{E}_{G^{\sigma}}$ could be made identically equal. This turns out to be too strong a requirement: it results in inconsistency. As we have stressed, however, the important function to be considered for possible invariance properties is the gravitas *margin* rather than gravitas itself. So it is pleasing to discover that the following strong axiom is in fact satisfiable:

Polarity Free Margin Invariance For every $A \subseteq M$,

$$Mar_{\mathcal{E}_{G^\sigma}}(A) = Mar_{G^\sigma}(A).$$

This concludes our list of axioms for the notion of gravitas.

4 Polarity Free Entropy (PFE)

In this section we define a measure, Polarity Free Entropy, or PFE, which satisfies all eight axioms for a notion of gravitas, G, described in the previous section, namely the Continuity, Locality, Voter Renaming, Monotonicity, Clone, Polarity Free, Classical Margin, and Polarity Free Margin Invariance axioms.

Definition 1 Given M, $A \subseteq M$ and σ,

$$PFE^\sigma(A) = -\sum_{\alpha \in D(A)} \frac{\sigma(\alpha \vee \overline{\alpha})}{2} \, log_2 \frac{\sigma(\alpha \vee \overline{\alpha})}{2} \quad \text{if } A \neq \emptyset$$

$$0 \qquad \text{otherwise.}$$

Note that for $A \neq \emptyset$ the definition is just one plus the usual Shannon entropy (to the base 2), but taken over the set of polarity free events $\alpha \vee \overline{\alpha}$. Another, perhaps more natural, interpretation of $PFE^\sigma(A)$ is as the Shannon entropy of a hypothetical distribution $h(\sigma_A)$ obtained from σ_A by, for every $\alpha \in D(A)$, redistributing the probability of each polarity free event $\alpha \vee \overline{\alpha}$ equally over the events α and $\overline{\alpha}$. Here one can think of the entropy of $h(\sigma_A)$ as being a measure of the uncertainty in σ_A if one "forgets," or discards as irrelevant, the available information about the *particular* manner in which the value of each $\sigma_A(\alpha \vee \overline{\alpha})$ is subdivided by σ_A between α and $\overline{\alpha}$.

It follows from the above definition that the polarity free expectation of PFE^σ function, $\mathcal{E}_{PFE^\sigma}$, as defined in the previous section, is given by

$$\mathcal{E}_{PFE^\sigma}(A) = 1 - \frac{1}{2} \sum_{\beta \in D(A^c)} \sigma(\beta \vee \overline{\beta}) \sum_{\alpha \in D(A)} \frac{\sigma(\alpha\beta \vee \overline{\alpha}\overline{\beta})}{\sigma(\beta \vee \overline{\beta})} \, log_2 \frac{\sigma(\alpha\beta \vee \overline{\alpha}\overline{\beta})}{\sigma(\beta \vee \overline{\beta})}$$

for $A \neq \emptyset, M$, with $\mathcal{E}_{PFE^\sigma}(A) = PFE^\sigma(A)$ otherwise.

It is now straightforward to verify that

Theorem 2 *The measure of gravitas PFE defined above satisfies the Continuity, Locality, Voter Renaming, Monotonicity, Clone, Polarity Free, Classical Margin, and Polarity Free Margin Invariance axioms.*

In addition PFE has the following two properties:

1. For any $A \subseteq M$ and any σ

$$\mathcal{E}_{PFE^\sigma}(A) = PFE^\sigma(M) - PFE^\sigma(A^c) + \delta_A$$

   where $\delta_A = 0$ if $A = \emptyset$ or M, and $\delta_A = 1$ otherwise.
   It is this equation, a translation by δ_A of the equation satisfied by the classical Shannon entropy, which ensures that the axiom of polarity free margin invariance holds: the result then follows just by writing the equations corresponding to $\mathcal{E}_{PFE^\sigma}(A)$ and $\mathcal{E}_{PFE^\sigma}(A^c)$ and subtracting.
   In the special case of the uniform distribution, it is trivial to verify that PFE satisfies a strong form of the classical margin axiom, namely

$$PFE^{unif}(A) = |A|.$$

We may remark here that although any translation of PFE^σ by the addition of a constant value K still satisfies the eight axioms of the previous section, if we also we require the property (2) above to be satisfied, then no nontrivial such translation is possible.

We should also note that if A is a singleton then $PFE^\sigma(A) = 1$ for any σ.

It is also worth remarking that if we define $G^\sigma(A)$ trivially to be $|A|$ for all σ, then this function satisfies all the axioms of the previous section except the clone axiom. It might therefore at first sight be concluded that the axioms are rather weak. This however is not at all the case: in the presence of the clone axiom the remaining axioms, and especially the polarity free margin invariance axiom, take on a far stronger meaning.

5 Conclusions and Open Problems

Much further research is necessary to elucidate the foundations of a theory of gravitas, together with the gravitas majority (or supermajority) decision rules which can be derived from the concept. The axioms suggested, in particular those involving the notion of polarity freeness, are by no means unchallengeable. In fact these axioms emerged because the author started by investigating a simpler notion of gravitas, consisting simply of the usual Shannon entropy of σ_A, namely $\sum_{\alpha \in D(A)} -\sigma(\alpha)\, log_2\, \sigma(\alpha)$. This definition has many pleasant properties and satisfies all the axioms given in Section 3 above except the polarity free axiom and the polarity free margin invariance axiom. However this last axiom should not really be counted as a failure since Shannon entropy *does* satisfy the simpler margin invariance axiom which can be formulated in the absence of the "polarity free" requirement, by replacing $\mathcal{E}_{G^\sigma}$ by the conditional Shannon entropy E_{G^σ}, defined by

$$E_{G^{\sigma}}(A) = \sum_{\beta \in D(A^c)} \sigma(\beta)\, G^{\sigma|\beta}(A)$$

where $\sigma|\beta$ denotes the conditionalisation of σ_A to the event β.

With this change we get the axiom of

Margin Invariance For every $A \subseteq M$, $Mar_{E_{G^{\sigma}}}(A) = Mar_{G^{\sigma}}(A)$.

In the case when G is interpreted as Shannon entropy $E_{G^{\sigma}}$ is simply the usual conditional entropy of A given A^c; then the analogous property to (1) of the previous section holds (i.e., with the constant term δ_A deleted) which immediately implies that the above axiom of margin invariance is satisfied. Furthermore Shannon entropy also possesses an at-first-sight attractive property which is not possessed by PFE: namely it is additive for the union of two disjoint sets of voters A and B in the case when the probability distributions over A and over B are independent of each other. Nevertheless, as a measure of gravitas Shannon entropy possesses some difficult counterintuitive properties. We can see this by looking at the example of a singleton A. Here the Shannon entropy varies between 0 and 1 depending on how close to $\frac{1}{2}$ the probability that the unique member of A votes yes is. This does not seem to make much sense as a measure of gravitas: in a two person committee we would surely not prefer, in the absence of other information, the judgement of a voter whose previous record indicated she was equally likely to vote yes or no, against a voter who previously almost always voted no, but on this particular occasion voted yes! This example is not really a problem for PFE however since, as noted above, PFE gives each individual voter an equal gravitas of 1.

Another reason to distrust Shannon entropy as a measure of gravitas is that the Shannon entropy of A satisfies a very strong symmetry property which we may call

Division Renaming $G^{\sigma}(M)$ is invariant under any permutation of $D(M)$ if the probability distribution σ is adjusted to reflect the permutation.[12]

For ease of notation we have stated the property for M, but it is clear that in the case when G is Shannon entropy (or in the presence of the Locality axiom) we could replace M by an arbitrary subset A. This property is far stronger than voter renaming: in a sense it makes the voters almost redundant to the calculation of G^{σ} since the divisions now all acquire identical status as abstract objects, and their original relationship to the voters of A appears to be irrelevant. This does not seem to have any intuitive justification as far as the notion of gravitas is concerned; furthermore it is inconsistent with the polarity free axiom, modulo only the trivial requirement that $G^{\sigma}(M)$ be not independent of variations of σ.

[12] That is, if π is a permutation of $D(M)$ and σ^{π} is defined by $\sigma^{\pi}(\pi(\alpha)) = \sigma(\alpha)$ for each $\alpha \in D(M)$, then $G^{\sigma^{\pi}}(M) = G^{\sigma}(M)$. Note that whereas a permutation of M always induces a permutation of $D(M)$ the converse is not true.

Turning to the rule R_{G^σ} discussed briefly in Section 2 as a motivation for the study of gravitas, it is worth noting that in a dynamic context where $\sigma(t)$ is changing over time, the clone and continuity axioms appear to ensure that such a rule would have the effect of strongly discouraging the formation of factions, by penalizing the voting power (or success) of any such faction: over time this would occur quite irrespective of whether the factions existed as formal entities.

To analyse this claim further let us define the *success rate* of a voter as the probability that that voter will belong to the winning camp when a new motion is presented and a vote is taken.[13] One of the most intractable problems arising from the use of conventional static voting rules is the fact that such voting rules are prone to instabilities caused by the increase in success rates which a set of voters may achieve individually by forming a faction which votes as a block, using an internal voting rule to decide which way all the members of that faction vote. The formation of one such faction in turn encourages the formation of other factions in a process which is inherently unstable unless one faction is large enough to constitute by itself a winning coalition, i.e., a majority dictatorship.[14] More serious than the instabilities, however, is the fact that voters are no longer voting honestly on the individual motions presented to them. In a political context the factions formed in the above manner are often given a formal status and called parties; we are now culturally so accustomed to this phenomenon that, even though the negative effects of party discipline on the discourse of politicians are well recognised, the existence of parties is regarded as an intrinsic part, or even a *sine qua non*, of the process of democratic decision making. Yet the possibility of a dynamic voting rule such as $R_{G^{\sigma(t)}}$ indicates that the phenomenon of dishonest voting can at least be actively discouraged. For if $R_{G^{\sigma(t)}}$ is taken to be the voting rule then as soon as a faction has formed for long enough for the effects of block voting to be partially reflected in the probability distribution $\sigma(t)$, the gravitas of any set of voters including that faction as a subset would tend to decrease; hence it seems likely that the formation of a faction would result in at least some members of that faction suffering a decreased success rate shortly after its formation, thus undermining the *raison d'être* of the faction. This can be seen as encouraging honest voting, and as a strong disincentive to the formation of factions. While this positive effect seems intuitively clear for a gravitas majority voting rule, rigorous mathematical results along these lines are likely to be hard both to formulate and to prove.

In the light of the above, it is interesting to consider Rousseau's observations concerning the problems arising from the formation of factions in a political context. According to Rousseau's notoriously ill-defined, but sometimes unfairly maligned, intuitive concept of "general will," the general will is always correct, but may well be at variance with the vote of the majority (see (Rousseau, 1762)). In Rousseau's conception the general will cannot be directly accessed, but while the opinion of the majority provides an indication of the general will, presumably in some probabilistic

---

[13] See (Laruelle et al., 2006) for an analysis of the concept of success in the context of voting systems.

[14] Effects of this kind have been studied in a number of recent papers on voting power; see (Felsenthal and Machover, 2002, 2008; Gelman, 2003).

sense, it can be "mistaken." Reasons given by Rousseau as to why such "errors" can occur include insufficient or incorrect information available for the formation of judgments, and especially distortions caused by the formation of factions:

> If, when the people, being furnished with adequate information, held its deliberations, the citizens had no communication one with another, the grand total of the small differences would always give the general will, and the decision would always be good. But when factions arise, and partial associations are formed at the expense of the great association, the will of each of these associations becomes general in relation to its members, while it remains particular in relation to the State: it may then be said that there are no longer as many votes as there are men, but only as many as there are associations. The differences become less numerous and give a less general result. Lastly, when one of these associations is so great as to prevail over all the rest, the result is no longer a sum of small differences, but a single difference; in this case there is no longer a general will, and the opinion which prevails is purely particular. It is therefore essential, if the general will is to be able to express itself, that there should be no partial society within the State, and that each citizen should think only his own thoughts .... (Rousseau, 1762, Book 2, Ch. 3)

There is in this quotation from Rousseau a serious conceptual problem which arises from the first sentence, and which has been much commented upon. It is difficult to understand how people could be both well-informed and have thoroughly considered a question, if they have no communication with one another.[15] It seems here as if Rousseau is grappling with an intractable difficulty, because while he desires an informed people who have fully deliberated the questions to be decided, he is painfully aware of the negative effects on the voting outcome which may be caused by factional substructures. However we can now see that the use of a gravitas majority rule could well cut through the difficulty which Rousseau was facing; under a gravitas majority rule, the influence of factions is likely to evaporate soon after they start to be formed, and the outcome of the voting rule can once again be considered, figuratively, the "sums of small differences" of independent opinions, even though there is nothing like a bijective correspondence between independent opinions and individual voters.

If indeed some sense can be made of Rousseau's idea of the general will, a question concerning which this author takes no position, then the notion of gravitas margin seems likely to provide a far more plausible indication of the general will than that provided by the classical margin. Indeed if the general will is interpreted as the limit of a probabilistic notion, then it may well be possible, using the notion of gravitas, to give the general will a more precise sense, which would be reasonably faithful to Rousseau's underlying idea.

For many reasons the ideas put forward in the present paper are likely to be met with a certain scepticism; hence it seems appropriate to end this paperby posing two precise philosophical challenges to those who would question whether a notion of gravitas majority can serve any useful purpose in the context of human systems of collective choice:

---

[15] The original French text of the first sentence, which is difficult to translate exactly, is as follows: "Si, quand le peuple suffisamment informé délibère, les citoyens n'avaient aucune communication entre eux, du grand nombre de petites différences résulterait toujours la volonté générale, et la délibération serait toujours bonne."

1. In defining a decision rule for collective choice, is there some convincing philosophical principle which would exclude as unreasonable the use of additional information about the abstract relationships between voters' previous choices such as that encoded by σ?
2. Suppose that it could be demonstrated by empirical methods as suggested in Section 2 that a particular gravitas majority rule R_{G^σ} performed systematically better than the simple majority rule when applied in contexts in which the correctness of M's decisions could be compared with an independently accessible objective truth. To what extent would such empirical evidence validate the use of the rule R_{G^σ} in contexts (a) where there exists an independent standard of objective truth or correctness which is not in general accessible, or (b) where there exists no independent standard of objective truth or correctness?

Acknowledgement I wish to express my thanks to Moshé Machover, Alena Vencovska, Hykel Hosni, Luc Bovens and Greg Holland for helpful comments on earlier versions of some of the ideas presented in this paper.

Bibliography

Arrow, K. (1963). *Social Choice and Individual Values.* Yale University Press, New Haven, CT.

Boland, P. J. (1989). Majority systems and the Condorcet jury theorem. *The Statistician,* 38:181–189.

Cohen, J. (1986). An epistemic conception of democracy. *Ethics,* 97:26–38.

Coleman, J. and Ferejohn, J. (1986). Democracy and social choice. *Ethics,* 97:6–25.

Condorcet, M. d. (1785). *Essai sur l'application de l'analyse à la probabilité des décisions rendues à la pluralité des voix.* de l'imprimerie Royale, Paris.

Dummett, M. (1984). *Voting Procedures.* Oxford University Press, Oxford.

Felsenthal, D. and Machover, M. (2002). Annexations and alliances: When are blocs advantageous a priori? *Social Choice and Welfare,* 19:295–312.

Felsenthal, D. and Machover, M. (2008). Further reflections on the expediency and stability of alliances. In Braham, M. and Steffen, F., editors, *Power, Freedom and Voting.* Springer, Berlin.

Gelman, A. (2003). Forming voting blocs and coalitions as a prisoner's dilemma: a possible theoretical explanation for political instability. *Contributions to Economic Analysis and Policy,* 2:1–14.

Grofman, B. N., Owen, G., and Feld, S. L. (1983). Thirteen theorems in search of the truth. *Theory and Decision,* 15:261–278.

Ladha, K. K. (1993). Condorcet's jury theorem in light of de Finetti's theorem. *Social Choice and Welfare,* 10:69–85.

Laruelle, A., Martinez, R., and Valenciano, F. (2006). Success versus decisiveness: Conceptual discussion and case study. *Journal of Theoretical Politics,* 18:185–205.

List, C. (2004). On the significance of the absolute margin. *British Journal for the Philosophy of Science,* 55:521–544.

List, C. and Goodin, R. (2001). Epistemic democracy: Generalizing the Condorcet jury theorem. *Journal of Political Philosophy,* 9:277–306.

Paris, J. (1994). *The Uncertain Reasoner's Companion.* Cambridge University Press, Cambridge.

Rousseau, J. (1762). *Du Contrat Social.* Available online at http://abu.cnam.fr/BIB/index.html. English translation by G.D.H. Cole avalailable at http://www.constitution.org/jjr/socon.htm

Shannon, C. and Weaver, W. (1964). *The Mathematical Theory of Communication.* University of Illinois Press, Urbana, IL.

Chapter 24
Scientific Knowledge and Structural Knowledge

Peter Clark

1 Introduction

There is a very general question in the philosophy of science. It is this: how do we think our theories represent the world? What is theoretical knowledge in the sciences, knowledge of? One very influential answer to this question is the thesis of Structuralism: theoretical knowledge is not knowledge of unobservable objects and the hidden relations among them but is always and only knowledge of structure.

There is no doubt that Structuralism is currently a very influential thesis. From its stronghold within the philosophy of mathematics it has recently burst forth into the methodology and the epistemology of science with a proposed solution to the pessimistic induction. The leading advocates of this view, John Worrall and Elie Zahar (see especially (Worrall, 1989, 2007; Zahar, 2001)) describe it as being the only viable solution to the problem of giving a (minimally) realistic interpretation of the natural sciences. Thereby they claim it meets the challenge of Constructive Empiricism, which is the claim that science does not aim at the truth but at empirical adequacy alone.

Not only this but also it has sallied forth into the metaphysics of science, as an interpretation of modern physics and virtually the whole of the special sciences. Here its champions are James Ladyman and Don Ross[1] whose recent book is aptly and humourously entitled "Every *Thing* Must Go" (the emphasis is mine). They claim two further motivations which go far beyond Worrall and Zahar's modest defensive operation. They see a need for "an ontology apt for contemporary physics, and a way of dissolving some of the metaphysical conundrums it presents" and "for a conception of how theories represent the world that is compatible with the role of models and idealisations in physics" (Ladyman and Ross, 2007, p. 131).

P. Clark (✉)
Proctor and Provost of St Leonard's College, Vice-Principal and Dean of Graduate Studies; Professor of Philosophy, University of St. Andrews, St. Andrews, UK
e-mail: pjc@st-andrews.ac.uk

[1] See (Ladyman and Ross, 2007). These are the main authors. The volume also contains contributions jointly written with James Ladyman and Dan Ross by David Spurrett and John Collier.

D. DeVidi et al. (eds.), *Logic, Mathematics, Philosophy: Vintage Enthusiasms*, The Western Ontario Series in Philosophy of Science 75,
DOI 10.1007/978-94-007-0214-1_24,

These are worthy aims indeed, but the battle cry with which they lead their assault is really revolutionary. They deny that "strictly speaking there are 'things'," they deny that "in the material world as represented by the currently accepted scientific structures, individual objects have any distinctive status." They argue that "some real patterns behave like things, traditionally conceived, while others behave like traditional instances of events and processes." But that "from the metaphysical point of view, what exists are just real patterns .... Science motivates no separate metaphysical theories about objects, events and processes" (p. 121). Their clarion call is "There are no things, Structure is all there is."

Now, one must not underestimate the seriousness of their claim. Their claim is not that there are some queer things about in modern physics, entities with startling and peculiar properties, for we can all agree on that, but rather it is that they want to do without entities of any sort whatsoever and rely merely on the existence of patterns (to use Resnik's phrase) and structures. It is a manifest fact that modern physical theory can be formalised in first-order set theory with individuals as urelements and consequently that in that formalisation there are universally and existentially quantified expressions, the quantifiers thought of as ranging over sets and the urelements. If to be asserted to be by a theory is to be the value of a bound variable occurring in that theory, then it would appear that modern physical theory is committed to all manner of entities, objects and things which serve as the range of those variables, in the standard univocal Quinean way. Ladyman and Ross et al. are fully aware of this and that is why they eschew that means of exhibiting the ontological content of theory and that Quinean view of theoretical commitment to existence.[2]

Before pursuing this matter further I should put my cards on the table. I am no structuralist, in mathematics, physics or metaphysics, though I am happy to accept the existence of all sorts of structures from ω through $\mathbb{C}$ (the field of complex numbers), to physical fields, and even, if it is required by metaphysics, to atomless gunk. But all these structures are structures of things. One can in my view talk of structures quite properly as themselves things, when one thinks of the structure as an equivalence class, as an abstraction over some suitable equivalence relation itself defined over classes of objects. But I think that structures are always structures of somethings.

Structuralism as a general view about the nature of mathematics really got started in a remark of Dedekind in his classic of 1887 *Was sind und was sollen die Zahlen?* (Quotations are from Dedekind (1963).) He famously remarks there:

> If in the contemplation of a singly infinite system N, ordered by a representation φ, we disregard entirely the peculiar nature of the elements, retaining only the possibility of distinguishing them, and considering only the relations in which they are placed by the ordering

[2] Thus they say explicitly, "It is not part of our realism that every time a scientist quantifies over something in formulation of a theory or hypothesis she is staking out an existential commitment .... Indeed, we will argue that, semantic appearances notwithstanding, we should not interpret science—either fundamental physics or special sciences—as metaphysically committed to the existence of self-subsistent individuals" (Ladyman and Ross, 2007, p. 119).

representation φ, then these elements are called *natural numbers* or *ordinal numbers* or simply *numbers*.

This manifesto for structuralism, that the natural numbers should be regarded as merely place holders in any ω-sequence, was replied to by Russell in 1903, before he himself temporarily adopted a structuralist view in (Russell, 1927). I have always thought that Russell's reply then was exactly the right one, not just to the specific claim but to the general structuralist thesis. It is worth quoting in full. He argues as follows:

> Moreover it is impossible that the ordinals should be, as Dedekind suggests, nothing but the terms of such relations as constitute a progression. If they are to be anything at all they must be intrinsically something: they must differ from other entities as points from instants, or colours from sounds. What Dedekind intended to indicate was probably a definition by means of the principle of abstraction .... But a definition so made always indicates some class of entities having (or being) a genuine nature of their own, and not logically dependent upon the manner in which they have been defined. The entities should at least be visible to the mind's eye; what the principle asserts is that, under certain conditions, there are such entities, if we only knew where to look for them. But whether when we have found them, they will be ordinals or cardinals, or even something quite different, is not to be decided off hand. And in any case, Dedekind does not show us what it is that all progressions have in common, nor give any reason for supposing it to be the ordinal numbers, except that all progressions obey the same laws as ordinals do, which would prove equally that any assigned progression is what all progressions have in common. (Russell, 1903, p. 249)

I will return to this issue later, we should begin by considering the methodological version of structural realism.

2 The Modest Defence of Scientific Structural Realism

Let us now turn to the modest defence of Scientific Realism proposed by Worrall and Zahar. What is it that they are defending and why does it require a defence at all? They are defending the claim of Scientific Realism which they, I think entirely rightly, characterise as a combination of a metaphysical claim and an epistemological one. The metaphysical claim asserts that there is a mind-independent reality of which scientific theories attempt to give true descriptions, and the epistemological thesis asserts that not only is this reality partially accessible to human discovery but that "it is reasonable to believe that the successful theories of mature science—the unified theories that explain the phenomena without ad hoc assumptions—have indeed latched on, in some no doubt partial and approximate way, to that structured reality, that they are, if you like, approximately true" (Worrall, 2007, p. 153–154). Why does such a claim need defending? It is essentially because of the pessimistic induction.

While at the empirical observational level of specific confirmed predictions science is cumulative, at the theoretical or explanatory level it is highly non-cumulative. Indeed the history of science reveals a sequence of dramatic revolutions in theory and that despite much, often very surprising predictive success by earlier theories, subsequent theories, while retaining as predictions those earlier predictive successes, have denied that the underlying structure of unobservable reality was anything like what earlier theories said it was. Up to the present then, all past theories have been proved to be false, Newtonian gravitational theory giving way to Relativity and classical electromagnetism giving way to quantum theory being prime examples, though the list is legion. Why do we not then reasonably infer that our current scientific theories are false? That is the pessimistic induction. But that induction further invites the question as to in what sense then were those admittedly predictively successful earlier theories even approximately true, given that such revolutions have occurred in successive theoretical accounts of the deep structure of the Universe? That is the central threat to scientific realism.

One standard realist defence against the pessimistic induction, the *no miracles argument*, is manifestly invalid. This very well known argument of Putnam and Boyd says that if we do not grant some notion of approximate truth to successive theories, then the explanatory success of science at the empirical predictive level would be a miracle (Putnam, 1975; Boyd, 1984). It would be a matter of the remotest chance, the remotest coincidence. But the invalidity of this argument is clear since it hinges upon what we take the prior likelihood of a false theory having a true and interesting consequence to be. It is only on the very specific assumption that the prior likelihood of that is incredibly low, that the argument has the slightest force. After all, any sequence of events can seem miraculous if one chooses a sufficiently low prior probability for outcomes of the required kind (see particularly (Howson, 2000), especially pp. 35–60).

Indeed further it is a requirement, if you like, of the practice of science that the predictive successes of earlier theories are carried over into succeeding theories as a necessary condition for the *adequacy* of those theories. The once much vaunted problem of "Kuhn loss," as it was called, turned out to be a myth.[3] The key idea in the structuralist defence of scientific realism is really contained in what is sometimes called the "correspondence principle." This is the claim that if we look at the history of major scientific revolutions then an adequacy condition on succeeding theories is that they yield the predecessor theory as a limiting case. The classic example of this is special relativity and Newtonian mechanics, where the Lorentz transformations of the former yield the Galilean transformations of the latter when the velocities considered are small with respect to that of light, and where the dynamical laws of the former yield as a limit the Newtonian Second Law when again the velocities are small with respect to that of light.

Now the key claim in general of Worrall and Zahar's position is that there is a strong notion of continuity which can be extracted from the history of theory change in science. It is continuity at the mathematical structural level, not at the ontological level.[4]

[3] Kuhn loss was the supposed historical phenomenon that the explanatory successes of earlier theories at the level of successful predictions would fail to be captured by succeeding theories.

[4] Van Fraassen in an excellent recent reply to structuralism (van Fraassen, 2006), gives a nice example of how observations reveal the structure of the phenomena behind them without revealing

As Worrall puts it:

> According to the account of theory change that underpins SSR [Structural Scientific Realism], successive theories in science have not only been successively more empirically adequate, but there has always been a reason, when viewed from the vantage point of the later theory, why the earlier theory achieved the degree of empirical adequacy that it did—namely that the earlier theory continues to look approximately structurally correct: its mathematical equations are retained modulo the correspondence principle. (Worrall, 2007, p. 143)

This seems to me to be a modest historical thesis, which one might well accept. The difficulty is however that it comes embedded with another claim, which seems to me to be completely fallacious as an account of theoretical knowledge in science. This claim is the view that a theory's full cognitive content is captured by its Ramsey Sentence, where the Ramsey sentence of a given theory T, results from T by replacing all the theoretical predicates occurring in T by second order variables and then existentially quantifying over them. So if T is an empirical theory with a theoretical vocabulary $\Theta_1, \ldots, \Theta_n$ and with observational vocabulary $O_1, \ldots, O_k$, then expressing T as $T(\Theta_1, \ldots, \Theta_n, O_1, \ldots, O_k)$ the Ramsey sentence of T, denoted by $R(T)$, is the claim $\exists X_1 \ldots X_n T(X_1, \ldots, X_n, O_1, \ldots, O_k)$. As is very well known, T and $R(T)$ have exactly the same observational consequences.[5] Now why is it the case that Structural realists want to identify the cognitive content of T with $R(T)$? The answer is because $R(T)$ says there are certain attributes, $X_1, \ldots, X_n$, whose relational structure is exhibited by $R(T)$ and that that structure imposes exactly the content on the observations that T does, (recall that T and $R(T)$ are observationally equivalent). So $R(T)$ exhibits the structural constraints that T imposes on the observations.

There is however a real difficulty with this view. It is the old problem pointed out by the topologist Max Newman (Newman, 1928) when criticising Russell's version of structural realism (as developed in (Russell, 1927), and that based upon his distinction between knowledge by acquaintance and knowledge by description). This objection has been greatly elaborated upon by William Demopoulos and

---

the qualities and inner nature of that hidden reality. His example is based on one provided by Russell in (Russell, 1927). Van Fraassen writes: "Listen to the radio, and hear the sounds which were produced in the studio many miles away. In between are the radio waves which have none of the qualities of sound. But we can infer that they must have structure which encodes the structure of this sound. Thus we know a great deal about those radio waves on the basis of observation: not what qualities they have or what they are like in themselves, but their structure. And it's precisely that, and only that, which science describes (as it happens of course what Maxwell's equations describe)"(p. 289).

[5] The proof is straightforward. If an observation sentence O is not a consequence of T then some interpretation I of T satisfies T but fails to satisfy O. But then that same interpretation will satisfy $R(T)$, since $R(T)$ is merely an existential generalisation of some predicates in T, but will fail to satisfy O. So if O is a consequence of $R(T)$ then it must be a consequence of T by contraposition. Similarly if O is not a consequence of $R(T)$, then some interpretation of the existentially generalised predicates of T satisfies $R(T)$ but fails to satisfy O. Precisely that interpretation of the predicates of T will then provide an interpretation of T itself which satisfies T but fails to satisfy O. Hence by contraposition if O is a consequence of T, O is a consequence of $R(T)$. So as far as O-consequences are concerned T and $R(T)$ are equivalent.

Michael Friedman (Demopoulos and Friedman, 1985; Demopoulos, 2003a, b, 2007). I will concentrate on Demopoulos' most recent elaboration of Newman's argument (Demopoulos, 2008; see also Ketland, 2004; Ainsworth, 2009) for I think his argument is very telling. Essentially the argument boils down to a model-theoretic argument of the following sort. Take a model, call it $\mathcal{M}$ in which all of the purely O-sentences of T hold. Then provided the cardinality of $\mathcal{M}$ is sufficiently great it can be expanded to a model $\mathcal{M}^*$ in which of course the O-sentences are satisfied, but in which all of the existential claims of $R(T)$ are also satisfied. This follows because of the richness of the set of subsets of the domain of $\mathcal{M}$. It is always possible, provided the domain of $\mathcal{M}$ is sufficiently large, to find subsets of the domain which will satisfy the existential claims of $R(T)$ exhibited above.[6] What this means then is that, if T is consistent, $R(T)$ expresses at most a cardinality claim on the universe, which is manifestly not what the theoretical claims of empirical science do. As Demopoulos puts it "modulo a logical assumption of satisfiability and an empirical assumption about cardinality, it follows that if the O-sentences are true, the T-sentences are true" (Demopoulos, 2008, p. 24). This result also holds if we do not, in forming the Ramsey sentence of the theory, allow for existential quantification over so-called mixed terms, that is those terms which apply both to theoretical and observational entities. So it cannot be right to identify the cognitive content of T with $R(T)$, which is what the structural realist insists we should do. This clearly presents a dilemma for the structuralist realist. Either identify the cognitive content of T with $R(T)$ with resulting trivialisation, or do not, but then what is to constitute the structural content of an empirical theory T?

Interestingly ontic structural realists like Ladyman and Ross do not identify the Ramsey sentence as the structural content of T, for they eschew altogether this form of analysis of structural content. So we should now turn to an analysis of their view.

3 Ontic Structural Realism

Ross and Ladyman and their co-authors clearly want metaphysics to fulfil an explanatory purpose, in particular they see it as performing an explanatory unification at the highest theoretical level. So, to take a straightforward example, one might genuinely regard the metaphysical theory of Perdurantism (see, for example, (Hawley, 2001)), perhaps enhanced by stage theory, as a good example of an attempt to unify the large scale picture of the four-dimensional block universe of Minkowski Space-Time in Relativity with ordinary facts about objects and how they change in our experience as codified in common sense theories. That particular model of Relativity could be treated independently of theories about ordinary objects which change and age but it is clearly helpful if an explanatory unification can take place.

[6] For proofs, see (Ketland, 2004; Demopoulos, 2008, 2003a). The background model theory is explained in (van Dalen, 1983).

Similarly one might recall seventeenth and eighteenth century struggles with the concept of matter, which sought to find a metaphysical concept of matter which could marry the concept of substance as a carrier for the primary qualities of matter as exhibited in mechanics (for example those of mass, inertia, impenetrability, and *vis viva*) to the other extraordinary property of matter, its response to the gravitational field and all that that apparently implied about action at a distance.

Now Ross and Ladyman lay down a constraint on metaphysical hypotheses which they label the Principle of Naturalistic Closure (PNC). This, modulo some technical stipulations, says:

> Any new metaphysical theory that is to be taken seriously at time *t* should be motivated by, and only by, the service it would perform, if true, in showing how two or more specific scientific hypotheses, at least one of which is drawn from fundamental physics, jointly explain more than the sum of what is explained by the two hypotheses taken separately, ... (Ladyman and Ross, 2007, p. 37)

Given this view about the role of metaphysics embodied in the PNC they argue for a positive and a negative thesis. They assert that:

> *The single most important idea we are promoting in this book is that to take the conventional philosophical model of an individual as being equivalent to the model of an existent mistakes practical convenience for metaphysical generalisation.* We can understand what individuals are by reference to the properties of real patterns. Attempting to do the opposite—as in most historical (Western) metaphysical projects—produces profound confusion. (Witness the debates about identity over time, identity over change in parts, and vagueness, none of which are PNC-compatibly motivated problems.) (Ladyman and Ross, 2007, p. 229)

So they use the PNC to dismiss some, if not all, of contemporary metaphysics. This seems to me unfair since much of modern metaphysics has been motivated by an attempt to bring a consistency proof forward for common sense views and fundamental physics (thereby surely satisfying the PNC). But I shall not dwell upon this point. Rather let us turn to their positive ontic structuralist thesis. That is the claim that all that exists is structure or patterns, all the way down. Though they never quite spell out what they mean by this view it is clear that they are committed to a reductionist view that objects, atoms, molecules, quarks and tables and chairs are to be thought of as "bundles" of structural universals.

Now it seems to me evident that whatever form of structuralism is involved the view cannot provide an answer to the pessimistic induction. For where before we had revolution in underlying ontology through scientific revolutions, we now have revolutions in structure. If there was a dramatic change in ontology from, say, the substance of heat, phlogiston, to the kinetic energy of molecules in the transition from early theories of heat and of Carnot's thermodynamics to late nineteenth century thermodynamics, then how is continuity to be restored by appeal to a transition from the structure of continuous media subject to a conservation law over cycles to a structure of discrete moving particles governed by rectilinear motion except at collision? Mathematical structure can change discontinuously just as traditional ontology can. So where is the advantage in this talk of structures?

But there is I think an even more serious objection, one that Russell made long ago and with which I began this paper. The objection is that the doctrine isn't coherent, if it is globalised so that everything is structure. Ladyman and Ross espouse the semantic view of theories; they wish to use all the techniques of model theory, domains, isomorphisms and embeddings, etc., to formulate and analyze the structural relations postulated by empirical theories. But they understand objects in a domain merely by the relational nexus in which they take part—objects are certain "higher order" relations holding among relations on the domain. But then which relations hold on the domain will depend upon the objects in the domain, and *that* depends upon which relations hold on the domain. I submit that this is viciously circular, not in the sense of understanding or definability alone, but in the ontological sense.

Perhaps the argument is better put by alluding to the Newman-Demopoulos argument mentioned above. That argument showed that structuralism does not yield an adequate account of theoretical knowledge because it does not yield any constraints on the World except ones of cardinality, which is certainly not what we take our theories to be telling us about the nature of reality. Now, there is no doubt that we can ascribe properties and relations to relations. We can say correctly that the relation less-than-or-equal-to is antisymmetric, or that the successor relation is one-one, or that the relation of isomorphism is part of the relation of ordinal similarity. But if all science does is speak of the relations holding among or of relations, it is perfectly consistent with the world being empty.[7] We are not given a coherent account of how the World can be unless some of the relata are individuals, but then structure can't be everything that there is.

So, despite their claims, nothing is gained by the structuralist account of theoretical knowledge either methodologically or ontologically, and the debate between the realist and the constructive empiricist over the interpretation of theoretical knowledge is left unresolved, by this move at least.

Bibliography

Ainsworth, P. M. (2009). Newman's objection. *British Journal for the Philosophy of Science*, 10:1–37.

Boyd, R. (1984). The current status of scientific realism. In Leplin, J., editor, *Scientific Realism*, pages 41–82. University of California Press, Berkeley, CA.

Dedekind, R. (1963). *The Nature and Meaning of Numbers*. Dover, New York, NY. Transl. W.W. Beman. Originally published 1887 as *Was sind und was sollen die Zahlen?*

Demopoulos, W. (2003a). On the rational reconstruction of our theoretical knowledge. *British Journal for the Philosophy of Science*, 54:371–403.

Demopoulos, W. (2003b). Russell's structuralism and the absolute description of the world. In Griffin, N., editor, *The Cambridge Companion to Russell*, pages 392–419. Cambridge University Press, Cambridge.

[7] Van Fraassen makes a closely related point. He puts the matter forcefully as follows: "if structure is not just there as mathematical or abstract entity, then it is not true that structure is all there is" (van Fraassen, 2006, p. 294).

Demopoulos, W. (2007). Carnap on the rational reconstruction of scientific theories. In Creath, R. and Friedman, M., editors, *The Cambridge Companion to Carnap*, pages 248–272. Cambridge University Press, Cambridge.
Demopoulos, W. (2008). Some remarks on the bearing of model theory on the theory of theories. *Synthese*, 164:359–383. (A special issue edited by Paolo Mancoscu dedicated to William Craig).
Demopoulos, W. and Friedman, M. (1985). Russell's *Analysis of Matter*: Its historical context and contemporary interest. *Philosophy of science*, 52:621–639.
Hawley, K. (2001). *How Things Persist*. Oxford University Press, Oxford.
Howson, C. (2000). *Hume's Problem: Induction and the Justification of Belief*. Oxford University Press, Oxford.
Ketland, J. (2004). Empirical adequacy and ramsification. *British Journal for the Philosophy of Science*, 55:287–300.
Ladyman, J. and Ross, D. (2007). *Every Thing Must Go: Metaphysics Naturalized*. Oxford University Press, Oxford.
Newman, M. H. A. (1928). Mr. Russell's causal theory of perception. *Mind*, 37:137–148.
Putnam, H. (1975). What is mathematical truth? In *Mathematics, Matter and Method: Philosophical Papers, Volume 1*, pages 60–78. Cambridge University Press, Cambridge.
Russell, B. (1903). *The Principles of Mathematics*. George Allen and Unwin, London.
Russell, B. (1927). *The Analysis of Matter*. K. Paul, Trench, Trubner, London. Reprinted by Dover Books in 1954.
van Dalen, D. (1983). *Logic and Structure*. Springer, Berlin.
van Fraassen, B. (2006). Structure: Its shadow and substance. *British Journal for the Philosophy of Science*, 57:275–307.
Worrall, J. (1989). Structural realism: The best of both worlds. *Dialectica*, 43:99–124.
Worrall, J. (2007). Miracles and models: Why reports of the death of structural realism may be exaggerated. In *Philosophy of Science*, volume 61 of *Royal Institute of Philosophy Supplement*, pages 125–54.
Zahar, E. G. (2001). *Poincaré's Philosophy: from Conventionalism to Phenomenology*. Open Court, Chicago and La Salle, IL.

Part VI
Aesthetics

Chapter 25
Musing on Music

Richard Feist

The man that hath no music in himself,
Nor is moved with concord of sweet sounds,
Is fit for treasons, stratagems, and spoils;
The motions of his spirit are dull as night,
And his affections dark as Erebus.
Let no such man be trusted. Mark the music.
The Merchant of Venice, Act V, Scene 1

1 Introduction: Boethius on Music

Anicius Manlius Severinus Boethius (ca. 480–524) would have been most puzzled by what Shakespeare's Lorenzo says. The reasons for his supposed perplexity lie in the background views on music that he appropriates from ancient Greek philosophy. Boethius' compendium on music, *De institutione musica* (*The Fundamentals of Music*), along with similar texts on arithmetic, geometry and astronomy, formed the medieval quadrivium. It is surprising that the scholastic philosophers, who were often deeply concerned with logical consistency and order, were unperturbed by the inconsistencies running rampant throughout *De institutione musica*. On the one hand, the work's first part is Pythagorean: music was inseparable from numbers, which governed the universe. Music exemplified the cosmic order. On the other hand, the work's latter sections contain anti-Pythagorean sentiments; music is not necessarily anything that expresses universal orders. So, it is difficult to say precisely what Boethius thought about music. Further complicating this is the fact that, with respect to music, Boethius was not always a terribly original thinker. He often borrowed from a (now lost) treatise on music by Nicomachus and from the first book of Ptolemy's *Harmonics*. Still, in the most original section of the book, the opening chapters, Boethius expresses a view of music that is relevant to the historical development of music and how thinkers addressed problems in the philosophy of music.

In these chapters Boethius divides music into three kinds. The first is *musica mundana* (cosmic music), which consists of the permanent and orderly numerical

R. Feist (✉)
Dean and Associate Professor of Philosophy, St Paul University, Ottawa, ON, Canada
e-mail: rfeist@ustpaul.ca

D. DeVidi et al. (eds.), *Logic, Mathematics, Philosophy: Vintage Enthusiasms*, The Western Ontario Series in Philosophy of Science 75,
DOI 10.1007/978-94-007-0214-1_25,

relations that could be "seen" in nature. Such relations would be planetary motions and seasonal changes. But there was another, natural music, *musica humana* (human music), which controls that obscure, yet supposedly common union: the body and the soul. Finally, there is music that exemplifies all these orderly relations, *musica instrumentalis*, which includes instrumental and vocal music. Boethius insisted that music was a key part of education; not only was it an important, albeit passive, object of study, but it influenced morals and served as an introduction, a platform of sorts, to more advanced philosophical studies.

An elegant discussion of music as a platform of education can be found in Plato's *Republic*:

> Then aren't these the reasons, Glaucon, that musical training is most important? First, because rhythm and harmony permeate the inner-most element of the soul, affect it more powerfully than anything else, and bring it grace, such education makes one graceful if one is properly trained, and the opposite if one is not. Second, because anyone who has been properly trained will quickly notice if something has been omitted from a thing or if that thing has not been well crafted or well grown. And so, since he feels distaste correctly, he will praise fine things, be pleased by them, take them into his soul, and, through being nourished by them, become fine and good. What is ugly or shameful, on the other hand, he will correctly condemn and hate while he is still young, before he is able to grasp the reason. And, because he has been so trained, he will welcome the reason when it comes and recognize it easily because of its kinship with himself. (401d4–402a4)

According to Plato music, not surprisingly, is not the absolute foundation of education, the forms are. Socrates compares the forms to letters; music entering the soul is like learning to read. Reason, however, comes afterwards. This notion of music first, followed by reason, is somewhat echoed in Boethius. Ultimately, the true musician according to Boethius was not the person who naturally sings or plays an instrument by ear, but the philosopher, the critic, "... who exhibits the faculty of forming judgments according to speculation or reason relative and appropriate to music" (Boethius, *De insitutione musica* 1.34, transl. Calvin M. Bower, quoted from Grout and Palisca (1963, p. 51)). Boethius would have regarded a human outside of music as outside of nature altogether, a "monster" in the traditional sense of that term.

A general and important theme here is the all-compassing nature of music, at least according to the ancients and, of course, Boethius. One could say that music originally determined, or at least highly influenced, the boundaries of sense. It was the primordial giver of order, of harmony. This thought has been echoed in our times by the great violinist Yehudi Menuhin:

> Music creates order out of chaos; for rhythm imposes unanimity upon the divergent; melody imposes continuity upon the disjointed, and harmony imposes compatibility upon the incongruous. (Menuhin, 1972, p. 9)

From relatively recent scientific literature, there is a similar notion of music as order-giving. Oliver Sacks, in his description of sufferers of post-encephalitic Parkinsonianism, describes one case as follows:

> By far the best treatment of her crises was music, the effects of which were almost uncanny. One minute would see Miss D. compressed, clenched and blocked, or jerking ticking and jabbering—like a sort of human bomb; the next, with the sound of music from a wireless

> or a gramophone, the complete disappearance of all these obstructive-explosive phenomena and their replacement by a blissful ease and flow of movement as Miss D., suddenly freed of her automatisms, smilingly 'conducted' the music, or rose and danced to it. (Sacks, 1981, p. 56–57)

Unfortunately, music has dark effects—in rare cases. For instance, it has been documented that one patient suffered a *grand mal*, a major epileptic seizure, simply by listening to a recording of Tchaikovsky's *Valse des Fleurs* (Critchley, 1977, p. 344). Even rarer are cases in which merely recalling a tune can cause seizures. Such deep and diametrically opposed effects of music on the soul are simply grist to Plato's mill.

For now, let us note that the basic idea is that, so to speak, it was once strongly held that there really was no meaning outside of music. We can see this even if we travel back to the origins of Western literature, as embodied in the Homeric poems. Some hold that these works are in fact the original works of Western literature.[1] To forget the role that music plays in these epics can lead to misunderstanding them. Because the ancients saw music as a mnemonic aid, they crafted their works according to metrical considerations. As the work of the great Homeric scholar, Milman Perry, established, the demands of the musical rhythm in many cases took precedence over the semantic content.[2] In many passages of the *Iliad*, for instance, Achilles is repeatedly referred to as "swift-footed," even in those scenes where he is seated. Certain scholars, while attempting to understand works like the *Iliad* are often needlessly perplexed by such repeated phrases.[3]

Nonetheless, by Shakespeare's time, it had become possible at least to think of a human existing outside of music. Such a marginalized human was still monstrous, being outside the moral order of the universe. But what I wish to stress is that music remained very much a part of the world, even though it may not be a kind of part that exemplifies the whole. In fact Shakespeare is most likely echoing some of the concerns over the development of music that had occurred since antiquity. Music was seen by many in the sixteenth century as having declined since antiquity. Basically, music had gone from being that which is deep within the soul, perhaps a mover of sorts, to being regarded as a formal sonic structure. This concern that music had become too formal can be found in a variety of thinkers, religious and non-religious, of the time historians refer to as "the Renaissance."

2 Formalized Music and the Renaissance

An example of one who reacted to the over-formalization of music is the religious leader Bernardino Cirillo (1500–1575), the archpriest of the Santa Casa of Loreto, a famous shrine and destination for pilgrimages. Cirillo complains that the polyphonic

---

[1] This is an overstatement, but it is often made; the epic of Gilgamesh predates those of Homer.
[2] For an overview of Milman's work, see (Powell, 2004, Ch. 1), "The Philologist's Homer."
[3] For an example of needless perplexity see (Smith, 2001, p. 7).

masses of the time were disappointing since they failed to move him. The masses paid too much attention to form. Cirillo insists that:

> ... when a Mass is sung in church, I should like the music to consist of certain harmonies and rhythms apt at moving our sentiments to religion and piety according to the meaning of the words. ... Today every effort and diligence is bent on making a work in strict fugue so that when one says "Sanctus," another pronounces "Sabaoth," while a third sings "Gloria tua," with certain wails, bellows, and bleating that at times they sound like cats in January ... (Feb. 16, 1649 letter to Ugolino Gualteruzzi, quoted from Grout and Palisca (1963, p. 153))

A similar, but more moderate reaction to formalization is in the secular writings of a prominent musician of the time, Gioseffo Zarlino (1517–1590). Zarlino regarded with pride the achievements in contrapuntal technique; however, he also agreed that music had fallen from its heights among the ancients. But, Zarlino insisted that music was currently undergoing a rebirth. Mere technique, he insisted, was not enough: music must move us to praise the glory of God. Indeed, Zarlino credits God with providing the world with Adrian Willaert (1490–1562), whom he refers to as the "new Pythagoras" who will restore music to the honour and dignity that it once had and ought to have. Basically, Zarlino credits Willaert with leading a new movement in music, where music is a vehicle of emotional expression. In other words, music was to be a language of the emotions.

The important point to draw out here is that viewing music formally has certainly been done—long before formalism as a general approach was applied to various disciplines. What lies behind this is the distinction between "that it can be done" and "should it be done?" Clearly, the Renaissance thinkers agreed that formalized views of music are possible; formalism in fact was staring them all in the face. But, they reacted, holding that a formalist take on music was misguided. In many ways, the formalist view of music strips it of a world. It is interesting to note that formalists today often replace that lost world by claiming that another one has taken its place. However, what this other world of music is, is often not at all explained.

But I wish to return to the claim that "music is a language of the emotions." This is a very old claim and, indeed, an ambiguous one. It has at least two possible readings. One could say that music expresses emotions or that it signifies emotions. Certainly music has great power. In fact, music is often claimed to have powers that outstrip any other human activity. To put any constraints on music's power has been seen by some, says contemporary philosopher of music Peter Kivy, to be an act of sacrilege. Still, one must ask, just what are the powers of music? What, indeed, can it really do? There are several questions to keep in mind while thinking about music. What is essential to it? Simply because we *can* interpret music in a highly abstract manner, does that mean that we *should*? Another, related, set of questions concern the supposed powers of music. What exactly, can it do? Peter Kivy offers an amusing but cautionary note, when considering the powers of music.

> ... long experience has taught me that whenever someone denies to music any power at all, or any important property, such denial tends to be seen as treachery or barbarism or some kind of musical insensitivity only to be expected from philosophical analysis or formalism, both of which I suppose that my work exemplifies. Had I denied that music can predict the future or remove warts, I am certain there would have been at least two responses in the

> literature to the effect that I had missed some sense in which it can predict the future or remove warts, although of course, one must not be quite so rigid or pedantic about what it means to "predict" or "remove" or what exactly the "future" or a "wart" might be. (Kivy, 1997, p. 140)

Kivy, whom I shall discuss later on, does not deny all power to music.

3 The Turning Point in Aesthetics: Representation

I now pass over the 1600s, to the years 1700–1750. During this time a number of important works dealing with philosophical reflections on art appeared in England and the continent. (For a list, see (Kivy, 1997, p. 4).) But as the historian Kristeller writes:

> The decisive step towards a system of the fine arts was taken by the Abbé Batteux in his famous and influential treatise, *Les beaux arts réduit à un meme principe* (1746). (Kristeller, 1992, p. 38)

What eventually emerged from Batteux's discussion could be called the *modus operandi* of the science of the arts, now called "aesthetics." The proposed view was that the arts are of a piece, united by their representative nature. The representative function, or *mimesis*, certainly played a role in the history of reflections on art and originates in the works of Plato and Aristotle. However, prior to Batteux, the general consensus was that the arts were all quite different; there was no clear notion of a "system" of the arts. They were more or less thought of as the disparate arts. The modern "system of the arts," as Kristeller calls it, consists of five arts. Three of them, painting, sculpture, and poetry, are not immediately problematic with respect to representation. Music and architecture are, at least to us today, more problematic with respect to the representative function. It was only around the mid-eighteenth century that representative function became that which was the shared property of all the arts.

If we shift back to the history of music, we note a parallel. Let us divide music into vocal and instrumental music. Clearly both had been in existence for centuries; the question is not one of existence, but emphasis. That is, prior to the mid-eighteenth century the emphasis, in quantity of production and what composers and their patrons valued, was vocal music. This type of music is, to a certain extent, much more easily subsumed under the representative function than instrumental music. Of course, instrumental music had existed and did so in sophisticated forms prior to the mid-eighteenth century. A quick consideration of Bach's instrumental works, written in the first half of the eighteenth century, illustrates the high levels instrumental works achieved. However, it remains an empirical, sociological fact that vocal music dominated. Vocal music was, once again, more valued than any other kind of music; vocal music commanded money whereas other musical types at the time generally did not—or at least not to the same degree. But all this changed in the latter half of the eighteenth century as it became possible to make a decent living composing purely instrumental music, absolute music. Now, once separated from any kind of vocal or literary associations and assuming a respected and popular status, music as absolute music becomes an object that is difficult to subsume under the representative function.

Standard examples of absolute music include Bach's *Art of Fugue* and Haydn's later symphonies. A typical text in music appreciation offers the following definition:

> Absolute music is music for which the composer has not indicated to us any nonmusical associations, whether of story, scene, or mood. Here the musical ideas are organized in such a way that, without any aid from external images, they give the listener a satisfying sense of order and continuity. (Machlis, 1970, p. 116)

Clearly this definition is intentionally-based; the foundation of the music/non-music connection is the intention of the composer. Of course, musical associations happen in a variety of ways. But, does this mean that music, via associations, can represent anything? This discussion turns on what one takes "representation" to mean.

In the early 1700s, one could say that the search for a clear definition of "representation," perhaps even any serious reflection on the notion of representation, is still a long way off into the future. At this time the mimetic function was a resemblance theory. In other words, "*X* represents *Y*" is unpacked as "*X* resembles *Y*." With this understanding of representation, at least three options loomed before aesthetic thinkers in the second half of the eighteenth century. First, they could simply exclude absolute music from the arts. Second, accept absolute music as a fine art and find a way to explain its representative nature. Third, accept absolute music as a fine art but construct a non-representative theory of the arts.

Without doubt the key thinker in aesthetics at this time is Kant. Indeed, Kant's *Critique of Judgment* sets the stage for the ensuing nineteenth century's aesthetical philosophy. Unfortunately, Kant does not clearly answer the question: "does absolute music represent or not?" Without going into the difficult notion of Kantian aesthetic ideas, one could say that Kant offers us a hybrid answer. In the sense of its form, absolute music is a fine art that does not represent; with the addition of poetry to it, the latter's evocation of the aesthetic ideas can cause the representative function to be engaged. Regardless of how one wishes to understand this, Kant in the end does not think that absolute music alone can represent anything. However, he still wants to have a general theory of art that has a representative function and can incorporate absolute music as an art. Whether or not he is successful—and it would appear that he is not—is not to be decided here.[4]

4 The Nineteenth Century

The giants of nineteenth century aesthetic thought are Hegel and Schopenhauer. Hegel emphasizes the dialectical nature of reality, which ultimately encompasses all forms of knowing. Without doubt he would have stressed the system of the arts

[4] For more on Kant see (Kivy, 1997, pp. 15–18), and (Bicknell, 2002).

as unified. Nonetheless, the sheer difficulty of understanding Hegel's reflections on music, along with his unclear references, make any serious discussion of his aesthetics beyond the scope of my work here. And so I turn briefly to Schopenhauer, who restores music to the status of an all-encompassing art form by returning to Platonic thought. According to Schopenhauer, art in general:

> ... repeats the eternal Ideas apprehended through pure contemplation, the essential and abiding element in all the phenomena of the world. According to the material in which it repeats, it is sculpture, painting, poetry, or music. Its only source is knowledge of the Ideas; its sole aim is communication of this knowledge. (Schopenhauer, 1966, p. 184)

Despite this, Schopenhauer insists that art does more than communicate knowledge of the Ideas. Art is a liberating mechanism, that is, it allows us an escape from our everyday world to one that is, at all times, nearby and yet difficult to remain within.

> There always lies so near to us a realm in which we have escaped entirely from all our affliction; but who has the strength to remain in it for long? As soon as any relation to our will, to our person, even of those objects of pure contemplation, again enters consciousness, the magic is at an end. We fall back into knowledge governed by the principle of sufficient reason; we now no longer know the Idea, but the individual thing, the link of a chain to which we also belong, and we are again abandoned to all our woe. (Schopenhauer, 1966, p. 198)

The non-musical arts aim at the Platonic Ideas by stimulating our minds to reach for such ideas.

> The (Platonic) Ideas are the adequate objectification of the will. To stimulate the knowledge of these by depicting individual things (for works of art are themselves always such) is the aim of all the other arts (and it is possible with a corresponding change in the knowing subject). Hence, all of them objectify the world only indirectly, in other words, by means of the Ideas. (Schopenhauer, 1966, p. 257)

Now we see the difference with music. Schopenhauer goes on to say that the visible world is simply an appearance of the Ideas. But music, he says, passes over the Ideas and is therefore independent of the world. That is, music could exist even if the world did not; this possible non-worldly existence of music cannot be attributed to any of the other arts. Schopenhauer then concludes:

> Thus music is as *immediate* an objectification and copy of the whole *will* as the world itself is, indeed as the Ideas are, the multiplied phenomenon of which constitutes the world of individual things. Therefore music is by no means like the other arts, namely a copy of the Ideas. For this reason the effect of music is so very much more powerful and penetrating than that of the other arts, for these others speak only of the shadow, but music of the essence.[5] (Schopenhauer, 1966, p. 257)

---

[5] It is interesting to note that Paul Hindemith has the exact opposite view of music. "The reactions music evokes are not feelings, but they are the images, memories of feelings. Dreams, memories, musical reactions—all three are made of the same stuff." Visual arts and poetry release direct emotions. Music is a trickster. "Paintings, poems, sculptures, works of architecture ... do not—contrary to music—release images of feelings; instead they speak to the real, untransformed, and unmodified feelings." (Hindemith, 1961, p.42)

Now, why would we accept that music is of the essence of anything? Music, when understood, is supposedly understood immediately.

> This close relation that music has to the true nature of all things can also explain the fact that, when music suitable to any scene, action, event, or environment is played, it seems to disclose to us its most secret meaning, and appears to be the most accurate and distinct commentary on it. (Schopenhauer, 1966, p. 262)

Schopenhauer goes on to state that when you listen to music, give yourself up to it, you see the events of life, but if you then reflect (presumably on the music) you cannot put the music and life into some kind of correspondence. Music, Schopenhauer repeats, just is not a copy of the phenomena.

> Accordingly, we could just as well call the world embodied music as embodied will; this is the reason why music makes every picture, indeed every science from real life and from the world, at once appear in enhanced significance, and this is, of course, all the greater, the more analogous its melody is to the inner spirit of the given phenomenon. (Schopenhauer, 1966, p. 263)

Still, Schopenhauer is quick to point out that there are strict limitations on these analogies between music and anything else.

> But we must never forget when referring to all these analogies I have brought forward, that music has no direct relation to them, but only an indirect one; for it never expresses the phenomenon, but only the inner nature, the in-itself, of every phenomenon, the will itself. Therefore music does not express this or that particular and definite pleasure, this or that affliction, pain, sorrow, horror, gaiety, merriment, or piece of mind, but joy, pain, sorrow, horror, gaiety, merriment, peace of mind *themselves*, to a certain extent in the abstract, their essential nature, without any accessories and so also without the motives for them. Nevertheless, we understand them perfectly in this extracted quintessence. (Schopenhauer, 1966, p. 261)

Much has been written about Schopenhauer's view of music. Without doubt it has been an influential view—especially in the early twentieth century. (See, for instance, (Yewdale, 1928).) Despite all of the criticisms of Schopenhauer, most of which reduce down to the accusation that he is a muddle-headed metaphysician, I think that he has much to say to modern philosophy of music. Indeed, Peter Kivy, one of the most influential of modern philosophers of music, despite what he explicitly says, has much in common with Schopenhauer. This now brings us to the twentieth century.

5 The Twentieth and Twenty-First Centuries

These centuries, in many ways, do not have a single giant in aesthetics akin to a Hegel or a Schopenhauer.[6] Nonetheless, perhaps some of the best work, at least in the analytic tradition, is that by Nelson Goodman, who offered numerous departure

---

[6] Clearly some would argue that Theodor W. Adorno (1903–1969) would be a suitable candidate for such status.

points for aesthetical investigations in his *Languages of Art*.[7] We have seen that the history of reflection on music illustrates, overall, an encroaching formalism in that music was originally held to be a broad framework of all meaning, which over the years, dwindled in its scope as content slowly drained from it. So now let us consider what happens after all content is in fact drained. In our time Peter Kivy is the key representative of the formalist view of music.

In several works, Kivy attacks the notion that music is "about anything," or that music can *mean* anything (Kivy, 1990, 1993, 1997). Still, Kivy insists that music can be profound. But its profundity cannot be unpacked in terms of hidden hermeneutic content, that is, in terms of what the music *says* or *means*, because, once again, music cannot say or mean anything. Then again, Kivy notes, even if music could say something about something, that alone is not enough to make it profound. In his attack on one particular attempt (Levinson's) to find musical profundity, Kivy concludes:

> It seems to me, then, that Levinson has failed to make out a plausible case for musical "aboutness." And without "aboutness" the case for musical profundity must also fail. But perhaps some might think this is a logical quibble or my concept of "aboutness" overly stringent and restrictive. So let us grant, for the sake of the argument, that Levinson has established the possibility of a musical aboutness and ask ourselves whether, even then, a case can be made for this kind of "aboutness." For profundity does not just require being able to say *something* about *something*; it requires being able to say something *profound*. I do not believe, even given the concept of musical aboutness as a gift, that Levinson can make his case for musical profundity. (Kivy, 1997, p. 169)

Now, in order to make his own case for musical profundity, Kivy discusses some of Schopenhaeur's themes. First, music differs from the other fine arts. Second, music represents the Will, while the other arts represent Platonic Ideas. Third, music liberates us from the world, while the other arts plunge us more deeply into it. Because Kivy rejects the notion of music representing anything, he rejects theme two. But he does accept one and three. His view is best summarized by his own words. Kivy writes:

> ... it is neither a "lack" in music that it possesses no content, nor a "lack" in the contentful arts that they possess no power to liberate. On the contrary, it is a defining virtue of the contentful arts that they do *not* liberate us from our workaday world but engage us, albeit in ways characteristic of the fine arts. And it is a defining virtue of absolute music, so I shall argue, that it does not engage us in our workaday world but liberates us from it. (Kivy, 1997, p. 204)

In sum, absolute music lacks content but has the power to deliver us from this world. Such a compound predicate would apply to a formalist view of mathematics. Kivy acknowledges this; and he says that such a common predicate does not bother him at all since it does not force us to equate mathematics and music (Kivy, 1997, p. 210).

---

[7] (Goodman, 1976). The first edition appeared in 1968 and caused quite a stir in the academic world. A standard text for many kinds of arguments in aesthetics, somewhat dated now but still quite useful, is (Beardsley, 1981).

That is quite true. However, a hot bath could possess such a compound predicate as well. In any case, Kivy's description of absolute music leaves a lot open by not being terribly precise about music itself. Still, one must be careful with Kivy's use of the term "content." Although he calls absolute music "organized but meaningless noise," he nonetheless calls it a "world" (Kivy, 1997, p. 206). Now in some ways this resembles Hilbert's famous formalist pronouncement: "No one shall drive us out of the paradise which Cantor has created for us" (Hilbert, 1991, p. 191). In any case, Kivy's musical world is some kind of object. In certain places he refers to it as an "intentional object" (Kivy, 1997, p. 209). This now resembles more of an intuitionist understanding of mathematics.

Nonetheless, a key distinction between mathematics and music is provided by G.H. Hardy, who states that most people enjoy mathematics, just as they do a musical tune. But music can stimulate mass emotion, whereas mathematics cannot (Hardy, 1969, p. 86). The importance of this distinction is that it cuts directly against Kivy's view of absolute music being liberating. In many ways, music ties us directly to this world in which we live.

In any case, Kivy does not hold that absolute music lacks all meaning; he deems his position "enhanced formalism" (Kivy, 1997, p. 216). This partial retreat from a "straightforward formalism" is not unique to Kivy—nor to aesthetics. Kivy's claims to formalism and then an actual partial retreat is an excellent example of Carnap's warning to be aware that philosophers say they are doing one thing and then actually do something else. Indeed, this retreat can also be found in the philosophy of mathematics.[8] But, we can also find this retreat earlier in the philosophy of music. Kivy's great 19th century predecessor, Eduard Hanslick, struggled with formalism in much the same way. However, Hanslick offers some deep insights into some age old philosophical problems. Hanslick writes:

> Reading so many books on musical aesthetics, all of which defined the nature of music in terms of the 'feelings,' and which ascribed to music a definite expressive capability, had long excited in me both doubt and opposition. The nature of music is even harder to fix within philosophical categories than painting, since in music the decisive concepts of 'form' and 'content' are impossible of independence and separation. If one wishes to attribute a definitive content to purely instrumental music—in vocal music content derives from the poem, not from the music—then one must discard the precious pearls of the musical art, in which no one can demonstrate a 'content' distinct from the 'form,' nor even deduce it. On the other hand, I readily agree that it is idle to speak of absolute lack of content in instrumental music, which my opponents accuse me of having done in my treatise. How is one to distinguish scientifically in music between inspired form and empty form? I had the former in mind; my opponents accused me of the other. (Hanslick, 1963, pp. 26–27)

As far as I know, Hanslick does not answer his own question, namely, "how does one seriously articulate, within absolute (instrumental) music, the distinction between 'inspired' and 'empty' form?" Nonetheless, it is quite interesting that he

[8] There are different kinds of formalist views of mathematics and often the formalist view is taken precisely to preserve some ontological view, such as Hilbert's formalism as a means to defend the consistency of mathematics. For a general overview of this, see (Shapiro, 2000, Ch. 6).

in fact posed such a question.[9] The point is that music is never truly understood as a purely sonic form. A purely sonic structure would be something like a drum rudiment repeatedly played on a table top.[10] Even those who claim to be formalists retain some aspect of content in absolute music. Eventually, then, the two are blurred. It seems that content in music, as Hanslick points out, just cannot be separated from form. This raises all sorts of difficult questions, but it is interesting to note that Hanslick was saying things about content and form with respect to music that W.V.O. Quine was to say a century later concerning content and form with respect to language.[11]

With this notion of thoughts on music as being a type of harbinger concerning particular philosophical reflections on language, I would now like to consider a troublesome theme in the philosophy of music, namely whether or not music can represent. Clearly past thinkers thought that it could. My suspicion, which I shall try to flesh out to some degree in the next section, is that many arguments against music as being able to represent cut against any symbolic system's claim to represent. The rest of the paper, then, is a small musing on this but does not claim to exhaust it in any way.[12]

6 Representation

Without doubt the term "representation" has vexed philosophers as well as other scholars. In philosophy we often worry about how sentences represent anything; in literature the problem of how a novel can represent anything is equally problematic, perhaps more so. In any case, one theme in the complex story of representation is that representative capacity has largely been assigned to sentential systems and denied to non-sentential ones. If one thinks about pictures and the origins of human communication, undoubtedly those from the caves of Lascaux come to mind. Pictures might have been the earliest form of human communication: the earliest way to say something to someone about something else.

Certainly the pictorial form of communication has been slowly, but surely, shunted aside in the history of mathematics. A picture or a diagram has long been seen as merely heuristic to understanding a proof, but not forming part of the proof itself. There are several reasons as to why this is the case. The most typically cited reasons are that pictures are misleading and subject to severe constraints on their expressive powers. This is, without doubt, quite true. However, one should also keep

---

[9] It is interesting to note that Kivy rejects the form-content identity in music simply because the latter has no content. (See his "On the Unity of Form and Content" in (Kivy, 1997).) But again, this makes his notion of "enhanced formalism" very difficult to understand.

[10] Would anyone call such an event a "piece of music?" Perhaps, and this raises the interesting, but difficult question of the ontology of music. I will return to this briefly at the end of this paper.

[11] This, I suggest, is a large topic and merits much further exploration. However, that is a task for another time.

[12] For an extended discussion of this question see (Robinson, 1994).

in mind that all forms of representation are constrained in some way or another. This was pointed out, albeit in a slightly different sense, by Wittgenstein, when he insisted, based on the notion that representations and their objects are part of the same world, that any representation, Y, of any object, X, involves a non-empty intersection of their forms (Wittgenstein, 1961, 2.17). The two sets may differ in some ways, but they cannot be disjoint. Without going into the details, suffice it to say that the nature of these constraints on representation, at least in the context of Wittgenstein's remarks, cannot ultimately be divorced from the logical atomism that underlies them. It would be reasonable to generalize and say that when we form judgments concerning the limits of representation—what can represent what—these judgments will reflect deeply held metaphysical pictures of the world. Needless to say, there is much more to be discussed in the history and philosophy of representation than I can enter into here.[13]

Metaphysical niceties aside, it should be noted that there is an everyday context for questions involving representation. If you are a confused tourist desperate to get to the airport to catch a plane, then what counts as a "good representation of the city" could be the sparse maps often found on restaurant place-mats. Highly detailed representations found in tourist guides will often provide quicker routes to the airport, but the time needed to digest the guide's information plus traveling the quicker route to the airport could be longer than the time needed to use the place mat plus traveling the longer route to the airport. However, if one is at home leisurely planning the trip, then the tourist guide is a much better representation of the city. So, if we ask for a "good representation of a city" the answer will depend on the context. More generally, when we ask if X represents Y we are often asking this question within a particular context Z.

I would like to put the question of the context to the side for a moment and simply ask whether music—in and of itself—can represent anything. As we have seen, in the past it was clearly assumed that it could, to the point of being the boundaries of meaning. What was running underneath all the previous musings on music was that "X represents Y" means "X resembles Y."[14] Let us begin with what appears as an easy case. No doubt musical instruments can emit certain sounds that resemble non-musical sounds; certain scales played on a clarinet resemble clocks and certain piano riffs can sound remarkably like running trains. However, most people hearing these scales and riffs do not automatically notice the resemblance to non-musical sounds, instead people often notice first how much these scales and riffs resemble other musical sounds. Still, if listeners are coached, advised that what they are hearing "sounds like" a clock or freight train, most listeners rapidly assent.

One might argue that the listener of an instrumentally-emitted sound should not require such coaching in order to recognize the sound as resembling a non-musical sound. It should be heard as resembling a non-musical sound. After all, open the

[13] For an examination of the implicit metaphysics involved in making judgments concerning representations, see (Greaves, 2002).

[14] For an extended critique of the resemblance theory of representation, see (Goodman, 1976, § 1), "Reality Remade."

door to coaching and, given a good coach, any musical sound could be said to resemble any other kind of sound. But this demand is not universal. Anyone who has seen an ultrasound or an X-ray realizes that seeing-Q-in-Z often requires a high level of training, which, arguably, is a kind of coaching. I doubt that anyone would ultimately reject the view that the ultrasound resembles the fetus. Nonetheless, very few individuals lacking the background training to read ultrasounds would immediately see a fetus, let alone specific gender markers. Finally, the fact that coaching is needed does not entail that a fetus is not represented in the ultrasound. Perhaps an even more extreme example would be Relativity Theory's field equations. They do represent a situation, yet who without specific training can read anything into them?

So, in addition to the problem of the context of degrees of representation, an external problem, there is an internal problem as well. Just because people require coaching to see that particular Q in a supposed representation, Z, of the "real Q," it by no means follows that Z does not actually represent the real Q.[15]

One might wish to reply that in specific examples such as the field equations of relativity or some other scientific representation of the world there is something within that representation that serves to point the interpretation in a particular direction, to a particular referent. That something internal to the scientific representation, call it IM (internal marker) is what sets a scientific representation aside from the question of musical representation.

This opens a huge debate, upon which I only make a brief comment. One could argue the case that there really is no such IM in scientific representations of the world. Theories, if formalized, can be given a variety of interpretations. (See (Putnam, 1983).) It would seem to be some kind of external criterion, EC, that selects a particular interpretation of a theory T out of a set of possible interpretations of T. This EC is most likely a pragmatic notion, such as success.[16] In any case, we are wise to heed Quine's words, "... reference is nonsense except relative to a coordinate system" (Quine, 1969, p. 48). Indeed, when we ask whether music can represent, and then deny that it can, we are doing this within a language. Language, even natural ones, do not represent in and of themselves.

But something similar might be going on with respect to music as well. As we have seen in the simple case of musical representation, the supposedly intended interpretation of a given musical set of sounds is often not selected by an uninformed listener. Coaching is required—and this is also required in many fields as well. Now consider a more involved issue—an actual composition of music, a quartet or symphony or what have you. Can this represent? These pieces, when played, often elicit many different interpretations. But, as we have seen with Schopenhauer, there is a sense of constraint on the interpretations. If one listens to a piece of music, such as Beethoven's "storm," one may not say, "that represents a storm." Or consider Schubert's "Die Forelle"—very few would notice that it represents a trout. But as

[15] I am barely scratching the surface of all the problems connected to representation. For a general discussion see (Goodman, 1976).

[16] This is another huge issue in the philosophy of science but one that I shall simply pass over here.

Schopenhauer stated, quite rightly I believe, there is a natural fit between music and the world. Military music around the world all sounds *military*. Sacred music, too, has a certain kind of feel. This, again, is merely to echo a common position to two thinkers as far apart as Hardy and Schopenhauer. Clearly the constraints on the set of possible interpretations of a given piece of music would be far weaker than the set of possible interpretations of a given piece of scientific theory, but that is a difference in degree, not kind. The main point is that if we are open to degrees of representation and do not see it as an all or nothing affair, then music, too, can represent.

Finally, there is another sense to the openness of representation. Not only are there better and worse representations within the particular context, but there is also a creative aspect. Goodman writes:

> To a complaint that his portrait of Gertrude Stein did not look like her, Picasso is said to have answered, "No matter, it will." (Goodman, 1976, p. 33)

7 Summary and Conclusion

Throughout history music has been seen as a kind of representation of the world. In the strongest form of this view, music was equated with the absolute foundations of the cosmos. However, music was slowly drained of all this content. This ultimate representation and boundary of meaning has, slowly but surely, retreated over the centuries until the twentieth and twenty-first centuries have come to the view that music cannot represent at all. In the end music has but a liberating effect. But, counter to this drift many have argued that to see music in a formal way is to strip it of something essential to it. In a sense, music stubbornly refuses to let go of this world.

I do not claim by any means to have established that music can represent and how in fact it does so. I have merely tried to motivate the view that if one seriously looks at some of problems of musical representation then the arguments that lead to the claim that music cannot represent can also be directed at other disciplines which people, at least many people, would argue *can* represent.

Finally, the removal of music from this world makes nonsense of the history of music itself. That is, if it is seen as some kind of abstract structure, then the idea of "Viennese music" or "Renaissance music" becomes problematic. This, again, strikes me as too formal of a view. Music is indeed part of this world and in its own way, indicates other parts of this world.

But I would like to conclude on what could be called a Socratic note. As is well-known, Socrates insisted on a definition of virtue prior to pondering whether it could be taught. In general, he asked for a definition of X before asking anything else about X. I have only briefly mentioned this issue, namely, the problem of saying what is and is not music, although it is central to such discussions as I have offered here. Many philosophers, however, shy away from it. In other words,what requires clarification in this context is the musical ontology being employed. Is a clarinet scale or a piano riff truly a piece of music? I can make all sorts of sounds on a clarinet, from random noises, mouse-like squeaks, to scales and what we might call "tunes." But when, exactly, have I passed from emitting sounds to playing music? So, before even tackling the question of a piece of music resembling anything, there would have to be an acceptable ontology of musical works.[17]

In this light one might ponder the nature of composer John Cage's *4′3″*, which is simply a time span with no notes of any kind played. This takes Claude Debussy's comment "music is the space between the notes" to an extreme. Perhaps *4′3″* qualifies as truly formal music. It is all form; there is no content. Or perhaps it frees music from all constraints, both form and content. Cage seems to offer such an interpretation of *4′3″* saying that it is not to be thought of as "empty space." Instead, it liberates the listener, not in Kivy's liberation from this world, but a liberation to more closely connect with this world and its depths, almost a Schopenhauerian twist. Cage writes:

> I think perhaps my own best piece, at least the one that I like the most, is the silent piece. ... I wanted my work to be free of my own likes and dislikes, because I think music should be free of the feelings and ideas of the composer. I have felt and hoped to have led other people to feel that the sounds of their environment constitute a music which is more interesting than the music they would hear if they went into a concert hall. (Quoted from Davies (2005, p. 14).)

I suggest that Boethius would have regarded *4'3"* as embodying, at least in spirit, the notion of his own *musica mundane*. In many ways, my musings have come full circle.

Bibliography

Beardsley, M. C. (1981). *Aesthetics: Problems in the Philosophy of Criticism.* Hackett Publishing Co., Indianapolis.

Bicknell, J. (2002). Can music convey semantic content? A Kantian approach. *The Journal of Aesthetics and Art Criticism*, 3:153–261.

Critchley, M. (1977). Musicological epilepsy. In Critchley, M. and Henson, R. A., editors, *Music and the Brain.* Heineman Medical Books, London.

Davies, S. (2005). The ontology of musical works and the authenticity of their performances. In *Themes in the Philosophy of Music*, pages 1–26. Oxford University Press, Oxford.

Goodman, N. (1976). *Languages of Art: An Approach to a Theory of Symbols.* Hackett Publishing Co., Indianapolis, IN, Second Edition.

Greaves, M. (2002). *The Philosophical Status of Diagrams.* CSLI Publications, Standford.

Grout, D. J. and Palisca, C. V., editors (1963). *A History of Western Music.* W.W. Norton, New York, NY.

Hanslick, E. (1963). *Music Criticism, 1846–99.* Penguin Books, Harmondsworth.

Hardy, G. H. (1969). *A Mathematician's Apology.* Cambridge University Press, Cambridge.

Hilbert, D. (1991). On the infinite. In Benacerraf, P. and Putnam, H., editors, *Philosophy of Mathematics: Selected Readings.* Cambridge University Press, Cambridge, second edition.

[17] This is a subject of much dispute and involves many more issues than I can go into here. See (Davies, 2005).

Hindemith, P. (1961). *A Composer's World.* Anchor Books, New York, NY.
Kivy, P. (1990). *Music Alone: Philosophical Reflections on Purely Musical Experience.* Cornell University Press, Ithaca, NY.
Kivy, P. (1993). *The Fine Art of Repetition: Essays in the Philosophy of Music.* Cambridge University Press, Cambridge.
Kivy, P. (1997). *Philosophies of Art: An Essay in Differences.* Cambridge University Press, Cambridge.
Kristeller, P. O. (1992). The modern system of the arts. In Kivy, P., editor, *Essays on the History of Aesthetics.* University of Rochester Press, Rochester, NY.
Machlis, J. (1970). *The Enjoyment of Music.* W.W. Norton & Company, New York, NY, 3rd edition.
Menuhin, Y. (1972). *Theme and Variations.* Stein and Day, New York, NY.
Powell, B. B. (2004). *Homer.* Blackwell Publishing, Oxford.
Putnam, H. (1983). Models and reality. In *Realism and Reason: Philosophical Papers, Volume 3,* pages 1–26. Cambridge University Press, Cambridge.
Quine, W. V. O. (1969). *Ontological Relativity and Other Essays.* Columbia University Press, New York, NY.
Robinson, J. (1994). Music as a representational art. In Anderson, P., editor, *An Introduction to the Philosophy of Music.* Pennsylvania State Press, University Park.
Sacks, O. (1981). *Awakenings.* Pan Books, London, revised edition.
Schopenhauer, A. (1966). *The World as Will and Representation,* volume 1. Dover Books, New York, NY. transl. E. F. Payne.
Shapiro, S. (2000). *Thinking About Mathematics.* Oxford University Press, Oxford.
Smith, N. D. (2001). Some thoughts about the origins of Greek ethics. *The Journal of Ethics,* 5:3–20.
Wittgenstein, L. (1961). *Tractatus-Logico-Philosophicus.* Routledge and Kegan Paul, London. transl. David F. Pears and Brian F. McGuinness.
Yewdale, M. S. (1928). The metaphysical foundations of pure music. *The Musical Quarterly,* 14(3):397–402.

Chapter 26
Inscrutable Harmonies: The Continuous and the Discrete as Reflected in the Playing of Jascha Heifetz and Glenn Gould

Joel Bennhall

We owe the Pythagoreans the revelation that the harmonies of music derive from number, that is, from the discrete. This must be seen as a triumph of inscrutability. Inscrutable indeed is the resulting subsumption of music within mathematics, a colourless, forbidding subject to most, indeed the polar opposite of music, whose gaudy, yet profound, epiphanies offer a striking contrast. The Pythagoreans are to be blamed for the fact that music came to find itself in the embrace of such unlikely bedfellows as arithmetic, geometry and astronomy, the other members of the mathematical *quadrivium*.

Still, this evidence might cause a Marxist to respond that, while the basis of musical *organization* is to be located in the discrete, its *means of production* originate in the realm of the continuous. For is not sound itself nothing more or less than a continuous vibratory excitation of the atmospheric envelope, whether induced by blowing, plucking, or striking a tensed string, beating a drum, blowing down a tube, or straining one's vocal cords? The Pythagorean discovery, at bottom, is an instance of *ex continuo discretum*.

Musical instruments may be classed as continuous or discrete according to the manner in which individual notes are sounded. Thus the voice, bowed string instruments, and slide trombones are naturally identified as continuous, while valved wind instruments such as the clarinet or oboe, plucked or struck string instruments such as the lute or dulcimer, and keyboard instruments such as the harpsichord or piano may be classified as discrete.

J. Bennhall (✉)
Musicologist and Composer

Editors' note: The musicologist and composer Joel Bennhall was briefly, when very young, a pupil of the Danish composer Dag Henrik Esrum-Hellerup. Later he studied with Schoenberg. His fascination with the relationship between music and mathematics led him to create topomusicological analysis, at one time a rival in musicological circles to Hans Keller's better known functional analysis. In his old age Bennhall met the man celebrated in this volume at a meeting of the London Melomaniacal Society, at which he read the present paper; sadly he was prevented from publishing it owing to his sudden demise (at the age of 98). The Editors are pleased to provide it with a suitable resting place.

D. DeVidi et al. (eds.), *Logic, Mathematics, Philosophy: Vintage Enthusiasms*, The Western Ontario Series in Philosophy of Science 75,
DOI 10.1007/978-94-007-0214-1_26,

Accomplished vocalists and players of continuous instruments have considerable freedom both in determining the quality of individual notes and in the shaping of the "line" engendered by the succession of notes. This freedom is manifested above all in the case of the violin. In the violin the continuous and the discrete are truly united. For while the violinist's bow is the source *par excellence* of continuous sound, of a variable intensity controlled by subtle alterations in pressure of the fingers on the strings, in the hands of a virtuoso that very bow is also employed to spectacular effect in engendering discreteness: witness, for example, *spiccato*, *staccato* and *col legno* bowing.

With the violinist's left hand the order of continuity and discreteness is reversed, since the violinist's digits are employed in the first instance to produce separate discrete notes through "stopping" the strings. But just as the bow can generate discreteness, so can the left hand generate continuity, e.g. through *vibrato*, the continuous minute oscillation of pitch of a single note[1]; the *portamento*, the subtle continuous movement from one note to another by gently gliding the finger along the string; and the *shift*, the violinist's equivalent of the mathematician's continuous change of coordinate system.

While the discrete instruments lack these refinements, they have one great advantage over their continuous counterparts, namely, their capacity to support *simultaneity*. A mere tyro on the guitar plays multiply voiced chords as a matter of course, while even a competent violinist may have difficulty in playing double-stops in tune. With the keyboard instruments this natural capacity to engender simultaneity has achieved its highest development in the polyphonic structures created by the composers of the Baroque period, and above all by J.S. Bach. Bach raises keyboard polyphony to undreamt-of heights, and certainly to a level far surpassing that achievable on any single stringed instrument. Even that most elaborate four-part fugue in the C major solo sonata cannot compare in complexity with the cyclopean edifice of the Art of Fugue!

Now let us turn to consider two modern masters of their respective instruments—Jascha Heifetz, the violinist whose technical command of the instrument is widely regarded as supreme—and Glenn Gould, the wizard of piano polyphony. The one, the master of the continuous, the other, the master of the discrete.

Jascha Heifetz—a name with which to conjure. For the present writer the name evokes a cluster of associations, each of which involves the continuous in one way or another. One recalls for example the famous exchange at the young Heifetz's New York debut between two members of an audience packed with musicians eager to hear the new *Wunderkind*—Mischa Elman, great violinist, and Leopold Godowsky, equally great pianist. To Elman's ingenuous observation, "It's hot in here, isn't it?", no wittier response is conceivable than Godowsky's "Yes, but not for pianists!" In this instance the discrete could afford to smile at the embarrassment of

[1] The *trill*—the rapid alternation of the main note with that a tone or semitone above it—is of course a discrete effect.

the continuous. (And yet it must be recalled that pianist-composers such as Liszt and Chopin were strongly influenced by the virtuosity of Paganini.)

Consider also Heifetz's celebrated definition of a Russian: one Russian—an anarchist; two Russians—a game of chess; three Russians—a revolution; four Russians—the Budapest String Quartet. Here one sees a striking progression from the pure discreteness of the unit to the continuity of stringed instruments.

Even Virgil Thomson's nastily dismissive description of Heifetz's repertoire as "silk underwear music" brings the continuous to mind.

But in truth Heifetz *was* the supreme master of silkiness, indeed of that ultimate form of continuity that mathematicians call *smoothness*. This quality is best heard in his recordings of the 1930s and 1940s, most arrestingly in encore pieces and lightweight concertos such as those of Korngold and Gruenberg. In these latter works the silky smoothness of Heifetz's tone—endlessly imitated but never duplicated by generations of violinists under his spell—almost surpasses belief.

Mathematicians have introduced the concept of a smooth topos, a mathematical "world" in which all correlations are arbitrarily many times differentiable, there are no jagged edges and in which, in Leibniz's words, *natura non facit saltus*. There is no question that Heifetz would be the canonical fiddler in such a world.

But Heifetz was also a master of the discrete effects achievable on the violin; above all he could play detached notes on his preferred Guarnerius with blinding celerity. A remarkable example of his facility in this respect is provided by his recording of the Sinding *Suite*, in which the first, *presto* movement is despatched with truly hair-raising speed and accuracy. There could not be a more striking contrast between this glittering flurry of notes and the smooth, yet sweetly earnest and heartfelt manner in which Heifetz delivers the second, *adagio* movement.

The present writer did not have many opportunities to see Heifetz on the concert platform, but they remain treasured experiences. Especially memorable was the recital at the Royal Festival Hall in the 1950s at which Heifetz was due to begin with the Vitali *Chaconne*. This piece, familiar to all violinists, begins with a G minor chord, so when Heifetz picked up his violin and struck a G major chord the ranks of violinists occupying the first few rows of seats fell back in shock. Heifetz nonchalantly went on to play the English national anthem in G major.

Fortunately there are in existence a handful of filmed performances of Heifetz. One of the most remarkable of these is the rendition of that famed encore piece the *Hora Staccato*, transcribed by Heifetz from a Rumanian original. The writer once had the experience of hearing this exacting morsel played by a gypsy fiddler in an Amsterdam restaurant. While adequate, the performance could not compare with that of Heifetz, who contrives to make the staccato effect *sound* discrete but *appear* to the eye as continuous.

Now let us turn to Glenn Gould. While he was of course a master of the keyboard, with an unexampled command of polyphonic technique, one suspects that he may have envied the string player and the singer their immediate contact with the continuous. Grounds for this surmise are provided by his admitted inability to suppress the vocalise which invariably accompanied his piano playing, and which was such a source of vexation to critics and listeners alike.

On the other hand, for Gould, in the final analysis, polyphony was all, and polyphony, on the piano at least, is achieved by the systematic exploitation of discreteness. So it is reasonable to suppose that at some point Gould made the conscious decision to celebrate the discreteness of the piano, to avoid the mimicking of continuity by what he saw as contrived and hackneyed effects such as overpedalling and the gratuitous use of legato. Thus he strove for a *secco, detaché* sound, with each individual note rejoicing in its separateness. This approach is extraordinarily effective in Baroque works; and also with the twentieth century composers Gould most admired: Schoenberg, Webern, Krenek,Hindemith. In the present writer's opinion, the approach is also effective in Beethoven, especially in the early works of that master. With Mozart, however, the result is, to this writer's ears at least, nothing short of disastrous—but this was, of course, exactly what Gould, who disliked Mozart's music (with the conspicuous exception of the early sonatas K. 279–284 and the Fantasy and Fugue K. 394) was trying to achieve. Here Gould carried discrete deconstruction to the point of destruction.

It is to Gould's transcendent performances of works by composers he esteemed that one turns again and again. And above all, of course, to the compositions of J.S. Bach. Although Gould was most famed for his recordings of the *Goldberg Variations*—a fame that led this work to be identified by his fans as the "Gouldberg" Variations—the composition of Bach's he revered above all others was the *Art of Fugue*. And yet Gould produced no complete recording of this supreme, but alas, unfinished, masterpiece of the polyphonist's art. For this writer the most exciting rendition of any part of this work is Gould's 1967 Canadian radio broadcast of Contrapuncti IX, XI and XIII. Here Gould's playing achieves what can only be described as an ecstatic seamlessness fusing discreteness and continuity in an almost Hegelian *Aufhebung*.

Finally, we must consider the question of how Heifetz and Gould would have sounded had they played together. Would these supreme exponents of continuity and discreteness have achieved a harmonious union?

The vast majority of Heifetz's duo recordings were made with contract pianists—able, but somewhat colourless. An exception is the magnificent recording of Brahms' op. 108 sonata Heifetz made in the 1950s with the brilliant American pianist William Kapell (who died tragically young). Here the power of the pianist's playing comes close to matching Heifetz's, driving the latter to peaks even he did not always attain with his usual accompanists.

As for Gould, he made only a handful of recordings with violinists. One recalls the curious *rencontre* with Yehudi Menuhin during which Gould persuaded the violinist to play the Schoenberg *Fantasy* op. 47, a work to which, like all of Schoenberg's output, Gould was partial, but which Menuhin later said he found totally incomprehensible. In this connection it is pertinent to recall Heifetz's similar antipathy to Schoenberg's *oeuvre*. Heifetz actually commissioned Schoenberg's Violin Concerto op. 36 but on seeing the score instantly rejected it, giving the scarcely credible excuse that to play it would require him to grow a sixth finger. "I can wait," Schoenberg is reputed to have replied.

Gould did record the Bach violin and keyboard sonatas with Jaime Laredo, a good violinist, but who takes a back seat to Gould. The one string player who really stood up to Gould in duet performance was the cellist Leonard Rose, who, in their recording of Bach's sonatas for cello and keyboard gives a robust performance fully matching Gould's powerful rendition.

The upshot is that we can only imagine the sound of a Heifetz/Gould recital. One's musical tongue waters at the idea of recordings by these two masters of the Bach, Beethoven, or Brahms sonatas. The nearest approach we can make to this ideal is to listen to the pair of Bach violin concertos (in E major and A minor) and their keyboard transcriptions (in D major and G minor) as recorded, respectively, by Heifetz and Gould. It is a rare treat to hear Bach's sublime lines played first continuously, and then with discrete elaboration.

If only Heifetz and Gould had collaborated! That would have been the ultimate synthesis of the continuous and the discrete.

Publications of John L. Bell

Books

1. (with A. B. Slomson) *Models and Ultraproducts: An Introduction.* North-Holland, Amsterdam, 1969. (3 printings.) Dover reprint, 2006.
2. (with M. Machover) *A Course in Mathematical Logic.* North-Holland, Amsterdam, 1977. 4th printing, 2004.
3. *Boolean-Valued Models and Independence Proofs in Set Theory.* Clarendon Press, Oxford, 1977. Preface by Dana Scott. (Oxford Logic Guides, Volume 4.)
4. *Boolean-Valued Models and Independence Proofs in Set Theory.* Clarendon Press, Oxford, 1984. Revised and expanded Second Edition of (3). Revised Preface by Dana Scott. (Oxford Logic Guides, Volume 12.)
5. *Toposes and Local Set Theories: An Introduction.* Oxford University Press, 1988. (Oxford Logic Guides, Volume 14.) Dover reprint 2008.
6. *A Primer of Infinitesimal Analysis.* Cambridge University Press, 1998.
7. *The Art of the Intelligible: An Elementary Approach to the Conceptual Development of Mathematics.* Western Ontario Series in Philosophy of Science, Kluwer, 1999.
8. (with D. DeVidi and G. Solomon) *Logical Options: An Introduction to Classical and Alternative Logics.* Broadview Press, Peterborough, ON, 2001.
9. *The Continuous and the Infinitesimal in Mathematics and Philosophy.* Polimetrica, Milan, 2005.
10. *Set Theory: Boolean-Valued Models and Independence Proofs.* Clarendon Press, Oxford, 2005. Third Edition of (3), revised and expanded version of (4), with the same Preface by Dana Scott. (Oxford Logic Guides, Volume 47.)
11. *A Primer of Infinitesimal Analysis.* Cambridge University Press, 2008. Revised and expanded Second Edition of (6).
12. *The Axiom of Choice*, volume 22 of the Series in Mathematical Logic and Foundations, College Publications, 2009.

D. DeVidi et al. (eds.), *Logic, Mathematics, Philosophy: Vintage Enthusiasms*,
The Western Ontario Series in Philosophy of Science 75,
DOI 10.1007/978-94-007-0214-1,

Articles and Chapters

1. "Weak Compactness in Restricted Second-Order Languages," *Bulletin de l'Academie Polonaise des Sciences* **18** (1970) 111–114.
2. (with F. Jellett) "On the Relationship between the Boolean Prime Ideal Theorem and Two Principles of Functional Analysis," *Bulletin de l'Academie Polonaise des Sciences*, **19** (1971) 191–194.
3. "Some Remarks on Current Mathematical Practice," in *Proceedings of the Bertrand Russell Memorial Logic Conference, Uldum, Denmark, 1971*, John L. Bell, J. Cole, G. Priest and A. Slomson, eds., Leeds, 1973.
4. "On the Relationship Between Weak Compactness and Restricted Second-Order Languages," *Archiv für mathematische Logik und Grundlagenforschung* **15** (1972) 74–78.
5. (with D. H. Fremlin) "The Maximal Ideal Theorem for Lattices of Sets," *Bulletin of the London Mathematical Society* **4** (1972) 1–2.
6. (with D. H. Fremlin) "A Geometric Form of the Axiom of Choice," *Fundamenta Mathematicae* **77** (1972) 167–170.
7. "Brève critique de la pratique mathématique actuelle," in *Pourquoi la Mathématique?* Editions 10/18, Paris, 1974.
8. "On Compact Cardinals," *Zeitschrift für mathematische Logik und Grundlagen der Mathematik* **20** (1974) 389–393.
9. "A Characterization of Universal Complete Boolean Algebras," *Journal of the London Mathematical Society (2)*, **12** (1975/1976) 86–88.
10. "Universal Complete Boolean Algebras and Cardinal Collapsing," *Zeitschrift für mathematische Logik und Grundlagen der Mathematik* **22** (1975) 161–164.
11. "A Note on Generic Ultrafilters," *Zeitschrift für mathematische Logik und Grundlagen der Mathematik* **22** (1976) 307–310.
12. "Uncountable Standard Models of ZFC + V = L," in *Set Theory and Hierarchy Theory: a memorial tribute to Andrzej Mostowski*, W. Marek, M. Srebrny, and A. Zarach, eds., *Springer Lecture Notes in Mathematics* **537**, Berlin, Springer, 1976, 29–36.
13. "Boolean Extensions as Toposes," *Bulletin de la société française de logique, méthodologie et philosophie des sciences* **6** (1979).
14. "The Infinite Past Regained: A Reply to Whitrow," *British Journal for the Philosophy of Science* **30** (1979) 161–65.
15. "Category Theory and the Foundations of Mathematics," *British Journal for the Philosophy of Science* **32** (1981) 349–358.
16. "Isomorphism of Structures in S-Toposes," *Journal of Symbolic Logic* **46** (1981) 449–459.
17. "Some Aspects of the Category of Subobjects of Constant Objects in a Topos," *Journal of Pure and Applied Algebra* **24** (1982) 245–259.
18. (with M. F. Hallett) "Logic, Quantum 'logic' and Empiricism," *Philosophy of Science* **49** (1982) 355–379.
19. "Categories, Toposes and Sets," *Synthese* **51** (1982) 292–338.
20. "On the Strength of the Sikorski Extension Theorem for Boolean Algebras," *Journal of Symbolic Logic* **48** (1983) 841–846.

21. "Orthologic, Forcing and the Manifestation of Attributes," in *Southeast Asian Conference on Logic: Studies in Logic and Foundations of Mathematics*, C.T. Chong, ed., North-Holland, Amsterdam, 1983, 13–36.
22. "Edward Hubert Linfoot," *Bulletin of the London Mathematical Society* **16** (1984).
23. "Orthospaces and Quantum Logic," *Foundations of Physics* **15** (1985) 1179–1202.
24. "A New Approach to Quantum Logic," *British Journal for the Philosophy of Science* **37** (1986) 83–99.
25. "From Absolute to Local Mathematics," *Synthese* **69** (1986) 409–426.
26. "Logic, the Paradoxes and the Foundations of Mathematics," *L.S.E. Quarterly I* (1987) 317–326.
27. "Infinitesimals," *Synthese* **75** (1988) 285–315.
28. "Some Propositions Equivalent to the Sikorski Extension Theorem for Boolean Algebras," *Fundamenta Mathematicae* **130** (1988) 51–55.
29. "Hilbert's ε-operator and classical logic," *Journal of Philosophical Logic* **22** (1993) 1–18.
30. (with W. Demopoulos) "Frege's Theory of Concepts and Objects and the Interpretation of Second-Order Logic," *Philosophia Mathematica* **1** (1993) 139–156.
31. "Hilbert's ε-Operator in Intuitionistic Type Theories," *Mathematical Logic Quarterly* **39** (1993) 323–337.
32. "Fregean Extensions of First-Order Theories," *Mathematical Logic Quarterly* **40** (1994) 27–30. (Reprinted in W. Demopoulos, ed., *Frege's Philosophy of Mathematics*, Harvard University Press, 1995, 432–437).
33. "Frege's Theorem and the Zermelo-Bourbaki Lemma," Appendix to William Demopoulos's "Introduction" in *Frege's Philosophy of Mathematics*, W. Demopoulos, ed., Harvard University Press, 1995, 21–26.
34. "Type-Reducing Correspondences and Well-Orderings: Frege's and Zermelo's Constructions Re-examined," *Journal of Symbolic Logic* **60** (1995) 209–221.
35. "Infinitesimals and the Continuum," *Mathematical Intelligencer* **17** (1995) 55–57.
36. (with R. Clifton) "QuasiBoolean Algebras and Simultaneously Definite Properties in Quantum Mechanics," *International Journal of Theoretical Physics* **34** (1995) 2409–2421.
37. "Logical Reflections on the Kochen-Specker Theorem," in *Perspectives on Quantum Reality*, R. Clifton, ed., Kluwer, 1996, 227–235.
38. (with W. Demopoulos) "Elementary Propositions and Independence," *Notre Dame Journal of Formal Logic*, **37** (1996) 112–124.
39. "Polymodal Lattices and Polymodal Logic," *Mathematical Logic Quarterly* **42** (1996) 219–233.
40. (with S. Gebellato) "Precovers, Modalities and Universal Closure Operators in a Topos," *Mathematical Logic Quarterly* **42** (1996) 289–299.
41. "Zorn's Lemma and Complete Boolean Algebras in Intuitionistic Type Theories," *Journal of Symbolic Logic* **62** (1997) 1265–1279.
42. "Boolean Algebras," *Routledge Encyclopedia of Philosophy*, 1998, 97.

43. "Boolean Algebras and Distributive Lattices Treated Constructively," *Mathematical Logic Quarterly* **45** (1999) 135–143.
44. "Frege's Theorem in a Constructive Setting," *Journal of Symbolic Logic* **64** (1999) 486–488.
45. "Finite Sets and Frege Structures," *Journal of Symbolic Logic* **64** (1999) 1552–1556.
46. "Infinitary Logic," *Stanford Encyclopedia of Philosophy*, 2000. (This may be found online at http://plato.stanford.edu).
47. "Sets and Classes as Many," *Journal of Philosophical Logic* **29** (2000) 585–601.
48. "Hermann Weyl on Intuition and the Continuum," *Philosophia Mathematica* **8** (2000) 259–273.
49. "Continuity and the Logic of Perception," *Transcendent Philosophy* **1** (2000) 1–7.
50. "The Continuum in Smooth Infinitesimal Analysis," in *Reuniting the Antipodes—Constructive and Nonstandard Views of the Continuum.* Symposium Proceedings, San Servolo/Venice, Italy, May 16–22, 1999. U. Berger, H. Osswald and P. Schuster, eds., Kluwer, 2001, 1–24.
51. "Observations on Category Theory," *Axiomathes* **12** (2001) 151–155.
52. "Time and Causation in Gödel's Universe," *Transcendent Philosophy* **3** (2002) 341–346.
53. "Some New Intuitionistic Equivalents of Zorn's Lemma," *Archive for Mathematical Logic* **42** (2003) 811–814.
54. "Hermann Weyl's Later Philosophical Views: His Divergence from Husserl," *Husserl and the Sciences*, R. Feist, ed., U. of Ottawa Press, 2004, 173–185.
55. "Observations on Mathematics," *Proceedings of* Mathematics as Story, *A Symposium on Mathematics Through the Lenses of Art and Technology, Faculty of Education, University of Western Ontario, 13–15 June, 2003*, George Gadanidis, Cornelia Hoogland and Kamran Sedig, eds., University of Western Ontario, 2004, 68–71;103–107.
56. "Russell's Paradox and Diagonalization in a Constructive Context," *100 Years of Russell's Paradox, Munich 2001*, Godehard Link, ed., Walter de Gruyter, 2004, 221–225.
57. "Whole and Part in Mathematics," *Axiomathes* **14** (2004) 285–294.
58. "Continuity and Infinitesimals," *Stanford Encyclopedia of Philosophy*, 2005 (`http://plato.stanford.edu/entries/continuity/`).
59. "The Development of Categorical Logic," *Handbook of Philosophical Logic*, Volume 12, D.M. Gabbay and F. Guenther, eds., Springer, 2005, 279–361.
60. "Divergent Concepts of the Continuum in 19th and Early 20th Century Mathematics and Philosophy," *Axiomathes* **15** (2005) 63–84.
61. "Oppositions and Paradoxes in Mathematics and Philosophy," *Axiomathes* **15** (2005) 165–180.
62. (With Geoffrey Hellman) "Pluralism and the Foundations of Mathematics," in *Scientific Pluralism*, Stephen Kellert, Helen Longino and C. K. Waters, eds., University of Minnesota Press 2006, 64–79.

63. "Choice Principles in Intuitionistic Set Theory," in *A Logical Approach to Philosophy, Essays in Honour of Graham Solomon*, D. DeVidi and T. Kenyon, eds., Springer, 2006, 36–44.
64. "Abstract and Variable Sets in Category Theory," in *What is Category Theory?*, Giandomenica Sico, ed., Polimetrica, 2006, 9–16.
65. "Cosmological Theories and the Question of the Existence of a Creator," in *Religion and the Challenges of Science*, William Sweet and Richard Feist, eds., Ashgate Publishers, 2007, 109–120.
66. "Cover Schemes, Frame-Valued Sets and Their Potential Uses in Spacetime Physics," in *Spacetime Physics Research Trends*, Albert Reimer, ed., Horizons in World Physics Volume 248, Nova Science Publishers, New York, 2007, 47–74.
67. "Incompleteness in a General Setting," *Bulletin of Symbolic Logic* **13** (2007) 21–30. "Corrigendum to 'Incompleteness in a General Setting' by John L. Bell," *The Bulletin of Symbolic Logic* **14** (2008) 122.
68. "Contribution to *Philosophy of Mathematics: 5 Questions*," V. Hendricks and H. Leitgeb, eds., Automatic Press, 2007, 13–26.
69. "The Axiom of Choice," *Stanford Encyclopedia of Philosophy* 2008, (`http://plato.stanford.edu/entries/axiom-choice/`).
70. "The Axiom of Choice and the Law of Excluded Middle in Weak Set Theories," *Mathematical Logic Quarterly* **54** (2008) 194–201.
71. "Hermann Weyl," *Stanford Encyclopedia of Philosophy* 2009, (`http://plato.stanford.edu/entries/weyl/`).
72. "Cohesiveness," *Intellectica* **51** (2009) 145–168.
73. "Types, Sets and Categories," in *Handbook of the History of Logic*, Elsevier, forthcoming.
74. "The Axiom of Choice in the Foundations of Mathematics," in a volume on Foundations of Mathematics, Giovanni Sommaruga, ed., in the Western Ontario Series in Philosophy of Science, Springer, forthcoming.

Book Reviews

1. "Review of *Simplified Independence Proofs: Boolean-Valued Models of Set Theory* by J.B. Rosser," in *Bulletin of the London Mathematical Society* **3** (1971) 117–118.
2. "Review of *Mathematical Logic* by S.W.P. Steen," in *British Journal for the Philosophy of Science* **23** (1972).
3. "Review of *Foundations of Set Theory* by Fraenkel, Bar-Hillel and Levy," in *British Journal for the Philosophy of Science* **26** (1975) 165–170.
4. "Review of *Set Theory* by F.R. Drake and *The Axiom of Choice* by T. Jech," in *British Journal for the Philosophy of Science* **27** (1975) 187–191.
5. "Gone to the Dogs," in *Times Literary Supplement*, **8** (July 1977).
6. "Review of *Handbook of Mathematical Logic*, edited by J. Barwise," in *British Journal for the Philosophy of Science* **30** (1979).

7. "Review of *Topoi: the Categorial Analysis of Logic* by R. Goldblatt," in *British Journal for the Philosophy of Science* **38** (1982) 95–96.
8. "Review of *Zermelo's Axiom of Choice*, by G. Moore," in *Bulletin of the London Mathematical Society* **15** (1983).
9. "Review of *Numbers, Sets and Axioms* by A.G. Hamilton," in *Times Higher Education Supplement* **13** (May 1983).
10. "Review of *Constructive Analysis* by E. Bishop and D. Bridges," in *Bulletin of the London Mathematical Society* **18** (1986) 509–511.
11. "Review of *A Theory of Sets* by A.P. Morse," in *Bulletin of the London Mathematical Society* **19** (1987).
12. "Review of *Stone Spaces* by P.T. Johnstone," in *Bulletin of the London Mathematical Society* **19** (1987) 195–196.
13. "Review of *Collected Works, Vol.I* by Kurt Gödel," in *Philosophical Quarterly* **38** (1987) 216–218.
14. "Review of *The Limits of Quantum Logic* by P. Gibbins," in *Philosophical Quarterly* **38** (1988).
15. "Review of *Reflections on Kurt Gödel* by Hao Wang," in *Philosophical Quarterly* **39** (1989) 115–117.
16. "Review of *Introduction to Higher-Order Categorical Logic* by J. Lambek and P.J. Scott," in *Journal of Symbolic Logic* **54** (1989).
17. "Review of *Relative Category Theory and Geometric Morphisms* by J. Chapman and F. Rowbottom," in *Bulletin of the London Mathematical Society* **25** (1993).
18. "Review of *Elementary Categories, Elementary Toposes* by C. McLarty," in *Journal of Symbolic Logic* **58** (1993).
19. "Review of *Ad Infinitum: The Ghost in Turing's Machine* B. Rotman," in *Philosophia Mathematica* **3** (1995) 218–221.
20. "Review of *Conceptual Mathematics: A First Introduction to Categories* by F.W. Lawvere and S. Schanuel," in *Minds and Machines* **5** (1995) 436–440.
21. "Review of *Categorical Logic and Type Theory* by B. Jacob," in *Studia Logica* **69** (2001) 429–431.
22. "Review of *Sets for Mathematics* by F.W. Lawvere and R. Rosebrugh," featured review in *Mathematical Reviews* (2003), 03E99.
23. "Review of *Bolzano's Philosophy and the Emergence of Modern Mathematics* by P. Rusnock," in *Philosophia Mathematica* **14** (2006) 362–364.
24. "Review of *Synthetic Differential Geometry (2nd edition)* by A. Kock," in *Bulletin of Symbolic Logic* **13** (2007) 244–245.

Other Publication

Editing

1. Editor (with J. Cole, G. Priest and A. Slomson) of *Proceedings of the Bertrand Russell Memorial Logic Conference, Uldum, Denmark, 1971* Leeds, 1973.

2. Guest Editor, "Categories in the Foundations of Mathematics and Language," a special issue of *Philosophia Mathematica* **2** (1994).

Translation

1. Translation of *La géometrie dans le monde sensible* by J. Nicod, published as "Geometry in the Sensible World," in *Geometry and Induction: Containing Geometry in the Sensible World and The Logical Problem of Induction*, with Prefaces by André Lalande and Bertrand Russell (from 1930) and a new Preface and Appendix by Sir Roy Harrod, Routledge & Kegan Paul, London, 1970.
2. Translation of *Groupes algébriques* by M. Demazure and P. Gabriel, published as *Introduction to Algebraic Geometry and Algebraic Groups*, North-Holland, Amsterdam, 1980.

Research Reports, Lecture Notes and Theses

1. *A Short Survey of Phrase-Structure Grammars*, Elliott Computers Technical Report 65/122, Sept.1965.
2. (With A. B. Slomson). *Introduction to Model Theory*, Mathematical Institute, Oxford, 1965.
3. *Infinitary Languages*, Diploma Dissertation, Oxford 1966.
4. *Completeness and Axiomatization Results for Weak Second-Order Languages*, Doctoral Dissertation, Oxford, 1969.
5. *Iterated Boolean Extensions and the Consistency of Souslin's Hypothesis*, Lecture Notes No.10, Dept. of Mathematics, National University of Singapore, 1982.
6. Notes circulated in mimeographed form:

    - Basic Model Theory
    - Set Theory
    - Constructivity in Mathematics
    - Foundations of Mathematics
    - General Topology
    - Measurable Cardinals
    - Notes on Logic

Miscellany

1. *Perpetual Motion: My First Thirty Years*, a memoir.
2. *Philosophy in Literature*.

Name Index